Metaphysics in Contemporary Physics

Poznań Studies in the Philosophy of the Sciences and the Humanities

VOLUME 104

The titles published in this series are listed at *brill.com/ps*

Metaphysics in Contemporary Physics

Edited by

Tomasz Bigaj
Christian Wüthrich

BRILL
RODOPI

LEIDEN | BOSTON

Cover illustration: Atomium, Brussels, Morguefile

Library of Congress Control Number: 2015955686

ISSN 0303-8157
ISBN 978-90-04-30963-0 (hardback)
ISBN 978-90-04-31082-7 (e-book)

This book is printed on acid-free paper.

CONTENTS

INTRODUCTION

The present collection assembles new work in the flourishing field of the metaphysics of physics, running the full gamut from the philosophical consideration of the foundations of contemporary physics to a scientifically informed analysis of traditional metaphysical concerns. Our desire to understand the innermost foldings of the world we inhabit has naturally brought physics and philosophy in close contact over the millennia; in fact, both disciplines have emerged out of the same systematic attempts to satiate this human zeal. Despite occasional dissonance and miscommunication, the nexus between the two fields was mutually beneficial for the most part, and forged the foundation of modern science in the first scientific revolution of the 16th and 17th centuries and was instrumental in initiating the second scientific revolution during the late 19th and early 20th centuries. The resulting unprecedented success of physics in predictive accuracy, explanatory abundance, and range of technological applications has led to a more asymmetric relation between physics and philosophy: as the former shines in its well-deserved acclaim as the kingpin of science, the latter struggles to remain relevant as foundational and philosophical questions are increasingly seen as arcane, inscrutable, and unnecessary. Lest the reader mistakes us to condone philosophy's supposed plight, we affirm that instead of waning in significance, foundational and philosophical work has acquired new urgency in the light of fundamental physics' continued struggle to even just formulate a complete quantum theory of gravity, let alone a comprehensive and unified foundation for all of contemporary physics.

This volume is concerned with specifically metaphysical issues that connect to physics. But even if the state of philosophy is altogether not that precarious, the prospects of metaphysics are routinely considered downright daunting and its standing has only very recently started to recover from the logical empiricists' onslaught almost a century ago. Its status and even its possibility have been the subject of protracted debates for long – in fact, long before the heyday of logical empiricism. Although we

In: Tomasz Bigaj and Christian Wüthrich (eds.), *Metaphysics in Contemporary Physics* *Poznań Studies in the Philosophy of the Sciences and the Humanities*, vol. 104), pp. 7-24. Amsterdam/New York, NY: Rodopi | Brill, 2015.

share the staunchly scientific spirit of the logical empiricists, we believe that the rehabilitation of metaphysics is long overdue and offer the following collection as evidence that naturalism and metaphysics can productively interact with one another.

1. Metaphysics and Its Subject Matter

So what, precisely, is metaphysics, and what possible intimate relations with cutting-edge scientific theories can it have? In a nutshell, metaphysics is the study of the fundamental structure of reality.

Let us try to be more specific. A discipline can be identified by its unique subject matter and by its specific methodology. Regarding the former, the subject matter of any field of inquiry is usually assumed to consist of a set of objects – the domain – and a set of distinguished properties and relations among these objects. Using Kit Fine's terminology (Fine 2013) we may say that elements of the subject matter for a given discipline can occur in it either objectually, or predicatively. One characteristic trait of metaphysics is that every object can in principle be an element of its domain of inquiry; another that it is customary to differentiate metaphysics from other branches of philosophy by excluding from its subject matter only the epistemic relation between the object of knowledge and the perceiving subject. This general trait is characteristic of physics too, with the additional restriction that the domain of physical inquiry is limited to material, spatiotemporal objects. Generally, metaphysics does not obey this restriction, as metaphysical considerations can, and often do, reach out into the realm of non-physical entities (abstracta, possibilia, values, etc.). However, it has to be admitted that there is a substantial overlap between the objectual parts of the subject matters for both physics and metaphysics.

The difference in the subject matter between the two disciplines becomes more conspicuous when we turn to the predicative part. While physics deals with fairly broad concepts, such as the notion of material objects, elementary particles, fields and interactions, metaphysics centers its analyses around even broader categories of objects, properties, identity and the like. However, it would not be correct to explicate the generality of metaphysical concepts simply in terms of the breadth of their scope. For instance, the concept of identity seems to be more universal than the concept of a (mereological) part, and yet the scope of the latter clearly includes the entire former category (as the numerical identity of objects x and y obviously implies that x is an (improper) part of y). In light of this observation Kit Fine (ibid.) proposes to spell out the requisite notion of generality in terms of invariance. Since the relation of identity is invariant under all permutations, while in the case of parthood a much narrower set of rearrangements of objects leave this relation unchanged, identity is less sensitive to the difference in descriptive character of objects than parthood, and therefore is considered more universal. With respect to its generality, metaphysics can be located between even more general – and topic-neutral – logic and decidedly less general science, including physics. It goes without saying that there are no clear cut-offs on the scale of diminishing generality that could precisely separate these fields of inquiry, and therefore some logical and scientific questions can, on this criterion, be plausibly categorized as borderline metaphysical.

It is often said, as we did above, that metaphysics is not merely a general study of objective reality, but a study of its most fundamental aspects. In other words, metaphysics is concerned with how things are by their very nature. Fine calls this feature of metaphysics *eidictic*, and he goes on to characterize eidictic theories as those whose propositions are true in virtue of their subject matter, where again the subject matter of a theory is assumed to consist of elements that occur objectually in its propositions, and elements that occur predicatively.[1] For instance, the logical proposition $\forall xy\ (x = y \rightarrow y = x)$ is true in virtue of the purely logical elements $\forall$ and $=$, and thus reflects the nature of these elements. However, the proposition $\exists x\ (x = x)$ is not true in virtue of the nature of the logical elements $\exists$ and $=$ only; we need the extralogical assumption (taken, for instance, from mathematics) that there is at least one object in the universe. According to Fine, metaphysics is not the only field of inquiry of an eidictic character – other eidictic theories include logic, mathematics and physics (for instance the statement ‘Electrons are fermions’ is arguably true by virtue of the nature of electrons and the property of being a fermion, i.e. the property of having a half-integer spin). However, the unique character of metaphysics is shown in that the concept of eidicity itself belongs to the (distinctive) metaphysical subject matter. Thus, only metaphysics contains propositions which explicitly use the expression ‘by nature’, as in the following example: ‘If a substance is composed of X, then it is composed of X by its nature’.

In Fine’s approach, the necessity of metaphysical truths is a direct consequence of their eidictic character. As there are eidictic fields of inquiry other than metaphysics, they too define their own categories of necessity.

[1] We are slightly simplifying Fine’s proposal here. His original characterization of eidictic theories involves the subtler notion of *distinctive* subject matters which reflects the fact that some elements of the subject matter of a given theory may be borrowed from another field of inquiry (viz. logical concepts used in metaphysics).

Thus there is for instance the concept of logical necessity, which applies to all propositions that are true in virtue of the logical subject matter. Metaphysical necessity characterizes all eidictic propositions involving its subject matter that are not logically necessary. Analogously, physically necessary propositions include those propositions true in virtue of the nature of the physical subject matter that are not metaphysically necessary. Metaphysical possibility is assumed to be narrower than logical possibility but broader than physical possibility. For instance, the existence of electrons that are bosons (i.e., have an integer spin) is metaphysically (and logically), but not physically, possible. On the other hand, it is logically, but presumably not metaphysically, possible for an object to possess two determinate properties of one determinable (e.g., two distinct colors).[2]

2. The Epistemic Status and Methods of Metaphysics

We have so far portrayed metaphysics as a field of inquiry which does not place any restrictions on the objects of its investigations other than their subject-independent existence, which deals with significantly broad concepts, and whose statements are true in virtue of the nature of participating elements. But what possible method can help us attain reliable knowledge in the area of study characterized in such a way? A typical response to the question of the epistemic status of metaphysics is that it is an a priori discipline, strictly separated from experience. However, it seems unclear what the source of the metaphysical aprioricity might be. Is metaphysics, as some suggest, based on intuitions, or intuitive insight into the nature of things? Such an account of the rationalist, non-empirical character of metaphysical truths looks implausible, since the intuitions seemingly underlying these truths are most likely acquired through learning and experience. As Don Ross, James Ladyman and David Spurrett (2007, pp. 10–15) correctly observe, our common sense and intuitions result from interacting with medium-size objects occupying a very restricted, small region within the spatiotemporal vastness of the universe. It is unreasonable to expect that intuitions developed in such a parochial way could underlie a metaphysical theory that would be impervious to the verdicts of refined scientific studies of the empirical world.

---

[2] However, we should note that some scientifically oriented philosophers are skeptical as to the existence of a viable concept of metaphysical necessity/possibility distinct from physical necessity/possibility (see Callender 2011, p. 44).

A different, if somewhat related, answer to the question of the epistemic source of metaphysics is that metaphysical truths are products of reflecting on our fundamental concepts by doing what is called conceptual analysis. In this approach metaphysics serves as a tool for systematizing and categorizing the conceptual framework in which we would like to describe the most fundamental aspects of reality. Consequently, metaphysical theses become something akin to the logical consequences of meaning postulates (terminological conventions) governing the use of the primitive concepts within a selected metaphysical framework. To those who are worried that this approach reduces metaphysical truths to trivial linguistic stipulations two responses may be offered. First, an analogous interpretation of mathematical theories and statements (as logical consequences of axioms which in turn serve the role of contextual definitions for the primitive terms of a given theory) does not seem to imply that mathematical facts are in any significant sense 'trivial' or 'merely linguistic'. Second, the choice of appropriate postulates characterizing the meaning of the terms of interest is by no means a trivial task. In selecting a particular conceptual framework we may be guided by several principles. One such principle may be the postulate to preserve some pretheoretical intuitions and snap judgments regarding a particular area of interest. However, the sway this principle holds on our philosophical analyses may be offset by a substantial increase in theoretical virtues, such as simplicity and parsimony. While theoretical virtues can offset pretheoretical intuition, the naturalism predicated of much of the work found in this volume demands this to be even more so for considerations arising from and connected to experience. For our current purposes it is of crucial importance to recognize that another important stimulus to develop a particular metaphysical framework can thus come from scientific theories. Due to its foundational character, physics seems to be best suited to offering guidance as to what such a metaphysical framework should look like.

The purported detachment from experience of metaphysics as traditionally conceived has been the source of well-known severe criticism by numerous philosophers, from the British empiricists through Immanuel Kant to the logical positivists. One possible reaction to this challenge is to embark on a research program that can be called naturalistic metaphysics. Alyssa Ney in her recent response to the neo-positivist critique of metaphysics (Ney 2012) points out that a naturalistic metaphysician can focus her attention on identifying entities, structures and principles that are present in every fundamental physical theory, and seem to be indispensable to our best scientific theories. Ney believes that in naturalistic ('neo-positivist') metaphysics there is still room for armchair methods. These rationalist methods can be used to elucidate the consequences of the metaphysical commitments indispensable for science, and to fill in the details of an incomplete metaphysics arrived at during the first, scientifically informed stage of inquiry. Similar ideas regarding scientifically oriented metaphysics are expressed by Craig Callender (2011). Callender remarks that a naturalistic metaphysics, while informed by our best science, does not have to be a 'handmaiden' to science. Scientific theories often leave serious gaps in the interpretation of their fundamental concepts, and this is where a serious and autonomous metaphysical analysis can prove its worth.

3. Between Metaphysics and Physics

How can we understand the expression 'metaphysics in physics'? In what sense can metaphysics be 'present' in physics? There are two broad options available here: either metaphysics can be somehow discovered as 'preexisting' in physics, or it can be put in there 'by hand'. Let's call the first option 'intrinsic' ('from within') and the second 'extrinsic' ('from without'). The 'from within' option attempts to extract the metaphysical implications of a physical theory, e.g. by developing a broad conceptual framework representing the fundamental features of a world in which the theory is stipulated to be true. To put it more succinctly, we are asking the question of what the world should look like in its most fundamental aspects for a particular physical theory to be true.

The main problem that this approach must face is the notorious phenomenon of the underdetermination of metaphysics by physics (see e.g. French 2011). A classic example of this phenomenon is provided by the quantum theory of many particles of the same type. Due to the symmetrization postulate imposed on the joint states of identical bosons and identical fermions, particles of the same type cannot be distinguished by appropriate reduced states. This arguably leads to the conclusion that particles of the same type violate the *Principle of the Identity of Indiscernibles*, and therefore their individualities cannot be grounded in differences in their properties. The original underdetermination thesis concludes that there are two metaphysical views compatible with this fact: one claiming that quantum particles possess so-called transcendent individualities (individualities grounded in non-qualitative features, such as *haecceities*), and the other denying that quantum particles are individuals in any metaphysically significant sense. It turns out that the situation may be even worse than that, as more metaphysical options have recently been added to the debate on the identity and individuality of quantum particles. These options include the widely debated proposal of grounding the numerical distinctness of particles in weakly discerning relations,[3] as well as an unorthodox approach to the problem of individuation and symmetrization which implies that fermions and bosons can be discerned by their properties after all (for more on this most recent suggestion see the contributions by Adam Caulton and Tomasz Bigaj in this volume).

Similar arguments in favor of the underdetermination of metaphysics by physics can be found in virtually all cases where attempts are made to draw metaphysical lessons from physical theories. To mention one more famous example: the initial hope that general relativity will break the impasse between the rival metaphysical positions of substantivalism and relationism in favor of the latter has been shattered by well-known arguments based on the existence of the vacuum solutions of Einstein's field equations (cf. the contribution by Matteo Morganti). Thus it seems that a physical theory can at best restrict the vast array of possible metaphysical frameworks by imposing on them the demand of the compatibility with its physical contents, without actually being able to select one that uniquely meets the requirements of the theory. In order to make such a selection, additional assumptions need to be accepted, and these assumptions will most likely come from outside the physical theory in question. For instance, philosophical arguments against haecceities (purported non-qualitative properties unique to all individual objects) can be brought into consideration when discussing the metaphysical consequences of the symmetrization postulate in quantum mechanics. But these arguments are not, strictly speaking, part of the considered physical theory – they have to be introduced *from without*.

This brings us to the second, extrinsic role that metaphysics can play in relation to physics. Two ways of introducing metaphysical theses into physics can be distinguished. One is related to the process of constructing a new fundamental theory. Sometimes certain metaphysical assumptions are explicitly built into a newly developed physical theory as part of its foundation. This method of incorporating metaphysical presuppositions into physical science is best illustrated by how Einstein approached the task of constructing his general theory of relativity. As is well known, Einstein's intention was to design a new theory which would satisfy certain general principles of a broadly metaphysical character, such as the principle of relativity, stating the fundamental physical equivalence of all coordinate systems. Similar foundational aspects are being taken into account by some physicists in their latest attempts to combine general relativity and quantum mechanics into a consistent theory of quantum gravity.

[3] Cf. Saunders (2006), Muller and Saunders (2008), and the recent critical overview by Bigaj (2015).

Some go as far as to claim that no successful theory of that kind can be built without a serious reconsideration of the fundamental metaphysical questions regarding the ultimate nature of reality.[4]

However, metaphysics can still influence physical theories even after they have been developed and empirically confirmed. The second way of introducing metaphysics from without has to do with the general task of providing an interpretation of the established mathematical formalism of a given theory, especially in the light of outstanding conceptual difficulties. Nowhere is this approach more perspicuous than in the case of quantum mechanics and its numerous interpretations. In spite of its tremendous empirical successes, since its inception quantum mechanics has been afflicted with fundamental conceptual problems, of which the measurement problem is the most prominent. In order to cope with these difficulties, several new additions to the standard formalism have been considered, each with its own metaphysical assumptions and implications. Thus the famous many-worlds interpretation comes with the bold metaphysical conjecture of the possibility of the existence of an infinite number of parallel universes which are created when the world splits into an array of copies of itself each with its own unique measurement outcome. The competing GRW theory includes an assumption of fundamental and irreducible randomness in the world, while Bohmian mechanics presupposes a world which at its most basic level is perfectly deterministic and dynamically complete. Some of the latest foundational works in quantum mechanics focus on the problem of how to interpret the concept of the wave function,[5] and this issue opens the door to even more metaphysics inside physics, including the widely discussed dispositional account of laws and properties (cf. the contribution by Mauro Dorato and Michael Esfeld).

4. The Contents of This Volume

The current volume offers but a sample of the extensive work that has recently been done at the border of metaphysics and the physical sciences. The reader can find here a rich variety of approaches and perspectives on the relations connecting metaphysical considerations with questions coming from our best physical theories. Some contributions put stronger

[4] See the recent special issue on principles in quantum gravity of *Studies in the History and Philosophy of Modern Physics* edited by Karen Crowther and Dean Rickles.

[5] For an excellent, up-to-date overview of this topic see Ney and Albert (2013). Cf. also Vincent Lam's contribution.

emphasis on the metaphysical side, while others pay more attention to the technical aspects of selected theories in contemporary physics. Together, they span the full spectrum from metaphysics to physics.

The opening chapter by Kerry McKenzie and Steven French sets the stage for all the subsequent contributions by tackling head-on the controversial question of the relation between metaphysics and science. As has been already noted, many contemporary philosophers of science denigrate mainstream analytic metaphysics for being woefully detached from scientific endeavors. While sympathetic to the scientifically-oriented motivations behind this general charge, McKenzie and French nevertheless attempt to rehabilitate at least some parts of contemporary metaphysical inquiries. Their main claim is that metaphysics can play a useful heuristic role by providing scientists and philosophers of science with a set of conceptual tools with which they can analyze and interpret their own theories. In their earlier analysis of this problem McKenzie and French expressed the hope that their heuristic approach to metaphysics can supply a criterion distinguishing between metaphysical projects that have the potential of being scientifically interesting from the projects that don't. Such a criterion can be given in the form of the compatibility principle, which states roughly that the constraints placed by a metaphysical theory on some entities should be compatible with at least some scientific theories that invoke these entities. McKenzie and French carefully discuss and repel a popular argument against the compatibility principle based on the premise that metaphysics deals with possible objects while science concerns itself with actual objects only.

However, the main challenge to the heuristic approach analyzed in the current contribution is that, as it appears, there is some tension between the compatibility principle and the fact that it is difficult, if not outright impossible, to predict the future development of scientific disciplines. If metaphysical theories are to be legitimized by their potential applications to the interpretative problems created in the wake of scientific progress, then it seems unreasonable to criticize even the most ill-conceived metaphysical speculations for fear that they might just happen to be useful in the context of some future scientific developments. McKenzie and French respond to this challenge by observing that the heuristic justification of metaphysics is highly conditionalized on the specific goings-on in science and natural metaphysics, and therefore it cannot offer a blanket recognition for any metaphysical speculation whatsoever. The fact that a given metaphysical project can fortuitously become a useful tool for future scientists and philosophers of science does not relieve the metaphysicians from their duty to engage with current science.

The ambitious goal of Douglas Kutach's contribution is to "demonstrate how a general metaphysical framework can be fruitfully integrated with contemporary fundamental physics to help advance our understanding of quantum ontology" (p. 55). The proposed framework of 'empirical fundamentalism' distinguishes between a fundamental reality – the actual world – and a 'derivative' one. Whether something is fundamental or not is a primitive fact about that thing that resists further analysis. The fundamental and the derivative realities are related to one another by a process of 'abstreduction', "an ontological reduction where the derivative quantity is identified as an abstraction from fundamental quantities" (p. 58). Once this framework is in place, an 'empirical surrogate', i.e. a formal representation of the phenomena, is introduced to bridge the gap between the fundamental and the derivative. Kutach then applies his 'empirical fundamentalism' to classical spacetime theories and to non-relativistic quantum mechanics. In both cases, a Machian spacetime containing point-particles carving out inextendible worldlines serves as empirical surrogate. An identification of what is fundamental then requires a delicately balanced trade-off between parsimony and the avoidance of "conspiratorial arrangements of attributes" (p. 65). This balance, Kutach maintains, leads the empirical fundamentalist to side with the space(-time) substantivalist and against the main interpretations of quantum mechanics as they are standardly understood.

Quantum physics offers very rich grounds for metaphysics and its frequent appearance in this volume should surprise no one. One of the recurring themes in this context is the question of just what ontology quantum physics recommends, requires, or rules out in the light of such challenges as the measurement problem and quantum non-locality. Vincent Lam argues that an ontic structural realist framework is what solves the main interpretative conundrum for an advocate of Bohmian mechanics trying to fit wave functions into her primitive ontology of elementary particles. The original challenge for the Bohmian camp insisting on a 'primitive ontology' of some fundamental material stuff floating around in three-dimensional space is to accommodate the all-important wave function, which is a denizen not of 3-space, but instead of a generally very high-dimensional configuration space. Lam discusses three Bohmian proposals of addressing this challenge, and settles for the third, which takes the wave function to be nothing but a codification of relations obtaining among the local beables, the Bohmian particles. On this understanding, the wave function encodes a holistic property of the entire configuration of these particles, a highly non-local relational complex.[6] This offers a natural entrée for ontic structural realism, which suggests that we thus interpret the wave function as a physical structure in 3-space (or 4-spacetime).

Lam continues that although ontic structural realism salvages the Bohmian insistence on a primitive ontology in the case of non-relativistic quantum mechanics (and perhaps in quantum field theory), matters get complicated in the more speculative realm of quantum gravity. On many approaches to quantum gravity, spacetime turns out to be emergent rather than fundamental.[7] If spacetime vanishes from the fundamental ontology in quantum gravity, then the structuralist salvation of the primitive ontologist's troubles with the wave function and its attendant manifest explanatory stratagem will no longer succeed, even though a purely structuralist framework remains a live ontological option, Lam argues.

In a similarly structuralist vein, Dean Rickles and Jessica Bloom defend the ontological view according to which on the fundamental level there are no things but irreducible relations only. Thus things ought to be reconceptualized in terms of primitive relations. In a clear exemplification of the approach described in the previous section as 'intrinsic', the authors argue that the overwhelming support for this claim is provided by modern physical theories, especially ones that Einstein referred to as "principle theories". Rickles and Bloom insist that their variant of ontic structural realism (which falls between the moderate structuralism of Michael Esfeld and Vincent Lam and the radical eliminativist structuralism of James Ladyman and Don Ross) has the potential to lead to new kinds of advancements in physics. If true, this contention would strengthen the claim of the 'from without' relationship between metaphysics and physical theories. However, most of the chapter is devoted to the discussion of four examples from contemporary physics that can either provide an object-free framework or receive a better explanation within such a framework. Among these examples are the application of category theory to the formalization of physical theories, the notorious case of quantum entanglement, and the phenomenon of duality present in many contemporary physical theories.[8]

The abandonment of the ontology of individual objects in the context of the quantum theory is similarly urged by Olimpia Lombardi and Dennis Dieks. Their proposal is to reinterpret quantum particles as bundles of properties, without positing any kind of substratum or haecceity.

---

[6] Cf. the contribution by Dorato and Esfeld for a similar claim.

[7] Cf. Huggett and Wüthrich (2013) and the contribution by Vassallo.

[8] It should be noted, however, that the question of which structuralist position is supported by the reformulation of physical theories in terms of category theory remains a controversial and debated issue (see Lam and Wüthrich forthcoming).

In their approach, different from the traditional bundle theory, bundles associated with individual quantum systems consist of type-properties (represented by self-adjoint operators) and case-properties (represented by eigenvalues). However, due to the quantum-mechanical limitations, such as the one imposed by the Kochen-Specker theorem, fully determinate bundles of actual case-properties are impossible. Lombardi and Dieks note that their version of the bundle theory can throw new light on the notorious problem of the indistinguishability of quantum particles. Two bundles of identical type-properties and case-properties are considered one whole with no individual components (this is best described as one bundle which is doubly instantiated). But in some cases a twice instantiated type-property may receive two different case-properties (for instance two different locations of a wave packet). Consequently, individuality turns out to be an emergent property, since its applicability appears to depend on contingent physical facts.

Tomasz Bigaj in his contribution focuses on the extrinsic aspect of metaphysical analyses in the interpretation of physical theories. The starting point of his discussion is the general question of how to identify fundamental physical objects in counterfactual situations. In response to this question Bigaj advances a metaphysical doctrine he calls "serious essentialism". In a nutshell, this position asserts that the only acceptable way of determining which possible objects can represent *de re* a given actual object is with the help of purely qualitative properties, typically referred to as 'essential'. The doctrine of serious essentialism can be used to evaluate some debates in the foundations of physical theories, such as the controversy over the status of spacetime in the light of the infamous hole argument, and the problem of the indiscernibility of quantum particles which results from the permutation invariance. Serious essentialism appears to give support to a moderate version of spacetime substantivalism, as well as to the claim that the absolute discernment of particles of the same type is possible, despite the symmetrization postulate.

The connection between the concepts of qualitativity and symmetry is scrutinized by Thomas Møller-Nielsen. The main goal of his article is to criticize the commonly accepted doctrine ("Received View") that physical symmetries indicate the superfluousness of certain non-qualitative structures. Møller-Nielsen argues that there are symmetries, such as the Galilean boost symmetry in Newtonian gravity (relating solutions that differ with respect to the absolute velocity of all matter) and the gauge symmetry of electromagnetism, which connect qualitatively discernible solutions. Moreover, the author points out that the view that symmetries should act as guides to redundant nonqualitative structures runs into two serious problems. One problem is that, by analogy with the famous Leibniz shift argument aimed at showing the non-existence of absolute space, the fact that permutations of intrinsically indiscernible particles are symmetries leads to the unpalatable conclusion that individual objects (particles) should be eliminated as well. The second problem arises in connection with Tim Maudlin's distinction between kinematic and static shifts in Newtonian gravitation theory. Plausibly, the argument leading from the existence of a symmetry connecting two qualitatively indistinguishable scenarios before and after a kinematic shift to the excision of the superfluous nonqualitative structure can be resisted, as we can use indexicals to distinguish the current location from the shifted one.

Matteo Morganti urges the reconsideration of a relational, as opposed to a substantivalist or an eliminativist, metaphysics of time. Arguing in a naturalistic mode, he bases his stance on Julian Barbour's 'Machian' program of formulating non-relativistic physics and outline of a quantum theory of gravity. He contends that Barbour's insights would better serve a relational interpretation of time, rather than the radically eliminativist gloss that Barbour himself offers. On this suggestion, Barbour's 'best matching' relations would constitute the temporal relations connecting the states of a system at different 'times'. Morganti proposes to mix this temporal relationalism with substantivalism about space as at least a serious interpretive option. This preservation of a Machian perspective on time without Barbour's attendant eliminativism, Morganti argues, promises a more direct recovery of time and consequently avoids the shortcomings of Barbour's 'error theory' to account for our phenomenology of temporality. Time, on Morganti's proposal, would thus rather straightforwardly emerge from the ordering of the fundamental ontology of spatial states. To explicate the emergence of time (and possibly space) from an ontology devoid of substantial time (and space) remains one of the most urgent tasks of research programs in quantum gravity.[9]

One paper to address the emergence of spacetime in quantum gravity is Antonio Vassallo's. Vassallo considers what it might mean to incorporate the lessons of general relativity into extensions of the theory. He amalgamates these lessons into what he dubs "GR-desideratum" – essentially the demand that the classical limit of a successor theory be a theory formulated on a d-dimensional pseudo-Riemannian spacetime with the d-dimensional diffeomorphisms as its gauge group. The main part of the paper is dedicated to an investigation of two 5-dimensional generalizations of general relativity, Kaluza-Klein theory and Paul Wesson's 'space-time-matter' approach it inspired. In the latter, the complete physics of our apparently

[9] Cf. Huggett and Wüthrich (forthcoming).

4-dimensional world, including, importantly, its matter and energy content, is induced by the fundamental 5-dimensional metric tensor. Vassallo finds that these extensions of general relativity only satisfy a relaxed version of the GR-desideratum and expresses his hope that such a relaxation might benefit other troubled issues such as the problem of time in canonical quantum gravity.

Ioan Muntean also considers the metaphysics of physics beyond the currently accepted theories – in his case string theory. String theory is an attempt at unifying the physics of the standard model of particle physics with general relativity, our best theory of the fourth force, gravity. In line with the outlook of this entire collection, he assumes that contemporary physics is a legitimate and important resource for building a metaphysical program. He contends, however, that this does not only apply to securely confirmed theories such as quantum mechanics and general relativity, but also to more speculative, 'good enough' theories such as string theory. Muntean argues that the resulting insights from 'string ontology' and 'string ideology' suggest a model-based and pluralistic metaphysics rather different from what one might expect from quantum field theory. In particular, though some weaker form of fundamentalism can be maintained, the 'duality-pairing' relation among models of string theory suggests a non-reductive, non-hierarchical ontology. Muntean concludes to a pluralism about fundamentality, grounding, parthood, and modality. We agree with Muntean that the novel conceptions of standard metaphysical notions in string theory, and in other programs of quantum gravity, should motivate any philosopher to take a closer look.

Mereology (the formal theory of parthood) is often considered a prime example of a bona fide metaphysical theory. Adam Caulton argues that the principles of mereology are not entirely immune to empirical falsification. To prove his case he uses fermionic composite systems as an example. The standard quantum theory imposes an important restriction, known as permutation invariance, on the joint states of identical fermions. Caulton insists that permutation invariance be interpreted as reflecting a representational redundancy of the standard formalism of quantum mechanics. This heterodox interpretation of permutation invariance prompts a revision of some quantum concepts, such as the notion of an entangled state. Focusing his attention on the assemblies of fermions which are non-entangled in this sense, Caulton shows that their states can be represented by subspaces of the single-system Hilbert space, while the relation of being a part of is best interpreted as the relation of subspacehood. Given the appropriate translation rules from the quantum-mechanical formalism to the language of mereology it can be proven that the principle of mereological fusion is violated, i.e. there are subsystems of the entire assembly that do not jointly constitute a whole satisfying the conditions of a mereological fusion (even though subsystems create *fermionic* fusions, which however lack the required Boolean structure). At the end of his paper Caulton suggests that mereology can be saved by admitting arbitrary mereological fusions which are not fermionic systems.

Mereological considerations intersecting with classical and quantum physics also take center stage in Andreas Hüttemann's contribution. He argues that a particular brand of physicalism – what he dubs "part-whole physicalism" – is not supported by considerations of classical physics and quantum mechanics. Part-whole physicalism asserts that the "properties of compound systems are the way they are in virtue of the properties of their parts", plus some relational facts including concerning how they interact. Crucially, part-whole physicalism includes a reference to the 'in virtue of' relation, a grounding relation of asymmetric determination. In other words, part-whole physicalism demands that for all wholes, there exist parts which, together with relational facts about these parts, asymmetrically determine the whole. On this conception, part-whole physicalism can only be true if the determination is asymmetric, i.e., if facts about a whole are partially grounded in facts about any of its specific part, but not vice versa.

Hüttemann then goes on to argue that in systems described by classical physics as well as in dynamics of multipartite quantum systems, while the properties of the whole are indeed determined by its parts, asymmetry of determination does not obtain: just as the whole is determined by its parts and relations among them such as laws of composition and interaction, a part is equally determined by the whole, the other parts, and relations among them. In other words, the partial determination between the whole and any of its parts is mutual and thus symmetrical. So unless the part-whole physicalist comes up with some way to break the symmetry of determination she cannot hope to succeed. Hüttemann does not see how such an asymmetry could be introduced, at least not as long as there is no additional case for insisting that all determinates of a determinable need to be at the same level (either all micro or all macro) and as long as the macrostate contains sufficiently 'fine grained' information.

Jessica Wilson's paper contains a comprehensive metaphysical analysis of the notion of emergence. All accounts of emergence should reflect its two aspects: synchronic dependence of higher-level entities on lower-level entities, and ontological and causal autonomy of higher-level entities. In spite of the enormous diversity of existing explications of emergent dependence and emergent autonomy, Wilson claims that there are actually only two schematic conceptions of higher-level metaphysical emergence, which she calls "Strong emergence" and "Weak emergence". They can be viewed as two possible responses to the problem of higher-level causation, articulated by Jaegwon Kim, which arise from the fact that certain plausible requirements regarding how special science entities cause effects cannot be jointly satisfied. Strong emergence solves the difficulty by denying physical causal closure, that is, by admitting that higher-level entities have novel powers which are not identical with lower-level powers. Weak emergence (or non-reductive physicalism), on the other hand, accepts overdetermination and assumes that the set of higher-level powers is a proper subset of the set of lower-level powers. The article shows how various and seemingly diverse accounts of emergent dependence and autonomy can be subsumed under the two broad schemas mentioned above.

One of the central questions of naturalistic metaphysics concerns the ontological status of the laws of nature. Primitivism and dispositionalism are two dominating non-Humean solutions to this problem. Mauro Dorato and Michael Esfeld compare these two metaphysical options available to anti-Humeans using two case studies: one from classical physics and one from quantum physics. Classical physics is founded on the Newtonian laws of motion, of which the first law is uninstantiated due to the impossibility of screening off gravitational interactions. Dorato and Esfeld argue that primitivism, in contrast to dispositionalism, has difficulties with accommodating uninstantiated laws of nature. Dispositionalism, in turn, implies that there is no possible world in which actual physical properties (such as mass) would be instantiated and yet the laws would be different. The second case study discussed by the authors is the 'primitive ontology' approach to quantum mechanics, as exemplified by Bohmian mechanics and two versions of the GRW theory. The main difference between the dispositionalist interpretations of the quantum and the classical cases is that in the quantum scenario laws encoded in the wave-function are grounded in global and holistic properties of matter, rather than local or intrinsic properties of individual particles. In spite of this setback, the authors maintain that dispositionalism is to be preferred over primitivism, since it can better accommodate the fact that the (nomological) wave-function develops according to the Schrödinger equation, and that there are many wave-functions that are compatible with the same dynamical laws.

The problem of determinism also belongs to the canons of scientifically motivated metaphysics. Marek Kuś considers the question of whether classical and quantum physics admit genuine randomness which is not a mere result of the limitation of our knowledge. Classical mechanics leaves some room for indeterminism, as the existence of systems with non-unique solutions for some initial conditions attests. However, quantum mechanics stands a much better chance of proving the existence of inherent randomness in the world. Experimental confirmations of the violation of Bell's inequalities are usually taken as indicative of the non-deterministic character of quantum processes, but this conclusion can be questioned on the basis of the fact that we have to assume first that the selection of measurement settings is genuinely random (cf. also Wüthrich 2011). However, as Kuś argues, a better argument is provided by the phenomenon of the amplification of randomness. Such a process starts with a sequence of bits of a given randomness (or even a perfectly deterministic one) and produces new sequences of an increasing degree of randomness. It has been proven that the amplification of randomness is impossible in the classical regime. However, recent investigations have revealed that by using a string of Bell-type experiments it is possible to achieve a genuine amplification of randomness, thus confirming that quantum mechanics outperforms classical physics in this respect.

There are some technical aspects of contemporary physical theories that arouse a particularly keen interest of philosophers of science. The notion of renormalization, used in the context of quantum field theory, belongs to that category. As Jeremy Butterfield and Nazim Bouatta explain in their extensive article, there are two main approaches to the problem of renormalization. According to the old school, renormalizability acts as a selection rule for theories. An acceptable quantum field theory has to be renormalizable, that is, it has to be possible to eliminate infinities occurring in this theory. The new approach, on the other hand, places emphasis on the general question why certain theories are renormalizable while others are not. This approach makes precise mathematical sense of the notion of a space of theories and a flow on this space, and using these concepts it offers an explanation of the renormalizability of some theories. Butterfield and Bouatta clarify in details the concept of a renormalization group flow central to this approach. At the end of the chapter they compare the concept of universality present in renormalization theories with the Nagelian conception of inter-theoretical reduction, and they argue that universality is a particular instantiation of the general philosophical idea of multiple realizability.

Tomasz Bigaj and Christian Wüthrich

REFERENCES

Bigaj, T. (2015). Dissecting weak discernibility of quanta. *Studies in History and Philosophy of Modern Physics* **50**, 43–53.

Callender, C. (2011). Philosophy of science and metaphysics. In: S. French and J. Saatsi (eds.), *The Continuum Companion to the Philosophy of Science*, pp. 33–54. London – New York: Continuum.

Crowther, K., Rickles, D. (2014). Principles of quantum gravity. *Studies in the History and Philosophy of Modern Physics* **46**, 135–326.

Fine, K. (2013). What is metaphysics? In: T. Tahko (ed.), *Contemporary Aristotelian Metaphysics*, pp. 8–25. Cambridge: Cambridge University Press.

French, S. (2011). Metaphysical underdetermination: Why worry?. *Synthese* **180**, 205–221.

Huggett, N., Wüthrich, C. (2013). The emergence of spacetime in quantum theories of gravity. *Studies in the History and Philosophy of Modern Physics* **44**, 273–364.

Huggett, N., Wüthrich, C. (forthcoming). *Out of Nowhere: The Emergence of Spacetime in Quantum Theories of Gravity*. Oxford: Oxford University Press.

Lam, V., Wüthrich, C. (forthcoming). No categorial support for radical ontic structural realism. *British Journal for the Philosophy of Science*.

Muller, F.A., Saunders, S. (2008). Discerning fermions. *British Journal for the Philosophy of Science* **59**, 499–548.

Ney, A. (2012). Neo-positivist metaphysics. *Philosophical Studies* **160**, 53–78.

Ney, A., Albert, D. (2013). *The Wave Function: Essays in the Metaphysics of Quantum Mechanics.* Oxford: Oxford University Press.

Ross, D., Ladyman, J., Spurrett, D. (2007). In defence of scientism. In: J. Ladyman and D. Ross (eds.), *Every Thing Must Go*, pp. 1–65. Oxford: Oxford University Press.

Saunders, S. (2006). Are quantum particles objects? *Analysis* **66**, 52–63.

Wüthrich, C. (2011). Can the world be shown to be indeterministic after all?. In: C. Beisbart and S. Hartmann (eds.), *Probabilities in Physics*, pp. 365–389. Oxford: Oxford University Press.

Steven French
Kerry McKenzie

RETHINKING OUTSIDE THE TOOLBOX: REFLECTING AGAIN ON THE RELATIONSHIP BETWEEN PHILOSOPHY OF SCIENCE AND METAPHYSICS

ABSTRACT. In a recent work, 'Thinking Outside the Toolbox', we mounted a qualified defence of analytic metaphysics in the face of ardent criticism. While sympathizing with other philosophers of science in decrying the lack of engagement of metaphysicians with real science when addressing central metaphysical problems, we also wanted to acknowledge the role that analytic metaphysics has played in providing useful tools for naturalistic metaphysicians. This double-edged stance compels us to identify what feature it is that marks out as problematic some, but not all, analytic metaphysics, and this we thought we could do by appeal to something we call here the *compatibility principle*. It now strikes us, however, that the approach we took in that earlier work is fundamentally unstable. After giving a streamlined presentation of our earlier argument, we will identify where we take the instability to lie. From there we shall make a more nuanced proposal for how naturalistic metaphysicians should regard the work of their analytic counterparts.

1. Introduction

A couple of years ago, we were fortunate enough to be invited to comment on the reflections of Michael Dummett on the state of contemporary analytic philosophy.[1] We were asked, in particular, to comment upon his lament regarding the present lack of engagement between philosophy and physics. As he put it,

[1] Michael Dummett was in turn invited to comment on our reflections in the same volume, but very sadly shortly after we finished writing our article he passed away.

In: Tomasz Bigaj and Christian Wüthrich (eds.), *Metaphysics in Contemporary Physics* (*Poznań Studies in the Philosophy of the Sciences and the Humanities*, vol. 104), pp. 25-54
Amsterdam/New York, NY: Rodopi | Brill, 2015.

> What is a genuine case of regret is the paucity of dialogue between philosophers and physicists. The generality of philosophers know too little physics to dare to venture to treat of the philosophical problems it raises, or to take due account of physical theories when addressing problems on which they bear... Never before, I believe, have philosophy and the natural sciences been so far apart. (Dummett 2012, p. 19)

We should emphasize that Dummett is similarly disparaging of 'scientistic' attitudes on the part of many scientists themselves – as he says, 'it is not from science that we know genocide is wicked, or that Michaelangelo was a great artist' – and also of the resultant 'shameful intimidation' of some philosophers, 'who hope that by humbling themselves before the sciences they will be entitled to share in some of their triumphalism'. But his frustration concerning the remove of analytic philosophy, and of analytic metaphysics in particular, from contemporary science is shared by many philosophers of physics, and we wanted to take the opportunity to consider how we ourselves stood with respect to the criticisms presented by our philosophy of physics colleagues.[2]

Our claim in that paper was that philosophers of physics – at least those who, like us, are interested in the *metaphysics* of physics – are not in any position to decry scientifically disengaged metaphysics *tout court*, because analytic metaphysics has proved a useful heuristic for philosophers of physics. Nevertheless, we also felt convinced that many of the examples philosophers of physics have cited in support of their anti-metaphysical stance raised genuine problems for metaphysics. What we therefore attempted to do was demarcate between the scientifically disengaged metaphysics that was *prima facie* somehow legitimate, and the scientifically disengaged metaphysics that we think ought to be condemned. Since then, however, we have come to regard the distinction as we drew it there as fundamentally unstable, and part of what we would like to do in what follows is to explain why.

In the ensuing, we'll therefore present a streamlined outline of the argument of our earlier paper, before going on to highlight the instability that we now perceive in it. In a nutshell, we have come to believe that the 'heuristic' justification we offered for (what we took to be) a subset of analytic metaphysics cannot but sanction all metaphysics whatsoever. However, while it might sound as though this casts analytic metaphysicians as (to speak crudely) the 'winners' of this debate – something that Dummett himself would no doubt have been unhappy with – we ourselves think it does no such thing. We think, rather, that this conclusion serves to highlight just

[2] Note that since our claim will be that the most extreme claims of both sides in this debate have to be tempered, what we have to say will also have critical ramifications for the avowed 'scientism' of some philosophers of science, such as Ladyman and Ross (2007).

how *conditionalized* the value of analytic metaphysics is from a naturalistic point of view. That conditionalized support for metaphysics, *conceived of as a tool for philosophers of physics*, may be compared with the criticisms we will by that point have mounted against metaphysics *conceived of as many analytic metaphysicians themselves do*. Although the picture that results is a complex one, we think the comparison reveals that, whatever the positive spin that can be put on it, the naturalistic backlash against analytic metaphysics remains well-motivated and that metaphysics needs to recommit to science if it is to succeed on its own terms.

Before we proceed with all that however some terminological and dialectical remarks are in order. By 'analytic' metaphysics, we will mean metaphysics that is 'non-naturalistic'; by 'naturalistic' metaphysics, we will mean metaphysics that somehow 'engages with', 'is continuous with', or is in some sense 'informed by' science.[3] While we appreciate that these are metaphors and that spelling out what they, and thus 'naturalistic metaphysics' itself, actually amount to is non-trivial (cf. Chakravartty 2013), for present purposes we will take it that there is good enough agreement at least on the extension of the term.[4] Furthermore, the idea that naturalistic metaphysics is a legitimate form of enquiry by virtue of its relative 'closeness' to the sciences is not one that will be questioned in this work. Though there are of course more empirically inclined philosophers of physics who would not see things that way, the principal motivation for the current paper is to investigate the claims naturalistic metaphysicians have made regarding the diminished status of analytic metaphysics relative to that of their own. Whether or not that latter status is itself legitimate is a matter for another day.

2. The Critical Background

As noted above, many contemporary philosophers of physics share the concerns that Dummett voiced regarding the insular nature of today's

[3] So, as an example of the former we include discussions of 'gunk' in mereology, and of the latter, we would include consideration of whether quantum mechanics supports monism; we will provide further examples below. Note that this distinction is made on the basis of the nature of the relevant considerations or discussion; one and the same metaphysician can work both sides of the divide. Note finally that if the reader is sceptical that there is a firm distinction to be drawn here, it is part of the *raison d'etre* of this paper to problematize precisely that assumption!

[4] The fact that metaphysicians tend to self-identify as one or the other of course lends support to this claim.

analytic metaphysics. These frustrations have been expounded in articles and works by a number of philosophers of physics – see, for example, Maudlin (2007), Callender (2011), and Price (2009) – but the classic statement of the view is without doubt Chapter 1 of Ladyman and Ross' *Every Thing Must Go*. As they put it, "one of the main contentions [of that work] is that contemporary analytic metaphysics, a professional activity engaged in by some extremely intelligent and morally serious people, fails to qualify as part of the enlightened pursuit of objective truth, and should be discontinued" (Ladyman and Ross, p. vii). Reading through their opening chapter as well as the other works cited above, one finds several grounds cited for making such scathing claims regarding the work of their colleagues across the hall.

2.1. *Metaphysics Is Frivolous*

Leaf through a handful of recent works in metaphysics and you will soon find yourself on trips to possible worlds populated by zombies (e.g. Chalmers 1996), disembodied spirits (e.g. Yoshimi 2007), unicorns (Lewis 1986, p. 88), dragons (ibid), trout-turkeys (Lewis 1991, p. 7), writer-cucumbers (Elder 2013, p. 75), gunk spheres (Sider 2003), and – in a chummy in-joke – the mereological fusion of David Lewis and a talking donkey (Hawthorne and Uzquiano 2011). Even the most unrepentant of analytic metaphysicians should be willing to concede that it at least *looks* bad that such paraphernalia is the stock-in-trade of today's metaphysicians, given their pretensions to be engaged in a noble intellectual pursuit. An obvious reply at this point would of course be that, if one were to rummage through the literature in philosophy of science then one could also pull out examples of such fantastical creatures as evil demons slamming doors open and shut and people with electron microscope eyes (Maxwell 1962). Nevertheless, one could plausibly claim that the use of such exotica in the latter case is merely to illustrate a thesis that could very well be stated without it; in the former, however, the idea that a gunk-sphere or a zombie is somehow very much a 'real' possibility does essential work in the arguments in which they are cited, since their very possibility is often taken to refute a rival thesis. As such, taking these preposterous entities ontologically seriously is crucial in the analytic context, and the seriousness that we feel able to impart to metaphysics correspondingly diminished.[5]

[5] Indeed, 'zombie' is one of the most cited terms in Chalmers' book, since his anti-reductionist thesis depends strongly on their possibility.

2.2. *Metaphysics Relies Too Much on Intuition Trading*

Rather than coherence with any body of theory outside itself, metaphysics often depends heavily on appeals to intuition in order to justify its claims. Nowhere to our mind is this better exemplified than in the debate around van Inwagen's 'special composition question'. When thinking about the general conditions under which a pair of objects could be said to form a whole, van Inwagen considers such options as stitching, gluing, and making contiguous, and asserts in each case that our intuition tells against regarding the resultant putative composite as a legitimate object (Van Inwagen 1990). These consultations of his intuitions moreover do the lion's share of the work in his argument for the notorious claim that there are no composites except composite living things. It should be underlined that few people in metaphysics buy into van Inwagen's theory: Ted Sider, for example, takes it to be refuted by his intuition that 'surely there is a gunk world in which some gunk is shaped into a giant sphere, and another where some gunk has the shape of a cube. Surely, there are gunk worlds that most of us would describe as containing objects much like objects from our world: tables and chairs, mountains and molehills, etc'.[6] In other words, Sider's intuitions are invoked to counter van Inwagen's intuitions, but whether the former count as any weightier than the latter is impossible for us to decide.

Again, we can concede that every theory, whether in philosophy or science, is going to have to rely on intuitions at some point. According to many accounts, scientists, for example, have hunches about what hypotheses to test, or which approximation methods might work – hunches that often prove very fruitful even if they ultimately cannot say why.[7] Similarly, many programs in naturalistic metaphysics often begin with intuitions that more received metaphysical pictures are not adequate for modern-day science (this is certainly the case with the structuralist metaphysics to be discussed below). But it seems that there is an asymmetry in the role of intuition in each case: in the scientific case, and arguably in the naturalistic case, the intuitions are functioning only as a starting point, a guide to what to try and justify by other means[8]; by contrast, in the van Inwagen case intuition itself has an essential justificatory role. Given that we no longer

---

[6] Sider (1993), p. 286. 'Gunk' is a term for matter that is resolvable into mereological parts ad infinitum.

[7] Although the role and overall significance of such hunches may be considerably less than such accounts presume, particularly given the role of heuristic factors discussed in numerous analyses of scientific discovery and pursuit.

[8] Here one might invoke some form of the discovery-justification distinction.

have God in the picture to underwrite the veracity of these intuitions, and given moreover the litany of errors that intuition has led us to, it is hard to avoid the conclusion that the reliance upon them for justification is a deeply problematic aspect of present-day metaphysics (cf. Putnam 1962).

2.3. *Metaphysics has Become Altogether Too Domesticated*

A curious feature of analytic metaphysics is that, over a period roughly contemporaneous with that in which it became decoupled from physics, it became preoccupied with the *ontologically fundamental.*[9] It was not that long ago that analytic philosophy was dominated by 'ordinary language' considerations, and ordinary objects in turn.[10] But for reasons that we won't attempt to chart here, the concern with ordinary objects was largely replaced with an express concern with the fundamental in particular.[11] Thus in the contemporary literature one finds assertions that the fundamental level can be resolved without remainder into a separable 'mosaic' of local matters of fact (Lewis 1986, Kim 1998); side-taking over Markosian's debate concerning whether the 'fundamental building blocks' of matter should be regarded as 'pointy' or rather 'maximally continuous extended' simples, akin to tiny bits of plasticine (see e.g. Markosian 1998); and debates over the modal implications of fundamental physics properties, such as quark color and flavor, played out in terms of whether or not quarks can be permuted among one another in space (Lewis 1986, p. 163). But the claim regarding the 'mosaic' is simply asserted as if quantum mechanics never happened[12]; the debate over the structure of fundamental entities is conducted as though that between Democritus and Anaxagoras remains fit to serve as the model; and the debate over the modal profile of the fundamental physics properties is conducted as though they and the laws they feature in are the same in all relevant metaphysical respects as their classical counterparts.[13] In sum, in each case it is simply *assumed* that the most fundamental regimes of the world can be regarded as a sort of 'doll's house' version of the world of everyday experience. But while few pretend to have a satisfactory positive picture of what fundamental reality

---

[9] See e.g. Paul (2012) for an explicit statement of this view.

[10] The 'descriptive' metaphysics associated with Strawson's *Individuals* is an example of what we have in mind.

[11] Callender (2011) gives some important parts of the story.

[12] Recall our point about the division between 'analytic' and 'naturalistic' metaphysics: Lewis of course did acknowledge that quantum mechanics might have an impact on the 'mosaic' account but the point remains that neither he nor many other metaphysicians explored the nature or extent of that impact.

[13] For commentary on this last debate, see McKenzie (2014).

is like, we *do* know that it is very hard to maintain that it is like the way that these classical pictures dressed up in modern physics clothing present it.[14] Even a passing acquaintance with the science pages of the newspaper would suffice to establish that.

Since it is this last set of criticisms that directly concern the relationship of metaphysics and physics, it is this set that we, as philosophers of physics, feel most confident in asserting. In what follows, therefore, we will take the fact that analytic metaphysics is overwhelmingly wedded to an outdated ontological picture to constitute the core criticism of it. Indeed, it is this feature which Ladyman and Ross themselves are most frustrated by. As they put, "mainstream contemporary analytic metaphysics" is 'no longer 'informed by real physics" and "has, like the nineteenth-century metaphysics against which Russell revolted, become almost entirely a priori". It is principally on these grounds that they hold it should be "discontinued".

This is fighting talk! But we should be absolutely clear at the outset that philosophers of physics such as ourselves, Maudlin, and Ladyman and co. are all likewise inclined to metaphysical speculation, albeit, we claim, of an avowedly 'naturalistic' sort. It therefore seems only fair to ask whether such philosophers of physics are really in any position to baldly assert that other approaches within the discipline ought to simply be drawn to a halt. To cut to the chase, our feeling is such a sweeping claim is ultimately unjustified. And we think that we can cite some facts about how philosophy of physics is done in support of that view.

3. The Heuristic Approach to Metaphysics

Our claim is that once we reflect on how philosophy of physics is produced in practice, we see that imposing a blanket ban on scientifically disinterested metaphysics would likely be counterproductive. As such, naturalistically inclined metaphysicians would be ill-advised to criticize metaphysics merely on the grounds of its disengagement from science.[15]

---

[14] Of course, different interpretations of quantum mechanics make reality look more and less classical. But quantum mechanics is not classical mechanics, and thus all of them will be non-classical in some respect.

[15] To be clear, our claim is based on how philosophy of physics is, as a matter of fact, 'done', and thus on facts about how we do things in practice; it is not based on a prescriptive claim about how we should do things, at least not in the first instance. Some philosophers of physics have claimed in response to our argument that the way we present metaphysics as being done is incredibly inefficient, and that what we have effectively shown is that all metaphysics

To flesh out this claim, we find it useful to explain how it is that we go about creating structuralist metaphysics of physics in particular. What makes this case so apposite – aside from the fact that it is the area in which we both work – is that structuralism is the metaphysical programme defended by Ladyman and Ross, the chief horsemen of the metaphysical apocalypse; and yet is a research program that is *up to its eyeballs* in all sorts of involved metaphysics. As such, it seems as appropriate ground for testing whether naturalistic metaphysicians such Ladyman are trying to have things both ways. So to begin, let us briefly introduce what we understand by the doctrine known as 'ontic structuralism'.

In a nutshell, ontic structuralism is the view that relational structure is ontologically fundamental. The doctrine proposes that if we take modern physics – principally, quantum theory and relativity – seriously, then the category of physical objects must be regarded as a derivative category, in contrast to the category of structure; or at the very least, that it can no longer be regarded as a category ontologically prior to that of relations and structure. It contends that the centrality of symmetry considerations in contemporary physics is a harbinger of deep ontological facts, that the identity conditions for both individuals and kinds are parasitic on structures in some essential way, and that global nomic concepts must replace more local, dispositional ones.

As even that cursory survey makes clear, ontic structuralism is characterized by a cluster of claims, any one of which is sorely in need of careful and sustained defence. Indeed, structuralists seem to have their work cut out just articulating exactly what it is that these claims mean in the first place. Thus in order to maintain their position, structuralists have had to say, first, exactly what it is that they mean by the categories of 'objects' 'structure', and 'relations'; they have also had to explain precisely what they understand by words like 'fundamentality', 'priority', 'derivativeness', and 'symmetry structure' in the context of physical ontology. With the meanings of these claims established (at least to some acceptable degree), they have then had to defend themselves against the gamut of objections that have been waged against them, including accusations of metaphysical incoherence, epistemic triviality, and their revival of a discredited Platonism. With so much work needing to be done, you might think, where did structuralists even begin?

should be 'made to order' and not simply taken 'off the peg' in the way that we present. We ourselves are sceptical that metaphysics would proceed better in this way, at least in all cases; but our argument in any case proceeds from how things are done, for better or for worse. In any case, we'll have more to say about this at the end.

The short answer to this question is that structuralists began by *looking at extant work in metaphysics*, and in our view that was as good a place as any to begin. To give some concrete examples, to articulate the core claim that structure is ontologically fundamental, structuralists have found it useful to draw on the work of Kit Fine, and in particular his work on ontological dependence (McKenzie 2013, French 2010). To articulate the relationship that they take to hold between symmetry structures, in particular, and the associated elementary particles, structuralists have found it helpful to borrow from work by Jessica Wilson on determinates vs determinables (Wilson 2012, French 2014). Ross Cameron's theory of truthmaking has been invoked to communicate how radical structuralists interpret physicists' talk about objects while denying that there fundamentally are any (French 2014, sect. 7.4.2.3; Cameron 2008). Simon Saunders has appropriated Leibniz's principle of the identity of indiscernibles, revamping it *à la* Quine to allow discernibility with respect to relations in the context of quantum mechanics (cf. Saunders 2003, Ladyman and Ross 2007, McKenzie 2013). And in the effort to defend structuralism against a well-known triviality objection, known as the Newman objection, David Lewis' notion of 'elite' or 'perfectly natural' properties has been taken to offer appropriate resources.[16]

There are many other examples that we could cite in this connection.[17] But the key point is that all these metaphysical packages that have proved useful to appropriate in structuralism were not only (and by definition) created independently of structuralism, but were moreover (by and large) developed independently of *any scientific considerations whatsoever*. Despite their usefulness in the fundamental physics context, neither Kit Fine nor David Lewis, for example, are exactly famed for their engagement with science – indeed in the latter case, often quite the opposite. Cameron's version of truthmaker theory was developed to understand talk about tables and chairs, and Leibniz' principle of the identity of indiscernibles was originally articulated several centuries too early to hope to incorporate the principles governing the quantum ontology that it subsequently helped to illuminate. We therefore see that scientifically disengaged metaphysics has, at least in many cases, provided us with a set of resources for doing the sort of metaphysics that resolutely *does* engage with modern

[16] See Melia and Saatsi (2006) for discussion, but also Saunders and McKenzie (2014).

[17] Outwith the context of structuralist philosophy of physics, we might mention how Meinard Kuhlmann (2010) has appropriated the trope ontologies of Keith Campbell and Peter Simons in the context of algebraic quantum field theory, and how Michael Esfeld, Mauro Dorato and others have appealed to the concepts of dispositional properties developed by Mumford and Bird to interpret the GRW approach to quantum mechanics (Dorato and Esfeld 2010).

physics. As such, it strikes us that we can and should view at least *some* constructions of analytic metaphysics as useful tools for shaping our own naturalistic accounts. This view of analytic metaphysics as the source of a set of resources that can be applied, appropriated, and generally used and abused by philosophers of physics in the process of developing naturalistic accounts, we have dubbed the 'heuristic approach' to metaphysics.

Indeed, in our view there is a neat analogy between, on the one hand, philosophy of physics and analytic metaphysics, and on the other, physics itself and pure mathematics. Just as it was useful to Einstein that the theory of non-Euclidean geometry was there for the taking when the moment arose, so it was useful to eliminative structuralists that a theory of dependence compatible with the elimination of the dependent entity had been developed. Likewise, just as it was useful for the development of particle physics that the theory of Lie groups was largely completed by the time the appropriately high-energy regimes could be probed, so it was beneficial to the defender of the Everett interpretation that a theory of personal identity that makes decision-making make sense in branching universes was already on the market.[18] And just as it was fortuitous that the theory of imaginary numbers was fit for use at the advent of the quantum revolution, so it has proved useful that various metaphysical packages were in place to provide possible frameworks for its interpretation, including Saunders' form of Leibniz's PII but also theories involving haecceities (see French and Krause 2006). Now, to be clear, nothing in this analogy is supposed to discourage the development of 'made to order' frameworks that engage (more or less) directly with the physics, such as the metaphysics of non-individuals and the associated formalism of quasi-set theory – any more than physicists should be discouraged from developing mathematics as and when new empirical situations arise (ibid). But nonetheless, just as areas of pure mathematics subsequently proved useful in physics it cannot be denied that empirically disengaged metaphysics has in the past proved useful to philosophers of physics. And given that the deliverances of 17th century, rationalist metaphysician have been usefully appropriated by the philosopher of quantum physics, it seems it would be folly to try to predict in advance what will and will not prove similarly useful in the course of time.

In our view, then, scientifically disengaged metaphysics can and has performed a useful function in naturalistic contexts, since it provides us with raw materials from which our own theories can be developed. And

[18] Of course, this is not to say that the relevant mathematics was developed entirely independently from the physical context (see Bueno and French forthcoming).

once that much is conceded, we think that it becomes very problematic to baldly assert that it should be 'discontinued'. It seems, rather, that doing so would be to simply bite the hand that feeds us.

4. Reining in the Metaphysics

The above considerations in support of analytic metaphysics undermine the most extreme claims regarding scientifically disengaged metaphysics. But it must now be acknowledged that there seems to be a tension in what we have said so far. We opened up this paper with a litany of grievances that philosophers of physics have had against analytic metaphysicians, and it seems to us that these remain strong grounds for deploring analytic metaphysics as currently practiced. We then said, however, that analytic metaphysics had played an important role in naturalistic metaphysics, and that it is to be valued for that reason. So are we with the analytic metaphysicians, or against them?

However – and not unusually for a dichotomy – this last ultimatum is much too simplistic. It should be obvious that disavowing blanket statements to the effect that *all* contemporary work in an area is worthless and should be abandoned is compatible with regarding *some* of that work in precisely that way; and it was such a differential attitude that we ourselves proposed in *Thinking Outside the Toolbox*. But if this is the attitude that one wants to take, then one is clearly obliged to say what it is about the offending cases that *makes* them offensive, and what it is about the acceptable cases that gets them off the hook. So given that the considerations of the last section suggest sanctioning some metaphysical projects, although we have as yet no clear reason to say all, let us make a normative distinction to siphon such projects into two classes, which we shall (somewhat artlessly) call 'Type I' and 'Type II':

Type I: metaphysics that is scientifically disinterested and that, at least *prima facie*, doesn't *need* to be so interested, or even that might *have* to be so disinterested;[19]

Type II: metaphysics that is disinterested but that *should not be*.

[19] In this paper, we are staying quiet on the issue of whether there is a body of metaphysics that can be regarded as legitimate enquiry but to which science could not contribute in principle, so that such metaphysics would *have* to be scientifically disinterested. This issue however is discussed in more detail in McKenzie, 'The Plurality of Priority' (in preparation). See also Bealer (1987).

Clearly, Type I metaphysics is the metaphysics that we want to protect, want to be regarded as *legitimate*, despite its disengagement from science; Type II is that which we wish to be cast to the flames. But while it seems clear that there is a normative distinction to be drawn here, the grounds on which the distinction is to be drawn are less so. How are the two types to be identified?

Since the aim, presumably, is to come to some sort of reflective equilibrium in our judgments, let's start off just trying to characterize the two types extensionally. Beginning with metaphysics of Type I, it seems clear that this category pertains to the 'good' metaphysics that we think can be defended, and if we go with what we've said about the role of analytic metaphysics in structuralism then it seems that *anything that has demonstrated its usefulness in naturalistic contexts* should be filed into this category. Thus into Type I go Leibniz's PII, Fine's theory of ontological dependence, and whatever it was that Lewis said about 'eliteness' that helped block the triviality objections to structuralism.[20] Into Type II, by contrast, will get filed the metaphysics that we vilified at the outset – so that, at the very least, Lewis' assertion that the fundamental level can be regarded as a 'mosaic' of local matters of fact, Markosian's debate over whether the fundamental entities are pointlike or continuous, and the debate in modal metaphysics over whether quarks can freely recombine, will all feature here.[21] These, recall, were regarded as problematic on account of the fact that they were not paying sufficient attention to science.

Whatever it is that ultimately grounds the distinction between two classes, it strikes us that the above examples should be classed as they are. So now we must ask *what it is* about, in particular, those examples classed as Type II that makes it the case that they *should* have engaged with some relevant science, even though they did not, given that we don't insist on any and all metaphysics doing so?

In a nutshell, the reason that these projects in particular strike us as the sort of thing that *should* engage with science even though they do not is simply that they *putatively refer to things that science itself is directly concerned with.*[22] After all, these projects are all taken to concern the *ontologically fundamental*, and given physicalism – commitment to which

[20] As we shall see below, however, the 'elite' properties are taken to have more features than this in Lewis' system, and not all the claims Lewis made about them will end up in being classified as Type I.

[21] Lewis' assertion that has of course come under withering attack by many philosophers of physics; see e.g. Maudlin (2007).

[22] We might say that they concern *physical* ontology in addition to what would normally be regarded as the *metaphysics* of that ontology.

"is about as close to a bit of orthodoxy as one will find in contemporary philosophy" (Hall 2010) – metaphysicians themselves will claim that the fundamental regimes of the world are going to be described by physics, or at least that they will be if they are to be described at all. But it seems obvious that one cannot simply *postulate* that things described in physics have such-and-such features: one has to actually *check* that they do in fact have those features, or at least that they can be reasonably claimed to, and moreover that one must be willing to give up on the idea that they do have those features if the physics seems to contradict it. Moreover – at least when we wrote *Thinking Outside the Toolbox* – it struck us as *entirely uncontroversial* that one should demand of metaphysicians that they incorporate the relevant findings of science regarding the entities they are interested in, *whenever there are such findings*. After all, here we are simply echoing Dummett's lament that 'the generality of philosophers [fail to] take due account of physical theories when addressing problems on which they bear'. And how could one possibly take issue with *that*?

To a first approximation, then, let us say that the problem with the metaphysics that results in its being classed as of the problematic, Type II sort is that it violates the *compatibility principle*:

The compatibility principle: the constraint that any metaphysical theory invoking entities x and deployed at some time t should be compatible with at least some independent, well-supported, overall 'serious' scientific theory that directly describes or that is otherwise relevant to those entities, should such a theory exist at that time.

To repeat, this principle (or something like it) should strike one as *prima facie* basically unobjectionable. But let us make a few further comments about it. Firstly, the principle is clearly to be regarded as a first approximation: we do not ultimately want to formulate a principle so strong that philosophers can only ever hope to be the supplicants at the door of science, never ever to be permitted to contradict received scientific views on what it is that science is telling us. Nevertheless, we think we can expect such cases to be the exception rather than the rule; so let us insist on adherence to the compatibility principle as formulated above in at least the vast majority of cases. Secondly, whether or not a project in metaphysics is legitimate or not – that is, is to be cast as Type I or not – is a feature that can change with time. That seems right: what was defensible metaphysics in the 18th century will not in general be defensible today. Thirdly, given the difficulties in interpreting physical theories, respecting the compatibility principle still leaves an abundance of space for metaphysicians to disagree on how to conceive of fundamental ontology. That is of course unfortunate from a certain point of view, but also strikes us as philosophically 'healthy'.

Fourthly, however – and most pertinently for current purposes – while we intend the compatibility principle to disqualify many metaphysical projects from being legitimate objects of serious debate, it is nevertheless in other respects generous. For example, if one could claim that no conceivable answer to the special composition question could be regarded as incompatible with the science that we currently have, then debate over this question may (at least thus far) be regarded as belonging in the legitimate, Type I class. If there are more robustly naturalistic metaphysicians who feel that that just means the compatibility principle, while excluding some things, does not exclude enough, then recall that we are here *trying* to make space for scientifically disinterested metaphysics, given our observations regarding the practices in philosophy of physics; those who think that is too lenient are of course welcome to develop more demanding proposals. Note, however, that the problems we raised for the special composition question concerned not the *incompatibility of science* with assumptions made about the debated ontology, but rather the *reliance on intuition* when conducting debates about it.[23] Thus the problems voiced above concerned not so much the assumptions made about *what* was debated, so much as the *way* in which the debate was conducted and the associated standards of evidence. And while the reliance on intuition certainly does strike us as problematic, we are also acutely aware that we do not have a well-developed epistemology of metaphysics in general (nor, indeed, of mathematics), and as such we fear that if we disqualify the debate over this question *merely* for its reliance on intuition, then we run the risk of throwing out the naturalistic baby with the analytical bathwater.[24] In any case, that discussion over what (epistemological) principles should be added to our (ontological) demand of compatibility with science is one for another day.

At this stage, then, we take the compatibility principle to disqualify many extant non-naturalistic metaphysical projects as legitimate ways to occupy one's time, while not taking it to disqualify all such projects. And just to repeat, although we could appreciate the view that in demanding

[23] This of course is not to say that there are no conceivable *ontological* objections that one could make to the debate around the SCQ; see for example the criticisms in Ladyman and Ross (2007, p. 21), and McKenzie and Muller (unpublished). Our point here is simply that the problems *we* cited above concerning the debate around this issue were not these same problems.

[24] Empiricists of course will be perfectly happy with this conclusion, but as naturalistic metaphysicians we are operating under the assumption that metaphysics that is somehow 'continuous' with science is in better shape.

mere compatibility we have not gone far enough, it is our *aim* to formulate a principle that rules out certain projects while being otherwise lenient. Some such generosity is deserved, we have argued, given our observations regarding the appropriation of plenty of scientifically disinterested metaphysics in the service of philosophy of physics, observations that prompt taking what we have called the 'heuristic approach' to metaphysics.

As stated at the outset, however, we are now worried that this 'half-way house' attitude to metaphysics is fundamentally unstable. In particular, we are worried that insistence on the compatibility principle is actually inconsistent with the heuristic approach to metaphysics. Since the compatibility principle strikes us as completely unobjectionable, and since (something like) the heuristic approach to metaphysics seems likewise unassailable given the history of philosophy of physics as practiced, this situation strikes us as verging on the paradoxical. But before we explain what we take this perceived instability to consist in, and what we think we should say in the face of it, it will be helpful to discuss how metaphysicians themselves have responded to the allegation that their work violates (something like) the compatibility principle, and that it is deeply problematic in consequence.

5. Metaphysicians Defend Metaphysics

A common response of metaphysicians to the claim that their work flies in the face of science is, in a nutshell, to simply deny that they *are* talking exclusively or even predominantly about the entities that are described in science, in spite of what may be initial appearances. According to them, while science can talk only of what is actual, what *they* are discussing are *possible* entities, and as such things of which science, as an investigation into the actual, knows only a tiny fragment.

Such a move is an expression of a general shift that has taken place in metaphysicians' own conception of metaphysics over the course of the 20th century: whereas the classical view of philosophy is as the search for what is necessarily the case, metaphysics is now more often characterized in terms of an investigation into possibility generally. That this is so is sometimes made explicit in places in the 'metametaphysical' literature: according to Conee and Sider, for example,

> Metaphysics is about *the most explanatory basic necessities and possibilities*. Metaphysics is about *what could be* and *what must be*. Except incidentally, metaphysics is not about explanatorily ultimate aspects of reality that are actual... (Conee and Sider 2005, p. 203).

Similarly, according to Lowe (2011, pp. 100, 106):

> metaphysics may [...] be characterized as the science of the possible, charged with charting the domain of objective or real possibility [...] All metaphysics is implicitly modal, because it is primarily concerned with kinds of things are possible or compossible, and only subsequently with what kinds of things are actual.

But if this is how analytic metaphysicians now conceive of their discipline, then it is easy to see how one may be led to believe that any apparent conflict with the compatibility principle may be effaced at a stroke. To be explicit: while today's metaphysicians are predominantly focused on the ontologically fundamental, and while the vast majority are physicalists when it comes to the actual world, when accused of conflict with actual science those metaphysicians may claim that the fundamental entities they are theorizing about *are entities of another world.* The net result of this, it appears, is that *nothing* discussed in metaphysics need ever fall foul of the compatibility principle, and all metaphysics is automatically recast as the legitimate, Type I class by our criterion.

How compelling is this move? Does the idea that metaphysics is 'the science of the possible' represent a get-out-of-jail-free card for analytic metaphysicians in the face of the complaints of Ladyman et al.?[25] We ourselves are pessimistic. Here we will enumerate just a few reasons why we remain distinctly unimpressed by this move.

(1) **It's unconvincing**. Lewis' assumption of locality, Markosian's debate over maximally continuous vs 'pointy' matter, and the debate over the recombinability of quarks all have one feature in common: they all assume manifestly classical concepts when debating what they regard as fundamental. But if metaphysics is all about possibility space generally, then *why does everything look so classical*?[26] Presumably, if we take possibility space seriously then somewhere in it there are entities at least as complicated as the Lorentz-invariant smorgasbords of probability functions that one finds in quantum field theory. Why, then, are entities of comparable complexity not discussed and debated? Could it be that metaphysicians are only *saying* that they are interested in possibility generally to mask their unwillingness to forfeit the classical assumptions that make their life so much easier?

[25] This phrase is first used, to our knowledge, in Russell (1919); given the earlier quote from Ladyman and Ross concerning Russell's revolt, this situation is somewhat ironic! For an example of the contrasting view, see Bealer (1987).

[26] And again we take the point – noted by a referee – that Lewis took his pointillism to be a contingent thesis. Nevertheless, as we have said, many metaphysicians have happily ploughed this particular furrow without taking into account that the thesis might not only be contingent but actually false.

(2) **The literature suggests that conceivability implies possibility.** Most of the cases cited as possibilities and taken seriously as such in analytic metaphysics – such as the existence of infinitely continuous matter or the existence of gunk – do not follow, or at the very least are not presented as following, from systematic modal assumptions. Rather, they are taken to be possibilities merely because they can be conceived. But to hold that whatever can be conceived of is possible is to assume the 'conceivability implies possibility' link that has been subject to much scrutiny, especially in the wake of the work of Kripke.[27] As such, we feel that the burden of proof is very much on the metaphysician who would claim that the mechanisms through which humans conceive things in thought may be relied upon to provide us with evidence for what is metaphysically, and not merely epistemically, possible. Furthermore, the very fact that Lewis made such an impact on modal metaphysics suggests that metaphysicians themselves would ideally like to be more systematic in their theorizing than they would be were they to merely exercise their imaginations, given that Lewis explicitly rejects the idea that "every seemingly possible description or conception of a world does fit some world" (as of course he must if there is to be a role for his theory).[28] But if Lewis is to be our model of how to be systematic in our modal theorizing, then that offers up yet another reason as to why the 'science of the possible' move does not relieve metaphysicians of having to attend to actual science, for the following reason.

(3) **Systematic theories of possibility space can be falsified by actual physics.** Lewis' possible world analysis is widely regarded as the best – indeed for some the only – systematic theory of possibility on the market. But the tenability of Lewis' system rests on some non-trivial assumptions about fundamental properties: in particular, the assumption that all the fundamental properties are *intrinsic*.

[27] See Bird (2007). (This objection is of course related to the problems of reliance on intuition in metaphysics.) There is, of course, an extensive discussion of the relation between conceivability and possibility and of the manner in which the former might be defeasible (cf. Chalmers 2002, Yablo 1993). The upshot of such considerations – or so it seems to us – is a whole range of different frameworks of possibility, each dependent on the afore-mentioned relation plus defeasibility factors, in terms of which the modal claims of analytic metaphysicians should be indexed. How that then might bear on our account is a subject for another essay.

[28] Lewis, p. 87. Lewis himself claimed that buying into the conceivability implies possibility link "indiscriminately endorses offhand opinion about what is possible" (ibid.), but given the detailed literature on the nature of conceiving in this context we can imagine many philosophers taking issue with that characterization of the relationship.

Such an assumption is crucial for Lewis, for only if properties are intrinsic will they be open to free recombination, and it is the principle of recombination applied to fundamental properties that is the generator of Lewisian possible worlds.[29] As such, the free recombinability of fundamental properties is a *sine qua non* of his whole system. But if all the fundamental properties are to be intrinsic and freely recombinable, that of course means that all the *this-worldly* fundamental properties in particular must be; and by physicalism, that means that all the fundamental *physics* properties have to have these features. That the fundamental physics properties do indeed have these features is something Lewis himself never investigates or makes any real attempt to justify.[30] But there is in fact good reason to think that the fundamental physics properties are *not* in general freely recombinable, since there is good reason to think that they are not intrinsic – at least not *qua fundamental properties*.[31] Our support of this claim must here be confined to a thumbnail sketch, but our argument is basically this.[32]

Our most fundamental framework for physics (at least at the moment) is quantum field theory (QFT).[33] In this framework, the magnitudes of physical properties, such as mass and electric charge, can change with the energy scale in a way that is described by the renormalization group equation (or 'Callan-Symanzik' equation). Furthermore, since spacetime is represented as continuous in QFT, according to this framework there is no limit

[29] It is because this principle is taken to be expressible in language devoid of modal concepts that is taken to secure the reductive character of his theory – the feature standardly understood to earn it the accolade 'best' (cf Sider 2003, Sec. 3.5). Note that intrinsicality is not sufficient for free recombination, making the latter the stronger assumption.

[30] Once again, we acknowledge the point that, in response to quantum mechanics, at least in part, he does contemplate the suggestion that there might be actual fundamental non-spatio-temporal external relations. Nevertheless, see what he says at Lewis (1983), p. 16; (1986), p. 61.

[31] In our previous paper we argued for this conclusion on the basis of considerations from gauge theory – considerations that a respondent argued simply begged the question at hand (see Livanios 2012). While that criticism was correct and legitimate with respect to the original presentation of our argument, we nevertheless think that our conclusion still stands. What was missing from our earlier argument was an emphasis on the constraints that are placed on *fundamental* properties in particular: it is fundamentality constraints that necessitate the connection between the fundamental constituents of matter and gauge bosons.

[32] This argument is discussed in more detail in McKenzie (ms).

[33] It should be pointed out as well that we do not think that focusing our discussion on laws and properties as they are represented in quantum field theory in particular – and thus not some other assumed 'possible' physical framework – need beg any questions. For discussion, see McKenzie (2014), Section 4.

to how high these energy scales can grow.[34] It follows that properties can be regarded as fundamental in this framework only if they stay mathematically well-defined, and thus finite in magnitude, in the infinite-energy limit. This turns out, however, to be an extremely demanding requirement, and there is reason to think that it is satisfied only if the property occurs in a *local gauge theory* containing *only a small number of fermion types*.[35] For example, it turns out that the colour charge on a quark will behave as a fundamental property if, but only if, (1) there exist gluons in addition to quarks, and (2) there are at most 16 distinct types, or 'flavours', of quarks in the theory (see e.g. Srednicki 2007, p. 485). Should there be more flavours present, the colour charge will diverge in the limit so that it can no longer be regarded as fundamental after all. It follows from all this that the fundamental physics properties cannot in general be regarded as intrinsic, at least not *qua fundamental properties*; for the very fundamentality of such properties can be sensitive to what exists in addition to any given bearer of them, in any world in which they occur.[36] As such, we cannot simply postulate a world with fundamental physics properties, add and subtract objects and properties at random, and *a priori* maintain that what we obtain is a new manifold of fundamental properties. But that each free recombination takes us from one manifold of fundamental properties to another such manifold is the central postulate of Lewis' world-building system. Quantum field theory, and the fundamentality considerations it engenders, thus seems to strike right at the heart of what many take to be our most successful modal system.

---

[34] The continuity assumption might of course be given up in a quantum theory of gravity. But for the moment QFT is the best we have, and naturalism enjoins us to take it seriously. There is also increasing optimism that gravity can be incorporated into the basic framework of QFT, though what exactly that entails for spacetime continuity is a complicated issue on which we won't speculate.

[35] This is because these properties are required in order for a theory to be asymptotically free. While there is a more general class of fundamental theories (namely, the asymptotically safe theories), this is only class that is tractable enough for us to investigate at present. Again, see McKenzie (ms) for discussion.

[36] Of course, in a fuller discussion we would commit to how exactly it is that we understand 'intrinsic' here: suffice to say for now that lone object-based analyses seem entirely inappropriate in this context and are more inclined towards the sort of dependence-based account expounded in Witmer et al. (2005). But all that is crucial for present purposes is that these facts about the renormalization group prohibit a conception of intrinsicality that would allow for free recombination: we cannot add arbitrarily many new flavors of quark to a world that is in other respects like this one and expect colour to remain fundamental. Thus if colour is fundamental, we cannot add or subtract objects from worlds in which it in instantiated in the way free recombination demands; and that is enough to prove the present point.

We think that this example makes salient the fact that *even if* we are happy to take metaphysicians at their word that they are engaged in 'the science of the possible', and *even if* we regard the investigation of metaphysical possibility space as a defensible academic enterprise in principle, it may yet be that the actual can veto crucial assumptions about what those possibilities are. As such, it remains that those metaphysicians who follow Lewis in engaging in systematic modal metaphysics have to pay attention to any respected, well-confirmed science that describes the actual portions of their modal ontology, since it may reveal those assumptions to be false; in other words, if they want their systems to be taken seriously then they should respect the compatibility principle.[37] So if the aim of appealing to 'the science of the possible' was to get around the need for compatibility, it seems that really nothing has been gained.

6. The Tension

This, then, is where we're at. We've said that some scientifically disinterested metaphysics should be protected from naturalistic criticism, on the grounds that it has proved useful in a naturalistic context. We've said that nevertheless some metaphysics – namely, that which falls foul of the compatibility principle – should by contrast be condemned. We've also underlined that metaphysicians' attempt to recast any compatibility-principle flouting metaphysics as merely 'the science of the possible' did not succeed in exonerating them from their failure to comply with the principle. What, then, is our worry?

In a nutshell, our worry is this. While we still deny that nothing in metaphysics is in principle incompatible with actual science, in the way that the 'science of the possible' move would hope, we worry that, *given our argument for taking the 'heuristic approach' to metaphysics, we are not actually in any position to demand compatibility in the first place.* The reason for this is that, ultimately, we have only the dimmest idea of what changes in physics lie ahead of us.[38] How, then, do we know that the

[37] It may be worthwhile noting at this point just how much weaker the requirement that our theory of possibility be consistent with physics is than the demand that all possibility is physical possibility: were it not the case that the fundamental physics properties were intrinsic, some variant of Lewis' recombinatorial thesis might have had a shot at structuring a possibility space with physically impossible worlds in it.

[38] This isn't of course to say that there are no principles we can expect to govern theory development: we should at the very least demand correspondence in the limit (cf. Post 1971). But satisfaction of that requirement of course still underdetermines a great deal.

current metaphysical models, even though they seem to be in contradiction with actual physics and problematic for that reason, might not themselves come to be useful in the course of time? And given that we have resisted the blanket condemnation of contemporary metaphysics by Ladyman and Ross on these heuristic grounds, how are we then not committed to sanctioning a blanket free-for-all in metaphysics, in which any metaphysics – as domesticated, juvenile, and intuition-driven as you like – is to be regarded as immune from criticism?

This conclusion leaves us somewhat aghast! And since it seems to us that one cannot reasonably deny *either* that philosophers of physics have utilized analytic metaphysics to their benefit, *or* that the compatibility principle is a reasonable requirement on theories, *or* indeed that what lies ahead in science is something that we cannot at this point predict, this conclusion also strikes us as somewhat paradoxical. Before turning to what exactly it is that we should say in the face of this seeming paradox, we sketch some responses the naturalistic metaphysician might offer to see off the metaphysical free-for-all that seems to beckon at this point. Disappointingly, however, we don't think that any of them really succeed.

The naturalist might first point out that

(1) **Analytic constructions never survive in philosophy of physics in the form they were originally given**. Consider again the PII. While arguably ruled out by quantum mechanics in its original form, it resurfaced through the work of Saunders: following Quine, he extended the principle to cover both the 'intrinsic denominations' of objects as well as their relations to one another, and in so doing significantly changed the dialectic in the debate over quantum individuality.[39] This illustrates the fact that analytic constructions typically only function as a starting point for naturalistic metaphysics, for they are then altered and adapted in various ways to suit the needs of the physical situation. This, it might be claimed, blocks the idea that *the analytic constructions themselves* are actually useful in naturalistic contexts, because they generally need to be significantly altered; and if that is the case, then this blocks the idea that they should be valued insofar as we value naturalistic metaphysics.

But of course, this fact that analytic constructions are typically altered in various ways is perfectly consistent with our heuristic approach, in which we value analytic constructions as tools for the development of more

[39] On how this Quinean form is not the same construction as the Leibnizian PII, see Ladyman and Bigaj (2010).

tailor-made theories. After all, the mere fact that a tool is useful as a starting point only does not make it any less of a tool. A much better objection to the idea that even compatibility-principle flouting metaphysics might prove useful in the future is the widely-held belief that:

(2) **Physics is likely only going to get less classical, not more.**[40] And should it do so, it is obviously going to move further and further away from the kinds of initial intuitions that motivate analytic constructions. Thus insofar as a big part of the problem with contemporary metaphysics is that it is so stubbornly classical, if what prompts the worry that we are committed to a metaphysical free-for-all is that we don't know what physics will throw at us in the future then we are worrying about nothing.

While this point seems broadly compelling, we ourselves are less convinced that things are so simple. First of all, we should be clear that we still lack a demonstration that gravity is amenable to quantum treatment, so that at this point, for all we know, classicality might be a fundamental feature of the world.[41] But even if fundamental physics should turn out to be pervasively non-classical, it remains that classical metaphysical concepts may be crucial for interpreting it. One obvious reason for this is that, insofar as the measurement problem has been the core conceptual problem in quantum theory, that conceptual problem concerns, in part, the relationship between quantum and classical ontology, and illuminating the nature of one term in a relationship can often illuminate the relation itself. Indeed, in this connection one need only think of the work of Wallace to appreciate how getting a better purchase on the nature non-fundamental, including classical, ontology can be illuminating in this way (see e.g. Wallace 2010).

It might be objected at this point, however, that this is a red herring in this context: no-one ever thought that there need be anything problematic in principle about a metaphysics describing the classical as long as it is explicit that that metaphysics is *intended* to be about non-fundamental ontology.[42] Thus one might object that there is nothing in Wallace's metaphysics of the non-fundamental that gives license to the sorts of metaphysics we

[40] We might mention in passing that David Bohm was of the belief that the world was structured in alternating layers governed by classical and quantum principles, although he provided little by way of support for this claim! See Bohm (1957), chapter 4.

[41] And of course, the different interpretations of QM present it as being dissimilar to classical physics in various respects and to varying degrees, so that no classical concepts whatsoever may well be useful in interpreting future quantum physics for that reason.

[42] While as we noted there has been a preoccupation with the fundamental in metaphysics, we ourselves do not think that an 'effective' metaphysics of the non-fundamental is in principle

cited at the beginning. Nevertheless, and even though that latter metaphysics has misguided ambitions to directly describe the fundamental and thus seems to flout the compatibility principle as a result, we *still* think that such compatibility principle-flouting metaphysics may well have a useful function in naturalistic contexts. To see this, consider again the objections that have been made to Lewis' separability assumption. By now everyone knows that one cannot blithely maintain, as Lewis did, that separability is a fundamental feature of the world, because it is arguably so at odds with the basic structure of quantum mechanics.[43] However, in learning that, do we not thereby learn something important *about quantum metaphysics*? Is it not the case, in point of fact, that we actually understand *a great deal* of the metaphysical content of quantum mechanics precisely by understanding what classical metaphysical concepts do *not* apply in that context, and on account of what principles? It seems to us at least that understanding that quantum physics is (arguably) *not* local and *not* separable in the way that classical metaphysics is is actually absolutely crucial to understanding the metaphysics of quantum physics, and it also seems to us that all but the most specialized philosophers of physics will struggle to fill in the details of a *positive* picture as to what the metaphysics of quantum physics is, beyond justifying and elaborating upon these negative claims. It therefore seems to us that while what philosophers of physics are ultimately aiming for is a *positive* picture of quantum reality, classical metaphysics can nonetheless furnish us with *negative analogies* that are crucial for understanding quantum metaphysics, and especially so while we are in lieu of a clear positive picture.[44] Therefore even though assertions such as Lewis's that the fundamental level exhibits separability fall foul of the compatibility principle, *recognizing that they do so* can be an important contribution to the metaphysical theories that are appropriate at the fundamental

unnecessary or illegitimate; indeed, we think that the embrace of merely 'effective' ontologies in physics at least invites us to embrace a merely effective metaphysics of it.

[43] Though of course the extent to which this is true depends on what interpretation of QM is adopted; see e.g. Miller (2013), Belousek (2003). This is of course not to say that one should regard separability as thereby vindicated; the point is that one cannot *blithely* maintain it, partly because doing so is replete with other physical implications. We note also that it an approach to quantum mechanics in which the wavefunction is taken to evolve in configuration space is widely held to restore separability. But we ourselves are deeply skeptical about the viability of such an approach, primarily because such a space requires particle number to be well-defined at all times and this is not the case relativistically; on this, see Myrvold (2015).

[44] Furthermore, given that our concepts were acquired in the same classical environments that metaphysicians treat as exhaustive of reality, perhaps there is a claim to be made that the classical will always have some sort of privileged role in our metaphysical understanding (a conjecture that of course recalls Bohr.) But we do not want to pursue this point here.

level. Thus, while clearly not every negative analogy stands a chance of being relevant and illuminating, it seems that even false metaphysics can in principle be useful in this sense. And that just seems to corroborate our worry that our heuristic justification can sanction even compatability principle-flouting metaphysics.

Finally, it might be objected that

(3) **The heuristic approach instrumentalizes metaphysics in a way that is patronizing to metaphysicians**. Perhaps. But seeing that contemporary metaphysicians seem somewhat desperate to have their discipline regarded as akin to the sciences (as the adoption of the 'science of the possible' moniker itself suggests), and given the lack of obvious alternative accolades for analytic metaphysics in comparison with other contemporary disciplines, we believe that metaphysicians would be very willing to embrace our justification of metaphysics in heuristic terms.[45] Furthermore, our stance preserves the autonomy of metaphysics in a way that the approach of Ladyman and others does not. All that metaphysicians have to accept is the occasional raiding party from philosophers of science, keen (we hope) to see what they're up to and what they can use for their own purposes; or, putting it once again in less confrontational terms, all that they have to put up with is the perspective – which they don't even have to be made aware of – that as far as philosophers of science are concerned, what they are doing is filling up the toolbox for us.

7. Evaluation

In the wake of this defense of even compatibility principle-flouting metaphysics on the grounds that even that might come in useful in the course of time, we find ourselves at a point that has notes of Lakatos – in that we are claiming that no proposition of metaphysics may categorically be pronounced dead. And insofar as we are defending analytic metaphysics in general on the grounds that it may prove a useful heuristic for the philosophy of science, our position also invokes Feyerabend in that it suggests

[45] Paul (2012) is another expression of the desire to see metaphysics as analogous to science. We might add that seeing as metaphysicians have arguably had an insecurity complex about mathematics dating back to the time of Plato (see e.g. Moore 2012, *passim*), we think that the analogy with pure mathematics is something they will be more than happy to embrace too.

that the imposition of normative constraints risks choking off progress down the line. At this point, then, the conclusion that analytic metaphysics is simply off the hook, free to get back on with business as usual, seems ineluctable, and as such that the criticisms of so many philosophers of science must simply be withdrawn.

We think, however, that a closer look at the situation reveals this to be the wrong conclusion. Reminiscent of how one's *modus ponens* can be another's *modus tollens*, we think that the fact that this conclusion is even mooted draws attention to just how precarious our heuristic justification of metaphysics is. While we do, to be sure, remain convinced that it is difficult for the naturalist to flatly condemn the work of analytic metaphysicians given the extent to which we have appropriated, and continue to appropriate, it in our own work, we think that the tension articulated in the previous section brings to light just how highly *conditionalized* that justification is. Thus note that insofar as any support can be given to either Type I or Type II metaphysics via the heuristic approach, that support is *conditionalized twice over*: it is conditionalized

(i) upon naturalistic metaphysicians continuing to take metaphysics down 'off the shelf' and instead of making metaphysics to order by themselves; and furthermore
(ii) upon those analytical constructions actually turning out to be relevant and useful to the interpretation of science as it evolves.

How likely is it, we must ask, that each of these conditions will be fulfilled? Regarding point (ii), we are not sure how much can be said given that whether or not it is fulfilled hangs on future scientific developments that we have already argued are difficult to foresee. And regarding point (i), it is clear that this too is going to hang on the trajectory of science, but we should note that it hangs on the trajectory of the philosophy of science as well. For whether or not (i) is fulfilled will depend on the extent to which utilizing extant packages instead of making everything to order is not a grossly inefficient way to go about things.[46] But whether or not it *is* grossly inefficient is going to be at least in part a function of the nature of the relevant future science, *and also* of our success in philosophy of science – for the extent to which highly classical constructions will be useful and relevant depends on how non-classical future physics will turn out to be, *plus* how successful philosophers of physics are in coming up with positive as opposed to purely negative interpretations of that physics (the

[46] Of course, if the packages are already there it would seem churlish not to use them. But that clearly cannot be cited as a justification for continuing to produce them in the first place.

latter, we have argued, being likely to be cashed out in terms of negative analogies with the classical).[47] And what the prospects are in either case is not something that we feel anyone is in much of a position to place bets on.[48]

We think it follows from this that, while our heuristic considerations do in principle lend some support to analytic metaphysics, whether of Type I or Type II, that support is highly conditional and contingent on goings on both in science and in a naturalistic metaphysics of it. But now contrast the *support* we have offered empirically disengaged metaphysics, *conceived of as a tool for philosophers of science*, with our *criticisms* regarding metaphysics *conceived of as it is within the contemporary discipline* – namely, as the 'science of the possible'. Recall that it was many analytic metaphysicians' stated concern with mere possibilia that was supposed to relieve its practitioners of any duty to engage with physics. We argued that such disengagement was not in fact sanctioned on that basis, for this conception puts modal metaphysics at the heart of metaphysics, and systematic theories of modality can be falsified by actual physics.[49] We think that this shows that *even if* one conceives of metaphysics in these terms, then that is not enough to absolve metaphysicians of the responsibility to engage with science in a fundamental way.

Putting everything together, then, the following picture emerges. While there is heuristic support for analytic metaphysics if the latter is conceived of as a tool for philosophy of science, that support is highly conditional

[47] It may be, as a referee has suggested, that for whatever reason, philosophers of science simply refuse to use any tools from analytic metaphysics and of course, there would then be a sense in which analytic metaphysics could be described as having failed to be useful. Perhaps, then, we should be considering the tools that philosophers of science *could* be employing or *ought* to be. But this we feel we cannot do. Think of some of the reasons why philosophers of science might turn their backs on metaphysics, and let's ignore the possibility that philosophers of science turn their backs on metaphysics out of disciplinary churlishness or even other broadly 'sociological' reasons. A more likely reason is that philosophers of science simply reach the point where the tools made available by metaphysics are not fit for purpose, whether through their inherent classicality or whatever. Under those circumstances, of course, the game, as it were, would be up, as would be the possibility of any further fruitful relationship between metaphysics and the philosophy of science. But in that situation, we can't talk about what tools we ought to be using either – or at least, not for now.

[48] Though if it is objected that this makes for a 'monkeys at typewriters' evaluation of metaphysics, we could say that this is the case, at least to some extent, for science as well!

[49] Given what we have said about the future of physics being unpredictable, should we therefore not say that our argument that the fundamental properties such as colour charge are not intrinsic likewise could be falsified, so that Lewis' theory, is, for all we know, still a live possibility? We ourselves think that such a move would be somewhat pathetic, but we are sure the reader can fill in the reasons why.

on contingent developments outside of it. If, however, we conceive of metaphysics as contemporary metaphysicians themselves do, then there are strong and seemingly categorical arguments for the idea that it has to engage with science. Whatever conditionalized support metaphysics gets from naturalistic metaphysics, then, it seems that metaphysicians must themselves concede that the systematic disregard of real science simply cannot continue if they are to take their own projects seriously. As such, it seems that the most central of the criticisms with which we opened up this paper remain as trenchant as they appeared then. Naturalistic metaphysicians were never telling anyone that they shouldn't do metaphysics. What we object to is only the idea that it should take place in a disciplinary vacuum.

But it has to be said that the picture we have painted is a complicated one, and that there are considerations pulling from both sides. As things stand, both those who would decry contemporary metaphysics and those who would defend it are doing so from crude defensive positions. What is needed is the development of more nuanced positions on the basis of which more productive engagement between the two factions might be achieved. We would hope that the perspective developed here and in our previous work will contribute to that engagement.

University of Leeds
School of Philosophy, Religion and History of Science
e-mail: s.r.d.french@leeds.ac.uk

University of California, San Diego
Department of Philosophy
e-mail: kmckenzie@ucsd.edu

ACKNOWLEDGEMENTS

We would like to thank Wayne Myrvold, Yann Benétreau-Dupin, an anonymous referee, the Bay Area Philosophy of Science group, students and staff of the Department of Philosophy, University of Bristol, and participants at the Metametaphysical Club, Rotterdam, for helpful feedback and criticism.

REFERENCES

Bealer, G. (1987). The Philosophical Limits of Scientific Essentialism. *Philosophical Perspectives* **1**, 289–365.

Belousek, D.W. (2003). Non-Separability, Non-Supervenience, and Quantum Ontology. *Philosophy of Science* **70** (4), 791–811.

Bird, A. (2007). *Nature's Metaphysics*. Oxford: Oxford University Press.

Bohm, D. (1957). *Causality and Chance in Modern Physics*. Routledge and Kegan Paul Ltd.

Bueno, O., French, S. (unpublished). *From Weyl to von Neumann: An Analysis of the Application of Mathematics to Quantum Mechanics*. MS, University of Miami and University of Leeds, in preparation.

Callender, C. (2011). Philosophy of Science and Metaphysics. In: S. French and J. Saatsi (eds.), *The Continuum Companion to the Philosophy of Science*, pp. 33–54. London: Continuum.

Cameron, R. (2008). Truthmakers and ontological commitment: Or how to deal with complex objects and mathematical ontology without getting into trouble. *Philosophical Studies* **140**, 1–18.

Chalmers, D.J., (1996). *The Conscious Mind: In Search of a Fundamental Theory*. New York and Oxford: Oxford University Press.

Chalmers, D.J., (2002). Does Conceivability Entail Possibility? In: T.S. Gendler and J. Hawthorne (eds.), *Conceivability and Possibility*, pp. 145–200. Oxford: Oxford University Press.

Conee, E., Sider, T. (2005). *Riddles of Existence: A Guided Tour of Metaphysics*. Oxford: Oxford University Press.

Dorato, M., Esfeld, M. (2010). GRW as an ontology of dispositions. *Studies in History and Philosophy of Science* Part B **41** (1), 41–49.

Dummett, M. (2012). The Place of Philosophy in European Culture. *European Journal of Analytic Philosophy* **8** (1), 14–23.

Elder, C.L. (2011). *Familiar Objects and Their Shadows*. Cambridge: Cambridge University Press.

French, S. (2010). The Interdependence of Structure, Objects and Dependence. *Synthese* **175**, 89–109 .

French, S. (2014). *The Structure of the World*. Oxford: Oxford University Press.

French, S., McKenzie, K. (2012). Thinking Outside the Toolbox: Toward a More Productive Engagement between Metaphysics and Philosophy of Physics. *European Journal of Analytic Philosophy* **8** (1), 42–59.

French, S., Krause, D. (2006). *Identity in Physics: A Historical, Philosophical, and Formal Analysis*. Oxford: Oxford University Press.

Hall, N. (2010). David Lewis's Metaphysics. In: E.N. Zalta (ed.), *The Stanford Encyclopedia of Philosophy* (Fall 2010 Edition). URL = <http://plato.stanford.edu/archives/fall2010/entries/lewis-metaphysics/>.

Hawthorne, J., Uzquiano, G. (2011). How Many Angels Can Dance on the Point of a Needle? Transcendental Theology Meets Modal Metaphysics. *Mind* doi:10.1093/mind/fzr004.

Kim, J. (1998). *Mind in a Physical World: An Essay on the Mind-Body Problem and Mental Causation*. Cambridge: MIT Press.

Kuhlman, M. (2010). *The Ultimate Constituents of the Material World: In Search of an Ontology for Fundamental Physics*. Frankfurt: Ontos-Verlag.

Ladyman, J., Bigaj, T. (2010). The Principle of Identity of Indiscernibles and Quantum Mechanics. *Philosophy of Science* **77**, 117–136.

Ladyman, J., Ross, D. (2007). *Every Thing Must Go: Metaphysics Naturalized.* Oxford: Oxford University Press.

Lewis, D. (1983). New Work for a Theory of Universals. *Australasian Journal of Philosophy* **61**, 343–377.

Lewis, D. (1986). *On the Plurality of Worlds.* Oxford: Blackwell.

Lewis, D. (1991). *Parts of Classes.* Hoboken: Wiley-Blackwell.

Livanios, V. (2013). Is There a (Compelling) Gauge-Theoretic Argument against the Intrinsicality of Fundamental Properties? *European Journal of Analytic Philosophy* **8** (2), 30–38.

Lowe, E.J. (1998). *The Possibility of Metaphysics.* Oxford: Oxford University Press.

Lowe, E.J. (2011). The Rationality of Metaphysics. *Synthese* **178**, 99–109.

Maudlin, T. (2007). *The Metaphysics Within Physics.* Oxford: Oxford University Press.

Maxwell, G. (1962). The Ontological Status of Theoretical Entities. In: H. Feigl and G. Maxwell (eds.), *Scientific Explanation, Space, and Time* (*Minnesota Studies in the Philosophy of Science*), pp. 181–192. Minneapolis: University of Minnesota Press.

McKenzie, K. (2013). Priority and Particle Physics: Ontic Structural Realism as a Priority Thesis. *British Journal for the Philosophy of Science*, doi:10.1093/bjps/axt017.

McKenzie, K. (2014). In No Categorical Terms: A Sketch for an Alternative Route to Humeanism about Fundamental Laws. In: M.C. Galavotti, S. Hartmann, M. Weber, W. Gonzalez, D. Dieks and T. Uebel (eds.), *New Directions in the Philosophy of Science.* Dordrecht: Springer.

McKenzie, K. and Muller, F.A. (unpublished). *Bound States and the Special Composition Question.*

Markosian, N. (1998). Simples. *Australasian Journal of Philosophy* **76**, 213–226.

Melia, J. and Saatsi, J. (2006). Ramseyfication and Theoretical Content. *British Journal for the Philosophy of Science* **57**, 561–585.

Miller, E. (2013). Quantum Entanglement, Bohmian Mechanics and Humean Supervenience. *Australasian Journal of Philosophy*, doi:10.1080/00048402.2013.832786.

Moore, A.W. (2012). *The Evolution of Modern Metaphysics: Making Sense of Things.* Cambridge: Cambridge University Press.

Myrvold, W. (2015). What is a Wavefunction? *Synthese*, doi: 10.1007/s11229–014–0635–7.

Paul, L.A. (2012). Metaphysics as modeling: The handmaiden's tale. *Philosophical Studies* **160** (1), 1–29.

Post, H. (1971). Correspondence, invariance and heuristics: In praise of conservative induction. *Studies in History and Philosophy of Science* Part A **2** (3), 213–255.

Price, H. (2009). Metaphysics after Carnap: The Ghost Who Walks? In: D.J. Chalmers, D. Manley and R. Wasserman (eds.), *Metametaphysics: New Essays on the Foundations of Ontology*, pp. 320–346. Oxford: Oxford University Press.

Putnam, H. (1962). It Ain't Necessarily So. *Journal of Philosophy* **59** (22), 658–671.

Russell, B. (1919). *Mysticism and Logic and Other Essays.* London: Allen & Unwin.

Saunders, S. (2003). Physics and Leibniz's Principles. In: K. Brading and E. Castellani (eds.), *Symmetries in Physics*, pp. 289–307. Oxford: Oxford University Press.

Saunders, S., McKenzie, K. (2014). *Structure and Logic.* In: L. Sklar (ed.), Physical Theory: Method and Interpretation, pp. 127–162. Oxford: Oxford University Press.

Sider, T. (1991). Van Inwagen and the Possibility of Gunk. *Analysis* **53**, 285–289.

Sider, T. (2003). Reductive Theories of Modality. In: M.J. Loux and D.W. Zimmerman (eds.), *The Oxford Handbook of Metaphysics*, pp. 180–208. Oxford: Oxford University Press.

Sredniki, M. (2007). *Quantum Field Theory.* Cambridge: Cambridge University Press.

Wallace, D. (2010). Decoherence and Ontology: Or how I learned to stop worrying and love FAPP. In: S. Saunders, J. Barrett, A. Kent and D. Wallace (eds.), *Many Worlds? Everett, Quantum Theory, and Reality*, pp. 53–72. Oxford: Oxford University Press.

Wilson, J.M. (2012). Fundamental Determinables. *Philosopher's Imprint* **12** (4).
Witmer, D.G., Buchard, W., Trogdon, K. (2005). Intrinsicality Without Naturalness. *Philosophy and Phenomenological Research* **70** (2), 326–350.
Yablo, S. (1993). Is Conceivability a Guide to Possibility? *Philosophy and Phenomenological Research* **53**, 1–42.
Yoshimi, J. (2007). Supervenience, Determination and Dependence. *Pacific Philosophical Quarterly* **88**, 114–133.

Douglas Kutach

ONTOLOGY: AN EMPIRICAL FUNDAMENTALIST APPROACH

ABSTRACT. I apply the philosophical program Empirical Fundamentalism to the topic of ontology. My advice is to represent reality using two related components: a model of fundamental reality and a model of what we can observe in principle called an 'empirical surrogate'. The empirical surrogate need not be fundamental but can abstract away from fundamental reality while reducing to fundamental reality. Our ideal guess at ontology is a model of fundamental reality that (1) is a reduction base for the appropriate empirical surrogate, (2) posits no conspiratorial structure, and (3) has no redundant structure. I contrast my approach to alternatives that appeal to beables or primitive ontology.

1. Introduction

My aim here is to demonstrate how a general metaphysical framework can be fruitfully integrated with contemporary fundamental physics to address issues like ontology in classical and quantum physics. The basic idea is that metaphysics concerns fundamental reality and how fundamental reality is related to reality. If we accept that fundamental reality includes fundamental physics, then the investigation of fundamental reality by physicists is one component of the overall undertaking of metaphysics. The distinctively philosophical components include (1) identifying how fundamental reality relates to reality, (2) clarifying our epistemic grasp of fundamental reality, and (3) comparing our answers to existing alternatives.[1] It would take considerable effort to accomplish each of these tasks fully, but in the following discussion, I hope to sketch one particular philosophical program and how it can address them. My goal is to provide a proof of concept

[1] A fourth task, too large to be taken up here, is to ascertain whether fundamental reality includes more than fundamental physics.

In: Tomasz Bigaj and Christian Wüthrich (eds.), *Metaphysics in Contemporary Physics* (*Poznań Studies in the Philosophy of the Sciences and the Humanities*, vol. 104), pp. 55-80. Amsterdam/New York, NY: Rodopi | Brill, 2015.

for a scientific metaphysics that is broader and more conceptually oriented than science itself, yet is capable of supporting the conclusion that the actual world is ultimately just fundamental physics.

I will first sketch my proposed philosophical program and its central distinction between fundamental and derivative reality. In order to support adoption of the program, I will then address tasks (1) and (2) by introducing a formal device called an 'empirical surrogate'. Its purpose is to idealize the empirical phenomena that scientific theories attempt to predict and explain so that the fundamental theory can explain all empirical phenomena by explaining the empirical surrogate. My 'empirical surrogate' is intended as an alternative to Wilfrid Sellars' (1962) 'manifest image' and to a more recently proposed concept, 'primitive ontology'. In sections 4 and 5, I will illustrate the fundamental/derivative distinction and 'empirical surrogate' by applying them to the classic debate over the ontological status of space in classical mechanics and to contemporary debate over quantum ontology. Finally, I will address task (3) by contrasting my approach to some recent advocacy for primitive ontology and to Bell's appeal to 'local beables'.

2. Fundamental Reality

The purpose of this section is to summarize the four components of the philosophical program known as Empirical Fundamentalism that are most relevant to the relation between physics and metaphysics. Brevity prevents a fully adequate defense of the overall Empirical Fundamentalist framework, but the program has already been discussed extensively in (Kutach 2011, 2013). The arguments provided here are intended to support Empirical Fundamentalism by showing how its conception of fundamentality clarifies what is at stake in debates about ontology, whether classical or quantum. Of course, many other successful applications of this philosophical system must be spelled out in detail before an adequate case can be made for its adoption as a general metaphysical framework.

Let us now take up task (1), identifying how fundamental reality relates to reality. Three broad observations are worth making.

First, according to Empirical Fundamentalism, reality is understood as the totality of what exists. Reality is partitioned into exactly two parts, fundamental and derivative. That is, every existent is either fundamental or derivative, and no existent is both. This conception of fundamentality differs from other conceptions by requiring no *levels* of reality, no *more-fundamental-than* relation, and no relations of *ontological dependence*, *ontological priority*, or *grounding*. These concepts suggest that we understand fundamentality in terms of some *relation*. In Empirical Fundamentalism, the two monadic categories 'fundamental' and 'derivative' do all the work.

The justification for enforcing this binary conception of reality consists of a long list of its salutary consequences. The main benefit is that it nicely separates questions of *ontology*, which concern only fundamental reality, from questions of *semantic reference*, which concern derivative reality as well. Equating ontology with fundamental ontology is meant to direct the reader's mind away from the standard identification of ontology with existence.[2] The guiding metaphor is that God creates everything fundamental and then rests; no further activity is needed to create derivative reality or to bind derivative existents to fundamental reality with grounding relations.

Semantic reference to *X* can be legitimate even if *X* is not part of the ontology because *X* can exist without being fundamental. This supports a general non-confrontational resolution to many stock philosophical debates – whether there is free will, whether the mind and brain are the same, whether time flows, and so on – by allowing us to analyze each issue into two halves. One half concerns whether fundamental reality contains the debated entities, for example, whether there are fundamental free volitions (of the sort imagined by free will libertarians). The answers to such questions are meant to be sought in the same (mostly speculative, but still scientific) way people might try to ascertain whether protons are fundamental. The second half addresses what our attitudes should be toward the debated entity, given that we have hypothetically settled whether it is fundamental. Here, one is typically free to be very accommodative, maintaining that regardless of the exact character of fundamental reality, people make free choices, have feelings, and grow older.

Another justification for the binary distinction between fundamental and derivative is that it allows us to define an all-purpose supervenience relation called 'abstreduction' that quantifies how derivative reality depends on fundamental reality without introducing any mystery about the nature or ontological status of supervenience relations.

For yet another justification, it allows us to bracket debates about the part-whole relation, infinite descent, and monism – metaphysical topics that have only a tenuous connection with physics.

---

[2] Readers who are in the habit of using 'ontology' to refer to what exists should note that their more liberal conception of ontology can be expressed in Empirical Fundamentalism as 'all existents' or 'all of existence' or 'reality'. In general, Empirical Fundamentalism does not *forbid* making traditional metaphysical distinctions like the one between what exists and what does not; it merely advises us to de-emphasize them in favor of the proposed fundamental/derivative distinction.

Second, Empirical Fundamentalism maintains that 'the actual world' and 'fundamental reality' are two names for the same thing. For illustration, imagine the actual world consists of nothing more and nothing less than a single block of space-time instantiating entities, properties, and relations of the kind routinely postulated by theories of fundamental physics. All these supposed components of the actual world are fundamental existents. Everything else that exists – colors, feelings, numbers, injustice – are derivative existents because they exist but are not part of fundamental reality.

Identifying the actual world with fundamental reality suggests a fruitful definition of metaphysically possible worlds:

> A metaphysically possible world is a (logically) possible fundamental reality.

This conception of metaphysically possible worlds has the immediate benefit of preventing a priori metaphysical arguments from demonstrating anything interesting about ontology. Any argument that it is incoherent or self-refuting to deny the existence of some alleged entity *X* cannot suffice to demonstrate that *X* exists fundamentally. For example, one can adopt the essentially Cartesian premise that any thought with the content «thought does not exist» is self-falsifying. This premise might be understood to justify an inference to the existence of immaterial souls or some other sort of fundamental mentality. The Empirical Fundamentalist can easily refute such arguments by pointing out that there is nothing self-falsifying about denying the *fundamental* existence of thought, and there is nothing about the *guaranteed* existence of thought – fundamental or derivative – that can bear on whether any thought is *fundamental*.

The point here is simply that there are advantages of Empirical Fundamentalism's conception of fundamentality that have nothing specifically to do with science. A theory of fundamental physics acquires free benefits by being situated within the framework established by Empirical Fundamentalism.

Third, derivative reality can and should be thought of as an *abstraction* from fundamental reality. In particular, derivative reality supervenes on and reduces to fundamental reality in the sense that all derivative existents and derivative quantities can be defined – if they are characterized with sufficient precision – purely in terms of fundamental quantities, usually supplemented with fundamentally arbitrary parameters such as spatio-temporal coordinates. (A fundamentally arbitrary parameter is any quantity whose value is set merely by convention.)

For example, we can assign a derivative quantity like temperature to collections of fundamental particles in terms of a *function* of their masses and speeds (and possibly other fundamental quantities) relative to a fundamentally arbitrary rest frame. This sort of functional relationship constitutes a superior form of supervenience than is usually countenanced in metaphysics because it can enforce a continuity between fundamental and derivative reality simply by employing a continuous function. Arbitrarily small alterations to the derivative quantities will then be guaranteed to arise from suitably small alterations of fundamental quantities. Supervenience by itself provides no such guarantee.

Let us label this sort of relation between fundamental and derivative an 'abstreduction'. An *abstreduction* is an ontological reduction where the derivative quantity is identified as an abstraction from fundamental reality. Providing an abstreduction for some quantity or existent consists of supplying a reductive explanation in two steps. The first step is the explicit provision of a mathematical function that maps fundamental quantities and fundamentally arbitrary parameters to a precisification of one's chosen derivative existent or quantity. The second step consists of assuring one's audience that the functional relationship helps to clarify why the derivative entity or quantity would be useful for people to countenance and that the various possible settings for the fundamentally arbitrary parameters have no observable consequences.

Let us now take up task (2), clarifying our epistemic grasp of fundamental reality.

According to Empirical Fundamentalism, our operational grip on fundamental reality comes by way of adopting the following framework assumption:

> Our ideal guess at fundamental reality is a model that best accounts for all empirical phenomena.

Any application of this inferential principle – from one's best estimate of the empirical phenomena to one's best model (or theory) of it – is called a 'global abduction'. The upshot is that fundamental reality, insofar as it is epistemically accessible to us, is isomorphic to our best complete model of the observable universe, a model that explains the totality of evidence, including how any non-fundamental evidence exists by virtue of fundamental reality. The principles used to evaluate what best accounts for the phenomena are drawn from contemporary science including ontological parsimony, inferential strength, and avoiding ad hoc hypotheses.

To clarify 'global abduction' enough to address extant criticism of inference to the best explanation would exceed the scope of this article, but I can offer several comments to help to communicate its content more precisely.

First, the sort of explanation appropriate for evaluating the quality of a global abduction is a "complete story" explanation whose aim is to detail how all the various fundamental attributes fit together. For example, an exemplary model of fundamental physics might include a complete specification of a space-time, particle world lines extending through that space-time, all particle properties and relations, and a dynamical law. The dynamical law does the explanatory work by specifying how a complete arrangement of fundamental attributes at one time determines, or fixes probabilities for, fundamental attributes at other times.

Second, in a global abduction, we ignore epistemic limitations such as whether the dynamical consequences of the fundamental laws can be inferred, verified, or computed. We also ignore alternative sorts of explanations such as causal explanations that identify the important or salient causes of a chosen event. Empirical Fundamentalism does not reject the legitimacy of these explanations; it just does not use them to assess a global abduction. Although inference to the best explanation might be criticized for being a defective *general* inferential technique, its application here is restricted to fundamental reality.

Third, the scope of 'empirical phenomena' is intended to encompass the particular layout of matter throughout space and time – the sounds, shapes, colors and arrangements of macroscopic objects – as well as any resilient general patterns that could be revealed by experiment. The restriction to *empirical* phenomena does not enforce a narrow range of epistemic access such as a requirement that it include only sense data or only what can be observed without instrumentation. What's more, further clarification of the scope of what is empirical is left open to future adjudication. The difference between observable and unobservable is treated flexibly so that the content of science is insulated from debates about what counts as empirical.

Fourth, the totality of empirical data will always underdetermine which model of fundamental reality is best, and our standards for assessing the success of any respectable model are inherently imprecise. For these reasons, Empirical Fundamentalism allows (but does not require) the ideal guess at fundamental reality to comprise a class of empirically equivalent models. This insulates the goal of learning about fundamental reality from the fact that the scientific virtues presupposed by global abduction may be too imprecise or infelicitous to capture all the structure of fundamental reality.

It ought to go without saying that in general there is no way to verify whether fundamental reality matches the models that Empirical Fundamentalism advises us to seek, nor to measure the accuracy of the heuristics it borrows from scientific practice. Reliance on global abduction to guide our understanding of fundamental reality is the leap of faith inherent in Empirical Fundamentalism.

3. Empirical Surrogates

Much of science involves evaluating scientific theories and models in terms of how well their pronouncements match observations, yet there is no commonly accepted framework for understanding the empirical phenomena that serve as touchstones for scientific enquiry. The wide range of philosophical approaches to *experience* and its role in guiding belief speaks to the lack of any uncontroversial delineation of what counts as empirical.

Empirical Fundamentalism seeks to alleviate problems raised by the indeterminacy of the boundary between the observable and unobservable by introducing a theoretical device called an 'empirical surrogate'. An *empirical surrogate* is a formal model standing in for the totality of empirical phenomena. Its purpose is to serve as an intermediary between a model of fundamental reality and our informal grasp of what is observable. It allows the abstreduction – the reduction of empirical phenomena to fundamental attributes – to proceed in two steps. The mathematically rigorous step of the explanation requires specifying the function from one's model of fundamental reality to the empirical surrogate. The less rigorous step of the explanation defends the claim that explaining all the quantities in the empirical surrogate suffices to explain the totality of empirical phenomena (see Kutach (2013) for details).

One good example of an empirical surrogate is a Machian space-time endowed with a specified arrangement of point particle world lines and nothing else. Machian space-time (Barbour 1974, Earman 1989, pp. 27–30) has enough structure to distinguish global planes of simultaneity and (spatial) distance relations within these planes, but it cannot distinguish temporal durations nor inertial from non-inertial motion. The distances between particles in this empirical surrogate serve as a small set of quantities adequate for capturing all observable phenomena. Any theory of classical physics could use an empirical surrogate of this kind as could a Bohmian version of quantum mechanics (Dürr, Goldstein, and Zanghì 1995).

A slightly different example of an empirical surrogate is a Machian space-time with a continuous mass density field specified everywhere and nothing else. Intuitively, the mass density field is dense wherever massive objects are located in space-time and extremely rarified wherever there is empty space. The GRW_m version of quantum mechanics (Allori et al. 2008,

2012, Goldstein, Tumulka, and Zanghì 2012) can be understood as specifying such an empirical surrogate.

There are several generic features of an empircal surrogate that deserve emphasis. First, an empirical surrogate needs no dynamical laws nor any specification of particle masses or charges; it only needs enough structure to make the trajectories of matter through time sufficiently determinate. What's more, it is far better for an empirical surrogate to exclude all dynamical laws and masses and charges because these elements would add redundancy without benefit. It is received wisdom that a specification of laws, masses, charges, and so on are needed for prediction and explanation, but in Empirical Fundamentalism, we accommodate these elements in *fundamental reality*, not in the empirical surrogate.

Second, an empirical surrogate is meant to serve as a sparse but adequate stand-in for everything that is empirically accessible or observable, but in what sense? The intended conception of observability should not be understood too dualistically, as if it were meant to require a determinate distinction between observer and observed, nor too perspectivally as if it were meant to require a determinate distinction between direct and indirect perception. Instead, formulating empirical surrogates should involve postulating some public quantities that are generic enough and responsive enough to fundamental reality for their evolution over time to serve as a reliable indicator of all (or nearly all) fundamental quantities.

An empirical surrogate stands in for the totality of what is observable in the following sense. Anything that can reasonably count as observable can be rendered in terms of contrasts between the ways things are and the ways things are not. The act of observation, moreover, can be rendered in terms of the nomic consequences of these contrasts. For example, the island of Hawaii is uncontroversially observable. Primarily, it is observable because (A) it exists and (B) just about any reasonable attempt to observe Hawaii would very likely develop into a condition different from the condition that would develop if Hawaii were non-existent or unobservable. Secondarily, there are a large number of associated contrast-relations that hold. For example, Hawaii is vegetated and a person examining Hawaii would very likely evolve into a condition different from the condition that person would be in Hawaii were barren. Observers would be in the kind of condition that lead them to report 'vegetated' when asked. Although these relationships are stated informally, it is uncontroversial that the observable conditions of observable things are typically correlated with the condition of observers.

In brief, what allows us to recognize a complete set of particle trajectories in a space-time structure as an acceptable type of empirical surrogate is that (1) differences among possible particle configurations encode differences between the way the observed existent is and the many ways it isn't, and (2) we have many successful theories about how particle configurations covary with observable entities and attributes, including theories of hearing and vision as well as theories of protractors, lenses, and voltmeters.

Why should we think a set of point-particle trajectories in a Machian space-time could serve as an adequate empirical surrogate for a model of fundamental reality realistic enough to include all the interactions, particles, and fields we know exist? If current physics is any guide, what happens at any space-time location depends on *everything* fundamental that happens at a previous time within its past.[3] For illustration, consider that a full specification of the fundamental state spanning at least the entire solar system at one moment will determine where on Earth some particular shoe is located twenty minutes later. The precise position of the shoe depends on virtually every specific value of all the fundamental fields and particles throughout that vast region. Alter the position of an electron on the other side of Jupiter, and its gravitational attraction on the shoe will be slightly different. Alter some of the electromagnetic fields and the motion of other particles will be affected, transmitting some effects to the motion of the shoe. This feature of physics is unlikely to be overturned by any future discoveries, because electromagnetism and gravitation together with the strong and weak interactions constitute a pervasive medium of influence that affects all ordinary matter. As a result, the precise distance relations among point particles serve together as a generic marker for virtually any difference a fundamental attribute might make.

Can spatial distances be replaced by some other structure of an empirical surrogate? Yes, and such replacement is well-motivated when we are considering models of fundamental reality using a Lorentzian space-time for an arena, but let's not over-complicate the example.

Summing up, the sort of observability captured by an empirical surrogate does not presuppose a distinction between subject and object or between direct and indirect perception. Instead, it represents observation by having trajectories of the matter (in the empirical surrogate) shadow the nomic (and thus counterfactual) dependencies in fundamental reality. The shadowing exists by virtue of the presumed abstreduction.

[3] Technically, 'previous time within its past' may be understood here as the intersection of some inextendible space-like hyperplane lying wholly to the past of the corresponding location in the arena of fundamental reality with every inextendible non-space-like path intersecting that location.

This framework can capture anything that deserves to be called 'empirically accessible'. Even one's present phenomenal properties will be registered in the future distances among particles. If our physical behavior nomically depends on our phenomenal properties – which it presumably must in order for our reports of feelings to count as evidence of phenomenal properties – the effects on our physical behavior will manifest themselves in particle distances. In this way, perspectival and subjective aspects of observation are assimilated without distinguishing between subject and object or between sense data and other sorts of perceived entities.

The observations made in this section hint that little structure is required of an empirical surrogate. At this point, I want to avoid imposing more restrictions than necessary, so I will leave several questions open: Do point-particle world lines serve better than field quantities for formulating an empirical surrogate? To what extent can we come to an agreement about what kinds of empirical surrogates are adequate without specifying the fundamental theories to which they are intended to abstreduce? To what extent does superfluous structure in an empirical surrogate count as a deficiency, as for example might occur if one postulates an empirical surrogate with both particles and fields? These questions deserve more attention than the space here allows, but fortunately we do not need to answer these questions in order to make significant progress in understanding how to identify the ontology of a theory of fundamental physics.

4. Application to Classical Physics

Finding an adequate model of fundamental reality is an extremely difficult task, in part because it requires a breakthrough in theoretical physics tantamount to formulating the much-hoped-for (empirically adequate) theory of everything (TOE). A more manageable task for us is to engage in further conceptual engineering by imagining that we have already found an adequate TOE and that we are interested in fine-tuning its ontology. This philosophical activity is intended to clarify some of the considerations that bear on the relations among fundamental reality, an empirical surrogate, and empirical phenomena. By applying the broad principles introduced in the previous section to classical mechanics we can illustrate these concepts in terms of an antecedently understood example.

For the sake of having a specific model to play with, let us pretend that classical physics has been so successful that some model of classical mechanics (or collection of empirically equivalent models) adequately accounts for every detail in some empirical surrogate that matches the historical layout of our universe's matter. For simplicity, let us also assume the empirical surrogate is of the kind already discussed: a Machian space-time with point-particle world lines and nothing else. Our task now is to ascertain more precisely which models of fundamental reality are best justified under the fiction that classical physics is completely empirically successful. Those models of fundamental reality will dictate what ontology should be inferred from classical physics.

One reasonable stock model of a classical fundamental reality includes a Galilean space-time, point particles whose world lines extend across all time, mass properties that adhere to the particles, a distance relation between every two corpuscles at every time, and a deterministic law governing how these attributes evolve over time.

With this provisionally adequate model of fundamental reality in hand, we can proceed to fine tune its ontology with two familiar rules of thumb:

Parsimony Principle: A model of fundamental reality is ceteris paribus preferable if it has less structure.

Conspiracy Principle: A model of fundamental reality is ceteris paribus objectionable if it incorporates conspiratorial arrangements of attributes.

These principles pull our ontological commitments in opposite directions. When models *G* and *H* are exactly alike except that *H* incorporates space-time structure that plays no role or a redundant role in the evolution of matter, such as an absolute state of rest, the Parsimony Principle advises us to prefer *G*. When models *K* and *L* are alike except that *L* incorporates a dynamical law that specifies how the complete fundamental state at one time determines its state at other times according a relatively simple differential equation, and *K* specifies the very same arrangements at the very same times but with no law or other constraint that would imply the determination, the Conspiracy Principle advises us to reject *K* on the grounds that its layout of matter merely by curious happenstance matches what the relatively simple fundamental law dictated for *L*.

It is easy to list worrisome features of the Parsimony Principle and Conspiracy Principle. There is the inherent imprecision in what it means for a model to have less structure and of what arrangements of attributes should count as conspiratorial. There is also no definitive scale for quantifying them so that they can be weighed against each other. (Note that there is a difference in that the Parsimony Principle speaks of a model being preferable, whereas the Conspiracy Principle speaks of a model being objectionable; this difference is meant to encode the reasonable scientific judgment that avoiding conspiracies is much more important than being maximally parsimonious.)

Although the Parsimony Principle and Conspiracy Principle are contestable, exemplary scientific practice would be hard to understand without something roughly like these two principles. So, rather than become paralyzed with uncertainty concerning their content and scope, let us explore how these principles apply to a standard example where seasoned scientific judgment can provide guidance.

One well known example where the Parsimony Principle and Conspiracy Principle are at odds is the classic relationist/substantivalist debate over the ontological status of space as conducted in the Leibniz-Clarke Correspondence (Clarke 1717), Newton's Scholium to the Eighth Definition (Newton 1686), and later commentary. Because my goal here is to clarify the difference between fundamental and derivative in order to illuminate what counts as an adequate global abduction, an historically sensitive characterization of the debate is beside the point, and we may entertain contemporary mathematical machinery and methodological commitments in order to fashion a modernized formulation of the debate.

Within the framework of Empirical Fundamentalism, the contest between relationists and substantivalists can be rendered as a debate over what spatial structure (if any) is included in fundamental reality.

Space derivatism asserts that space itself exists as part of derivative reality, though some spatial relations among bits of matter may be fundamental. *Space fundamentalism* asserts that space exists as part of fundamental reality.

This distinction roughly tracks the traditional terminology. The relationist about space contends that there are fundamental distance relations between material bodies at any time but no fundamental spatial structure in which they are embedded. The relationist thinks we can abstract away from the fundamental distance relations by positing a (derivative) three-dimensional space in which these distances are embedded, but once we have accepted the existence of the fundamental particle distances, the extra posit of fundamental space adds ontological structure with no compensating value to our understanding of the apparent motion of material bodies. So the relationist uses the Parsimony Principle to infer that we should not posit fundamental space.

Space-derivativism is not exactly relationism. For one difference, relationism traditionally emphasizes the difference between matter and space and would presumably be hostile to the hypothesis that although 3-dimensional space is derivative, there is some other fundamental space, for example, the 11-dimensional arenas countenanced in string theory. For another difference, it is nowadays commonplace to conduct the relationist/substantivalist debate in terms of space-time rather than space. A substantivalist could reasonably hold that *space-time* is fundamental while *space* is derivative, but that would hardly be a victory for relationism. In any case, it is possible to consider whether space is a derivative existent or a fundamental existent while bracketing the issue of whether *space-time* is derivative or fundamental. So for simplicity, let's not worry about how the fundamentality of space-time bears on whether space should be considered derivative.

The space fundamentalist and space derivativist can in principle commence debate under the presumption that some particular (empirically adequate) empirical surrogate has been abstreduced to a model of fundamental reality that includes fundamental 3D Euclidean spaces at each time linked together by fundamental temporal relations. The abstreduction is easily constructed by preserving (with an identity function) the Machian structure of space-time as well as the particle trajectories while ignoring (and thus eliminating) the affine structure of Galilean space-time, the temporal metric, the dynamical laws, and the particle masses.

The subject of dispute is whether the space derivativist is justified in positing the same model of fundamental reality except that the fundamental spaces have been excised while retaining the fundamental spatial distances. In an opening maneuver, the space fundamentalist may contend that eliminating the 3D spaces from this model of fundamental reality greatly weakens the explanatory value of the model. The simplest version of this argument can press the objection that without the fundamental 3D spaces, there is no reason why the fundamental distance relations being posited by the space derivativist should have just the right magnitudes to be embeddable in a 3D Euclidean space. There is a conspiratorial arrangement of fundamental distances in the space derivativist's model of fundamental reality because with no constraints, a collection of n particles should have $1 + 2 + 3 + \ldots + (n - 1)$ independent distance relations, but the observed distance relations among particles in the 3D Euclidean space (of the empirical surrogate) has many fewer independent distance relations when n is large. Hypothetically introducing a new particle into an already populated 3D space by specifying its distance relations to four or five randomly chosen particles typically suffices for its distance to each of the remaining particles. By ontologically whittling away the fundamental spaces, the space derivativist is left to posit that the fundamental distances are *just so* arranged as to satisfy the precise constraints one would expect if they were always located in 3D space. Because there is a vast difference between the number of distance relations one would expect to be evident (via expectations of how matter will behave) if particles were located in fundamental 3D spaces and the number of distance relations one would expect if there were no fundamental 3D spaces, the space derivativist's arrangement of distances is prima facie conspiratorial. Hence, there is an explanatory obligation for the space derivativist that has not yet been discharged.

The space-fundamentalist here has just applied what we can designate as an '*undisposability argument*'. Space cannot be disposed from the ontology – in the sense of being shifted out of (an empirically adequate model of) fundamental reality into derivative reality – without saddling the resulting model of fundamental reality with an intolerable defect, in this case a conspiratorial arrangement of fundamental quantities. Therefore, space should be retained as part of one's model of fundamental reality.

An undisposability argument only applies to fundamental existents and should not be conflated with traditional *indispensability* arguments where indispensability is judged in terms of what is real or what exists. For example, a stock argument for Platonism concerning mathematical objects proceeds by alleging that they are indispensable for conducting science. This sort of indispensability is identified by examining the explanatory practices of science: for example, by observing that scientists sometimes refer to numbers when demonstrating that their predictions have been confirmed, and by arguing that such reference to numbers cannot be avoided without undermining the explanatory success. By contrast, when judging *disposability*, any reference to numbers in one's derivations or explanations is irrelevant. In the Empirical Fundamentalist framework, one does not assess disposability in terms of whether an explanation can survive a modification of the model that converts an existent into a non-existent, but whether it can survive the modification of a fundamental existent into a derivative existent. Scientific practice may well require the existence of numbers, but it does not require the *fundamentality* of numbers in order to make successful reference to numbers and physical magnitudes. Indispensability arguments for mathematical Platonism thus cannot establish that numbers are part of fundamental reality. Yet, a superficially similar undisposability argument can successfully establish that space should be retained as part of fundamental reality.

By virtue of the Parsimony Principle, the space derivativist's conception of fundamental reality would be superior to the space fundamentalist's if all other things were equal. However, they are not equal. Stripping space from one's model of fundamental reality without identifying some remaining fundamental structure that explains the observed 3D Euclidean geometry constitutes a conspiratorial posit, a deficiency not outweighed by the lesser virtue of ontological parsimony.

Note that it would amount to cheating for the space derivativist to appeal to the mere *existence* of a 3D space in order to explain the 3D Euclidean spatial relations among macro-objects. What is needed is a reductive explanation of 3D space in terms of *fundamental reality alone*. The relationist must explain why (in classical physics) a collection of such a vast number of *fundamental* distance relations should turn out to be embeddable in a 3D Euclidean space.

The main lessons so far are as follows. In order to assess a theory's ontological implications, we first need some way to characterize empirical phenomena in terms that can be explicitly connected to the theory's models. Otherwise, there would be no way to make sense of prediction, confirmation, and scientific explanation. In Empirical Fundamentalism, this role is played by an empirical surrogate. Second, we need one of the theory's models of fundamental reality to give rise to the empirical surrogate. In Empirical Fundamentalism, this 'giving rise to' is spelled out in terms of an abstreduction of the empirical surrogate to fundamental reality. The required function abstracts away from fundamental reality alone to arrive at all the structures and magnitudes in the empirical surrogate. (In this section's example, the 'abstracting away' consisted simply of ignoring the laws, masses, temporal metric and affine structure.)

5. Application to Quantum Mechanics

In this section, we can further explore Empirical Fundamentalism by investigating the ontological implications of quantum mechanics. There is a sizable literature on quantum ontology, but much of it is conducted as if it were intended to answer the question, "How should quantum theory be interpreted?" The term 'interpretation' is misleading, though, because it suggests the intended goal is to understand the content of quantum theory rather than the Empirical Fundamentalist's goal of discerning what quantum theory indicates about fundamental reality and fundamental reality's relation to reality.

If we adopt the stance that quantum mechanics has implications for the structure of fundamental reality, then Empirical Fundamentalism invites us to specify a functional relationship between some model of fundamental reality and an empirically adequate empirical surrogate that exhibits distinctively quantum phenomena such as interference patterns in electron diffraction experiments. Let us now sketch how one could connect fundamental reality to an empirical surrogate.

The initial task is to identify an appropriate empirical surrogate. Because one of the design criteria for a good empirical surrogate is that its essential character be insulated from the various candidate theories of fundamental reality that attempt to account for it, we might think that one can adopt the same empirical surrogate used for illustrating classical mechanics. So far as I can tell, there is no problem with using an empirical surrogate consisting of point-particle world-lines in a Machian space-time, but in doing so it is advisable to consider only non-relativistic quantum mechanics. The distinctively quantum effects are present in non-relativistic models, so bracketing concerns about relativity should be acceptable for current purposes.

One potential criticism of the choice to represent matter with point-particles is that it flouts the received wisdom that in quantum mechanics particles are inherently incapable of having perfectly precise positions. Several responses to this charge are reasonable. First, the received wisdom goes beyond the predictive content of the theory and so is more an element of ideology than fact, as evidenced by the coherence of Bohmian mechanics. Second, Empirical Fundamentalism allows an empirical surrogate to have precise particle locations even if fundamental reality does not. One could, for example, designate the location of the particle in the empirical surrogate with something like a center of mass function of an inherently spread out wave. Third, in any case, it is not difficult to adjust our chosen empirical surrogate, if needed, to make its entities spatially fuzzy. For simplicity, let's just retain our previous classical empirical surrogate.

The next task Empirical Fundamentalism requires of us is to identify an adequate function from the possible histories of fundamental states to the corresponding empirical surrogates. These functions are allowed to incorporate (in their domain) fundamental quantities as well as fundamentally arbitrary parameters like a choice of an inertial reference frame and a choice of (wave function) phase. The domain of the function does not permit any derivative existents.

At present, standard interpretations of quantum mechanics do not attempt this task, much less accomplish it. So, from the perspective of Empirical Fundamentalism, we cannot assume that the usual cast of characters – modal interpretations, Everettian, spontaneous collapse interpretations – are competing to explain what quantum mechanics indicates about fundamental reality (as understood in Empirical Fundamentalism). These need to be refurbished to specify how their models relate to an empirical surrogate. In some cases, this task may be trivial. In other cases, the interpretations will require extensive renovation. It would take too much space here to reformulate all existing versions of quantum mechanics and to survey all the implications. Instead, let us attend to the previously discussed tension between the Parsimony Principle and the Conspiracy Principle in order to highlight one tiny portion of the philosophical activity that has yet to be undertaken.

In order to explore how the Parsimony Principle and the Conspiracy Principle apply to quantum ontology, let us consider *quantum state fundamentalism*, the thesis that the complete history of the universe's quantum state should be regarded as fundamental. In favor of this thesis, we can point to the lack of any existing models of how the observed behavior of entangled particles could supervene on an entanglement-free fundamental physics. That is, we have no abstreduction of wave functions. Against this thesis, we can observe the lack of any existing arguments that the quantum state *must* be fundamental. I suspect that with most physicists being hostile towards versions of quantum mechanics that supplement the quantum state with additional material ontology, quantum state fundamentalism is a relatively popular default position, but because almost all discussion of quantum mechanics proceeds without the concept of fundamental reality imposed by Empirical Fundamentalism, it may be contentious to attribute quantum state fundamentalism to any particular person.[4]

Adopting quantum state fundamentalism provisionally for the sake of argument, we can contrast two existing views of what it implies for the ontological status of space and the ordinary material objects it contains.

On the one hand, there is a seed of a suggestive argument that space and spatially located objects are entirely derivative.[5] Drawing from our existing practices for evaluating theories of fundamental reality, we have a prima facie argument that the arena of quantum theory should be isomorphic to a direct sum of n copies of 3D Euclidean space plus one temporal dimension, where n is the number of particles in the universe. The reason is that an arena is by definition a container space for the material content of fundamental reality, and there are $3n$ dimensions required for a natural specification of a generic universal quantum state capable of a deterministic temporal development. The qualification 'natural' is included to make the claim about dimension non-trivial by ruling out the kinds of topological pathologies exhibited by space-filling curves and discontinuous mappings of higher-dimensional spaces into lower-dimensional spaces. Another way to restate the motivation for adopting the $(3n + 1)$-dimensional arena is that the set of all possible quantum states (at any one time) comprises exactly those states that can be formulated as a suitably integrable complex-valued

---

[4] Ney (2013) characterizes the position called "wave function realism" as the view that "the wave function of quantum mechanics is a real, fundamental entity", but it is unclear to me whether her 'wave function' is the same as my 'quantum state' because Ney appears to hold that wave function realism presupposes (or perhaps just implies) that the arena of quantum mechanics is a temporally-extended quantum configuration space. Because debate over the kind of arena to attribute to quantum theory goes beyond debate over whether the quantum state is fundamental, wave function realism appears to be different from quantum state fundamentalism. Also, wave function realism is advocated by Albert (2013) without his relating it to fundamental reality.

[5] This conclusion has been presented as the claim that space "is somehow flatly illusory" (Albert 1996, p. 277).

or spinor-valued function in a $3n$-dimensional space (taking into account an overall phase degree of freedom). Such an argument is one way to defend the conclusion proposed by Albert (2013).

On the other hand, there are coherent versions of quantum mechanics that treat space-time and its material contents as fundamental. These include Bohmian mechanics, and the mass-density and flashy versions of the spontaneous collapse interpretation (Allori et al. 2008, Allori et al. 2012, Tumulka 2007, Goldstein, Tumulka, and Zanghì 2012). I believe it is currently too unclear how to interpret Everettian versions of quantum mechanics, but some recent versions (Wallace and Timpson 2010, Wallace 2012) appear to treat space-time as fundamental and spatially located macroscopic objects as non-fundamental.

The reasons behind these competing assessments of the ontological status of space and space-time, I suggest, arise in great part from the tension between the Principle of Parsimony and the Principle of Conspiracy. Without taking sides in this debate, I contend that the kinds of considerations invoked for and against the fundamentality of spatial or spatio-temporal structure as well as the quantum state are largely the same as in classical mechanics.

Peter Lewis (2013) argued correctly, I think, that in order for a model of quantum mechanics to be empirically adequate without (in effect) positing a conspiracy, it needs more structure than what is provided by the $(3n + 1)$-dimensional arena. Especially, one needs an adequate explanation of why the Hamiltonians defining the temporal evolution of the quantum state take the specifically three-dimensional form that they do. For one example, a Coulomb interaction between two particles is treated in orthodox textbooks (Bohm 1951) as a potential function in a six-dimensional configuration space that exactly corresponds to the degrees of freedom of two distinct particles in three dimensional space. For another example, in textbook applications of quantum mechanics to single atoms, one routinely assumes a spherically symmetric potential generated at a single spatial location to approximate that a nucleus is a compact, massive source of positive charge around which the much lighter electrons circulate. What is prima facie conspiratorial in such examples is that existing interpretations of quantum mechanics do not posit any fundamental point-particles inhabiting a 3D space whose positions could serve in a law quantifying the determinate interaction potentials that play a role in the evolution of the quantum state. Quantum theories either treat the fundamental particles' positions as being exhaustively characterized by the (typically non-pointy) quantum state, or (in Bohmian versions) there are no means by which a particle's point-like position contributes to the evolution of the universal quantum state. If an atomic nucleus is instantiated fundamentally as a spread-out quantum state (with no point-like position), then there is no structure in fundamental reality that defines the distance needed to make determinate the magnitude of the Coulomb potential.

In all models of quantum mechanics where the arena is taken to be a $(3n + 1)$-dimensional space, the form of the Hamiltonian has not been derived purely from fundamental quantities. If such a derivation is lacking, it would count as a conspiracy for us to posit such a Hamiltonian (with all the familiar three-dimensional symmetries).

It is important to recognize that insofar as quantum mechanics is being wielded merely as a scientific theory, this conspiracy I am alleging has no significance. It is only a problem for the hypothesis that quantum theory accurately describes not only reality, but *fundamental reality*.

Thus we have arrived at the crux of the problem of quantum ontology. How can we explain entanglement, which (at least superficially) appears to presuppose a very high dimensional mathematical space in order to contain (naturally!) all lawfully permitted quantum states, while at the same time fully explaining how particles wiggle around in an empirical surrogate? On the one hand, the observed motion of particles (as represented by the empirical surrogate) appears to require that their interactions (as encoded by a Hamiltonian operator) be constrained so that all particles (at a single time) behave as if embeddable in three dimensions, but no theory with a $(3n + 1)$-dimensional arena has justified this constraint in terms of how *fundamental reality* is structured. On the other hand, we could posit a fundamental 3D space as the arena for quantum mechanics with some structure to serve as the source of interaction potentials, but the observed entanglement appears to require that we need a larger than 3D arena to contain the space of possible states (in the natural way mentioned above), and no one has proposed an acceptable derivation from fundamentally unentangled stuff in 3D space to the full (entangled) quantum state.

It does no good (for achieving the required abstreduction) to point out that quantum theory itself does not supply the correct form of the Hamiltonian and leaves us free to establish an appropriate Hamiltonian on a case-by-case basis in light of what we know about classical physics and experimental outcomes. The goal is to recover the 3D-ish behavior of the matter in the empirical surrogate *purely* from fundamental reality, unsupplemented by what is (presumably) derivative: the behavior of twentieth century laboratory equipment.

Gordon Belot (2012) surveys some options relevant to the problem I am describing and identifies one possible resolution: positing an n-dimensional multi-field in a $(3 + 1)$-dimensional arena. But no one has yet presented any details of how such a multi-field would operate in a way that would mitigate the tension between entanglement and the 3D-ish character of the interaction Hamiltonian.

Another attempted resolution (Wallace and Timpson 2010) appears to adopt space-time as fundamental and to accept that quantum states can be represented as superpositions of locally-well-defined operator-valued functions. As far as I can tell, this amounts to denying that entanglement motivates us to adopt a higher-dimensional arena and to accept that our pre-quantum-theory intuitions about how to identify a proper arena are inadequate for convicting quantum theory of smuggling conspiratorial structures into its model of fundamental reality.[6] This tactic of simply dismissing what I am claiming is an appeal to a conspiratorial form for the Hamiltonian is one that I cannot completely reject out of hand because perhaps our usual heuristics for judging conspiracy are defective. However, before accepting such a dismissal, we ought to have an explicit defense of the practice and to contrast this option with others that deserve exploration. In particular, we should attempt to ascertain whether it is possible to formulate some sort of fundamental state that does not incorporate entanglement but on which entangled quantum states supervene.

I do not have a preferred resolution to defend here. Instead, I merely want to emphasize that the service provided by Empirical Fundamentalism is (1) to identify the most important task in understanding the implications of quantum mechanics: ascertaining how the various versions of quantum mechanics address the tension between the effects of entanglement and the 3D-ish (or 4D-ish) nature of the four known interaction-types, and (2) to provide a structure for answering this challenge: an abstreduction of an adequate empirical surrogate to fundamental reality.

6. Comparisons

In this final section, I will review how the concepts 'fundamental reality' and 'empirical surrogate' as precisified in Empirical Fundamentalism might relate to concepts like 'primitive ontology' and 'local beable'.

John Bell (1976) invoked 'local beables' in an effort to expose deficiencies in the usual ways physicists like to talk about quantum mechanics, especially references to irreducible 'measurements' and 'observables'.

---

[6] I suspect that this kind of response may be common among practicing physicists, but even if there were a consensus, it would count for little because for almost all scientific purposes, it is irrelevant whether the fundamentality of the quantum state implies that the interaction Hamiltonian takes a conspiratorial form. These factors are only important insofar as we are concerned about fundamental reality.

Although I have no major disagreement with Bell's pronouncements, his discussion of 'beables' is so brief and provides so few examples or constraints that it can hardly be said to be in competition with Empirical Fundamentalism.[7] He rightly emphasizes the importance of identifying at least some 'local beables' for understanding the empirical adequacy of quantum mechanics, but he does not indicate what sorts of relations they ought to bear to non-local beables, nor does he argue that it is impossible to interpret the quantum state so that it can be said to exist locally in 3D space, for example, as discussed by Belot (2012) and Wallace and Timpson (2010). (Bell's arguments concerning quantum non-locality are irrelevant to this point because they concern non-local *influence*, not non-local *existence*.) Empirical Fundamentalism, by providing more detail than Bell and significantly tighter constraints on the adequacy of fundamental theories, can serve as a more precise framework for addressing questions about quantum ontology.

Much the same can be said of the more recent distinction between ψ-epistemic and ψ-ontic interpretations of the quantum state in the literature surveyed by Leifer (2014). This framework relies on insufficiently articulated conceptions of 'reality' and 'epistemic' and does not address the crux of the problem of quantum ontology, how to eliminate the tension between entanglement and the 3D-ish or 4D-ish character of ordinary physical interactions.

Finally, we can consider some recent advocacy for primitive ontology. The expression 'primitive ontology' first appears in "Quantum Equilibrium and the Origin of Absolute Uncertainty" (Dürr, Goldstein, Zanghì 1992) and is defined as "the basic kinds of entities that are to be the building blocks of everything else" except the wave function (p. 10). The intended interpretation of 'primitive ontology' is unfortunately opaque. They claim that whatever is "more primitive" is "more familiar and less abstract", and they also speak of what is "*physically* primitive" without distinguishing 'physically primitive' from 'primitive'. They also never cash out the 'building block' metaphor.

Here are some questions that deserve to be answered before a proper evaluation of 'primitive ontology' can be made. Is primitive ontology meant to serve the epistemological role I have designated for an empirical surrogate? Can the primitive ontology be spatio-temporally indeterminate? Can a primitive ontology consist of something a twentieth-century scientist has never heard of? Can it consist of less than is needed to account for all scientific observations? Can a theory posit multiple choices for its

[7] See Norsen (2011) for a sympathetic explanation of Bell on this issue.

primitive ontology, each of which is alone satisfactory but incompatible with the others? To what degree is it a deficiency for a primitive ontology to include redundant structure? Must the predictions of a theory must be derivable solely from the complete history of the primitive ontology?

Because so many important questions about primitive ontology have not yet been answered, I think no one is in a position to argue that Empirical Fundamentalism provides a better framework than one based on primitive ontology. (I left many questions about empirical surrogates unanswered here as well!) My hope is that the distinctions I have started to draw in this article can serve as a foil against which the concept of primitive ontology can be further clarified.

In any case, the more recent work of Allori (2013a) supplies some additional details about primitive ontology that may allow a few constructive comparisons to be drawn:

> Any fundamental physical theory must always contain a metaphysical hypothesis about what are the fundamental constituents of physical objects. We will call this the *primitive ontology* of the theory: entities living in three-dimensional space or in space-time, which are the fundamental building blocks of everything else, and whose histories through time provide a picture of the world according to the theory.

From this definition, I surmise that primitive ontology is meant to be understood as one form of *fundamental ontology* and that it constitutes an empirical surrogate in the sense that it stands in formally for the empirical phenomena that we use to confirm and disconfirm theories. Consider further evidence of this interpretation (Allori 2013a):

> [E]ven if the primitive ontology does not exhaust all the ontology, it is the one that makes direct contact between the manifest and the scientific image. Since the primitive ontology describes matter in the theory, we can directly compare its macroscopic behavior to the behavior of matter in the world of our everyday experience. Not so for the other non-primitive variables, that can only be compared indirectly in terms of the way in which they affect the behavior of the primitive ontology.

If the correct way to interpret 'primitive ontology' is that it is a special case of what I have designated as 'fundamental existents', there are several simple arguments available for why primitive ontology is less preferable than an empirical surrogate. First, there is no reason (given by its proponents or that I can see) why the stuff playing the role of an empirical surrogate needs to be fundamental. The observable stuff in 3D space could be non-fundamental as in string theory and loop quantum gravity. This possibility is explicitly recognized by Allori (2013a). Second, in quantum theory, it is easy to convert a theory with a primitive ontology into a theory with a not-entirely-fundamental empirical surrogate. One can simply introduce additional fuzziness in the positions of particles in the empirical surrogate, thereby rendering them non-fundamental without damaging the empirical adequacy of the surrogate. Third, there need not be one single choice of primitive ontology for a theory.

One point that advocates for primitive ontology get right by the lights of Empirical Fundamentalism is that interpretations of quantum theory ought to specify at least one kind of empirical surrogate as a means of defining its empirical consequences. This is especially pertinent to Everettian interpretations, where the relationship between the ridges in the history of the universal quantum state – what Everettians call 'worlds' – and the totality of empirical phenomena falls far short of an abstreduction.

Two more features of the primitive ontology approach identified by Allori (2013a) are worth considering:

> Any fundamental physical theory is supposed to account for the world around us (the manifest image), which appears to be constituted by three-dimensional macroscopic objects with definite properties. To accomplish that, the theory will be about a given primitive ontology: entities living in three-dimensional space or in space-time. They are the fundamental building blocks of everything else, and their histories through time provide a picture of the world according to the theory (the scientific image). The formalism of the theory contains primitive variables to describe the primitive ontology, and nonprimitive variables necessary to mathematically implement how the primitive variables will evolve in time. Once these ingredients are provided, all the properties of macroscopic objects of our everyday life follow from a clear explanatory scheme in terms of the primitive ontology.

The two features in this description I want to draw attention to are (1) its borrowing the distinction between the scientific and manifest image and (2) its claiming that the primitive ontology serves in a "clear explanatory scheme" of mundane interactions among macroscopic objects.

Concerning (1), it is valuable to distinguish how Wilfrid Sellars' (1962) distinction between the scientific image and manifest image differs from the distinctions used in Empirical Fundamentalism. First, fundamental reality is not strictly equated with models of fundamental physics. Empirical Fundamentalism does not rule out of hand the possibility that the best model of fundamental reality might require Cartesian souls, phenomenological properties, or Platonic numerical existents. Second, Sellars distinguishes the manifest image from the scientific image in terms of the inferential techniques used to ascertain its structure. The scientific image, by definition, incorporates an abduction, whereas features of the manifest image are inferred through other means such as enumerative induction. By contrast, Empirical Fundamentalism does not prevent the use of abduction for drawing inferences about derivative reality. Third, the manifest image is a much richer structure than an empirical surrogate and much more impoverished than derivative reality. In the manifest image, entities like trees are treated as "truncated persons", which is a maneuver on Sellars' part to suggest that an adequate reductive explanation of the motion of the tree's parts (in conjunction with its environment) is not necessarily an adequate reductive explanation of the tree. Sellars leaves open what sort of explanation is needed to relate the tree to fundamental particles and fields. Empirical Fundamentalism, by contrast, selects one particular kind of reduction, abstreduction, to be sufficient for an adequate link between the tree and fundamental physics. That said, the concept of the empirical surrogate is intended to serve as an improvement on the concept of the manifest image, and the sense in which they are playing the same role is that they are both idealized versions of our common sense conception of public reality.

Concerning (2), a primitive ontology does not by itself provide the kind of explanation relevant to an undisposability argument and thus to ontology as understood in Empirical Fundamentalism. (What sort of explanatory scheme primitive ontology is meant to support is unclear to me.) To show that a certain specification of fundamental attributes is sufficient for an abstreduction of an empirical surrogate (and that any supplementation to that ontology is disposable), one needs to show that the specified fundamental attributes and laws (together with a designated set of fundamentally arbitrary parameters) suffice for determinate values for all the quantities in the chosen (type of) empirical surrogate. In the quantum context, the primitive ontology is insufficient for this sort of derivation because it does not include the quantum state, which is needed for ascertaining what any fundamental state implies concerning later quantum states.[8] At least, no one yet has provided any scheme for deriving (probabilities for) final conditions from initial conditions that successfully predict the signature effects of particle entanglement without incorporating a quantum state as part of the fundamental attributes. Hence, the crux of the problem in quantum ontology – deriving empirically adequate phenomena purely from attributes that are taken to be fundamental and innocent of conspiracies – is unresolved until the quantum state and the dynamical interaction contained in the Hamiltonian can be declared free of any conspiratorial arrangements. Nothing in the literature on primitive ontology appears to bear on this problem because no one has been able to show how the quantum state can be derived from (non-entangled) fundamental relations among elements of a primitive ontology.

University of Pittsburgh
Center for Philosophy of Science
e-mail: requests@sagaciousmatter.org

[8] Belot (2012) and Ney and Phillips (2013) make this observation.

REFERENCES

Albert, D. (1996). Elementary Quantum Mechanics. In: J.T. Cushing, A. Fine and S. Goldstein (eds.), *Bohmian Mechanics and Quantum Theory: An Appraisal*, pp. 277–284. Dordrecht: Kluwer.

Albert, D. (2013). Wave Function Realism. In: A. Ney and D. Albert (eds.), *The Wave Function*, pp. 52–57. New York: Oxford University Press.

Albert, D., Ney, A. (2013). *The Wave Function.* New York: Oxford University Press.

Allori, V., Goldstein, S., Tumulka, R., Zanghì, N. (2012). *Predictions and Primitive Ontology in Quantum Foundations: A Study of Examples.* arXiv:1206.0019v2.

Allori, V., Goldstein, S., Tumulka, R., Zanghì, N. (2008). On the Common Structure of Bohmian Mechanics and the Ghirardi-Rimini-Weber Theory. *The British Journal for the Philosophy of Science* **59** (3), 353–389.

Allori, V. (2013a). Primitive Ontology and the Structure of Fundamental Physical Theories. In: A. Ney and D. Albert (eds.), *The Wave Function*, pp. 58–75. New York: Oxford University Press.

Allori, V. (2013b). The Metaphysics of Quantum Mechanics. In: S. Le Bihan (ed.), *Précis de Philosophie de la Physique*, pp. 116–140. Paris: Vuibert.

Barbour, J. (1974). Relative-Distance Machian Theories. *Nature* **249**, 328–329.

Bell, J.S. (1976). *The Theory of Local Beables, Epistemological Letters.* Reprinted in J.S. Bell, (2004), *Speakable and Unspeakable in Quantum Mechanics.* Cambridge: Cambridge University Press.

Belot, G. (2012). Quantum States for Primitive Ontologists. *European Journal for Philosophy of Science* **2**, 67–83.

Bohm, D. (1951). *Quantum Theory.* New York: Dover.

Clarke, S., ed. (1717). *Leibniz-Clarke Correspondence.* London: Printed for James Knapton at the Crown in St. Paul's Church-Yard.

Dürr, D., Goldstein, S., Zanghì, N. (1992). Quantum Equilibrium and the Origin of Absolute Uncertainty. *Journal of Statistical Physics* **67**, 843–907.

Dürr, D., Goldstein, S., Zanghì, N. (1995). *Bohmian Mechanics and the Meaning of the Wave Function.* arXiv:quant-ph/9512031.

Earman, J. (1989). *World Enough and Spacetime.* Cambridge: MIT Press.

Goldstein, S., Tumulka, R., Zanghì, N. (2012). *The Quantum Formalism and the GRW Formalism.* arXiv:0710.0885

Kutach, D. (2011). Reductive Identities: An Empirical Fundamentalist Approach. *Philosophia Naturalis* **47–48**, 67–101.

Kutach, D. (2013). *Causation and Its Basis in Fundamental Physics.* New York: Oxford University Press.

Lewis, P. (2013). Dimension and Illusion. In: A. Ney and D. Albert (eds.), *The Wave Function*, pp. 110–125. New York: Oxford University Press.

Leifer, M. (2014). *Is the quantum state real? A review of ψ-ontology theorems.* arXiv:1409.1570.

Newton, I. (1686). *Philosophiae Naturalis Principia Mathematica.* London: George Brookman.

Ney, A. (2013). Ontological Reduction and the Wave Function Ontology. In: A. Ney and D. Albert (eds.), *The Wave Function*, pp. 168–183. New York: Oxford University Press.

Ney, A., Phillips, K. (2013). Does an Adequate Physical Theory Demand a Primitive Ontology? *Philosophy of Science* **80**, 454–474.

Norsen, T. (2011). J.S. Bell's Concept of Local Causality. *American Journal of Physics* **79**, 1261–1275.

Sellars, W. (1962). Philosophy and the Scientific Image of Man. In: E. Colodny (ed.), *Frontiers of Science and Philosophy*, pp. 35–77. Pittsburgh: Pittsburgh University Press.

Tumulka, R. (2007). The 'Unromantic Pictures' of Quantum Theory. *Journal of Physics A: Mathematical and Theoretical* **40**, 3245–3273. arXiv:quant-ph/0607124.

Wallace, D., Timpson, C. (2010). Quantum Mechanics on Spacetime I: Spacetime State Realism. *British Journal for the Philosophy of Science* **61**, 697–727.

Wallace, D. (2012). *The Emergent Multiverse: Quantum Theory According to the Everett Interpretation.* Oxford: Oxford University Press.

Vincent Lam

QUANTUM STRUCTURE AND SPACETIME

ABSTRACT. The aim of this paper is twofold. In the first part, it clarifies the nature of the wave function within the framework of the primitive ontology approach to quantum mechanics using the tools of ontic structural realism. In the second part, it critically discusses the primitive ontological move of postulating from the start matter localized in spacetime as the ultimate referent for quantum theory, in particular in the case where this latter is applied to the general relativistic gravitational field.

1. Introduction

There is a recently much discussed approach to the ontology of quantum mechanics (QM) according to which the theory – and possibly any fundamental physical theory – is ultimately about entities in 3-dimensional space (or 4-dimensional spacetime) and their temporal evolution. Such an ontology postulating from the start matter localized in 'usual' physical space or spacetime (by contrast to e.g. a wave function on a high dimensional configuration space as within the framework of 'wave function realism') is called 'primitive ontology' in the recent literature on the topic. According to the proponents of the primitive ontology approach to QM, it is the best (realist) way to avoid the main difficulty of wave function realism: there is no 'illusion' or 'appearance' of matter in 3-dimensional space to be explained, since this fact is simply postulated from the start as the referent of the theory (i.e. as what the theory is fundamentally about). Of course, this 'postulate' is not the whole story: the theory in question has to specify how matter is instantiated in 3-dimensional space (particles, fields, strings, loops, ...) and how it evolves in time. And that's where things get interesting: in all the proposed primitive ontologies for QM (the paradigmatic examples are of course Bohmian mechanics (BM) and versions of

In: Tomasz Bigaj and Christian Wüthrich (eds.), *Metaphysics in Contemporary Physics* (*Poznań Studies in the Philosophy of the Sciences and the Humanities*, vol. 104), pp. 81-100. Amsterdam/New York, NY: Rodopi | Brill, 2015.

the theory of Ghirardi, Rimini and Weber (GRW)), the wave function plays a central and crucial role in the time evolution of the primitive ontology and in the account of quantum non-locality. An important and difficult task for the primitive ontologist is therefore to elucidate what the nature of the wave function can be in this context.

The aim of this contribution is twofold. First, it aims to clarify the nature of the wave function within the framework of the primitive ontology approach to QM using the tools of a recently much debated conception in the metaphysics of contemporary fundamental physics, namely ontic structural realism (OSR). Second, this article aims at discussing the very ontological move of postulating from the start matter localized in spacetime as the ultimate referent for any fundamental physical theory. In particular, this move seems debatable within the framework of the development of a quantum theory for the general relativistic gravitational field, where the nature of spacetime and its relationship to matter constitute central open issues. We will also more generally discuss the role of spacetime in the primitive ontology approach and in the OSR conception.

2. The Main Motivation for a Primitive Ontology

A primitive ontology for QM specifies explicitly what the theory is fundamentally about, i.e. what there is in the world according to QM, in terms of material entities localized in 3-dimensional space (or 4-dimensional spacetime) and their dynamics. There are different such specifications, in particular within two of the three standard realist conceptions of QM that take the measurement problem seriously: particles following continuous deterministic trajectories within the framework of BM, a continuous stochastic matter density field or stochastic point-like events ('flashes') within GRW, giving rise to GRWm and GRWf respectively (see Allori et al. 2008 and references therein). A primitive ontology in the context of the Everett (or 'many-worlds') framework can possibly be defined (for instance in terms of a deterministic matter density field, see Allori et al. 2011) but its meaning is less transparent.

The main motivation for specifying a primitive ontology for QM is rather straightforward: it provides a powerful and generic explanatory framework within which familiar macroscopic objects localized in 3-dimensional space and their (classical) behavior can be understood in terms of the behavior of (possibly fundamental) microscopic entities that are also localized in 3-dimensional space (in particular, there is an explicit connection between the behavior of these microscopic entities and what can be observed at the macroscopic level, for instance in terms of measurement outcomes). Obviously, the details of this account depend on the specific primitive ontology under consideration; the point is that such an account in terms of a primitive ontology does not have to bridge substantial explanatory gaps, for instance between macroscopic objects that are (or seem to be) localized in 3-dimensional space and fundamental microscopic entities that are not (as it seems to be the case within the framework of wave function realism; for recent discussions on primitive ontology and its contrast to a wave function ontology, see the contributions in Ney and Albert 2013).

The primitive ontology approach to QM finds part of its roots in Bell's notion of 'local beables', which was introduced in the context of his reflections on non-locality and the measurement problem (see the papers collected in Bell 1987). To some extent, a primitive ontology is made up of local beables – that is, of entities that "can be assigned to some bounded spacetime region" (Bell 1987, p. 53) – which can be directly related to the behavior of familiar material objects, such as a measurement apparatus for instance, so that the measurement problem simply does not arise within the framework of a primitive ontology for QM (in this sense, the main motivation for a primitive ontology for QM is that there is simply no quantum measurement problem in this context).

3. The Central Role of the Wave Function

However, the local beables alone (in 3-dimensional physical space) have little explanatory power; the material entities that are localized in 3-dimensional physical space and that constitute the primitive ontology have little explanatory power alone. The explanatory power of the primitive ontology approach to QM stems from the local beables together with their temporal development or dynamics, which crucially relies on the wave function. As a consequence, in this context, it seems unavoidable to accept the wave function on top of or as 'part of' the primitive ontology, in some sense to be clarified (note that in principle it does seem possible to consider the primitive ontology move within a purely Humean framework where all quantum features, including the wave function and the features related to quantum non-locality, supervene on the entire spacetime distribution of local beables, see Miller 2013 – however, the standard difficulties related to the explanatory power of Humeanism seem especially salient in the quantum case).

Let us consider BM as an illustration. Indeed, BM embodies the paradigmatic example of a primitive ontology for QM and as such will serve us as a very convenient study case throughout this contribution. The Bohmian particles constitute the primitive ontology (they obviously are local beables since they always have a definite position in 3-dimensional physical space), but the temporal evolution of the total configuration of the Bohmian particles crucially relies on the universal wave function through the Bohmian guiding equation or equation of motion. According to this equation, Bohmian particles continuously evolve along determinate trajectories, but such that the velocity of each particle depends on the positions of all the other particles: strictly speaking the velocity of each particle is a functional of the universal wave function defined on the whole configuration. In particular, the role of the wave function in this huge dynamical interdependence is central to the Bohmian account of quantum non-locality, that is, to its explanatory power (more on that below). Clearly, this crucial role of the wave function is shared by the other main primitive ontologies of QM (e.g. GRWm and GRWf); the wave function is an irreducible part of what Allori et al. (2008) have identified as the 'common structure' of all the conceptions within the primitive ontology approach to QM. The 'common structure' between GRWm, f and BM is that the considered theory is fundamentally about matter in spacetime, in contrast to, e.g., a wave function in a high-dimensional configuration space; however the point here, which is rather clear among the proponents of a primitive ontology for QM, is that the wave function cannot be entirely dropped from the ontological picture (see however the investigations in Dowker and Herbauts 2005). The theoretical, explanatory, ontological importance of the wave function therefore creates a tension within the primitive ontology approach to QM: it raises the (old) issue of the ontological status and the metaphysical nature of the wave function within the familiar ontological picture offered by the primitive ontology framework, that of matter localized in 3-dimensional space and evolving in time. Let us now consider what are the standard options for understanding the wave function in this context.

4. Three Categories for Understanding the Wave Function

There are mainly three realist ways to understand the wave function as part of the ontology of QM (see Belot 2012), each appealing to a different philosophical category and two of which being rather common within the primitive ontology approach. The first one is the most straightforward: to consider the wave function as a physical object on its own. It is also the most problematic from the point of view of the primitive ontology approach. Indeed, within the framework of BM, it amounts to recognize the wave function as a physical object, possibly not living in 3-dimensional space but in high-dimensional configuration space, in addition to the Bohmian particles in 3-dimensional space, thereby considerably inflating the ontology. The explanatory strength and simplicity of the primitive ontology approach would then be significantly weakened, and some of the difficulties of wave function realism – against which the primitive ontology approach was originally designed – would reappear (in particular, the link between the local beables in spacetime and the wave function existing in a different space would have to be clarified). It is therefore not surprising that this option is commonly rejected within the framework of the primitive ontology approach to QM.

The second understanding appeals to an entirely different philosophical category: it suggests considering the wave function as a law-like, nomological entity, that is, not as an additional substantial, physical entity in space and time. This interpretation of the wave function is favored by some of the most prominent current proponents of BM, who take as a heuristic argument the analogy with the common interpretation of the Hamiltonian on phase space within the framework of classical mechanics (see e.g. Dürr et al. 1997). So, within this understanding and the Bohmian context, the wave function is taken as an aspect of the Bohmian law of motion (guiding equation). However, this nomological interpretation of the wave function faces an important difficulty: the wave function can be time-dependent – a non-standard feature for a law-like entity, which requires some clarifications. A related difficulty concerns the status of the Schrödinger equation: what is the status of a law (the Schrödinger equation) that determines the temporal evolution of law-like entity (the wave function)? In order to deal with these difficulties, the proponents of the nomological understanding of the wave function within BM have deployed a strategy which contains three main components. First, the crucial distinction between the universal wave function, i.e. the wave function of the universe (the wave function corresponding to all the Bohmian particles in the universe), and the effective wave functions corresponding to (Bohmian) subsystems of the universe. If the latter are epistemically crucial (they are the ones that are dealt within standard QM as well as for predictive and operational purposes), only the former is ontologically fundamental strictly speaking (effective wave functions only possess a 'derivative' status compared to the universal wave function). Second, and most importantly for the later parts of this article, the expectation, based on the fundamental timelessness of the Wheeler-De-Witt equation in canonical QG, that the universal wave function is static (and possibly unique). Third, the (informed) conjecture that the time-dependent Schrödinger evolution is not fundamental, but only effective in the sense of only arising for subsystems and their effective ('time-dependent') wave functions. Even if supported by good heuristic arguments, the second and third component of this strategy remain speculative, making the purely nomological interpretation of the wave function somewhat less attractive (the important point to underline at this stage is the fact that the proponents of the nomological understanding of the wave function in the context of the Bohmian primitive ontology appeal to the QG domain to ground their conception, see e.g. Dürr et al. 1997, sect. 12). Moreover, it seems that the exact ontological picture resulting from the nomological understanding of the wave function depends on one's metaphysical stance with respect to laws, e.g. Humean or dispositional. There is no need to enter this venerable metaphysical debate here. Suffice it to note that if a Humean approach in this context is clearly not incoherent (leading to what could be called 'quantum Humeanism'), it can be argued that it would considerably weaken the explanatory power of the conception, which is one of the main motivations for the primitive ontology move in the first place. For instance, within this Humean framework, the wave function and crucial quantum features such as quantum non-locality that are encoded in the wave function would merely supervene on the whole distribution of the relevant local beables in the entire spacetime, rather than being anchored (and therefore 'explained' in some sense) in the nature or properties of these local beables postulated by the primitive ontology (see the comment above in section 3; see also the discussion in Esfeld et al. 2013).

The third understanding of the wave function precisely aims to do that: the idea is to interpret the wave function in terms of the properties of – more precisely, the relations among – the local beables. This understanding is appealing in the primitive ontology context: indeed, for example, within the framework of BM, the wave function determines through the Bohmian equation of motion (guiding equation) the temporal development of the local beables, that is, the velocities of the Bohmian particles. In this perspective, it is perfectly sensible to think of the wave function as describing a fundamental property of the Bohmian particles that determines their motion (this description possibly not being one-to-one, see Belot 2012, pp. 78–80 and the discussion in Esfeld et al. 2013). The main worry for this understanding comes from the fact that, in this context, the wave function encodes quantum non-locality, so that the fundamental property described by the wave function is rather peculiar. In the Bohmian case, the (universal) wave function is defined on the whole configuration of all Bohmian particles in the universe at a given time, so that the temporal development (the velocity) of each particle depends strictly speaking on the positions of all the other particles at that time through the (universal) wave function. Therefore, the wave function actually describes a kind of holistic property of the whole configuration of particles (at a given time). We discuss in the next two sections how ontic structural realism can help to clarify the nature of such a property, that is, the nature of the wave function in the primitive ontology approach.

5. Ontic Structural Realism and Primitive Ontology

Ontic structural realism (OSR) is a recently much debated conception in the metaphysics of contemporary fundamental physics, in particular quantum theory. As a metaphysical conception and interpretative framework for fundamental physics, its development has been mainly motivated by various fundamental relational physical features, in particular background independence and gauge-theoretic diffeomorphism invariance in the general relativistic domain (see e.g the contributions by Pooley, Rickles and Stachel in Rickles et al. 2006 as well as Esfeld and Lam 2008) and permutation invariance (together with other symmetry considerations), entanglement and non-locality in the quantum domain (see e.g. French and Ladyman 2003, Esfeld 2004, Ladyman et al. 2007, ch. 3, Kantorovich 2009, Muller 2011a, Lam 2013). The broad ontological thesis of OSR that is motivated by these relational features can be expressed in the following way: what there is in the world at the fundamental level (or in the cases where OSR is relevant) are physical structures, in the sense of networks of concrete physical relations among concrete physical objects (relata), whose existence depends in some sense on relations in which they stand and on structures they are part of (see French 2010 for a discussion of the relevant notion of 'existential dependence' in this context; see also recently Wolff 2012 and McKenzie 2013).

As mentioned above, it has been argued in the literature for some time now that OSR provides a general interpretative framework for the generic relational features of quantum entanglement and quantum non-locality as encoded in the violations of Bell-type inequalities. Since (of course) quantum non-locality has to be accounted for within the primitive ontology approach, it is no wonder that OSR is relevant in this context. Indeed, on the one hand, a primitive ontology for QM (such as BM) provides an ontological framework within which the general OSR understanding of quantum non-locality can be specified (in particular the relata of the relevant quantum structures can be specified). On the other hand, OSR provides the primitive ontology approach to QM with a convincing way to interpret the wave function and its encoding of quantum non-locality in this context (see Esfeld 2014 for a similar point of view).

6. The Wave Function as a Physical Structure

We have seen above that there is a tension about the wave function within the primitive ontology approach to QM (section 3): if it clearly plays a central role in the explanatory scheme of the primitive ontology approach (in particular, in the account of non-locality), it is not easy to anchor the wave function in the 3-dimensional physical space (or 4-dimensional spacetime) in which the relevant local beables live – as one would expect within the primitive ontology framework. OSR precisely provides such a spacetime anchorage for the wave function in the primitive ontology approach to QM: in this context, the wave function can be understood as a physical structure in spacetime whose relata are the local beables of the primitive ontology under consideration (this spacetime anchorage is crucial from the primitive ontology point of view; we will come back to this point below).

Let us illustrate how this account works in our study case, BM. The wave function is understood in terms of a concrete, physical structure instantiated by the Bohmian particles. This huge network of physical correlations (as described by the wave function) constitutes the physical ground for (the explicit) quantum non-locality in BM; in this fundamental quantum structure, each particle is strictly speaking related to all the others (in a way described by the wave function), so that its temporal development (its velocity) depends on the positions of all the other particles (hence the BM account of the violation of Bell-type inequalities in terms of the violation of parameter independence); note that, as mentioned above in section 4, the notion of effective wave function captures the operationally relevant aspects of such a huge dependence.

There are two aspects that jointly make this quantum structure, which is described in the quantum formalism by the wave function, a structure in the OSR sense. First, the quantum relations connecting all the Bohmian particles do not supervene on any intrinsic properties of the particles; therefore, these quantum relations and the corresponding quantum structure are fundamental and irreducible in the sense that they cannot be merely understood in terms of (they cannot be 'reduced' to) the intrinsic properties of the relata, namely the Bohmian particles. Second, even if some intrinsic individuality and identity can possibly be ascribed to them (e.g. in virtue of their spacetime location, if one accepts that it can be taken as an intrinsic feature), there is a sense in which Bohmian particles dynamically depend on the structure they are part of, through the dependence on the positions of all the other particles. Unlike the case of Newtonian gravity (where some structuralist dependence among all Newtonian particles also obtains), this dependence is strictly speaking not affected by spatial distance. So, in a sense, the very existence of Bohmian particles dynamically depend on the structure they are part of.

Furthermore, one could characterize Bohmian particles in terms of some dynamical (diachronic) identity that depends on the whole configuration of particles, that is, in terms of some non-intrinsic (structural, contextual) identity (about the notion of non-intrinsic identity in the context of OSR, see Lam 2014).

The tension between this structural identity and the above mentioned intrinsic identity based on the spacetime location is only apparent: besides the fact that its 'intrinsicness' can be put into question (it ultimately relies on the spacetime structure), this latter identity is dynamically inert, whereas the former plays a crucial dynamical and explanatory role, in particular in the account of quantum non-locality. Indeed, in this context, quantum non-locality is accounted for in terms of the dynamics of the (relevant part of the) quantum structure, within which the relata (i.e. the Bohmian particles) are interdependent; this dynamical interdependence is here precisely encoded in the notion of dynamical (or structural, contextual) identity. In this perspective, it seems to make sense to claim that for each Bohmian particle, the fact of being this very particle, which includes its own trajectory and dynamical features, depends on the structure it is part of. Moreover, this structuralist understanding receives further support from the permutation invariant formulation of BM, called 'identity-based' BM (see Goldstein et al. 2005a, Goldstein et. al 2005b), which highlights the fact that Bohmian particles can be genuinely understood as lacking intrinsic properties altogether.

This quantum structure instantiated by the Bohmian particles possesses a fundamental and inherently modal nature, as described by the wave function on configuration space. In particular, this modal nature grounds all the possible temporal developments of the interdependent Bohmian particles through the guiding equation. The modal nature and the dynamical role of the quantum structure is crucial to the structuralist understanding of the wave function: this latter should not be understood in terms of a bunch of dynamically and modally inert relations holding among Bohmian particles (or in terms of a plethora of non-actual relations), but rather in terms of a concrete, physical structure instantiated by the actual particle configuration that relates all the particles in determining their temporal development. The universal wave function and the guiding equation fully describe what sort of physical structure this quantum structure is and how it constraints as a whole the temporal development of each particle – in particular, correlating the temporal development of each particle with all the other particles.[1]

So, from the metaphysical point of view, the structuralist account of the wave function proposed here provides a clear metaphysical basis for the holistic aspects mentioned within the framework of the property understanding at the end of section 4; in this sense, OSR helps to clarify the nature and the status of the wave function – more precisely: what is

[1] Thanks to an anonymous referee for pressing me on this point.

represented by the wave function – within the primitive ontology approach to QM. OSR provides a way to understand the wave function in spatio-temporal terms, as a concrete physical structure instantiated by local material entities (beables) in spacetime. As mentioned above, this spacetime anchorage of the wave function is crucial to the primitive approach to QM, since it is postulated from the start that the fundamental ontology is about matter (and its properties/relations) in spacetime. In the next sections, we look more closely at this spacetime postulate that is at the heart of the primitive ontology approach to quantum theory, in particular when this latter is applied to the gravitational (general relativistic) domain.

7. Primitive Ontology and Spacetime

As we have discussed in section 2, the main motivation for postulating a primitive ontology for QM, that is, for postulating an ontology of material entities localized in spacetime is the explanatory power that such an ontological background provides. In particular, it allows for a classical explanatory scheme, that is, an explanatory scheme similar to the one at work in classical physics, where the behavior of familiar, macroscopic objects that appear localized in 3-dimensional space (or 4-dimensional spacetime) can be explained in terms of microscopic, fundamental constituents that are localized in 3-dimensional space too (we put aside possible questions related to the general reductionist framework, they are not directly relevant here). In this context, the common spacetime arena constitutes a crucial link between the manifest and the scientific image.

This specific ('spacetime') link, which is central to the classical explanatory scheme, is not available to an ontology of QM that does not postulate material entities localized in 4-dimensional spacetime at the fundamental level, such as within wave function realism (or, more precisely, wave function monism; the distinction does not fundamentally alter the issue here). According to this latter conception, what there is at the fundamental level is the wave function, understood as a real, substantial entity ('field') living not in 4-dimensional space-time but in (on) a high-dimensional space (isomorphic to what is usually called 'configuration space'). To the extent that the wave function is crucial to all the three standard realist interpretations of QM (i.e. Bohm, GRW, Everett; see section 3), these latter can all be ontologically understood along the lines of wave function realism. It is interesting to note that the debate between a primitive ontology approach to QM and wave function realism is a debate between different ontological frameworks underlying the three standard realist interpretations of QM (rather than a debate between these interpretations; see also Ney and Philips 2013, §2). If, as discussed above in section 2, it does seem that wave function realism has to bridge an important explanatory gap that is absent from the primitive ontology explanatory scheme, it is also important to underline that the proponents of a wave function ontology are not without explanatory resources (see e.g. Albert 1996, Albert 2013, Ney 2012). However, in the case of QM, it can be rather convincingly argued that the primitive ontology approach has a clear explanatory (and maybe ontological) advantage, be it only in terms of simplicity, where spacetime is a crucial component (similarly, spacetime implicitly plays a crucial role in the motivation for Bell's local beables).

Now, we would like to stress that the role of spacetime in elaborating an ontology for quantum theory applied to the gravitational field as described by the general theory of relativity (GTR) is not as clear-cut as it is in the case of QM (including standard quantum field theory). Indeed, in the latter case, spacetime is not part of the dynamical physical systems described by the theory, where a given fixed, non-dynamical (possibly curved) spacetime structure is therefore postulated from the start (the "stage" over which the physics unfolds, as Rovelli 2001 puts it), so that it seems coherent for the corresponding ontology to do as much. In the former case, things possibly are entirely different. Indeed, under a common understanding of GTR, the gravitational field and the spacetime structure are aspects of the same physical dynamical entity, which is described by the theory and its dynamical equations, the Einstein field equations (so that the spacetime "stage" becomes an "actor"). A crucial feature in this context is that there is no fixed, non-dynamical physical entity with respect to which physical systems and their dynamics can be considered in a privileged way; more precisely, the (metric-)gravitational field cannot be decomposed uniquely into an inertial (non-dynamical) part plus a gravitational (dynamical) part (this is a rough characterization of what is called 'background independence', a feature many think to be essential for a fundamental theory of QG, see e.g. Smolin 2006; more precise discussions of background independence can be found in Giulini 2007 and Rickles 2008 for instance). To the extent that this feature of background independence is fundamental (and again, many think it is), one can legitimately expect that it is reflected in the ontology in some way or another, so that we can already see at this classical stage that a tension with the spacetime postulate of primitive ontology may possibly arise.

Indeed, if one tries to develop a quantum theory of the gravitational field – that is, a quantum theory of the spacetime structure itself – within this background independent framework, it does not seem to be justified to postulate from the start that the ontology of such a QG theory is fundamentally about a spacetime arena within which and with respect to which physical systems are localized. The very nature and status of spacetime and its relationship to matter are actually central open issues, which are the focus of ongoing QG research programs. Within the framework of many QG candidate theories, such as the ones based on the canonical quantization of GTR, various ontological conceptions about spacetime have been argued for, including conceptions according to which spacetime is not fundamental in some sense, and the whole debate is still very much ongoing (the detailed arguments need not be exposed here, for a recent overview see Huggett and Wüthrich 2013). The important point is that the very status of spacetime is under debate within these candidate theories of QG. A primitive ontology approach just does not seem to be able to do justice to this very important debate and the very methodology underlying this approach seems flawed in this context (at least to the extent that it amounts to assuming from the start a certain ontological conception of spacetime in the very debate on the ontological nature and status of spacetime; see section 9 below for further reflections in this context on the role and status of spacetime in string theory and Bohmian approaches to QG).

There is for sure a venerable philosophical tradition according to which ontological claims about space and time (and their relationship to matter) can be made a priori, without relying on our best physical descriptions that are available. Without discussing the merits and the difficulties of such an epistemic stance, suffice is to note that it is certainly not in line with the 'naturalized metaphysics' framework according to which our fundamental ontology of nature should be grounded in our experimentally most successful physical theories and within which most of the current debates on the ontology of quantum theory (including the primitive ontology approach) take place. As a side remark, let us further note that there definitely are important and interesting debates around this naturalized approach to analytical metaphysics (for a recent overview, see the collections of essays in Ross et al. 2013) and in particular the role and status of spacetime within such a metaphysics of science, but they constitute different (sometimes related) topics of investigations from the one the primitive ontology conceptions (and this paper) are originally concerned with (namely, the ontology of quantum theory).

8. Ontic Structural Realism and Spacetime

The role of OSR in the understanding of the quantum wave function within the primitive ontology framework is double. First, it aims to encode the fundamental relational features of the wave function as manifested in quantum non-locality. Second, it allows to anchor the wave function in spacetime, as a physical structure in spacetime among the relevant local beables constituting the primitive ontology under consideration. As discussed in the last section, this spacetime anchorage may not be relevant anymore in the QG domain, where the very status of spacetime is an open issue. However, at least within canonical QG, the wave function (or the relevant quantum states) might still possess fundamental relational features grounding non-locality at the QM level, so that OSR might still provide a relevant interpretative framework in the QG context, even if it can be argued that the very notion of non-locality (as well as the related notions of entanglement and non-separability in their common understanding) make only sense, strictly speaking, within a spatio-temporal framework (see Lam and Esfeld 2013). One worry seems to be that in order to be entitled to use the notions of non-locality and entanglement, one has to spell out a differentiation in terms of a plurality of entities that are entangled (or non-locally correlated) with each other. The difficulty lies in the fact that such concrete differentiations are often made in spatio-temporal terms, which may not be available in the QG domain (similarly, the notion of non-separability in terms of which quantum entanglement can be understood seems to presuppose spatio-temporally separated entities).

However, spatio-temporal differentiation need not be the only way to obtain a plurality of fundamental objects. Indeed, within the OSR framework, the numerical diversity of fundamental relata is often considered as a primitive fact – this primitive numerical diversity is then said to be 'contextual' in the sense of a plurality of objects whose existence depends on the structures they are part of (as a consequence, and in contrast to a primitive diversity of 'isolated' individuals with intrinsic identity, primitive contextual diversity does not imply haeccesitism, see Ladyman 2007). In this structuralist context, there is actually a number of convincing arguments in favor of such a primitive understanding of numerical diversity (see Lam 2014, §5).

First, it obviously dissolves the possible worries around the controversial issue of the status of numerical diversity within the purely relational framework of OSR (in particular, it avoids the circularity concern that is sometimes voiced against OSR). It is important to note that the issue of the status of the numerical plurality of the fundamental entities in our ontology has to be faced in any case – whether or not spacetime is fundamental; in the context of a fundamental spacetime background providing the above mentioned standard spatio-temporal differentiation, the issue is then about the numerical diversity of spacetime points or regions themselves (indeed, motivated by GTR diffeomorphism invariance, Pooley 2006 and Esfeld and Lam 2008 suggests to take the contextual diversity of spacetime points as a primitive fact). Second, primitive numerical diversity enables OSR to clearly distinguish between issues related to the numerical identity of the relata and issues related to their (in)distinguishability; more specifically, it explicitly frees OSR from any commitment to the controversial Principle of the Identity of Indiscernibles (PII) – while remaining compatible with it, in particular in its weak version (see Ainsworth 2011 for a discussion of the links between OSR and PII) – and it allows for possible contextual but completely indiscernible objects, in the sense of not even weakly discernible objects.

So, without delving into the details of this issue, it does seem that taking the numerical diversity of the fundamental OSR relata as a primitive fact constitutes a coherent and convincing move (for a recent discussion on this issue, see Lam 2014 and references therein).

The important point here is that primitive contextual diversity shows that spatio-temporal differentiation is not necessary for having a multiplicity of fundamental objects and for attributing certain (relational) features to them, in particular for defining an OSR structure. In this sense, it does not seem that OSR is committed to spacetime; and in this sense, OSR might still provide a relevant interpretative framework for the QG domain, whatever the fundamental status of spacetime turns out to be (as already mentioned above, there are good reasons to think that a structuralist framework might still be relevant in the QG domain, see the essays in Rickles and French 2006).

The possibility of non-spatio-temporal OSR structures, possibly suggested by certain QG theories, raises however an important question. The OSR conception is about physical ontology, that is, about what there is in the concrete physical world, and as such it is concerned with concrete physical structures instantiated in the concrete physical world, in contrast to the abstract, mathematical structures of mathematical structuralism for instance. Now, what makes the structures OSR claims are out there in the world physical structures in contrast to abstract mathematical structures? One standard way to answer this question precisely relies on spacetime and is at the basis of the whole primitive ontology framework: physical structures are in spacetime, whereas mathematical ones are not. Of course, this criterion does not cover the case of the spacetime structure itself (as described by GTR for instance) or the possible non-spatio-temporal structures mentioned above in the QG context. In the face of this difficulty, a possible move is to simply reject the distinction between abstract and concrete entities or structures (Ladyman et al. 2007, §3.6). However, providing abstract mathematical structures with the same ontological status as (what one would commonly think as) concrete physical structures does not seem very satisfactory. It seems indeed more promising to consider the other standard way to characterize concrete entities in contrast to abstract ones: causal efficacy. Concrete physical structures as opposed to abstract mathematical ones can be considered to be causally efficacious in some modal sense, where the relevant modality is encoded in the structure through different metaphysical strategies, involving causal powers, dispositions or some inherent, objective modality (see above section 6; see also Esfeld 2009, Ladyman et al. 2007, ch. 2–3; Berenstain and Ladyman 2012 and recently French 2014, ch. 8–10). Such modality might characterize concrete physical structures possibly without relying on a spatio-temporal framework, and so would fit the QG context. Trying to spell out this modality in non-spatio-temporal terms within specific QG cases is an interesting and important task for future research.

9. Conclusion and Perspectives

In the first part of this contribution, we have highlighted the difficulties that arise about the status and nature of the wave function within the framework of the primitive ontology approach to QM, that is, within the realist interpretations of the theory – realist solutions to the measurement problem – according to which QM is ultimately about (material) entities localized in 3-dimensional physical space and evolving in time. We have then suggested how a structuralist understanding of the wave function in the sense of OSR might solve these difficulties. We have mainly taken BM – the paradigmatic example of a primitive ontology for QM – as a convenient study case for illustrating the interpretative relevance of this structuralist understanding. In particular, the explicit non-locality that is at the heart of BM (and to some extent, at the heart of QM in general) is naturally understood in terms of the relational features of the relevant quantum structure. The upshot is that the primitive ontology approach and OSR can be nicely combined to form a powerful interpretative framework for QM and its fundamental feature of quantum non-locality.

In the second part of this contribution, we have discussed the limits of this framework. In particular, the crucial role spacetime plays in the primitive ontology context seems at odds with certain approaches to QG, although the OSR conception itself does not rely on a spatio-temporal background. These considerations from the QG domain may actually seem too speculative, and the proponents of the primitive ontology approach may want to argue that we should first deliver a coherent ontology for QM – an experimentally extremely successful theory, in contrast to QG theories for the time being – what their approach convincingly does. However, the primitive ontology move is often expected to be universal in the sense of being valid for any fundamental physical theory (see e.g. Allori 2013a). Moreover, when it comes to ground a possible nomological status for the wave function, the proponents of the primitive ontology approach to QM readily appeal to the (canonical) QG domain and in particular to the timelessness of the Wheeler-DeWitt equation (see e.g. Goldstein and Zanghì 2013; see also section 4 above).

Allori (2013b, §8) very briefly mentions the status of spacetime in the QG domain (namely, string theory) within the framework of a discussion on primitive ontology. She identifies a potential difficulty in the fact that, similarly to the 'configuration space' of wave function realism, but to a much lesser extent, the relevant space of string theory is not a 3- or 4-dimensional space but rather a higher dimensional space. But then she quickly argues that the compactification of the extra dimensions within sting theory makes the primitive ontology approach and its explanatory scheme still relevant in this context.

Without entering into the details, the various string-theoretic 'dualities' actually make the situation much more complicated than that – in a way that quite explicitly conflicts with the primitive ontology approach (see recently Huggett and Wüthrich 2013, §2.4 for a brief overview). Depending on the dualities considered, even topology and dimension do not seem to be determinate, fundamental features of the world in the string-theoretic context. Of course, this latter context requires a cautious attitude, but the important point here, as already discussed in section 7, is that, whatever its merits and difficulties, it constitutes a further example in the QG domain where the very status of spacetime and its relationship to matter are open issues, so that postulating an ontology of material entities (local beables) localized in 4-dimensional spacetime does not seem like the appropriate interpretative move in this framework (in particular, with respect to these dualities, the 'strings' of string theory are no local beables in any way).

The basic strength of the primitive ontology approach is the fundamental requirement underlying the whole move that for any physical theory it should be clear what the theory is fundamentally about. Primitive ontologies understood in this broader sense have been put forward in the context of Bohmian approaches to canonical QG (see e.g. Goldstein and Teufel 2001). But it should be clear that primitive ontology in this broader sense is nothing but ontology tout court understood in a serious way. Within this broader framework, the role of local beables becomes less central: as such, a serious ontology need not be made up of local beables. But a serious ontology without local beables at the fundamental level is not a primitive ontology in the original sense; most importantly, the powerful (and simple) explanatory scheme of the primitive ontology involving local beables is not available to such an ontology (this is especially salient in various Bohmian approaches to quantum field theory and canonical QG, but this is a topic for another paper). In any case, the primitive ontology approach cannot have it both ways: it cannot stick to its successful explanatory scheme and at the same time provide an ontological framework for QG.

University of Lausanne
Department of Philosophy
e-mail: vincent.lam@unil.ch

ACKNOWLEDGMENTS

I am grateful to the Swiss National Science Foundation (Ambizione grant PZ00P1_142536/1) for financial support.

REFERENCES

Ainsworth, P.M. (2011). Ontic structural realism and the principle of the identity of indiscernibles. *Erkenntnis* **75**, 67–84.

Albert, D.Z. (1996). Elementary Quantum Metaphysics. In: J. Cushing, A. Fine, and S. Goldstein (eds.), *Bohmian Mechanics and Quantum Theory: An Appraisal*, pp. 277–284. Dordrecht: Kluwer.

Albert, D.Z. (2013). Wave Function Realism. In: A. Ney and D.Z. Albert (eds.), *The Wave Function: Essays in the Metaphysics of Quantum Mechanics*, pp. 52–57. New York: Oxford University Press.

Allori, V. (2013a). Contre les ontologies de la fonction d'onde: défense des ontologies primitives. In: S. Le Bihan (ed.), *Précis de philosophie de la physique*, pp. 116–140. Paris: Vuibert. English version at philsci-archive.pitt.edu/9343.

Allori, V. (2013b). Primitive Ontology and the Structure of Fundamental Physical Theories. In: A. Ney and D.Z. Albert (eds.), *The Wave Function: Essays in the Metaphysics of Quantum Mechanics*, pp. 58–75. New York: Oxford University Press.

Allori, V., Goldstein, S., Tumulka, R., Zanghì, N. (2008). On the Common Structure of Bohmian Mechanics and the Ghirardi-Rimini-Weber Theory. *British Journal for the Philosophy of Science* **59**, 353–389.

Allori, V., Goldstein, S., Tumulka, R., Zanghì, N. (2011). Many Worlds and Schrödinger's First Quantum Theory. *British Journal for the Philosophy of Science* **62**, 1–27.

Bell, J.S. (1987). *Speakable and unspeakable in quantum mechanics*. Cambridge: Cambridge University Press.

Belot, G. (2012). Quantum states for primitive ontologists. *European Journal for Philosophy of Science* **2**, 67–83.

Berenstain, N., Ladyman, J. (2012). Ontic Structural Realism and Modality. In: E. Landry and D. Rickles (eds.), *Structural Realism*, pp. 149–168. Dordrecht: Springer.

Dowker, F., Herbauts, I. (2005). The status of the wave function in dynamical collapse models. *Foundations of Physics Letters* **18**, 499–518.

Dürr, D., Goldstein, S, Zanghì, N. (1997). Bohmian mechanics and the meaning of the wave function. In: R. Cohen, M. Horn, and J. Stachel (eds.), *Experimental Metaphysics*, pp. 25–38. Dordrecht: Kluwer.

Esfeld, M. (2004). Quantum entanglement and a metaphysics of relations. *Studies in History and Philosophy of Modern Physics* **35**, 601–617.

Esfeld, M. (2009). The modal nature of structures in ontic structural realism. *International Studies in the Philosophy of Science* **23**, 179–194.

Esfeld, M. (2014). *How to account for quantum non-locality: ontic structural realism and the primitive ontology of quantum physics*. Synthese, doi: 10.1007/s11229–014-0549–4.

Esfeld, M., Lam, V. (2008). Moderate structural realism about space-time. *Synthese* **160**, 27–46.

Esfeld, M., Lazarovici, D., Hubert, M., Dürr, D. (2013). The Ontology of Bohmian Mechanics. *British Journal for the Philosophy of Science* **64**, doi:10.1093/bjps/axt019.

French, S. (2010). The Interdependence of Structure, Objects and Dependence. *Synthese* **175**, 89–109.

French, S. (2014). *The Structure of the World.* Oxford: Oxford University Press.

French, S., Ladyman, J. (2003). Remodelling Structural Realism: Quantum physics and the metaphysics of structure. *Synthese* **136** (1), 31–56.

Giulini, D. (2007). Remarks on the Notions of General Covariance and Background Independence. In: E. Seiler and I.-O. Stamatescu (eds.), *Approaches to Fundamental Physics: An Assessment of Current Theoretical Ideas*, pp. 105–122. Berlin: Springer.

Goldstein, S., Taylor, J., Tumulka, R., Zanghì, N. (2005a). Are all particles identical? *Journal of Physics A: Mathematical and General* **38** (7), 1567–1576.

Goldstein, S., Taylor, J., Tumulka, R., Zanghì, N. (2005b). Are all particles real? *Studies in History and Philosophy of Science Part B: Studies in History and Philosophy of Modern Physics* **36**, 103–112.

Goldstein, S., Teufel, S. (2001). Quantum Spacetime without Observers: Ontological Clarity and the Conceptual Foundations of Quantum Gravity. In: C. Callender and N. Huggett (eds.), *Physics Meets Philosophy at the Planck Scale*, pp. 275–289. Cambridge: Cambridge University Press.

Goldstein, S., Zanghì, N. (2013). Reality and the Role of the Wave Function in Quantum Theory. In: A. Ney and D.Z. Albert (eds.), *The Wave Function: Essays in the Metaphysics of Quantum Mechanics*, pp. 91–109. New York: Oxford University Press.

Huggett, N., Wüthrich, C. (2013). Emergent spacetime and empirical (in)coherence. *Studies in History and Philosophy of Modern Physics* **44**, 276–285.

Kantorovich, A. (2009). Ontic Structuralism and the Symmetries of Particles Physics. *Journal for General Philosophy of Science* **40**, 73–84.

Ladyman, J. (2007). On the identity and diversity of objects in a structure. *Proceedings of the Aristotelian Society Supplementary* **81** (1), 45–61.

Ladyman, J., Ross, D., Spurett, S., Collier, J. (2007). *Every Thing Must Go: Metaphysics Naturalized. Oxford*. Oxford University Press.

Lam, V. (2013). The Entanglement Structure of Quantum Field Systems. *International Studies in the Philosophy of Science* **27**, 59–72.

Lam, V. (2014). Entities without Intrinsic Physical Identity. *Erkenntnis* **79**, 1157–1171.

Lam, V., Esfeld, M. (2013). A Dilemma for the Emergence of Spacetime in Canonical Quantum Gravity. *Studies in History and Philosophy of Modern Physics* **44**, 286–293.

McKenzie, K. (2013). Priority and Particle Physics: Ontic Structural Realism as a Fundamentality Thesis. *British Journal for the Philosophy of Science*, doi:10.1093/bjps/axt017.

Miller, E. (2013). Quantum Entanglement, Bohmian Mechanics, and Humean Supervenience. *Australasian Journal of Philosophy*, doi:10.1080/00048402.2013.832786.

Muller, F.A. (2011). Withering Away, Weakly. *Synthese* **180**, 223–233.

Ney, A. (2012). The Status of our Ordinary Three Dimensions in a Quantum Universe. *Noûs* **64**, 525–560.

Ney, A., Albert, D.Z., eds. (2013). *The Wave Function: Essays in the Metaphysics of Quantum Mechanics*. New York: Oxford University Press.

Ney, A., Philips, K. (2013). Does an Adequate Physical Theory Demand a Primitive Ontology? *Philosophy of Science* **80**, 454–474.

Pooley, O. (2006). Points, Particles and Structural Realism. In: D. Rickles, S. French and J. Saatsi (eds.), *The Structural Foundations of Quantum Gravity*, pp 83–120. Oxford: Oxford University Press.

Rickles, D. (2008). Who's Afraid of Background Independence? In: D. Dieks (ed.), *Ontology of Spacetime. Philosophy and Foundations of Physics Series. Vol. 2*, pp. 133–152. Amsterdam: Elsevier.

Rickles, D., French, S. (2006). Quantum Gravity Meets Structuralism: Interweaving Relations in the Foundations of Physics. In: D. Rickles, S. French and J. Saatsi (eds.), *The Structural Foundations of Quantum Gravity*, pp. 1–39. Oxford: Oxford University Press.

Rickles, D., French, S., Saatsi., J. (2006). *The Structural Foundations of Quantum Gravity.* Oxford: Oxford University Press.

Ross, D., Ladyman, J., Kincaid, H., eds. (2013). *Scientific Metaphysics*. Oxford: Oxford University Press.

Rovelli, C. (2001). *Quantum spacetime: What do we know?* In: C. Callender and N. Huggett (eds.), *Physics Meets Philosophy at the Planck Scale*, pp. 101–124. Cambridge: Cambridge University Press.

Smolin, L. (2006). The Case for Background Independence. In: D. Rickles, S. French and J. Saatsi (eds.), *The Structural Foundations of Quantum Gravity*, pp. 196–239. Oxford: Oxford University Press.

Wolff, J. (2012). Do Objects Depend on Structures? *British Journal for the Philosophy of Science* **63**, 607–625.

Dean Rickles
Jessica Bloom

THINGS AIN'T WHAT THEY USED TO BE. PHYSICS WITHOUT OBJECTS

ABSTRACT. In this paper we draw attention to some problems with the view that physical theories are fundamentally about individual entities (= things). We highlight several ways in which sense can be made of a physics in which such an assumption is abandoned.

It doesn't mean a thing to me
It doesn't mean a thing to me
And it's about time you see
Things ain't what they used to be

Parsons and Ellington

1. These Foolish Things

A world of things or a world of relations?[1] *Things* have traditionally exposed scientific development to trouble: think phlogiston. Think luminiferous ether. *Things* have been at the root of many problems in the conceptual foundations of physical theories. Think spacetime points. Think quantum particles. Yet things also seem to be necessary for physical relations (found in the laws of physics) to make any kind of sense. How could there be a world *without* things; without particular objects or individuals? How could there be relations (spatiotemporal distance, for example) without things *standing* in those relations to 'bring them about'? So long as there exists

[1] "Thing" is simply our catch-all term for concrete, particular objects.

In: Tomasz Bigaj and Christian Wüthrich (eds.), *Metaphysics in Contemporary Physics* (*Poznań Studies in the Philosophy of the Sciences and the Humanities*, vol. 104), pp. 101-122. Amsterdam/New York, NY: Rodopi | Brill, 2015.

a universe of things, it appears to safe to say that there is also a universe of relations. Moreover, it seems sensible to say that these relations are *logically dependent* on things being related, so that wherever there are relations there must be things too. Surely these categories cannot be disentangled? Mustn't they come as a package deal?

That the world is fundamentally made up of things is an example of a belief so ingrained that (with some exceptions) it has become an unquestioned dogma (though a perfectly understandable one, for the reasons alluded to above). Recall Feynman's famous declaration on the importance of the atomic hypothesis: "Everything is made of atoms. That is the key hypothesis" (2011, p. 20). In this paper we will argue, contrary to this common line of thought, that the assumption that the world is *fundamentally* composed of 'things' (atoms or otherwise) is wrong (or at least problematic in the context of physics) and that jettisoning it might plausibly lead to novel advances in physics.[2]

The search for things (of an ever more 'elementary' kind) has been fruitful in many ways, characterising the way physics ought to be conducted for a very long time. It will presumably always have a role to play given that things provide our 'epistemic access' to the world in a very general sense. However, not only have specific choices for the fundamental things repeatedly been overturned throughout the history of science, aspects of our most fundamental theories (and those destined to become such in the future) imply a picture in which the fundamental ontological level (the 'ontological basement', to borrow Paul Davies' apt phrase) is occupied by primitive, *irreducible* relations. What is called for, we argue, is a reconceptualisation of our 'thing-world' in terms of a fundamentally relational ontology, which is perhaps better called 'structural physics' to avoid the troubling slide to the supposed ontological primacy of relata.[3]

Viewing the world as structurally constituted by primitive relations (not grounded in specific thing-based constructions) has the potential to

---

[2] Of course, Feynman's position was nothing like the idea that the world 'really is' made of atoms at a fundamental level. He was concerned with the massive 'information content' stored in the atomic hypothesis, unifying all manner of apparently disconnected phenomena. That we don't wish to dispute, nor do we wish to dispute the enormous utility of conceptualizing the world in terms of individual things. Our point is, rather, that at a certain level of analysis, the hypothesis that the world is fundamentally constituted by things is placed under pressure.

[3] Though structures too are most often conceived of as being *composed of* and *dependent on* a domain of elements. The key point is that we are arguing for a picture of the world in which relational structures are *prior* to things that might 'emerge' from such structures. There is unfortunately no pre-existing term to describe this state of affairs, as so we shall continue to employ 'structure', 'relations', and 'relata', with the above proviso in mind.

lead to new kinds of research in physics, and knowledge of a more stable sort. Indeed, in the past those theories that have adopted a broadly similar approach in terms of *methodology* – along the lines of what Einstein labeled 'principle theories' – have led to much needed advances in situations where thinkers were 'cognitively obstructed' and had to let go of some of their most deeply ingrained assumptions to make progress. Principle theory approaches often look to general 'structural aspects' of physical behavior over 'thing aspects' (what Einstein labeled 'constructive'), promoting invariances of world-structure to the status of general principles.[4] As Einstein puts it himself:

> These employ not the synthetic, but the analytic method. The starting point and basis are not constituted out of hypothetical constructional elements, but out of empirically discovered, general characteristics of natural processes (Principles), from which follow mathematically formulated criteria that the individual processes or their theoretical models have to satisfy. (A. Einstein, Time, Space, and Gravitation. *Times (London)*, 28th November, 1919.)

However, the notion that the principles have to be 'filled in' by individual processes (of thing-like character: Einstein's 'hypothetical constructional elements') is where the trouble starts. It is just such elements that tend to be overturned in scientific revolutions, and just such elements that have a tendency to be underdetermined by the observational data. Einstein had followed such principle-theory approaches in both his special and general theories of relativity because he wasn't convinced by the available (electromagnetic and quantum) filling; nor did he have a better alternative to hand. In place of a 'realistic', construction in terms of quantum particles (or singularities in a classical field or whatever) Einstein was forced to employ decidedly *unrealistic* rigid rods and clocks:

> *Sub specie aeterni* Poincaré, in my opinion, is right. The idea of the measuring rod and the idea of the clock coordinated with it in the theory of relativity do not find their exact correspondence in the real world. It is also clear that the solid body and the clock do not in the conceptual edifice of physics play the part of irreducible elements, but that of composite structures, which must not play any independent part in

[4] David Deutsch's recent work on 'constructor theory' (2012) is clearly inspired by the distinction between constructive and principle theories and, we think, constitutes a good example of a structural physics. In fact, we would say that it lies even closer to aspects of Arthur Eddington's philosophy of science, based on what Edmund Whittaker called "postulates of impotence" (1951, p. 18) – namely high-level, qualitative (empirical) assertions that express *prohibitions* on certain kinds of empirical observation, most often expressing the *independence* of physical quantities from certain kinds of transformation, standards, or labelling: "the statement of a conviction that all attempts to do a certain thing, however made, are bound to fail" (ibid). As we shall argue, such physical quantities must be conceived as structurally constituted.

> theoretical physics. But it is my conviction that in the present stage of development of theoretical physics these concepts must still be employed as independent concepts; for we are still far from possessing such certain knowledge of the theoretical principles of atomic structure as to be able to construct solid bodies and clocks theoretically from elementary concepts. (Einstein, Geometrie und Erfahrung. Erweiterte Fassung des Festvortrages gehalten an der Preussischen Akademie der Wissenschaften zu Berlin am 27. Januar 1921. Berlin: Julius Springer, 1921 – translation by Don Howard.)

We can agree that rods and clocks are not primitive, but that does not mean filling in by some kind of *elementary* things. Measurements of time and distance, in order to be invariant, must be of an irreducibly relational form, correlating (gauge-variant) quantities.[5] While the gauge variant quantities have a 'thingish' quality about them, they are strictly unphysical and realised only via a choice of gauge (which, again, involves a primitive correlation between them). This gauge fixing sets of a gauge-invariant correlation that can be imbued with physical significance. But to view this physical correlation as 'composed' out of pre-existing *unphysical* relata (the gauge variant quantities) does not make good physical sense.

While the basic idea defended here (a fundamental ontology of 'brute relations') can be found elsewhere in the philosophical literature on 'ontic structural realism',[6] we have yet to see the idea used as an argument for *advancing* physics, nor have we seen a truly convincing argument, involving a real construction based in modern physics, that successfully evades the so-called "incoherence objection" that there can be no relations without *first* (in logical order) having things so related.[7] We will sketch a position in this paper that provides such a construction, and is sufficiently general to apply to fundamental physics as a whole. The position we outline also provides a novel way of understanding the place of objects in ontic

[5] This is related to the well-known argument based on diffeomorphism invariance: physical quantities at points don't have fundamental physical meaning (are gauge-dependent), because of the freedom to perform diffeomorphisms. As Peter Bergmann colourfully writes, "A world point (identified by its coordinate values or, perhaps, by its geometric properties) gets mapped on a definite world point only if the field is fixed; clothed with different fields it will get mapped on different world points" (Bergmann 1971, p. 53). But *relations* between 'clothed' quantities can localise other quantities in a purely relative way. We will argue that these ought to be viewed as the fundamental basis for physics.

[6] See especially (Ladyman and Ross 2007) for an exceptionally detailed, kindred defence.

[7] A possible exception is Silberstein, Stuckey, and Cifone's "relational blockworld" approach to the foundations of physics, based on what they call a "non-separable geometric ontology of relations" (2008, p. 3). However, their position views spacetime relations as fundamental (hence, it is a variety of geometrodynamics), giving rise to particle and field phenomena. Our approach uses gauge-invariant structure, which is still more fundamental. A much older kindred work is (Cassirer 1923).

structuralism.[8] Further work is needed to establish the coherence of the scheme and its ability to reconstruct the thing-like appearance of the world.

2. All the Things You Are

Clearly, eliminating things as the fundamental ontological category (out of which the world is composed) will not really alter our everyday interactions with the world: we will still drink tea out of teacups, and not some flimsy structural counterparts thereof. Moreover, at least in terms of our direct experimental dealings, it will be things that form the objects of discourse (results of observations and experiments) and thing-language that is used to speak about them. However, when, e.g., the experimentalist speaks of a click in a counter or a spark on a screen or a number on a dial, he is really speaking elliptically about a component in a relation (that is, a relation in disguise). But before we begin our case for clearing things out of the ontological basement and moving them to the ground floor, we'd better firm up just what we have in mind by our terms 'thing', 'relation', and 'structure'.

What are things? The question lies, in many ways, at the heart of most of the central debates in metaphysics. "Thing" usually refers to something *concrete*: an individual entity with a definite location in space and time that takes on properties, and so on. In other words: a *subject*.[9] We will understand the notion in this sense. Things *qua* subjects are quantified over in models characterising a theory's possible solutions. Hence, we might find $\langle \mathcal{M}, A, B, C \rangle$ (where A, B, C are functions of $x \in \mathcal{M}$ and will map to some value space, e.g. $\mathbb{R}$, or some more complex space). Of course, quantum mechanical no-go theorems, such as Kochen and Specker's, already place considerable pressure on this notion of thinghood (at least as something that applies to all objects at all times in a *non-contextual* way), since complete, consistent simultaneous value assignments to the properties of

---

[8] Our position can be seen as falling somewhere between Ladyman and Ross' (2007) extreme, eliminativist structuralism (doing away with objects entirely), and the moderate structuralism of Esfeld and Lam (2010) (in which objects are retained, though in a much weakened sense). Our claim is that objects (things) and object-talk is *unphysical*, in a precise sense to be elucidated below.

[9] We should point out that philosophers have (thanks to Hume) a notion of a thing that does not require any 'thick' or substantial notion of a thing *underlying* properties: instead one can view things themselves as constituted by a *bundle* of properties held together by a 'compresence' relation. However, even in this case the standard subject-predicate form of description will apply.

a quantum system are not possible given certain reasonable (functional) constraints.

We are used to the notion of a *relative* or relational property via the distinction between primary and secondary qualities. A secondary quality (of a surface for example), such as being green, is just a property that demands a relationship between the coloured object and the observer (including background 'viewing conditions'): greenness involves appearing green *under the right conditions*. However, relational properties are still properties that things possess, as subjects, even if multiple such things are demanded.

Our approach might appear at first sight to be a defence of plain vanilla relationism, which classically (in the context of space-time theories) boils down to the view that space and time are dependent on configurations of material objects – implying that a space-time devoid of objects is impossible. The metric field in general relativity seems to support this perspective given that it is defined over the entire manifold, shares certain characteristics associated with matter (albeit uneasily), and determines the shape of the universe (*via* the mass distribution).[10] However, we wish to go further than this. Whilst the relationist position does treat some of the world's furniture relationally (as in the case of material objects over the existence of an independent spacetime), it still relies heavily on things to do this. It is things that provide the terminus in the reduction of relational (in this case, spatiotemporal) structure. This we take to be characteristic of *all* relationalist positions. In our view, such positions do not go far enough. Or rather they go *too far* in reducing down relational structure in a way that introduces unphysical structure (the gauge dependent entities that relationists suppose must underpin the physical, gauge independent structure).

Structuralism as we are proposing it, does away with this kind of reductive move. Whereas relationalism will seek to reduce one type of structure to the relational properties of things, we treat the relational structure as *irreducible* (in all but an 'unphysical' sense corresponding to gauge freedom, to be explained below). Rather than 'objects' and 'structure', we have only the structure which can be 'gauged' in various ways in to representations corresponding to things. This position has several advantages: it neatly sidesteps the (moribund) debate between relationism and

[10] A claim about the true, physical models is also thought to go along with relationalism. Namely that the relationalist must identify as the physical configuration an equivalence class of configurations related by the geometric symmetries. Structuralist physics would agree with this identification of the physical configuration. However, relationist and structuralist physics diverge on how this equivalence class is to be analysed: relationists will invoke some primitive objects (including fields), while structuralists will invoke brute correlational quantities.

substantivalism; resolves the incoherence objection to ontic structural realism, fits several strands of interpretively difficult contemporary physics; and it also provides a context in which to forge a new approach to physics.

3. We'll Be Together Again

Most, if not all, of the revolutionary episodes in physics have happened when some piece of the theoretical framework was found not to have a counterpart in reality. What was thought to be a substantial thing, was found not to be observable (even in principle), or could be eliminated without disrupting empirical aspects – that this overturning will *continue* to occur is, of course, the 'pessimistic meta-induction'. The stability of relations over relata (central to the principle theory approach mentioned earlier) has formed one path to structural realism, but this leaves us with the problem mentioned at the outset: how can there be relations without related *things*? It would be like having a handshake without the hands! Formally: if $R(x, y)$ then x and y are clearly bundled up in the relation. For example, let R be the taller-than relation. Then this is given by the set of ordered pairs of things, such as ⟨Magic Johnson , Danny DeVito⟩, ⟨Everest , K2⟩, ⟨Empire State Building , Tower of Pisa⟩, and so on. This brings us finally to our main case for rejecting 'object-oriented physics' (a phrase we borrow from Steven French).

It is becoming almost a truism amongst those working on quantum gravity that the relational revolution ignited by special relativity's dethroning of uniform motion has spread to include location too. But still the relationality is defined with respect to *elements*. For example, even when Lee Smolin (that paragon of background independence) speaks of relational variables "created by the system itself, as it evolves" that "do not exist *a priori*, but are defined in a context of relationships created by the dynamics of the system" (2003, p. 1082), he nonetheless understands the system to be composed of "a population of diverse elements" (ibid.). But the central point we wish to make is that the notion of elements (which, in addition, constitute an example of *absolute* elements) here ends up looking like that philosopher's nightmare: the 'bare particular'! A something we know not what waiting in an ontological limbo to receive its relational properties, whereupon it magically becomes a part of a physical structure. Instead, taken to its logical conclusion, the thesis of relational localisation (based in the gauge and diffeomorphism invariance of our theories) implies that at a fundamental level, things (such as Smolin's population of elements) do not exist. If anything, things are epiphenomena, by-products of more fundamental purely relational quantities. This amounts to doing away with another layer of absolute structure, or adding another layer of background independence.[11]

Our first principle is that observables characterise what our theories are about: whether we can actually perform an operation to determine their values or not, they provide the ontological foundation telling us what kinds of quantity exist in worlds that satisfy the theory and what operations and transformations they can satisfy. In a physically realistic theory, we should not wish to include 'external' background components (other than for convenience). When we do make observations, they are inevitably local (or approximately so), for reasons of practical necessity. But also, these observables have to be of a form that renders them invariant under general coordinate transformations. As has been discussed already, quantities defined at spacetime points will not be of this form, so some other means of localising is demanded and given that we wish to avoid external, unphysical elements, this should be done using physical degrees of freedom.[12] Hence, we see that the observables will be of a form involving relations between physical quantities. Now, though we speak of related quantities, this should not be taken to presuppose the priority of the individual quantities. Rather, the relation is the *sine qua non* of the quantities, but not vice versa. The reason for this is simple enough, though seems not to have been appreciated: the quantities qua independent entites are not invariant (that is, they are unphysical). To imbue them with robust ontological status (on a par with the gauge invariant quantity) is to dabble in something like a physicist's version of a bare particular.[13] We can perhaps diagnose this in

---

[11] Sir Arthur Eddington makes what looks like a similar statement in his discussion of Weyl's early unified theory of electromagnetism and gravitation: "any conception of structure (as opposed to substance) must be analysable into a complex of relations and relata, the relata having no structural significance except as the meeting point of several relations, and the relations having no significance except as connecting and ordering the relata" (1921, p. 121). However, we diverge from this 'package deal' viewpoint, arguing that the relata are unphysical in a way that the relations are not.

[12] The idea is an old and venerable one: Einstein postulated a "reference mullusc", DeWitt a "reference fluid" (consisting of an elastic medium holding a field of clocks), and more recently Rovelli suggested a "material reference frame" ("a 'realistic' local material reference system" also consisting of a field of clock-carrying particles). These all aim for the same basic goal: an invariant way to localise quantities. In each case, however, the 'physical' reference system is only elevated a little way above the external, absolute objects they are seeking to replace in that the reference frames themselves are treated as external material objects (things!).

[13] There is a curious parallel here with the measurement problem in the context of EPR experiments on spin-singlets: before a measurement is made, the components of the singlet are in a kind of limbo. But the crucial difference in the gauge theory case is that the limbo is not one of probabilistic uncertainty over which value will be found; rather it concerns unphysical

terms of the language we use which, based so strongly on subject-predicate form, forces us into speaking in terms of correlations, coincidences, relations, and so on, and this leads us to fall into the trap of thinking in terms of things standing in such relationships. Hence, traditional relationalism can't help us (and, indeed, can only hinder us) in the interpretation of the relational observables we find in spacetime theories since the things grounding the relations are themselves unphysical.[14]

The only other person we know of that has considered the deep implications of the kind of structural, correlational view that gauge invariant spacetime theories appear to imply is John Earman (2006, pp. 15–16). As he points out, the notion of physical observable we described above does not sit well with either of the usual interpretive suspects: neither absolutism/substantivalist nor the relational alternatives seem able to cope. The root of the problem, again, is that both substantivalism and relationalism invoke the traditional distinction between subject and predicate, viewing spacetime points (or regions) or material bodies respectively as subjects (things) bearing various properties. He goes so far as to suggest that the gauge-invariant content of GR might demand the introduction of a new ontological category that he labels a "coincidence occurrence" that encodes the idea of co-realizing values of pairs of gauge-variant quantities. He concludes:

> my feeling is that spacetime theories satisfying [substantive general covariance] are telling us that traditional subject-predicate ontologies, whether relational or absolute, have ceased to facilitate understanding" (p. 16–17).

We heartily concur! However, we feel he doesn't go quite far enough, since he speaks of the co-realization of the values of dynamical (i.e. physical) quantities – seemingly buying into the 'no relations without relata' objection. But this ignores the problem that one can't make physical sense of the quantities independently of the correlation that they form so that it is wrong to speak of them as dynamical variables. They are, rather, formal variables, coupled along the lines of a gauge-theoretic analogy of non-separability.

Just as coordinates provide a useful gauge allowing us to identify spacetime locations, with no physically objective reality, so *things* and

structure in the sense that no reality (other than a purely formal one) can be ascribed to it whatsoever.

[14] We might also add, at this point, that the idea that quantum gravity involves a (traditional) relationalist stance has become something of a recent dogma. We have become used to questioning the absoluteness of spacetime stucture, and also of continuum concepts. Yet we continue to impose the assumption of a thing-ontology. Our proposal is intended to counteract this imbalance.

thing-talk act as a kind of 'gauge fixing of world structure'.[15] If we might be permitted to change gear from already rather speculative to radical, it seems possible that evolutionary mechanisms determined the specific gauge fixing that characterises the human cognitive framework. It does not seem too outlandish to suppose that alternative cognitive frameworks (and perceptual faculties) would 'gauge fix' in a different way, while still preserving the same relational structures.[16] Hence, the choice of things seems to be more variable than the structure exhibited by things.

Rather than having an undifferentiated blooming, buzzing confusion, by fixing on things we can navigate the world. The thing-gauge thus acts as an identification map between a dictionary of thing words or concepts and aspects of world structure.[17]

---

[15] Technically speaking, a gauge will be a function of a parameter space (changing as one varies the location in this parameter, or base space): the appropriate base space for the 'thing gauge' would, we can assume, depend on the physics under consideration.

[16] Interestingly, this line of thought has impressive pedigree in the form of Poincaré (where he speaks of "enunciations" in place of our gauge fixing via coordinates): "Since the enunciation of our laws may vary with the conventions that we adopt, since these conventions may modify even the natural relations of these laws, is there in the manifold of these laws something independent of these conventions and which may, so to speak, play the role of *universal invariant*? For instance, the fiction has been introduced of beings who, having been educated in a world different from ours, would have been led to create a non-Euclidean geometry. If these beings were afterwards suddenly transported into our world, they would observe the same laws as we, but they would enunciate them in an entirely different way. In truth there would still be something in common between the two enunciations, but this is because the beings do not yet differ enough from us. Beings still more strange may be imagined, and the part common to the two systems of enunciations will shrink more and more. Will it thus shrink in convergence to zero, or will there remain an irreducible residue which will then be the universal invariant sought?" (2001, p. 334). His response to his question is also along the lines we suggest here: "What now is the nature of this invariant it is easy to understand, and a word will suffice us. The invariant laws are the relations between the crude facts, while the relations between the 'scientific facts' remain always dependent on certain conventions" (ibid, p. 336).

[17] This view clearly presupposes that 'the real world' (which we are understanding as a structural entity) does not have distinguished striations carving it up into 'natural' things. This pluralism seems more in keeping with recent trends in philosophy of science. We might also note here that our position may have some bearing on Lee Smolin's recent work on the reality of time. Smolin, taking his cue from Galen Strawson and Thomas Nagel, argues that any naturalist framework must take account of the existence and nature of and qualia and our experience of the world. We agree in the sense that it must find an explanation in the general physical picture one provides. That does not necessarily imply that one 'work backwards' from the qualia and experience to try and locate structures in the world that correspond to them. Smolin sees a realism about becoming (or what he labels 'temporal naturalism') as being "the only way to accommodate qualia and experience as intrinsic qualities of events in nature" (2013, p. 4). By contrast, we suppose that a world of irreducible structure (which we should add is considered to be timeless: time evolution constituting a gauge transformation) can nonetheless reproduce

Lest the reader suppose that a gauge fixing couldn't possibly have the kind of physical significance we are suggesting here, we would point to the case of the (arbitrary) gauge fixing used to pick a prime meridian or zero on the Earth, rendering longitude a (global) gauge symmetry.[18] Once selected by George Airy in 1851, Greenwich *physically* determined the evolution of multiple aspects of human life – it continues to do (albeit now using the International Reference Meridian) as the zero for GPS technology. Given its physical arbitrariness, one can easily imagine different choices for the meridian, but that would have led to slightly different human behavior. If one did choose to do this, then one would of course have to specify one's numbers relative to a choice of prime meridian for them to be physically meaningful.

4. There'll Be Some Changes Made

We don't expect the physics of things to vanish, as we mentioned above, but we do believe structural physics has a useful role to play in more fundamental physics involving what are usually thought of in traditional relationalist terms (such as relative locality). What kind of physics can be expected to flow from a physics without things? In the final section we don't propose to offer any new framework in which to conduct such physics; instead, we briefly discuss four examples of pre-existing physics that might be taken to either provide pieces of such a framework, or else receive a more meaningful explanation from the perspective provided by such a framework:

(1) *Categorical Physics*: The first example is bound up with the ongoing attempt to revise the mathematical foundations that might link up with the physical picture we have offered. In particular, category theory offers an alternative foundation to mathematical structure (and therefore, a potential alternative means of representing physical theories). The idea is to

appearances. Whereas Smolin wants a faithful mapping from our subjective experience of flow to *real* objective flow, we view the experiential states as illusory artefacts of arbitrary gauge fixing (emergent features of the gauge-invariant correlations) – the relevance to time and change is discussed in more detail in Rickles 2012; Huggett and Wüthrich 2013 also contains much that is relevant to the issue of 'recovering' a thing-populated, changing world from the physics.

[18] The process is a little more complicated than this suggests, since one must map the Earth's surface to a standard unit 2-sphere and then pick an orientation, a zero-point pole, and *then* a prime meridian.

view such structures (usually viewed as set theoretic entities composed of *and specified* by their elements) as objects in a category characterised by the morphisms (arrows) between objects. In this way, it is the arrows (or morphisms) to and from objects that receive prime billing and serve to specify objects.[19] As John Baez and Mike Stay point out, in the context of physics "the objects are often physical systems, and the morphisms are processes turning a state of one physical system into a state of another system – perhaps the same one". (2012, p. 95). Earlier, Baez (2006) has shown how one can set up appropriate categories for GR and quantum field theory by invoking the appropriate morphisms. The categories are *nCob* and *Hilb*, *n*-dimensional manifolds and mappings (cobordisms) between them and Hilbert spaces and mappings (linear operators) between them. As Michael Atiyah demonstrated, one can understand a topological quantum field theory (a diffeomorphism invariant quantum field theory, without local degrees of freedom) as a morphism between these categories (known as a 'functor').

The functor, joining categories, has particular importance to the proposal of this paper since it provides a means to question the fundamentality of the distinction between object (or system) and morphism (or process). One can use the functor concept to ground a notion of equivalence of categories (say A and B), give by a pair of functors $X: A \rightarrow B$ and $Y: B \rightarrow A$ satisfying some isomorphism conditions ($Y \circ X \cong \mathrm{id}_B$ and $X \circ Y \cong \mathrm{id}_A$). The interesting cases for us concern higher-dimensional categories (or 'multicategories' in which both the morphisms and objects of one category (the entire category itself) are the objects of a 'higher' category – see Leinster 2004). For example, Cat is the 2-category that has categories (objects + morphisms) as objects and functors as morphisms. That this procedure can be generalised to *n*-categories shows that the division into objects and morphisms is not an absolute one since one can switch to a higher category in which their roles change.[20]

We mention this example to point out that there is a mathematical foundation available with some features that have the potential to become aligned with the physical ideas we have presented, along with some initial forays into formulating the corresponding physics. Jonathan Bain (2011) has given a philosophical examination of the notion that category theory (and sheaf theory) and structuralism might make good bedfellows, arguing that there is indeed some level of support. Lam and Wüthrich (2013) have

[19] Indeed, Yoneda's lemma states that the information encoded in such morphisms can fully determine the object (uniquely, up to isomorphism).

[20] Our thanks to Ben Dribus for bringing this point to our attention.

argued that structuralists cannot expect to receive any support by shifting to category theoretic foundations and that category-theoretic representations do not involve an elimination of relata (things). That is, they maintain that it is impossible, even within a categorical formalisation, to uphold a purely structuralist philosophy independent of set-theoretic ideas. Both relations and their attendant relata are said to be physically meaningful only within set theory, which is irreducible to category theory, and vice versa. Bain's argument hinges on the category of sets, with the relevant category theoretic morphisms taking the form of functions between sets. Lam and Wüthrich demonstrate that an attempt to rid such a philosophical construction of relata succeeds in making the relations themselves meaningless.

We agree with Lam and Wüthrich that one cannot so simply escape from objects in a categorical description; however, we take the machinery of multicategories to at least point to an issue with treating either of process or object as fundamental notions. Bain's argument may not definitely establish that there is a mathematical formulation that completely eliminates relata. It does, however, indicate the tenuous hold that relata may have on future mathematics: whilst not replacing relata at this stage, the progress of theory is in the direction of an overall structural package.

(2) *Entanglement*: Though often heavily interpetation-dependent, quantum mechanics offers up several instances of examples pointing away from an ontology of physics and towards an ontology of structure.[21] We already mentioned the impact of Kochen and Specker's theorem above. Also importantly, quantum entanglement has some history of being bound up with issues of relationism and holism. It involves the notion that the total quantum system is not reducible to the intrinsic properties of its subsystems, thus exhibiting a kind of 'relational holism' – *cf.* §6 of Healey's entry on Holism and Nonseparability in Physics: http://plato.stanford.edu/entries/physics-holism. Formally, an entangled state (for an N-dimensional system) is 'non-factorizable' in the sense that $\Psi_{1,\ldots,N} \neq \Psi_1 \otimes \cdots \otimes \Psi_N$. Hence, there is a suggestion even in standard quantum mechanics that the world might not submit to being carved up in terms of individual things. Even here, however, some sense of spacetime frame is involved in the definition of the quantum theory, and so if one argues that quantum subsystems are not things, the underlying spacetime must seemingly be so understood in

[21] Though we don't mention it here, the theory's permutation invariance has most famously been seen to suggest a doing away of individual objects (see e.g. French and Krause 2006).

such terms in order to make sense of the notion of distant correlations and of the conservation of probabilities.[22]

Esfeld and Lam (2010) argue that such relations in entangled systems can be understood as brute and un-reducible to the intrinsic subsystem properties – and from this they construct a form of structural realism. In these cases too there is an external spacetime supporting the relations (and certain state-independent properties of the particles). But then they do not believe that any part of current physics warrants the elimination of objects:

> metaphysics should not be more revisionary than is required to account for the results of science, and in that respect, we do not see a cogent reason to abandon a commitment to objects (p. 148).

In this sense, they claim to tread a 'moderate' path towards structural realism. Ultimately, they propose that the 'object-property' distinction (where 'property' includes relations and structure) is not a fundamental ontological one, but a conceptual one, which sounds to our ears much like Earman's suggestion to dispense with the fundamentality of the subject-predicate distinction. A move, that we argued, recommends our further proposal to treat things as a kind of gauge fixing effect. Esfeld has more recently argued for something a little stronger, though still retaining a commitment to an ontology of things:

> quantum entanglement by no means implies that we have to abandon an ontology of objects (i.e. substances such as particles) in favour of an ontology of structures (as claimed by French and Ladyman). Any of the known proposals for a quantum ontology of matter in space-time is committed to objects. However, on any of these proposals, what determines the dynamics of these objects are not local, intrinsic properties, but a global or holistic property instantiated by all the objects together – that is, a structure that takes all the objects in the universe as its relata. (Esfeld forthcoming, p. 1)

He goes on to conclude that "as far as contemporary fundamental physics is concerned, there is no cogent reason to abandon the Aristotelian ontology of substances and properties" (ibid., p. 2). That is, so long as one adopts the stance of holism. The root of Esfeld's resistance to substance

---

[22] For this reason, David Mermin's so-called Ithaca interpretation, though seemingly along the same lines as our proposal (after all, he argues that "[c]orrelations have physical reality; that which they correlate does not", Mermin (1998, p. 753), is really very different. It is, quite explicitly, cast as a stance specific to quantum mechanics. Our thanks to Steven Weinstein for drawing our attention to the similarities between Mermin's position and our own. In fact, we don't see any real need for a non-relational spacetime framework to make sense of distant correlations and unitarity in quantum mechanics: a relational approach could presumably take it's place, so long as the fluctuations of whatever relative frame were employed were also taken into consideration.

elimination appears to lie in his belief that one needs objects to instantiate the structures of physics (so that they are individuated independently of the structure) in order to account for measurement outcomes and temporal development of quantities. However, there are relative time and relative localization proposals that evade such worries. These proposals take place in the context of quantum gravity, of course, and though he briefly mentions the Wheeler-DeWitt equation (and the absence of a background space and a background time), he fails to appreciate the impact of gauge invariance in this broader context. For this reason, he claims that the Wheeler-DeWitt equation "can still be regarded as referring to a configuration of elementary objects, such as a configuration of elementary parts of space (or a configuration of elementary parts of space-cum-matter)" (ibid, p. 13). Yet such a configuration would be an example of an unphysical degree of freedom. The elementary *physical* parts must be irreducibly relative structures.

On a more general note, we take some issue with Esfeld and Lam's proposed upper bound on the revisionary nature of metaphysics, particularly as regards quantum mechancics. Whilst it is certainly true that metaphysics must be firmly grounded in physics, or risk becoming theological, it seems clear that a revision of consensus metaphysics is often necessary to open the way for new science. Further, it is impossible to separate the quest to unify quantum and relativistic science from the need to find common ground between competing metaphysical positions. The conceptual limitations of each are some of the primary barriers to unification. Developing a structuralist understanding of quantum mechanics makes sense in this context, particularly considering that the background of the highly relational theory behind entanglement is already highly suggestive of the potential of such a move.

(3) *Dualities*: One of the most conceptually interesting concepts of contemporary physics is that of 'duality', that is an invertible mapping between a pair of theories that preserves all physical observables – see Rickles (2011) for a recent review. The concept of duality itself is nothing new. There is, of course, the well-known wave-particle duality (related to the momentum/position duality). However, there have more recently appeared a series of dualities with, potentially, quite remarkable ontological and epistemological implications. As Robert Savitt expresses it, "[a] dual theory in general does not have the same structure as the original theory" (Savitt 1980, p. 454). Though much of the focus has tended to be

on spacetime dualities,[23] there are also examples that affect more general aspects of thing-ontology, often mixing with spacetime aspects, rendering both spacetime and things existing within spacetime non-fundamental.[24]

Responsible for a large part of the hold that an ontology of things has on us is the idea of reductionism, and the related notion that any such reduction must terminate at some basic entities (converting composite entities into multiple elementary objects or excitations). This gives us a division between composites and elementary or fundamental things, which are usually thought to exist in a hierarchy of increasing complexity. Some dualities can demolish this division into 'composite' and 'elementary', at least at a fundamental (absolute) level: the choice becomes largely pragmatic, like a choice of coordinates. One can see this quite clearly in the case of the electric-magnetic duality of David Olive and Claus Montonen (1977), in which one has freedom of choice with respect to a pair of actions exchanging magnetic and electric degrees of freedom:

> In the original Lagrangian, the heavy gauge particles carry the $U(1)$ electric charge, which is a Noether charge, while the monopole solitons carry magnetic charge which is a topological charge. In the equivalent "dual" field theory the fundamental monopole fields, we conjecture, play the rôle of the heavy gauge particles, with the magnetic charge being now the Noether charge (and so related to the new $SO(3)$ gauge coupling constant). (1977, p. 117)

The exchange involves an S-duality mapping $e \longrightarrow 1/e$ (where e is the square root of the fine structure constant), also interchanging the 'elementary' excitations (visible in perturbation theory) and the non-perturbative 'solitonic' (composite) excitations (that is, the electric and magnetic charges as in the above quotation). Hence, there is no invariant notion of elementary object in theories that admit S-duality transformations. In which case, it becomes hard to get any grip on the notion of object at all.

The most radical duality is perhaps the AdS/CFT correspondence, linking a 4-dimensional quantum gauge theory on Minkowski space and a diffeomorphism invariant gravity theory (namely string theory) in

[23] E.g., *T-duality* relating pictures that differ in the radii of certain compactified coordinates (with theories being defined with radii of reciprocal lengths), and its generalisation *mirror-duality* relating pictures that differ with respect to deeper topological features, such as the complex structure of the manifolds. In these cases, one has isomorphism 'at the level of the observables' without having isomorphism at the level of the Riemannian manifolds forming the backgrounds for the theories.

[24] With specific reference to string theoretic dualities, Huggett and Wüthrich write that "dualities eliminate local beables from the basic furniture of the world" (2013, p. 281), retaining them as 'unphysical' ("surplus representational structure") along the lines we suggested above, in which the (local) thing ontology amounts to choosing a gauge, part of the unphysical representational scheme.

5-dimensions.[25] In other words, we have dual descriptions for spaces of different dimensions. But there is more to this: even the symmetry groups that one might think characterise theories are dual. For example, one theory has the Poincaré group and the other has the diffeomorphsm group. Given that we think of theories possessing the latter group as background independent, and the former as background dependent, we also have a duality between these other two categories. The upshot is the same as with the other dualities, however: thing-like features (objects located in spacetime) are not fundamental, but are instead part of the unphysical representational apparatus. Whether there is some deeper thing-structure underlying the AdS/CFT is not yet clear, since the degrees of freedom of string theory are still not known. However, it seems likely that in order to have general relativity emerge at the appropriate limits, it will have to depend on structural/relational degrees of freedom, since there is no background against which to define individual, local degrees of freedom.

(4) *Detectors First*: Finally, we note that Giovanni Amelino-Camelia's "detectors-first" methodology described in his essay for the 2012 FQXi essay competition (2012), also (suitably generalised) provides an interesting potential example of the kind of structural physics we have in mind. Amelino-Camelia proposes that locations be given a relational definition in terms of "detection at a given detector", with detection *times* likewise given by a material clock established at the detector site. Spacetime observables are then to be couched in such terms so that the notion of an independent spacetime framework becomes redundant (though often convenient, as with our prime meridian example). He proses that we "build all of our description of physics, including the so-called 'spacetime observables', using as primitive/most-fundamental notions the ones of material detectors and clocks" (ibid., p. 2). This suggests a general strategy for searching for quantum gravity theories, in which spacetime concepts are expected to be non-fundamental: "look for candidate theories of the exchange of signals among (Physical, material) emitters/detectors, now allowing for such theories an interplay between $\hbar$ and G_N, and without insisting on the availability of a spacetime abstraction suitable for organizing exactly all

[25] The duality in fact involves a dictionary between a four dimensional supersymmetric Yang-Mills theory and a superstring theory in ten dimensions (with the ten dimensions being broken down into a pair of five dimensional manifolds: anti-de Sitter space and a 5-sphere) such that the Yang-Mills theory 'lives on' the conformal boundary of the anti-de Sitter space. The AdS/CFT duality would require far too much by way of technical setting up to present here, to we simply give the results. For the details, including those relating to the arguments of this paper, see Rickles 2011.

such exchanges of signals" (ibid., p. 5). Of course, this still leaves the detectors and clocks in need of an invariant physical definition, lest they simply replicate Einstein's unphysical rods and clocks. As we argued, to do so requires that we don't assign them physical reality independently of the irreducibly relational construct in which they are implicated, since so viewed they will not have physical locations or properties of their own.

Amelino-Camelia mentions a variety of examples from particle physics in support of his view, showing that absolute spacetime concepts (detection times, locations, distances, etc.) can be eliminated from all emission-detection-type measurements – these will likewise form supporting cases for our view, though, again, combined with the proviso that the things (here 'detector' and 'detected system') not be invested with independent existence: they are defined by the more fundamental invariant correlation they represent.[26]

Finally, as one of us [DR] has discussed before, perhaps the clearest example of a potential physics without fundamental objects can be couched in Carlo Rovelli's scheme of partial and complete observables (2002). Partial observables are gauge-dependent (unphysical) degrees of freedom in a formal description of a system. They are the relata that would, if they had an unproblematic existence, be associated with the fundamental objects in a thing-ontology. However, according to Rovelli's way of defining them they are *physical* quantities to which one can associate a measurement leading to a number. This is incorrect, as argued in (Rickles 2012). The physical structure is given by the gauge-independent *complete* observables, defined as quantities whose values (or probability distributions) can be determined by the theory's laws given a specification of the initial conditions. The partial observables coordinatize the extended configuration space $\mathcal{Q}$ of a system, while the complete observables coordinatize the associated reduced space Γ_{red}, corresponding to the predictable correlations between partial observables. The physical dynamics is then given in terms of *relations between partial observables*, which amounts to a world involving only the relative evolution of (non-gauge invariant) variables as functions of each other. The relata are merely formal in this scheme and so we have a reversal of the usual metaphysical picture of individual objects (and their properties) providing a subvenience basis for all aspects of world-structure. In

[26] Moreover, their properties will depend on aspects of the spacetime structure (the symmetries and so on).

this alternative picture (in which gauge invariance plays the determining role in physical content) there are no relata without relations.[27]

5. Until the Real Thing Comes Along?

Can we really be so confident that a world of such primitive undecomposable structures will be the final story? Is there something *necessary* about it, or does it simply provide a useful path? We have proposed that thing-talk is a convenient fiction: irreducible structures are the fundamental thing. We motivated this using invariance principles, but one can also see it perhaps more clearly by considering the role of observation in physics. Take the observation of the distance moved by some object, or the time elapsed of some process. To establish this requires that points be specified between which the lengths and time are defined. However, a purely formal specification of families of four numbers (the coordinate values) won't do (because of general coordinate invariance): one needs to *physically* locate the end points containing the lengths and intervals. This involves 'marking' the points using some other object or process, and thus forming a relation, allowing for the required determination.[28] But to view this as marking some pre-existing space is wrongheaded: the spacetime framework is built from such correlations. If physics continues to be based on this kind of framework – and there is surely something to the Kantian view that *experience* (an ineliminable stage of physics) demands a spatiotemporal framework – then a physics of primitive relations like this is surely required.

---

[27] We take this to be the real root of Earman's problematic over 'subject-predicate'-style ontologies: the worlds described by the kinds of theories we and Earman discuss do not contain (Physical) independent subjects (things) that bear properties and the enter into relations. The gauge-dependent correlations can't be viewed as physically *analysable* into fundamental relata for that would imply that world-structure supervenes on unphysical quantities.

[28] Likewise, in quantum field theory one measures 'distance' from the vacuum state by using a similar physical setup of correlations, in this case collisions of various kinds. (One can specify general characteristics via the S-matrix, but even in Heisenberg's hands this S-matrix was supposed to be filled in by some physics of things, rather than viewing the S-matrix structure as a relational structure of which things can offer a representation.) This correlational form is presumably a general feature of empirical science, and underlying it is some notion of a spatiotemporal structure allowing for divisions between the object of study and the external means of testing or observing. But given diffeomorphism invariance, this usable spacetime structure must itself be bootstrapped into existence from physical quantities, which in order to make sense must have spatiotemporal properties. Thus the irreducibility of the relational structure in this case.

University of Sydney
Unit for History and Philosophy of Science
e-mail: dean.rickles@sydney.edu.au

University of Sydney
Department of Physics
e-mail: jessica.bloom@sydney.edu.au

REFERENCES

Amelino-Camelia, G. (2012). *Against Spacetime.* FQXi essay competition: http://www.fqxi.org/community/forum/topic/1442.

Baez, J. (2006). Quantum Quandries: A Category-Theoretic Perspective. In: D. Rickles, S. French, and J. Saatsi (eds.), *The Structural Foundations of Quantum Gravity*, pp. 240–265. Oxford: Oxford University Press.

Baez, J.C., Stay, M. (2012). Physics, Topology, Logic and Computation: A Rosetta Stone. In: B. Coecke (ed.), *New Structures for Physics*, pp. 91–166. Berlin, Heidelberg: Springer.

Bain, J. (2011). Category-Theoretic Structure and Radical Ontic Structural Realism. *Synthese* **190** (9). 1621–1635.

Bergmann, P. (1971). Foundations Problems in General Relativity. In: M. Bunge (ed.), *Problems in the Foundations of Physics*, pp. 49–55. New York: Springer.

Cassirer, E. (1923). *Substance and Function and Einstein's Theory of Relativity.* New York: Dover Publications.

Deutsch, D. (2012). *Constructor Theory.* arXiv: 1210.7439.

Earman, J. (2006). The Implications of General Covariance for the Ontology and Ideology of Spacetime. In: D. Dieks (ed.), *The Ontology of Spacetime*, pp. 3–23. Amsterdam: Elsevier.

Eddington, A.S. (1921). A Generalisation of Weyl's Theory of the Electromagnetic and Gravitational Fields. *Proceedings of the Royal Society of London. Series A* **99** (697), 104–122.

Esfeld, M. (forthcoming). The Reality of Relations: The Case from Quantum Physics. In: A. Marmodoro and D. Yates (eds.), *The Metaphysics of Relations.* Oxford: Oxford University Press.

Esfeld, M., Lam, V. (2010). Ontic Structural Realism as a Metaphysics of Objects. In: A. Bokulich and P. Bokulich (eds.), *Scientific Structuralism*, pp. 143–159. Dordrecht: Springer.

Feynman, R.P. (2011). *Six Easy Pieces.* New York: Basic Books.

French, S., Krause, D. (2006). *Identity in Physics: A Historical, Philosophical, and Formal Analysis.* Oxford: Oxford University Press.

Huggett, N., Wüthrich, C. (2013). Emergent Spacetime and Empirical (In)coherence. *Studies In History and Philosophy of Modern Physics* **44** (1), 276–285.

Ladyman, J., Ross, D. (2007). *Every Thing Must Go: Metaphysics Naturalized.* Oxford: Oxford University Press.

Lam, V., Wüthrich, C. (2013). *No Categorial Support for Radical Ontic Structural Realism.* arXiv:1306.2726.

Leinster, T. (2004). *Higher Operads, Higher Categories.* Cambridge: Cambridge University Press.

Maudlin, T. (2007). Completeness, Supervenience and Ontology. *Journal of Physics A: Mathematical and Theoretical* **40**, 3151–3171.

Mermin N.D. (1998). What is Quantum Mechanics Trying to Tell Us? *American Journal of Physics* **66**, 753–767.

Montonen, C., Olive, D.I. (1977). Magnetic Monopoles as Gauge Particles? *Physics Letters* **72B**, 117–120.

Poincaré, H. (2001). *The Value of Science: Essential Writings of Henri Poincaré*. New York: The Modern Library.

Rickles, D. (2011). A Philosopher Looks at String Dualities. *Studies In History and Philosophy of Modern Physics* **42** (1), 54–67.

Rickles, D. (2012). Time, Observables, and Structure. In: E.M. Landry and D.P. Rickles (eds.), *Structural Realism: Structure, Object, and Causality*, pp. 135–145. Dordrecht: Springer.

Rickles, D. (2011). AdS/CFT Duality and the Emergence of Spacetime. *Studies In History and Philosophy of Modern Physics* **44** (3), 312–320.

Rovelli, C. (2002). Partial Observables. *Physical Review D* **65**, 124013–124018.

Savitt, R. (1980). Duality in Field Theory and Statistical Systems. *Reviews of Modern Physics* **52**, 453–487.

Smolin, L. (2013). *Temporal Naturalism*. arXiv:1310.8539v1.

Smolin, L. (2003). The Self-Organization of Space and Time. *Philosophical Transactions of the Royal Society A* **361**, 1081–1088.

Silberstein, M., Cifone, M., Stuckey M. (2008). Why Quantum Mechanics Favors Adynamical and Acausal Interpretations Such as Relational Blockworld Over Backwardly Causal and Time-Symmetric Rivals. *Studies in History and Philosophy of Modern Physics* **39**, 736–751.

Stuckey, M., Silberstein, M. (2008). *Time, Space and Matter in the Relational Blockworld: A New Approach to the Problems of Time*. http://www.fqxi.org/data/essay-contest-files/Stucke_Stuckey_Silberstein.pdf.

Whittaker, E. (1951). *Eddington's Principle in the Philosophy of Science*. Cambridge: Cambridge University Press.

Olimpia Lombardi
Dennis Dieks

PARTICLES IN A QUANTUM ONTOLOGY OF PROPERTIES

ABSTRACT. We propose a new quantum ontology, in which *properties* are the fundamental building blocks. In this property ontology physical systems are defined as *bundles of type-properties* (specified by algebras of observables in a Hilbert space). Not all elements of such bundles are associated with definite *case-properties*, and this accommodates the Kochen-Specker theorem and contextuality. Moreover, we do not attribute an *identity* to the type-properties, which gives rise to a novel form of the bundle theory. There are no "particles" in the sense of classical individuals in this ontology, although the behavior of such individuals is mimicked in some circumstances. This picture leads in a natural way to the symmetrization postulates for systems of many "identical particles."

1. Introduction

Although talk of particles is part and parcel of everyday practice in quantum physics, it is generally recognized that it is less than clear what quantum particles *are*: quantum mechanics makes it difficult to think of them as independent and localized entities, in the way of classical physics. Typical non-classical features that are responsible for this problematic status of particles in quantum theory are contextuality, indistinguishability and non-separability. These are recognized novel characteristics of quantum theory, but most of the philosophy of physics literature treats them as more or less independent of each other and no unifying ontological picture has been proposed in which they all find a natural place. The present paper is part of a project that aims at filling this lacuna: we propose to develop a new quantum ontology in terms of which a general characterization of quantum systems can be given.

In: Tomasz Bigaj and Christian Wüthrich (eds.), *Metaphysics in Contemporary Physics* (*Poznań Studies in the Philosophy of the Sciences and the Humanities*, vol. 104), pp. 123-144. Amsterdam/New York, NY: Rodopi | Brill, 2015.

The perspective that guides our work is that *properties* constitute the fundamental ontological building blocks that form physical systems. As we will argue, in a *quantum property ontology* the notorious quantum peculiarities emerge as natural aspects of physical systems. In this article we will focus on contextuality and indistinguishability and explain how these features naturally fit into our properties perspective, and why this has the consequence that the concept of a "particle," with its classical connotations, cannot be taken as fundamental. We will also explain under what circumstances and with what limitations talk of particles can nevertheless be retained.

2. Quantum Systems as Bundles of Properties

What is an individual? The classical philosophical concept of an individual is inspired by the 'things' or 'objects' of everyday experience. An individual object is something that can be identified here and now, is different from other individuals, and continues to be what it is as time goes on.

A classical individual is an indivisible unity in the sense that it either cannot be divided at all or, if it can be divided, that the results of the division are different from the original. Moreover, an individual is subject to the Kantian category of quantity (unity-plurality): individuals are either one or many. In the latter case, they may form aggregates, in which they can be counted individually. These features distinguish the category of individual from the category of "stuff," which can be divided into portions without losing the stuff identity, and whose portions, when put together, cannot be individually counted (see Lewowicz and Lombardi 2013). As Henry Laycock puts it, the key to the character of the general category of individual "*evidently rests in the notions of unity and singularity – and thereby perhaps, more generally, in the concepts of number and countability.*" (Laycock 2010, p. 8).

Individuals can be given names and fall under definite descriptions. An individual possesses properties (possibly including relations as *n*-adic properties); linguistically this is captured by predicates applied to a subject. The ontology of objects possessing properties is basic to classical thinking and has generated the fundamental subject-predicate structure of language. This mirror relation between the ontological category of individual and the linguistic category of subject was highlighted by Peter Strawson in his classical book, *Individuals*, in which he states that an individual is "[a]*nything whatever can be introduced into discussion by means of a singular, definitely identifying substantival expression*" (1959, p. 137), and "*anything whatever can appear as a logical subject*" (1959, p. 227). Ernst Tugendhat expresses the same idea as follows:

> There is a class of linguistic expressions which are used to stand for an object; and here we can only say: to stand for something. These are the expressions which can function as the sentence-subject in so-called singular predicative statements and which in logic have also been called singular terms (1982, p. 23).

The usual systems of logic follow this pattern and make use of constants and variables, subject to predication, and thus are tailored to represent classical individuals. For instance, in first order logic, the sentence '*Pa*' says that the property corresponding to the predicate '*P*' applies to the individual denoted by the individual constant '*a*'; likewise, in the expressions '$\forall xPx$' and '$\exists xPx$', the range of the variable *x* is understood to be a domain of individuals. To quote Wittgenstein:

> the variable name '*x*' is the proper sign of the pseudo-concept object. Wherever the word 'object' ('thing', 'entity', etc.) is rightly used, it is expressed in logical symbolism by the variable name. For example in the proposition 'there are two objects which…' by '$\exists x, y$'. (1921, Proposition 4.1272).

As Wittgenstein thus makes clear, "object" is not a concept that is *defined* within a logical language, but rather is a category that is presupposed by a language and *shown* by its structure: it can be read off from the use of constants and variables.

The essential role of individual constants and variables is not limited to traditional logic: the vast majority of systems of logic, even extensions of traditional logic and deviant systems (see Haack 1974, 1978), use them; an ontology populated by individual objects is thus universally presupposed. The appropriate background theory for all these logics is set theory: '$a \in A$' expresses that the element '*a*' belongs to the set of individuals represented by '*A*'. In short, it is a universal and basic characteristic of traditional logic, and traditional thought in general, that there are individuals that can possess properties and can be represented by a constant or variable subject to predication.

The concrete identification of individuals requires a criterion, a "principle of individuation," in order to distinguish each individual separately (synchronic individuation) and to re-identify it over time (diachronic individuation). The first, synchronic problem can be expressed as: What makes an individual to be that individual and not another? Following Leibniz's Principle of the Identity of Indiscernibles, which says that two individuals cannot have exactly the same properties, it seems natural to respond that we may base synchronic individuation on the individual's properties. However, when we also address diachronic individuation we have to take into account that in general the properties of an individual change. Descartes in his Second Metaphysical Meditation discusses the example of a piece of wax: it has many sensory properties – it is white, has a certain smell, makes a certain sound when one raps it with one's finger, is hard, and has a certain taste – but it may lose them all when placed by the fire. If properties thus change drastically, what allows us to say that we are dealing with the same individual both before and after the change? Funes, the main character of one of Jorge Luis Borges's short stories (1942), "*was disturbed by the fact that a dog at three-fourteen (seen in profile) should have the same name as the dog at three-fifteen (seen from the front).*" The example of the Ship of Theseus, whose planks where replaced one by one until finally it was composed of entirely different planks, also illustrates this problem of identity over time. What makes an individual the same at different times?

A traditional response to these questions is that there is an underlying unchanging *bearer* of properties, a *substratum* or property-less *substance* that is the seat of individuality. In this case each individual is distinguished from the others by its own substance. In this way it is justified to think of the same individual, even if all its properties change over time.

The word '*substantia*' is an (unfortunate) Latin translation of the Greek term '*ousia*', and etymologically means "what stands (*stare*) under (*sub*)." In the history of philosophy, the notion of substance has developed into one of the most complex notions of metaphysics. Aristotelian *primary substance* (*prôtai ousiai*) corresponds to 'objects' or 'individuals'; it is composed of matter and form (form is the *secondary substance*) and is fundamental in Aristotelian ontology. But in modern philosophy the notion of substance has come to refer to a substratum ("bare particular"), so that an individual consists of substance plus properties. This is famously made explicit by Locke when he writes:

> The idea then we have, to which we give the general name substance, being nothing but the supposed, but unknown, support of those qualities we find existing, which we imagine cannot subsist sine re substante, without something to support them, we call that support substantia; which, according to the true import of the word, is, in plain English, standing under or upholding. (Locke 1690, Book II, Ch. 23).

This is the notion against which Hume directed his devastating criticism (more about this later).

This concept of substance can be characterized by a number of core characteristics (see Robinson 2013):

- ontological fundamentality: substances are the ontological principles that metaphysically sustain everything else;
- the ability to bear properties;
- permanence in spite of change of attached properties;
- ground for individuation and re-identification.

This concept of substance does not sit well with present-day science, however: it represents an element of reality that is unobservable *by definition*. The name *substance* conjures up the idea of ordinary physical or chemical substances – but this is a misleading analogy since ordinary substances possess physical or chemical properties, whereas the substance we are discussing here has no physical properties on its own: it is the mere *possibility* for a system to possess properties. Nevertheless, the arguments that we have reviewed seem to make it plausible to accept some substance-like principle; how else could we make sense of predication and individuality? The scholastic notion of *haecceity* ("primitive thisness," from the Latin *haec*, this) is such a substance-like notion that occurs even in recent philosophy of quantum mechanics.

Such a mere possibility remains mysterious and one wonders whether one cannot do without it. From a scientific viewpoint it is natural to wonder whether it is not possible to work directly with the physical properties themselves that characterize a system. The situation in quantum mechanics reinforces this question. For example, the problems surrounding "identical particles" in quantum mechanics give us a hint that quantum systems may be very unlike classical objects: there is at least one tradition in the philosophy of quantum mechanics saying that quantum particles are not individuals at all. This suggests that even if the substance-plus-properties picture is completely adequate for the treatment of classical systems, a quantum system may be better analyzed differently. It is therefore appropriate to pay attention to a rival of the substance view, namely the *bundle theory* according to which physical systems are just collections of properties, without a substance underlying them.

The idea of dismissing the category of substance from the ontology is anything but new in philosophy and dates back from far before any quantum challenges. In fact, many philosophers with an empiricist bent of mind have objected to substance because of its empirical unobservability in principle, following Hume's classical criticism. This stance has led them to suggest that individuals are just bundles of properties, so that properties obtain ontological priority over individuals and become the fundamental items of the ontology.

The question of whether an object is a substratum supporting properties or merely a bundle of properties has been and still is one of the big controversies in metaphysics (Loux 1998). In this classical debate the decision which side to choose has more or less remained a question of metaphysical taste (see Benovsky 2008). But quantum mechanics changes this situation. The traditional view of individuals as substances-plus-properties now more than before begins to show *scientific* limitations: limitations that are open to empirical investigation. In particular, the empirically well confirmed central principle of quantum mechanics that the total state of a system of "identical particles" must be symmetrical or anti-symmetrical hints that no physical meaning attaches to the notion of an exchange between "quantum particles," which seems to suggest the absence of a substance-like principle of individuality.

Before proceeding, a word of caution is in order. It is true that if one is determined to retain the idea of a classical particle one can do so without inconsistencies, like in Bohm's theory. The peculiar quantum statistical results can then be explained by supposing that correlations between measurements results are the consequence of peculiar initial or boundary conditions on particle states (see discussion in Dieks 1990, van Fraassen 1991), or by assuming that quantum particles exert "exchange forces" on each other (repulsion between fermions and attraction between bosons; see Mullin and Blaylock 2003 for a critical discussion). The evaluation of such proposals and comparison with the more standard ideas about quantum mechanics that we are discussing here is intricate. However, it does not merely depend on metaphysical taste: we are dealing with problems of scientific choice and scientific methodology, and although conclusive arguments for one position over another will certainly remain out of reach, empirically informed discussion about the pros and cons of the various alternatives is possible (see Acuña and Dieks 2014). If anything, arguments from the quantum realm point in the direction of problems with the classical notion of an object, and much more so than in classical physics are we driven into the direction of a properties ontology. In the next section we are going to introduce the general structure of such an ontology and explain in more detail why it accords well with quantum mechanics.

3. The Structure of the Quantum Properties Ontology

The idea of a quantum ontology of properties lacking substantial individuals was introduced in the context of modal interpretations of quantum mechanics (see the overview in Lombardi and Dieks 2013). In these interpretations definite values of physical quantities are ascribed to physical systems, according to criteria that depend on the specific interpretation. The idea was later developed (da Costa, Lombardi and Lastiri 2013, da Costa and Lombardi 2014) with the aim of exploring the general structure of the quantum domain, independently of the decision about the specific rule of definite-value ascription. But the applicability of the notion is not restricted to modal interpretations: also in other interpretations one can speak about properties of physical systems, albeit relativized to a (measurement) context – in particular, this is also true for the Copenhagen interpretation.

In quantum theory the descriptive concepts used in experimental practice (physical quantity, value of a physical quantity, state produced in a preparation procedure, etc.) have mathematical counterparts in the Hilbert space formalism (self-adjoint operators on a Hilbert space, eigenstates and eigenvalues of an operator, vectors in the Hilbert space, etc.). The general strategy we are going to follow is to endow this mathematical-physical language with ontological content, even outside the context of measurements in the usual sense. In this, we attempt to avoid adding ontological categories that have no empirical counterparts. Generalizing standard interpretational ideas, we establish the following semantic relations:

- *Observables* (self-adjoint operators on a Hilbert space) represent *type-properties* (like "electron energy"; these can themselves be seen as instances of *universal* type properties, like "energy").
- *Eigenvalues of an observable* represent the possible values of an observable, i.e. instances of the corresponding type-property; they stand for the possible *case-properties.*
- *The quantum state* (mathematically, a vector in the Hilbert space or, more generally, a functional on the space of operators) yields the *probabilities* for actualization of possible case-properties.

We have here made use of the distinction between type-properties and their instances. The question about the ontological status of (universal) type-properties leads us back to the problem of universals, which bedevils philosophy since Plato's *Parmenides.* We will only remark that our proposal is meant to be neutral with respect to this general question: we will not enter into the question of whether the type-properties are primitive or built up from their instances. A *realist* interpretation of universals should be compatible with our proposal, but it would not essentially change under many different forms of *nominalism*, such as predicate nominalism, concept nominalism, or class nominalism (see Rodriguez-Pereyra 2011). The essential point for us is that any type-property can be multiply instantiated. Moreover, a distinct feature of our proposal is that we will *not* assume that properties or their instances possess a form of individuality (except of course for the differences between numerically distinct eigenvalues). Our proposal does therefore not accord with views in which there is such an individuality, e.g. a tropes view in which the tropes possess a primitive identity (see, e.g., Ehring 2011).

The difference between type-properties and case-properties runs parallel to the distinction between *determinables* and *determinates* (Johnson 1921, Prior 1949). For instance, the property color is a determinable, a universal type-property. Redness and whiteness are determinate instantiations of this universal type-property. Similarly, mass and a mass of 1 kilogram are a determinable universal type-property and an instantiation of it, respectively (Sanford 2013). Redness, whiteness and mass of 1 kg are type-properties themselves, and can be instantiated in many cases. As a quantum example, the type-property "energy of the hydrogen atom" (itself an instance of the universal property "energy") has the particular energy values of the hydrogen spectrum as its possible case-properties.

On the basis of these fundamental distinctions we now build up our ontological structure:

- *Bundles of properties* define physical systems. The *type* of the system is determined by a bundle of *type-properties* (represented by an operator algebra on a Hilbert space); the concrete *case* is a collection of *case-properties*. Note: the fact that we use the familiar term "quantum system" does not imply that we are assuming that quantum systems are individuals (this point will be discussed below).
- An *atomic system* is a system that cannot be decomposed into smaller bundles. Its physical correlate is an elementary particle. Mathematically we should think of an irreducible representation of the theory's symmetry group; for example, the Galilean group in non-relativistic quantum mechanics.

Our ontological starting point is thus the idea of multiply instantiable type-properties, without a principle of individuation, and their case-properties. So we start with a realm that does not contain substantial objects: there is no substratum or other principle of individuation.[1] At the basis of our ontology is a tree-like categorical structure, in which universal type-properties (like "energy") have many instances (like "electron energy") and

[1] There are partial similarities here with what a number of other authors, some of them also motivated by the ontological challenges of quantum mechanics, have proposed. In particular, in "structural realism" (Worrall 1989, Ladyman 1998), at least in its ontic version, there is an ontological shift from substantive objects to places in a network of relations (French and Ladyman 2003). This follows Cassirer (1936) in an older claim that elemental objects are not old-style individuals but rather "points of intersection" of certain relations: physical objects are "*reduced to mere 'nodes' of the structure, or 'intersections' of the relevant relations*" (French 2006, p. 173). This is similar to our proposal to the extent that it is a modern form of the bundle theory and tries to do without the a priori concept of an individual. These authors also note the difficulties in even expressing this position, because of "*the descriptive inadequacies of modern logic and set theory which retains the classical framework of individual objects represented by variables and which are the subject of predication or membership respectively*" (French and Ladyman 2003, p. 41). This echoes our earlier remarks about the relation between standard language and the classical notion of individuality.

in which each such instance has, in general, many (possibly uncountably many) possible case-properties. The type-properties characterizing a quantum system are represented by an operator algebra on a Hilbert space. The question is how this ontology, to be further explained below, will allow us to face the quantum challenges, for example contextuality and indistinguishability. This will be the subject of the two following sections.

4. Contextuality in an Ontology of Properties

One of the first reactions to the probabilistic character of quantum theory was the attempt to interpret it as a statistical theory, in the style of classical statistical mechanics, so that the probabilities can be explained as frequencies in ensembles of systems with definite but "hidden" values of their observables. The *coup de grace* for such classical-like statistical interpretations was the Kochen-Specker theorem (1967), which proves the impossibility of ascribing precise values to *all* the observables of a quantum system simultaneously, while preserving the functional relations between commuting observables. It follows that which observables can be ascribed precise values must be contextual, i.e. situation-dependent (e.g., dependent on the measurement context).

This contextuality of quantum mechanics has far-reaching ontological consequences. Contextuality implies the violation of the *principle of omnimode determination*, a principle that has been generally accepted in modern philosophy. For example, it appears in the works of Wolff,[2] in the famous treatise on the calculus of probabilities by Bernoulli,[3] and is also repeated several times by Kant in his lectures on metaphysics.[4] The idea, which is almost self-evident in the context of pre-quantum thinking, is that in every individual all determinables are determinate: if the determinable "color" applies to an object, it necessarily has some determinate color, say red, independently of its other determinate properties, and also independently of our knowledge about what that determinate color is.

[2] "*Apparet hinc, individuum esse ens omnimode determinatum*" ("*Hence it appears that an individual is a completely determined being*") (Wolff 1728, p. 152).

[3] "*Sed quicquam in se et sua natura tale esse* [viz. *incertum et indeterminatum*], *non magis a nobis posse concipi, quam concipi potest, inde simul ab Auctore naturæ creatum esse et non creatum*" ("*That anything is uncertain and indeterminate in itself and by its very nature is as inconceivable to us as it would be inconceivable for that thing both to have been created and not created by the Author of nature*") (Bernoulli 1713, p. 227).

[4] "*Alles, was existirt, ist durchgängig determinirt*" ("*Everything that exists is continuously determined*") (1902, AA 18:332, 5710; AA 18:346, 5759; see also LM XXVIII 554).

Quantum mechanics, on the contrary, tells us that a system will generally be associated with determinables that are not determinate. For example, we can have a physical system to which the type-property position applies but which nevertheless has no definite position.

This feature of quantum mechanics has often been considered an interpretative problem in need of a solution. One approach has been to accommodate contextuality by adapting the logic: starting from the fact that contextuality relates to the non-Boolean structure of elementary quantum propositions, a non-classical propositional logic can be formulated in terms of the non-distributive, orthocomplemented lattice of the theory (see, e.g., Jauch and Piron 1969, Piron 1976, Beltrametti and Cassinelli 1981). From a more physical perspective, other authors have dealt with quantum contextuality by selecting a context, via an interpretive assumption (see, e.g., Bub and Clifton 1996, Dieks 2005), or by a physical process as decoherence (see, e.g., Zurek 1982, 2003). However, the general problem of what a physical system *is*, and what structure the quantum ontology should have in order to make contextuality natural has not been answered in a systematic way.

When the contextuality of quantum mechanics is considered from the viewpoint of our property ontology, it appears as a limitation regarding actual case-properties. Since the quantum system is a bundle of type-properties, with their corresponding *possible* case-properties, it is immune to the challenge by the Kochen-Specker theorem, since this theorem imposes no restriction on possibilities and on type-properties. The theorem states that the idea of a fully determinate bundle of *actual* case-properties for *all* the type-properties of the bundle cannot work in the quantum world: it is not possible to ascribe definite case-properties to all the type-properties of a bundle in a non-contradictory manner. In other words, the Kochen-Specker theorem places restrictions on which possible case-properties of a bundle can enter actuality: not every type-property of the bundle have an actual value for one of its possible case-properties. For example, in elementary quantum mechanics of systems without internal degrees of freedom all systems have both momentum and position as characteristic type-properties, in spite of the fact that these two quantities cannot both enter the realm of actuality in the form of definite case-properties.

In the light of contextuality it is interesting to notice the relation between the determinable/determinate distinction and the possible/actual distinction. The traditional principle of omnimode determination makes the distinction between determinables and determinates superfluous, since all determinables are determinate. But in the quantum case, given the Kochen-Specker theorem, it makes sense to consider the realm of possibility, since not all determinables can be determinate in a given system:

there are determinables such that none of their possible determinates are actual.[5]

The limitations imposed by the contextuality of quantum mechanics lead us to reflect on the difference between the traditional bundle theory and the present proposal. In the traditional version of the bundle theory a bundle is a combination of case-properties such that all the type-properties corresponding to that bundle are unequivocally determined. For instance, a particular billiard ball is the combination of a definite value of position, a definite shape, namely roundness, a definite color, say white, etc. So, in the debates about the metaphysical nature of individuals, the question is whether an individual is a substratum to which a definite position, roundness and whiteness belong, or whether it is rather a substance-less bundle of these same case-properties. In both cases all type-properties taken into account are actual.

It has been argued in the literature (Benovsky 2008) that in this case the difference between the substratum theory and the bundle theory is only verbal: in the bundle theory the bundling per object is done by a "compresence relation" that is specific for the object in question, and this relation fulfils exactly the same purposes as substance in the traditional theory. However, in our quantum proposal not all the type-properties that define a system have an actual case-value, and the status of a quantum system is consequently that of a bundle of *type-properties*, specified by an algebra of observables. The bundling, on this level, does not require a relation of compresence that is specific for a particular system: all the algebras for a system of, e.g., electrons are the same. There really is only *one* algebra that is multiply instantiable; we do not suppose a compresence relation that bestows individuality on any particular electron.

The traditional bundle theory aimed at reproducing the concept of an individual without reliance on the concept of substance or haecceity. A crucial ingredient needed to make this program work is the assumption that each thus-reconstructed individual can be characterized by at least one property that makes it unique and which allows re-identification over time (a form of Leibniz's Principle of the Identity of Indiscernibles). In classical physics *position* is the standard property that fits this bill: different particles have different positions and the continuous trajectories of particles make it possible to follow them over time. Therefore, in the context of classical physics the bundle theory leads to the same "surface" picture

[5] Because this might sound strange in the context of traditional discussions about determinables versus determinates, we have chosen to talk about *type-properties* and possible *case-properties*, among which not more than one can become actual.

as the substance theory: although the substance itself is lacking, its role is taken over by position. By contrast, in the quantum case the example of identical particles suggests that in general we cannot expect that systems can be individuated and re-identified over time by case-properties: quantum systems may lack the basic characteristics of individuals. The variation on the traditional bundle theory that is presented here embodies these features: *bundles are not individuals*, they have no "principle of individuality" that makes them to be a particular individual and not another. Accordingly, our quantum bundles do not aim at reproducing the same general picture as the traditional substance theory: whereas the substance theory is meant to ground an ontology of individuals, our quantum bundle theory leads to an ontology of properties, without individuals. As we already noted, this non-individual version of bundle theory does not make use of a notion of compresence that is just a substitute for substance or bare particular (Benovsky 2008).

The concept of an individual system does not fit comfortably in our quantum property ontology. In our proposal, to be further developed in the next section, properties will in general not build up individuals: there will just be properties, in general multiply instantiated, without generally valid individuating characteristics. This feature relates directly to the subject of indistinguishability.

5. Indistinguishability in an Ontology of Properties

It is a peculiarity of quantum mechanics that states of "*n* identical particles" are invariant (except for a possible change of sign) under any permutation of "particle labels." Therefore, permutations of these labels do not lead to any differences in the probabilities for measurement outcomes and consequently do not give rise to any empirical differences at all. This raises questions concerning Leibniz's Principle of the Identity of Indiscernibles (PII), according to which there cannot be two different completely indistinguishable objects: according to PII two indiscernible candidate-objects, i.e. objects with exactly the same properties, are in reality one and the same object. The principle admits different versions, depending on the set of properties over which we quantify. Quantum indistinguishability raises problems for even a (logically speaking) weak version of Leibniz's principle, according to which it is not possible for two distinct individuals to have *all physical* properties and relations in common. This seems to indicate that the notion of individual quantum particles can only be reconciled with a form of PII if we include non-physical ingredients (like haecceity or substance) in the definition of individuals. This is a disturbing situation: it is certainly against the spirit of modern physical science to introduce such metaphysical and unobservable-in-principle things at a fundamental level.

Various solutions have been proposed to this problem of quantum indistinguishability. Following a suggestion from the work of Quine (1976), Simon Saunders (2003, 2006) has argued that two fermions in an anti-symmetric state are *weakly discernible*. For example, in the singlet state two electrons stand in the relation of "having opposite spins," and this makes it possible to individuate the particles by means of PII after all, in a weak sense (i.e., without making it possible to refer, by means of a physical description, to a specific individual; PII here only says that there must be *two* particles, but cannot effectively distinguish them). Muller and Saunders (2008) and Muller and Seevinck (2009) have made (controversial) attempts at extending this analysis to the case of bosons. There is a threat of *petitio principii* here: the argument may beg the question to the extent that it relies on the idea of a multiplicity of entities from the start. As French and Krause (2006, pp. 170–171) remark: "*the worry is that in order to appeal to such* [irreflexive] *relations, one has already had to individuate the particles which are so related and the numerical diversity of the particles has been presupposed by the relation which hence cannot account for it*." The objection has been developed by Dieks and Versteegh (2008; see also Dieks 2014), a recent defense is proposed by Muller (2014); the subject remains controversial.

Among the attempts to salvage individuality along more traditional lines, we find the suggestion that the quantum restriction to symmetrical and anti-symmetrical identical particle states is not law-like, but rather due to contingent initial conditions; and suggestions that the quantum states describe traditional individual particles in an incomplete, statistical way, e.g. in the way proposed by Bohm. As French (1989, 1998) points out, strategies of this kind can keep the quantum mechanical formalism formally consistent with the traditional ontological view of identical particles as individuals. One can object that these attempts at salvaging the notion of individual particles have an *ad hoc* character, and add structure to the formalism of quantum mechanics without adding to the predictive content of the theory. Or like Michael Redhead and Paul Teller (1992) put it, this way of retaining the notion of individual can be accused of introducing *surplus structure* in the formalism; the individuals are put in by hand.

We do not wish to add to this well-known methodological debate here, however. Rather, we want to work out a new picture, in terms of our property ontology, in the hope that interpretational puzzles will be dissolved in this new framework. As Teller (1998, p. 122) says: "*I suggest that belief in haecceities, if only tacit and unacknowledged, plays a crucial role in the felt puzzles about quantum statistics*." Suggestions that a (silently) assumed classical ontology of individual particles is responsible for the conceptual problems surrounding indistinguishability can also be found with other authors, like Post (1963) and Maudlin (1998). However, these suggestions that an alternative ontology might be more suitable have not been developed in a systematic way. A possible exception is formed by proposals for new kinds of set theories: the semi-extensional quasisets theory developed by Newton da Costa and Decio Krause (1994, 1997, 1999; see also Krause 1992, and da Costa, French and Krause 1992) and the intensional quasets theory, developed by Maria Luisa dalla Chiara and Giuliano Toraldo di Francia (1993, 1995), describe collections of objects having cardinality but no ordinality. In these theories quantum particles are objects that are intrinsically indistinguishable; but they still are treated as individuals and, as such, are referred to by individual variables just as in classical set theory. The ontological significance of this formal description has not yet been clarified. Another drawback is that the problem of contextuality is left untouched: quantum systems violate the principle of omnimode determination, and this fact is not accounted for by the new set theory.

Our own positive proposal is not formal but is based on reflections about fundamental ontology. In our picture, quantum systems are not individuals but rather *bundles of properties* that can merge with each other and form new wholes, without individual components. Our claim is that this type of description is only natural, both within quantum mechanics and the framework of our property ontology.

Let us consider a system that corresponds to two elemental bundles of the same kind (i.e., generated by the same algebra of observables). We do not assign individuality to these bundles: we are dealing with a doubly instantiated algebra of observables, without additional distinguishing characteristics. We refrain therefore from using terminology like "bundle 1," "bundle 2," "each bundle," and so on. Rather, we wish to introduce such "individuality" concepts *a posteriori*, in cases in which this is possible, on the basis of the physical details of the situation.

So we say of any type-property A that it can be *twice* instantiated, with its own twice-instantiated case-properties. A twice-instantiated case-property of the twice-instantiated A may consist of two different values a_1 and a_2, or twice the same a value. Obviously, in the first case there is a physical distinction that makes it possible to speak about different bundles, namely the bundle containing a_1 and the bundle containing a_2, respectively. For ease of reference we may call them bundles 1 and 2; but we should not take this as showing that there is some metaphysical principle of identity that grounds these labels.

In the second case, the one in which *a* occurs twice, we cannot introduce distinguishing labels and we should therefore now say that we have *one* bundle in which *a* is *doubly instantiated.* If this situation arises during temporal evolution, we are entitled to say that two different original bundles have "merged." Within the framework of our property ontology there is now just one whole, in which the original bundles can no longer be identified. Indeed, the resulting *a-a* system is defined in such a way that it makes no sense to hold that one *a* comes from bundle 1 and the other from bundle 2, or *vice versa.*

The crux here is that individuality is not something given *a priori,* in terms of substance, haecceity, or system-specific compresence, but needs to be defined on the basis of differences in physical case-properties. In this sense individuality may be said to be *emergent,* since it is a notion whose applicability depends on physical facts, namely the values of the case-properties (we will come back to this point in the following section).

This ontological picture does not so much offer a solution to the problem of indistinguishability, but rather *dissolves* it. Indeed, the difficulties in standard discussions come from considering indistinguishability as a relation between particles whose individuality is already assumed to exist; and one would then like to relate this individuality to physical differences, via PII. By contrast, from the point of view introduced here the problem of indistinguishable individuals does not arise because now there simply can be no individuals with the same properties. A "merger" of two bundles, in the manner discussed above, produces one new whole. In the final situation there is no violation of Leibniz's principle since we are not dealing with two items, but with a single item to which the application of the principle makes no sense.

In order to represent these ontological ideas mathematically for a system of "identical particles," we may consider operators A_1 and A_2, both representing instances of the universal type-property A, acting on isomorphic Hilbert spaces $\mathcal{H}_1$ and $\mathcal{H}_2$, respectively. Note here that the indices 1 and 2 are not supposed to refer to individual particles (this would be question begging!), but only have a mathematical function, namely of referring to two copies of the same Hilbert space.

All observables must be symmetrical in 1 and 2, like the operator $A = A_1 \otimes I_2 + I_1 \otimes A_2$ or the operator $A_1 \otimes A_2$ (acting on the Hilbert space $\mathcal{H} = \mathcal{H}_1 \otimes \mathcal{H}_2$). Indeed, "twice-instantiated" (said of the algebra) means that the order of 1 and 2 must be without physical significance (see da Costa, Lombardi and Lastiri 2013). This symmetry is anchored in the ontological picture: there is not a physical type-property 1 and a physical type-property 2, but only A that can be instantiated more than once. In fact, this agrees with standard wisdom that all observables of an identical particle system have to be symmetrical in order to preserve the symmetry properties of the states over time. But in the usual treatment this is a requirement that has to be imposed, in addition to and independently of the symmetrization of the states. By contrast, in our approach we ground this property in the structure of the properties ontology; and we are going to deduce the symmetrization properties of the states from this symmetry of the observables.

It is clear from the above that in the mathematical language that we use we cannot easily dispense with indices, because the usual language of mathematics itself is, as that of the classical set theory on which it is based, a language that operates with the notion of individuals. However, it is essential to note here that although the differently indexed operators may be seen as different *mathematical* objects (even though they are identical copies of one mother object), we do not assume that they refer to different *physical individuals*. The indices are here employed as mere mathematical tools without physical and ontological significance.

When the idea of a property ontology without traditional individuals is taken seriously, it no longer is self-evidently natural to represent it by a formalism whose primitive symbols are variables referring to individual objects. An alternative possibility for handling such a property ontology could be a logic of relations in the spirit of the "calculus of relations" proposed by Tarski (1941), where individual variables are absent. This strategy was suggested by Lombardi and Castagnino (2008), and has begun to be worked out by Krause (2005).

An important point is that the ontology of properties not only provides a justification for the symmetry of observables, but also makes it possible to derive the symmetry and anti-symmetry postulates of quantum mechanics. Summarizing this derivation very briefly, we start with the observation that any operator can be decomposed into a symmetric and an anti-symmetric part. Then, when a state-operator ρ is used to assign a value (expectation value) to a symmetric observable operator, its anti-symmetric part plays no role: the value assigned by the state operator ρ is completely determined by its symmetric part (see Lombardi and Castagnino 2008, da Costa, Lombardi and Lastiri 2013). From this it follows that in the particular case of pure states the state can be expressed as either a symmetric state vector or an anti-symmetric state vector. Therefore, that state vectors of identical particles can only be symmetric or anti-symmetric is not the consequence of an *ad hoc* symmetrization or anti-symmetrization rule, but has an ontological background: these symmetry properties of the states are a consequence of the symmetry of the observables of the aggregate, and this symmetry is, in turn, a consequence of the fact that properties and not individuals are fundamental.

6. What Are Quantum Particles?

As we have argued in the previous sections, there are no individuals in our fundamental quantum ontology. However, the practice of physics is rife with talk about particles, and particles seem individuals *par excellence*. How can we understand this apparent conflict? In the previous section we have already indicated the essential answer, but let us be more specific here.

In everyday language the concept of an individual is central, and this is justified by the fact that this concept can be used very well to describe ordinary situations, also in the practice of experimental physics. It thus has become more or less self-evident to assume that even in the formalism of fundamental quantum mechanics individuals should be represented, and this is behind the almost universally accepted notion that the labels in a state of many "identical particles" refer to these particles as individuals – one could say that the labels are interpreted as haecceities. However, as we have explained above, this is an interpretation that is not unavoidable or even cogent: the labels can be taken to refer to different mathematical objects, different Hilbert spaces, and need not at all be considered as "particle names" (Dieks and Lubberdink 2011).

To see a simple illustration of this point, consider an Einstein-Podolsky-Rosen type of situation in which two identical particles are involved. Usually EPR experiments are discussed in terms of two particles at a large distance from each other, on which measurements are performed. According to this customary view, there is a left-side particle L and a right-side particle R, both treated as individuals that differ from each other in their positions. However, it is important to realize that the correct quantum mechanical state of the total system must be (anti-)symmetric. As a consequence, the indices 1 and 2 that occur in the quantum state *cannot* correspond to the wave packets on the left and right, respectively. These labels refer to Hilbert space 1 and Hilbert space 2, and in each of these Hilbert spaces we find a mixture of the L and R wave packets. This shows that the labels in the standard formalism in fact do not correspond to what we intuitively call particles. Rather, our intuitive notion of a particle is linked, in this example, to the two localized wave packets L and R. This shows that our ordinary intuitions actually follow the idea that what defines a particle is a set of physical *properties*, and not a metaphysical notion of identity. The notion that such an identity is inherent in the Hilbert space indices is simply wrong.

Elaborating on the remarks in the previous section about individuating case-properties, we note that under certain circumstances (decoherence is important here) narrow wave packets may occur in the description of a many particles system (see, e.g., Zurek 2003). Moreover, in classical limiting situations, such wave packets can remain narrow and more or less localized during a relatively long time. In this way, particle-like behavior can emerge: wave packets can represent approximately definite positions and can follow an approximately definite trajectory. In terms of case-properties, we thus have (approximate) positions and trajectories, and these features can be used to define and individuate particles. Note, however, that these particles do not correspond to the Hilbert space labels (as was just illustrated for the EPR case). Rather, their individuality resides in the distinctness of the case-properties that define them.

In the case of "identical particles" systems it may occur that there are no distinct case-properties; so in general we cannot expect that there are individual particles defined by these properties. In this sense particles as we know them from classical physics are *emergent*: the concept of a particle becomes applicable only in special circumstances, and the fundamental ontology is one of properties that do not possess inherent individuality.

7. Conclusions

Traditionally, contextuality and indistinguishability have been discussed as unrelated problems. Here we have proposed an encompassing framework in which the basic ontology is an ontology of properties, and in which physical systems of a specific kind are represented by sets of observables, not all of which need to take definite values. Actual states of affairs are represented by case-properties of the (contextual) subset of observables that is definite-valued. In this way both contextuality and indistinguishability become natural elements in one ontology, that of quantum properties.

An additional bonus of our approach is that the distance between non-relativistic quantum mechanics and quantum field theory becomes smaller. It is well-known that in quantum field theory the concept of a particle, as a fundamental entity, is problematic. In quantum field theory, as in the approach we have sketched, the particle picture is only emergent and approximate. The property ontology of quantum mechanics that we have proposed here thus forms a bridge to more general quantum theories.

Our discussion here has focused on the general features of the property ontology that are relevant for indistinguishability and contextuality. We believe, however, that these ideas about the ontology of quantum theory can also shed new light on other typical features of quantum mechanics, in particular non-separability. This will be the subject of another publication.

University of Buenos Aires
CONICET
e-mail: olimpiafilo@arnet.com.ar

Utrecht University
History and Foundations of Science
e-mail: d.dieks@uu.nl

REFERENCES

Acuña, P., Dieks, D. (2014). Another look at empirical equivalence and underdetermination of theory choice. *European Journal for Philosophy of Science*, online DOI 10.1007/s13194–013-0080–3.

Beltrametti, E., Cassinelli, G. (1981). *The Logic of Quantum Mechanics.* Boston: Addison-Wesley.

Benovsky, J. (2008). The bundle theory and the substratum theory: deadly enemies or twin brothers? *Philosophical Studies* **141**, 175–190.

Bernoulli, J. (1713). *Ars Conjectandi, Opus Posthumum. Accedit Tractatus de Seriebus Infinitis, et Epistola Gallice Scripta de Ludo Pilae Reticularis.* Basel: Thurneysen.

Borges, J. L. (1942). *Ficciones*. English translation, New York: Grove Press, 1962. Reprinted, New York: Alfred A. Knopf/Everyman, 1993.

Bub, J., Clifton, R (1996). A uniqueness theorem for interpretations of quantum mechanics. *Studies in History and Philosophy of Modern Physics* **27**, 181–219.

Cassirer, E. ([1936] 1956). *Determinism and Indeterminism in Modern Physics.* New Haven: Yale University Press.

Castagnino, M., Lombardi, O. (2003). Self-induced decoherence: a new approach. *Studies in History and Philosophy of Modern Physics* **35**, 73–107.

Castagnino, M., Lombardi, O. (2005). Self-induced decoherence and the classical limit of quantum mechanics. *Philosophy of Science* **72**, 764–776.

da Costa, N., Krause, D. (1994). Schrödinger logics. *Studia Logica* **5**, 533–550.

da Costa, N., Krause, D. (1997). An intensional Schrödinger logic. *Notre Dame Journal of Formal Logic* **3**, 179–194.

da Costa, N., Krause, D. (1999). Set-theoretical models for quantum systems. In: M.L. dalla Chiara, R. Giuntini and F. Laudisa (eds.), *Language, Quantum, Music*, pp. 171–181. Dordrecht: Kluwer.

da Costa, N., French, S., Krause, D. (1992). The Schrödinger problem. In: M. Bitbol and O. Darrigol (eds.), *Erwin Schrödinger: Philosophy and the Birth of Quantum Mechanics*, pp. 445–460. Paris: Editions Frontières.

da Costa, N., Lombardi, O., Lastiri, M. (2013). A modal ontology of properties for quantum mechanics. *Synthese* **190**, 3671–3693.

da Costa, N., Lombardi, O. (2014). Ontology without individuals. *Foundations of Physics*, forthcoming.

dalla Chiara, M. L., Toraldo di Francia, G. (1993). Individuals, kinds and names in physics. In: G. Corsi, M. L. dalla Chiara and G. C. Ghirardi (eds.), *Bridging the Gap: Philosophy, Mathematics and Physics*, pp. 261–328. Dordrecht: Kluwer.

dalla Chiara, M. L., Toraldo di Francia, G. (1995). Identity questions from quantum theory. In: K. Gavroglu, J. Stachel and M.W. Wartofski (eds.), *Physics, Philosophy and the Scientific Community*, pp. 39–46. Dordrecht: Kluwer.

Dieks, D. (1990). Quantum statistics, identical particles and correlations. *Synthese* **82**, 127–155.

Dieks, D. (2005). Quantum mechanics: an intelligible description of objective reality? *Foundations of Physics* **35**, 399–415.

Dieks, D. (2014). Weak discernibility and the identity of spacetime points. In: V. Fano, F. Orilia and G. Macchia (eds.), *Space and Time: A Priori and a Posteriori Studies*, pp. 43–62. Berlin: Walter de Gruyter.

Dieks, D., Lubberdink, A. (2011). How classical particles emerge from the quantum world. *Foundations of Physics* **41**, 1051–1064.

Dieks, D., Versteegh, M. (2008). Identical quantum particles and weak discernibility. *Foundations of Physics* **38**, 923–934.

Ehring, D. (2011). *Tropes: Properties, Objects, and Mental Causation.* Oxford: Oxford University Press.

French, S. (1989). Identity and individuality in classical and quantum physics. *Australasian Journal of Philosophy* **67**, 432–446.

French, S. (1998). On the withering away of physical objects. In: E. Castellani (ed.), *Interpreting Bodies: Classical and Quantum Objects in Modern Physics*, pp. 93–113. Princeton: Princeton University Press.

French, S. (2006). Structure as a weapon of the realist. *Proceedings of the Aristotelian Society* **106**, 167–185.

French, S., Krause, D. (2006). *Identity in Physics: A Historical, Philosophical and Formal Analysis.* Oxford: Oxford University Press.

French, S., Ladyman, J. (2003). Remodelling structural realism: quantum physics and the metaphysics of structure. *Synthese* **136**, 31–56.

Haack, S. (1974). *Deviant Logic.* Cambridge: Cambridge University.

Haack, S. (1978). *Philosophy of Logics.* Cambridge: Cambridge University Press.

Jauch, J.M., Piron, C. (1969). On the structure of quantal propositional systems. *Helvetica Physica Acta* **42**, 842–848.

Johnson, W.E. (1921). *Logic, Part I.* Cambridge: Cambridge University Press.

Kant, I. (1902). *Gesammelte Schriften.* Berlin: Herausgegeben von der Preußischen Akademie der Wissenschaften (Bde. 1–22), der Deutschen Akademie der Wissenschaften zu Berlin (Bd. 23), und der Akademie der Wissenschaften zu Göttingen (Bde. 24, 25, 27–29).

Kochen, S., Specker, E.P. (1967). The problem of hidden variables in quantum mechanics. *Journal of Mathematics and Mechanics* **17**, 59–87.

Krause, D. (1992). On a quasi-set theory. *Notre Dame Journal of Formal Logic* **33**, 402–411.

Krause, D. (2005). Structures and structural realism. *Logic Journal of IGPL* **13**, 113–126.

Ladyman, J. (1998). What is structural realism? *Studies in the History and Philosophy of Science* **29**, 409–24.

Laycock, H. (2010), Object. In: E.N. Zalta (ed.), *The Stanford Encyclopedia of Philosophy* (Winter 2011 Edition), URL = <http://plato.stanford.edu/archives/win2011/entries/object/>.

Lewowicz, L., Lombardi, O. (2013). Stuff versus individuals. *Foundations of Chemistry* **15**, 65–77.

Lombardi, O., Castagnino, M. (2008). A modal-Hamiltonian interpretation of quantum mechanics. *Studies in History and Philosophy of Modern Physics* **39**, 380443.

Lombardi, O., Dieks, D. (2012). Modal Interpretations of Quantum Mechanics. In: E. N. Zalta (ed.) *The Stanford Encyclopedia of Philosophy* (Fall 2013 Edition) URL = <http://plato.stanford.edu/archives/fall2013/entries/qm-modal/>.

Loux, M. (1998). *Metaphysics. A Contemporary Introduction.* London-New York: Routledge.

Maudlin, T. (1998). Part and whole in quantum mechanics. In: E. Castellani (ed.), *Interpreting Bodies. Classical and Quantum Objects in Modern Physics*, pp. 46–60. Princeton: Princeton University Press.

Muller, F. (2014). The Rise of Relationals. *Mind*, forthcoming.

Muller, F., Saunders, S. (2008). Discerning fermions. *British Journal for the Philosophy of Science* **59**, 499–548.

Muller, F., Seevinck, M. (2009). Discerning elementary particles. *Philosophy of Science* **76**, 179–200.

Mullin, W. J., Blaylock, G. (2003). Quantum statistics: is there an effective fermion repulsion or boson attraction? *American Journal of Physics* **71**, 1223–1231.

Piron, C. (1976). *Foundations of Quantum Physics.* Reading: W.A. Benjamin.

Post, H. (1963). Individuality and physics. *Listener* **70**, 534–537.

Prior, A.N. (1949). Determinables, determinates, and determinants. *Mind* **58**, 178–194.

Quine, W.V. (1976). Grades of discriminability. *Journal of Philosophy* **73**, 113–116.

Redhead, M., Teller, P. (1992). Particle labels and the theory of indistinguishable particles in quantum mechanics. *British Journal for the Philosophy of Science* **43**, 201–218.

Robinson, H. (2013). Substance. In: E.N. Zalta (ed.), *The Stanford Encyclopedia of Philosophy* (Summer 2013 Edition), URL = <http://plato.stanford.edu/archives/sum2013/entries/substance/>.

Rodriguez-Pereyra, G. (2011). Nominalism in Metaphysics. In: E.N. Zalta (ed.), *The Stanford Encyclopedia of Philosophy* (Fall 2011 Edition), URL = <http://plato.stanford.edu/archives/fall2011/entries/nominalism-metaphysics/>.

Sanford, D. H. (2013), Determinates vs. determinables. In: E.N. Zalta (ed.), *The Stanford Encyclopedia of Philosophy* (Spring 2013 Edition), URL = <http://plato.stanford.edu/archives/spr2013/entries/determinate-determinables/>.

Saunders, S. (2003). Physics and Leibniz's principles. In: K. Brading and E. Castellani (eds.), *Symmetries in Physics: Philosophical Reflections*, pp. 289–307. Cambridge: Cambridge University Press.

Saunders, S. (2006). Are quantum particles objects? *Analysis* **66**, 52–63.

Strawson, P. (1959). *Individuals. An Essay in Descriptive Metaphysics.* London: Methuen.

Tarski, A. (1941). On the calculus of relations. *The Journal of Symbolic Logic,* **6**, 73–89.

Teller, P. (1998). Quantum mechanics and haecceities. In: E. Castellani (ed.), *Interpreting Bodies. Classical and Quantum Objects in Modern Physics*, pp. 114–141. Princeton: Princeton University Press.

Tugendhat, E. (1982). *Traditional and Analytical Philosophy: Lectures on the Philosophy of Language.* Cambridge: Cambridge University Press.

van Fraassen, B. (1991). *Quantum Mechanics: An Empiricist View.* London: Oxford University Press.

Wittgenstein, L. (1921). *Logisch-Philosophische Abhandlung. Annalen der Naturphilosophie* **14**. English version: *Tractatus Logico-Philosophicus*, 1922, C. K. Ogden (trans.). London: Routledge, Kegan Paul.

Wolff, Ch. (1728). *Philosophia Rationalis Sive Logica.* Reprint of the 1740 edition with introduction, notes and index by Jean École (ed.), New York: Georg Olms, 1980.

Worrall, J. (1989). Structural realism: the best of both worlds? *Dialectica* **43**, 99–124.

Zurek, W. H. (1982). Environment-induced superselection rules. *Physical Review D* **26**, 1862–1880.

Zurek, W. H. (2003). Decoherence, einselection, and the quantum origins of the classical. *Reviews of Modern Physics* **75**, 715–776.

Tomasz Bigaj

ESSENTIALISM AND MODERN PHYSICS

ABSTRACT. In the first part of the paper I develop and defend a metaphysical position regarding the interpretation of *de re* modalities which I call "serious essentialism." Subsequently, I show how this doctrine can be helpful in solving perennial interpretive problems that arise as a result of the general covariance of general relativity, and the permutation invariance of quantum mechanics.

1. Introduction

Historical accidents can sometimes reveal deeper and unexpected connections. When the pupils and followers of Aristotle coined the term *ta meta ta physica,* they meant to use it as nothing more than a way of cataloging their master's works on "first philosophy" together with other writings. But soon the artificial blend *meta-physics* took on a life of its own. While metaphysics no longer describes "what comes after physics" in the literal sense of the term, the suggested associations between metaphysics and physics are surprisingly accurate and go well beyond the mere adjacency on the shelves of an ancient library.

It is not hard to find cases supporting the claim that metaphysical discussions should be and in fact are infused with an influx of new ideas from contemporary physics.[1] Two well-worn examples are provided by how Einstein's relativity changed the philosophical concepts of space and time, and how quantum mechanics forced philosophers to reconsider the notion of causality. Of course many more cases like these can be found

[1] Some authors strongly urge to abandon arm-chair speculative metaphysics altogether in favor of a scientifically-informed one. (Maudlin 2007) is representative of this approach.

In: Tomasz Bigaj and Christian Wüthrich (eds.), *Metaphysics in Contemporary Physics* (*Poznań Studies in the Philosophy of the Sciences and the Humanities*, vol. 104), pp. 145-178. Amsterdam/New York, NY: Rodopi | Brill, 2015.

in the context of more recent physical theories. However, in this article I would like to focus on the somewhat more controversial side of the mutual relationships between metaphysics and physics; that is on the possible influences that the former can have on the latter. I am far from suggesting that arm-chair divagations of metaphysicians have the potential to lead to new and groundbreaking discoveries in physics. My claim is much more modest: I believe that rigorously formulated metaphysical doctrines and concepts may be used as a guide in discussing interpretive questions that arise in the context of fundamental physical theories.

As a particular illustration of this general thesis I have selected discussions related to the metaphysics of modality and objecthood. I will try to show how certain choices regarding *de re* representations of individual objects in modal contexts can impact the debates on the status of fundamental entities in two key physical theories: general relativity and quantum mechanics. I will start with a brief presentation of the connections between modalities and essential properties.

2. Metaphysics of Essential Properties

2.1. *De Re Modalities and Essences*

This is a well-known problem in quantified modal logic: how to account for modal statements that refer directly to individual objects rather than to their kinds? Consider the universal statement "Necessarily, all humans are mortal." Acceptance of this rather incontrovertible modal truth does not help us decide whether the singular sentence "Socrates is necessarily mortal" is also true, even though Socrates is clearly a human. The reason for this is that we don't know whether the property of being human characterizes Socrates as a matter of necessity or only accidentally. The difference in truth conditions between universal and singular modal statements can be best explained using the nowadays customary framework of possible worlds.[2] To assess the truth of the first statement we have to compare the

[2] In this paper I am adopting the standard possible-world semantics for modal logic. However, I wish to remain neutral with respect to the metaphysical status of possible worlds. Even though I will often use phrases of the sort "object x existing in a possible world w," I do not require that they be interpreted literally, as in modal realism. It is entirely possible that an actualist who believes that only actual objects exist literally can nevertheless produce plausible paraphrases of the above-mentioned expressions (for a recent proposal along these lines see Stalnaker 2012). I also acknowledge that there are philosophers who are deeply suspicious about the usefulness of the framework of possible worlds for interpreting modalities *de re*

extensions of the properties of manhood and mortality in every possible world to make sure that the first is included in the second. The evaluation of the second statement, on the other hand, requires that we identify the referents of the singular name "Socrates" across possible worlds. If there is a possible world in which the referent of "Socrates" is not human (he is for instance a god or an alien), he may very well turn out immortal there.

Thus, in order to make sense of modal *de re* statements regarding a given individual (e.g. Socrates) in the broad framework of possible worlds, we need to be able to identify possible objects that represent *de re* the selected individual. There are several well-known approaches to this problem. Here we can sketch the main options, some of which will be extensively scrutinized later in the text. According to extreme haecceitism, every possible object can in principle represent *de re* any actual individual, regardless of their qualitative properties. Hence an extreme haecceitist must admit the possibility that Socrates might for instance be a poached egg. Similarity-based counterpart theory, on the other hand, uses the counterpart relation to determine representation *de re*. A counterpart of an actual individual *a* is an object which is sufficiently qualitatively similar to *a*. The notion of similarity is intentionally left vague and context-dependent to ensure greater flexibility. The main proponent of counterpart theory is David Lewis, who actually developed its two versions. In the earlier 1968 version of Lewis's counterpart theory it is assumed that no object can have counterparts in its own world. This restriction is lifted in the 1986 approach in order to preserve certain intuitions regarding counterfactual switching. The later version of counterpart theory is sometimes called "cheap haecceitism," for reasons that will become clear later.

An alternative account of representation *de re* is offered by Saul Kripke. Kripke famously insists that possible worlds are not distant lands that we discover using powerful telescopes but our own creations (Kripke 1980, p. 44). Which objects represent *de re* a given individual is determined by our stipulations. The only restriction imposed by Kripke on our freedom of stipulation is that certain properties of the original individual should be preserved. These properties are commonly known as essential, i.e. such that without them an individual cannot retain its identity. A more radical version of essentialism maintains that each object is characterized by its own unique set of essential properties (so-called individual essences).

(Skow 2008, 2011). While I don't necessarily share these sentiments, I will try, whenever possible, to relate statements about objects and their identities across possible worlds to more metaphysically neutral claims involving modalities *de re*.

Under this assumption representation *de re* is reduced to possessing all the essential properties of a selected individual.

In subsequent discussions I will mostly ignore the metaphysical position of extreme haecceitism, focusing instead on the remaining theories of representation *de re* whose common denominator is that *some* qualitative characteristics of an individual should be relevant to what objects can represent it *de re*. I admit that I don't have a knock-down argument against extreme haecceitsm, other than my deep conviction that it is fundamentally inconsistent with the empirical character of scientific theories. In order to understand the remaining approaches better we have to make some terminological distinctions, starting with the central notion of an essential property.

The standard definition of an essential property of an object a is such that it is a property whose possession by any x is strictly necessary for x to be a (see Robertson and Atkins 2013, Mackie 2006, chapter 1). A general modal characterization of this concept, written in a language that does not presuppose the framework of possible worlds, can be formulated as follows:

(1) φ is an essential property (EP) of an individual a iff $\Box\forall x\ (x = a \rightarrow x$ has $\varphi)$

In order to explicate the meaning of formula (1) in terms of possible worlds we need to introduce a relation which will pick out those entities in other possible worlds that are supposed to represent *de re* our selected object a. Symbolizing this relation by I we can write down the following reformulation of (1):

(2) φ is an essential property (EP) of an individual a iff for all possible worlds w it is the case that $\forall x$ (x is in w and $Ixa \rightarrow x$ has φ in w)

The formal properties of relation I depend on the adopted approach to representation *de re*. For instance, according to Lewis's counterpart theory I is not transitive (as it is a similarity relation), and is such that many objects in one possible world can represent *de re* a given individual. The difference between the 1968 and 1986 variants of the theory is that according to the former I is reduced to numerical identity when restricted to the actual world, while the latter rejects this assumption. The individual essence theory, in contrast, defines relation Ixy as "x possesses the same individual essence as y." Given this characterization, I is an equivalence relation. Usually it is assumed in this case that in each world there is at most one object standing in relation I to the selected individual. However, technically the notion of individual essence can be used for representation *de re* even if this condition is not met, although the use of the adjective "individual" may be objectionable in that case (see sec. 2.3 for discussion).

An individual essence of an object a is a property (or a set of properties) such that possessing it is strictly necessary and sufficient for being a. More precisely:

(3) φ is an individual essence (IE) of a iff $\Box\forall x\,(x = a \leftrightarrow x \text{ has } \varphi)$,

or, equivalently,

(4) φ is an individual essence (IE) of a iff for all possible worlds w it is the case that $\forall x\,(x \text{ is in } w \text{ and } Ixa \leftrightarrow x \text{ has } \varphi \text{ in } w)$

In order not to trivialize the notion of an individual essence we should make it clear that the range of the property-variable φ is limited to "natural" qualitative properties,[3] and thus we exclude such non-standard properties as haecceities or world-indexed properties (see Mackie 2006, pp. 20–21). Given the assumption that every actual object possesses an individual essence, all *de re* modal statements regarding a particular object a can be translated into equivalent *de dicto* statements regarding any objects exemplifying the essential properties of a.

We can now formulate two related metaphysical theses regarding the notion of individual essences. The first one is that all actual objects possess individual essences (Individual Essences Claim – IEC):

(IEC) $\forall x\,[x \text{ is in } @ \rightarrow \exists\varphi\,(\varphi \text{ is an IE of } x)]$

[3] There are well-known problems with finding a satisfactory, non-circular characterization of qualitative properties. Typically, qualitative properties are specified as those properties whose ultimate definition does not contain any reference to individual objects (Adams 1979, pp. 7–8; Fara 2008; Swoyer and Orilia 2014). This explication is unsatisfactory for many reasons. For instance, does the expression "the tallest man ever" count as qualitative? Our intuition says yes, but doesn't it make reference to a (presumably) unique individual? On the other hand, the distinction between proper names and descriptions which could help us here is hard to make without implicitly relying on the notion of qualitative properties. A particularly scathing attack on the qualitative/nonqualitative distinction can be found in (Stalnaker 2012, pp.59–62). While I understand Stalnaker's frustration at the lack of a sharp distinction between qualitative and non-qualitative properties, I disagree with some of his specific remarks. In particular, I believe that his argument against dispositional properties is mistaken. The modal characterization of a disposition in terms of its stimulus and manifestation does make reference to an individual, i.e. the bearer of the property. But such reference is clearly indexical (it should be formalized by a variable, not a proper name), and therefore does not undermine the qualitative character of dispositions. I don't want to pretend that I have a solution to the perennial debate of how to define qualitative properties; nevertheless I will continue to make use of this important concept.

Under the assumption that no actual object can be represented *de re* in its world by entities other than itself, (IEC) implies the weaker claim (called the Distinctive Essences Claim) that all actual individuals are characterized by distinctive essences (there are no two actual individuals that are characterized by the same sets of essential properties):[4]

(DEC) $\forall x, y$ [x is in @ and y is in @ and $x \neq y \rightarrow \exists\varphi \neg(\varphi$ is an EP of $x \leftrightarrow \varphi$ is an EP of y)]

However, (DEC) does not entail (IEC). One counterexample to this implication involves two individuals a and b satisfying the following conditions: a is an actual object with essence φ, while b exists in a possible world and possesses φ, and yet $\neg Iab$. This scenario does not *per se* violate (DEC), but it clearly makes (IEC) false, since the equivalence given in (4) is evidently false. Another, perhaps less obvious example is a situation in which an actual object b possesses all essential properties of another actual object a but not essentially.[5] Such a situation does not make (DEC) false, but is impossible when (IEC) is true.

2.2. *Troubles with Individual Essences*

But is (IEC) a reasonable claim? Many authors express their skepticism regarding the existence of individual essences. This skepticism can be backed by some lessons from physics. Consider for instance two electrons. If they differ at all with respect to some qualitative properties[6], this difference will for certain involve accidental properties: their individual positions, momenta, energies, components of spin in a particular direction, etc. On the other hand, those properties of an electron that are the best candidates for constituting its essence, i.e. the so-called state-independent properties such as mass, electric charge, total spin, are shared by all electrons.[7] Thus

[4] Proof of this implication is straightforward: suppose that (DEC) is false, and thus there are two distinct actual objects a and b with the same essence φ. But φ can be the individual essence of neither a nor b, since by assumption they don't represent *de re* each other (hence possessing φ doesn't guarantee standing in relation I to a or b). When it is permitted for one actual object to represent *de re* another, (IEC) might be true while (DEC) is false. However, this move has some unintuitive consequences, of which we will say more in sec. 2.3.

[5] I owe this observation to Nat Jacobs.

[6] In this section I will ignore complications which arise as a result of the symmetrization postulate that applies to all particles of the same type, including electrons. We will return to this issue in section 3.

[7] Typical examples of essential properties considered in the metaphysical literature are: sortal properties (Properties associated with belonging to a particular natural kind) and origin properties (see Robertson and Atkins 2013). While sortal properties are rather unlikely to produce individual essences, the origin of a particular electron is an even more improbable candidate

it seems obvious that (DEC) is violated in the case of elementary particles, and this implies that the individual essence claim (IEC) is violated too. Similar arguments can be presented with respect to other entities postulated in physical theories, such as spacetime points. If we agree that the metric properties of a spacetime point constitute its essence, we have to accept that in universes with flat geometry and in universes characterized by some symmetries there will be points sharing their essential properties.

Unfortunately, there is a price to be paid for eschewing individual essences. Recently, a number of general metaphysical arguments in favor of the individual essences claim have been proposed.[8] These arguments are based on some natural assumptions, some of which have to be abandoned if we insist that there are objects with no individual essences. The arguments themselves are rather intricate, but their main ideas are very simple and easy to grasp. In a nutshell, typical arguments in support of individual essences describe certain unpleasant scenarios which are impossible to avoid if some objects are not characterized by their unique individual essences. Some of these setups are known by rather suggestive names: the role-switching scenario, the multiple occupancy scenario, and the reduplication scenario. I will briefly discuss each of these examples below.

Role-switching

Suppose that the actual world contains two distinct objects *a* and *b* possessing the same set of essential properties φ. In such a case we can envisage an alternative possible world in which *a* and *b* exchange all of their accidental properties. But such a world would be qualitatively indiscernible from the actual one, as the only difference between both worlds would lie in different identifications of objects *a* and *b*.

Multiple Occupancy

In this scenario the actual world looks exactly like in the role-switching case, but now we are considering a possible world in which there is only one object *c* with the set of properties φ. The problem we are facing now is whether *c* represents *de re a* or *b* (intuitively it seems that it can't be both,

---

for its essential property, since the history of an individual electron is not even a well-defined characteristic in quantum mechanics. After an interaction of two electrons, it is usually impossible to trace back their unique and separate histories (see also sec. 4.2).

[8] Mostly in (Forbes 1985, 1986). Penelope Mackie gives a thorough reconstruction of Graeme Forbes's arguments for individual essences in chapter 2 of her book (2006). My subsequent presentation of "troublesome scenarios" is loosely based on Mackie's interpretation of Forbes's arguments.

since this would somehow imply that *a* and *b* are identical – we will later see that this argument is faulty). Again, as in the previous case, the answer to this question cannot be given on the basis of qualitative facts only, hence we have a potential example of distinctness without qualitative difference. Alternatively, we could admit that *c* represents *de re* neither object *a* nor *b* (this is admissible, since possessing essential properties is *necessary* but not sufficient for representation *de re*). However, in that case the lack of representation *de re* would be a brute fact, not grounded in any qualitative feature of the possible world in which *c* exists.

Reduplication

This is actually a "reversal" of multiple occupancy, in which we start with the actual world containing only one object *a* with the essence φ (understood as the set of all essential properties), but then we consider an alternative scenario in which there are two objects *b* and *c* possessing the same essence φ. The question of which object *b* or *c* is identical with *a* apparently cannot be answered on the basis of the qualitative facts, hence we end up considering two distinct but qualitatively indiscernible possible worlds which only differ with regard to the representation *de re* of object *a* by *b* and *c*.

All three above-mentioned examples presuppose that the distinctive essences claim (DEC) is false, which implies the falsity of (IEC). However, it is possible to create yet another unsettling scenario in which (DEC) is saved and only (IEC) is abandoned. We have already mentioned a situation like that: it involves two non-*I*-related individuals *a* and *b* with the same essential properties occupying two different worlds. The fact that one object does not represent *de re* the other is just a brute fact, not explicable by reference to any qualitative property of either object.

It may be argued that a common element in all the discussed cases is the presence of some form of haecceitistic (i.e. non-qualitative) difference between otherwise indistinguishable situations. Admitting the possibility of haecceitistic differences is the hallmark of the metaphysical doctrine generically referred to as haecceitism. However, several non-equivalent reconstructions of haecceitism are possible, and we will have to specify them more precisely and probe deeper into how they relate to the available approaches to *de re* modality sketched in sec. 2.1. The most fundamental intuition behind haecceitism is that it admits that things might be different than they actually are without any difference in qualitative facts. Speaking metaphorically, haecceitism implies that we can "shuffle" objects around creating a situation that is distinct, yet qualitatively indistinguishable from the original one. One way of cashing out this intuition is as follows (see Fara 2009, Skow 2008):

(H1) There is an object *a* possessing property *P* such that *a* might not possess *P* while all qualitative statements retain their actual truth value.

Alternatively, we may express this thought by saying that non-qualitative descriptions of reality do not supervene on its complete qualitative description. It may be tempting to interpret (H1) in terms of possible worlds as follows:

(H2) There are two possible worlds which are qualitatively indistinguishable (i.e. which make exactly the same qualitative statements true) and yet differ with respect to how they represent *de re* some actual objects.

However, as will soon become clear, (H2) is not equivalent to (H1). Furthermore, one might think that (H2) can be alternatively expressed in the following way:

(H3) There are two distinct possible worlds which are qualitatively indistinguishable.

Yet (H3) would be equivalent to (H2) only if we excluded the existence of distinct but perfectly duplicate possible worlds (identical with respect to both qualitative facts and representations *de re*).[9]

It should be obvious that extreme haecceitism, defined as in sec. 2.1, implies all three theses (H1) – (H3). Individual essentialism in the form of claim (IEC), on the other hand, is committed to rejecting (H1) and (H2) (and (H3) as well, if we accept the above-mentioned assumption regarding the distinctness of possible worlds). But the case of Lewis's counterpart theory is more complex. To begin with, Lewis rejects haecceitism (H2) by endorsing the claim that what a possible world represents *de re* supervenes on its qualitative character (1986, p. 221). And in keeping with his modal realism he professes agnosticism regarding thesis (H3). As for the (H1) variant of haecceitism, its status depends on the specific version of counterpart theory. In the earlier 1968 version, thesis (H1) comes out false. This is so, because in this interpretation a counterpart of *a* that does not possess one of *a*'s properties must exist in a possible world different from

[9] In the actualist framework in which possible worlds are properties that the actual world might have this assumption is naturally satisfied (see Stalnaker 2012, pp. 58–59 for a discussion).

the actual world (the only counterpart of *a* that exists in its world is *a* itself). Thus the only way to make (H1) true in this case is to find a possible world which is qualitatively indistinguishable from the actual one, and yet represents *a* as not possessing *P*. But this evidently validates (H2) which was supposed to be rejected.

The 1986 version of Lewis's counterpart theory, on the other hand, makes it possible to retain (H1) while rejecting (H2). As we admit the possibility that an actual object *a* can have its distinct counterparts in the actual world, the statement "*a* might not possess *P*" can be made true by such an actual counterpart. And of course all qualitative statements remain unchanged, since we are still in the same actual world. Lewis argues in favor of this proposal, because it makes certain intuitive counterfactual statements (e.g. "I might have switched identity with my twin brother") true without committing us to haecceitism (H2). The price of this solution, however, is that the metaphysical distinction between what is actual and what is only possible gets blurred. The standard possible-world approach to modality is that the ways the actual world might be are just other worlds (or proper parts thereof). But now we have to accept that the ways the actual world might be can themselves be actual. This means that some actual situations lead a double life: they are both the ways the actual world is (as themselves), and the ways the actual world might be (as something else).[10]

In the next subsection we will add one more metaphysical position to the roster of available interpretations of *de re* modality. This new position will combine some elements of individual essentialism with counterpart theory. The main goal will be to eliminate all vestiges of haecceitism (clearly visible in Lewis's latest theory) without committment to the Distinctive Essences Claim (DEC).

[10] Delia Graf Fara moves this objection to a higher level by noting that Lewis's proposal leads to unacceptable consequences when we introduce an actuality operator into our language (Fara 2009). In spite of the unquestionable ingenuity of Fara's argument, I believe that Lewis has a promising strategy of defense against it. The argument is based on two premises: (1) if object *a* has an actual counterpart possessing property *P*, then it is possible for *a* to actually possess *P*, (2) if it is possible that actually *S*, then actually *S*. Acceptance of both (1) and (2) leads to a contradiction; however, it may be questioned whether there is an unequivocal interpretation of an actuality operator under which both premises are valid. A counterargument can be presented to the effect that premises (1) and (2) are supported by different construals of an actuality operator, and therefore the entire argument is based on equivocation. In sec. 2.3 I sketch my own argument against Lewis's cheap haecceitism which in my mind is decisive.

2.3. *Serious Essentialism to the Rescue*

I would like now to formulate several desiderata which will jointly constitute a metaphysical position I dub "serious essentialism" (the reader may want to use the alternative name "hard-nosed essentialism," or even "foolhardy essentialism"). Then we will see how this doctrine compares with its main competitors (in particular how it can avoid the problems described in the previous section while still not committing itself to individual essences). The first two desiderata specify how we are supposed to describe possible worlds.

(1) Possible worlds should be characterized purely qualitatively (*contra* Kripke). This means that a description of each possible world should be given in a language (common to all worlds) whose primitive predicates refer to a specified set of qualitative properties and relations. The set of selected basic properties and relations may vary depending on the context in which the modal apparatus is employed. Thus if we are for instance interested in considering the problem of identity and individuality of spatiotemporal points, the relevant properties will be metrical and physical properties of spacetime, as described in the General Theory of Relativity. It has to be stressed that no individual names whose reference is fixed across possible worlds (i.e. no rigid designators) are allowed at this stage. It is admissible to use individual names restricted to a particular possible world (introduced for instance with the help of definite descriptions), but these names are not to be "transferred" to the characterizations of other worlds. Thus identifications of objects across possible worlds are not allowed as parts of our descriptions.

(2) Two possible worlds whose (complete) qualitative descriptions are identical are to be identified. More precisely, if there is a bijective mapping from one world to another which preserves all qualitative properties and relations (i.e. an isomorphism), then these two worlds are considered one and the same.

The main rationale behind the first two postulates is to not prejudge the issue of the representation *de re* of objects independently from their given qualitative characteristics, and to eliminate any haecceitistic differences not grounded in qualitative facts. The next desiderata deal with the issue of how to deduce the identity facts from the qualitative facts about individuals in various possible worlds. The key to this is the concept of essential properties.

(3) All fundamental objects possess some essential properties. This thesis can be formally stated, in an analogous way to how we introduced claims (IEC) and (DEC), as follows:

(EC) $\forall x$ [x is in @ and $x \in F \rightarrow \exists \varphi$ (φ is an EP of x)],

where symbol F represents the distinguished set of fundamental objects. We don't have to assume a stronger form of essentialism according to which all objects have to be equipped with essences. After all, an attempt to identify essential properties of complex objects: bacteria, trees, asteroids, galaxies, etc., encounters serious difficulties. However, objects that constitute the ontological foundation of a given scientific theory ought to have certain properties essentially. This assumption is motivated by the perhaps overly optimistic conviction that the fundamental kinds of objects introduced by our best scientific theories should be precisely delineated in terms of well-defined properties. An example supporting this conviction can be supplied by the known types of elementary particles, each of which is clearly defined in terms of physical properties such as mass, charge, isospin, and so forth. Such definitional properties can then be reasonably claimed to be possessed not accidentally, but as a matter of (Physical) necessity. But even fundamental objects are not required to possess individual essences, as we have already argued in sec. 2.2.

(4) All representations *de re* are to be done with the help of the relation of possessing the same essence. This implies that all objects that represent *de re* a given individual *a* must possess its essential properties. This assumption is actually a logical consequence of the way we defined essential properties. But we are adding to it the additional requirement that possessing an object's essential properties is not only necessary but also sufficient for its representation *de re* (this is precisely what I mean by being a "serious" essentialist – we accept the fact that once we relinquish haecceitism we have no further means to distinguish which possible objects possessing the essence of a given individual represent it *de re* and which don't; therefore we should treat them all equally). Because we do not assume the existence of individual essences, such a relation will not have the properties of an identity relation, and therefore we will call it, following Lewis, the counterpart relation.[11] Thus an object x in

[11] This informal characterization may give rise to the suspicion that the vicious circle fallacy is committed here. For we use the notion of essential properties in order to define the

world w_1 is a counterpart of an object y in w_2 (which is assumed to be distinct from w_1) iff x and y possess the same essential properties. On the other hand, the only counterpart of any object in the actual world is the object itself.

(5) All *de re* modal statements are to be interpreted in the following way with the help of the counterpart relation. If a formula $\Phi(a)$ does not contain identities involving constant a, the modal sentence $\Box\Phi(a)$ is translated as: "In all possible worlds w it is the case that $\forall x$ (x is in $w \wedge Cxa \rightarrow \Phi(x)$)," where C denotes the counterpart relation. On the other hand, all identities of the form $x = a$ within the scope of a modal operator are to be directly replaced with Cxa. Thus, for instance, we will interpret the modal sentence $\Box\forall x(\Phi(x) \rightarrow x = a)$ as "In all possible worlds w, it is the case that $\forall x$ (x is in $w \rightarrow (\Phi(x) \rightarrow Cxa)$)," and not as "In all possible worlds w, it is the case that "$\forall y$ (y is in w and $Cya \rightarrow$ "$\forall x$ ($\Phi(x) \rightarrow x = y$)). Note that the latter "translation" clearly does not express the intended meaning of the original sentence, which is that the condition Φ is strictly (i.e. in all possible scenarios) sufficient for being identical with a given object a. Instead, the supplied paraphrase asserts that for each counterpart of a in any possible world it is the case that whatever satisfies condition Φ in this world is identical with this counterpart of a (which logically implies that if there is at least one object at a given possible world satisfying Φ, there is at most one counterpart of a at this world). But conditions for being identical with counterparts of a is not the intended subject of the original statement to begin with. Rather, we are stating what conditions must be satisfied by an object in an alternative scenario in order to be (i.e. to represent *de re*) a.

I would like to pause for a while and discuss in more detail some implications of the way serious essentialism characterizes the counterpart relation (Postulate 4). In particular, the question may be asked why we are reverting to the older version of Lewis's counterpart theory, rather than embracing his later proposal. Let us start to analyze this tangled and controversial issue by noting that the way we characterize the counterpart relation

counterpart relation, which in turn figures in the definition of essential properties (in the form of the relation I). But the problem disappears once we start proceeding more cautiously. We should first identify a distinguished set of properties φ for a given object a, and then we can define the set of the counterparts of a as comprising all possible objects possessing φ plus a itself. That way we can guarantee that φ indeed satisfies the definition of an essential property. However, φ is not guaranteed to be an individual essence of a, since there may be actual objects other than a that possess φ (by stipulation they are not counterparts of a).

within the actual world has some interesting consequences regarding our understanding of individual essences. The way we have defined the counterpart relation – as possessing the same distinguished set of properties as the original individual – may suggest on a superficial reading that we are reintroducing individual essences here. This is not the case; the condition of possessing the essence of a selected individual is sufficient for being its counterpart only if applied to an object in a non-actual world. In the actual world a distinct object with the same essence is not a counterpart of the original one. But what if we followed Lewis's 1986 theory? Then we would have individual essences on the cheap, as the strict equivalence between being an object and possessing its essence would be guaranteed by the adopted interpretation of the counterpart relation. However, as we have already noted in sec. 2.1, in this case the Individual Essences Claim (IEC) does not imply the Distinctive Essences Claim (DEC). That is, we may clearly have more than one object in the actual world with exactly the same set of essential properties. Speaking in such a situation about individual essences is a bit unnatural, and the responsibility for this terminological confusion should be placed squarely on Lewis's insistence that objects can have distinct counterparts in their own world.

But the problems with Lewis's cheap haecceitism may go even further than giving rise to a confusing terminology. Let Φ denote the essence of a given object a, and let us assume that more than one actual object possesses Φ. Given the introduced definition of the counterpart relation, we should accept the modal statement $\Box \forall x\, (\Phi(x) \rightarrow x = a)$, as it is equivalent to saying that in all possible worlds (including the actual one) possessing property Φ is sufficient for being a counterpart of a. But now we can select an actual object b distinct from a for which $\Phi(b)$. It seems that we have just violated the following, unquestionable principle of modal logic: if $\Box \forall x\, (P(x) \rightarrow Q(x))$, and $P(a)$, then $Q(a)$. The dangers of using inhabitants of one world as representing *de re* other inhabitants of the same world have now become apparent.[12]

I hope enough has been said to support the decision to limit the counterpart relation to numerical identity within the actual world. Now let us briefly discuss how serious essentialism deals with the troublesome

[12] I am sure that Lewis would be able to wriggle out of this tight spot, for instance by using the stratagem already applied in his "Postcripts to Counterpart Theory and Quantified Modal Logic" (Lewis 1983), where he scorned his critics: "don't guess, but read the counterpart-theoretic translations." If the counterpart-theoretical interpretation of an apparent logical rule implies something that is false, so much worse for the rule. The problem is, though, that counterpart-theoretic translations are not sancrosanct; they are useful only insofar as they express what we want them to express.

scenarios described earlier. The role switching scenario is an impossibility for the serious essentialist: we cannot exchange two objects possessing the same essences, since the result is the same situation as the original one. Here Lewis's cheap haecceitism delivers an entirely different verdict. The switching of two essence-sharing objects *a* and *b* does not create a new possible world (this would support haecceitism H2, which Lewis wants to reject) but it creates a new *possibility*, represented by the pair ⟨*b*, *a*⟩. Object *a*, being a counterpart of *b*, represents the possibility that *b* could possess all non-essential properties of *a*, and *vice versa*. But this is clearly haecceitism, cheap or not. If someone tells me that this particular electron could literally be identical with that one over there (I assume that all electrons have the same essential properties), I don't know how to make sense of this claim other than to assume that there is some non-qualitative "kernel" of identity that can be transferred from one object to another. The serious essentialist should frown upon any interpretation of the rigid designator "this electron" that involves anything over and above the essential properties of the electron, and therefore should deny that there is any genuine possibility, distinct from the actual situation, that two electrons could swap their roles.

In a similar manner, it doesn't make sense for the serious essentialist to ask which possible object in the reduplication scenario "really" represents *de re* the original one. The *de re* representation is done here by both "copies" of the original individual simultaneously, thanks to the assumption that the counterpart relation may be one-to-many even if we limit its co-domain to one possible world. That much is assumed by Lewis's 1986 theory; there is no disagreement here. The multiple occupancy case is defused by noting that one possible object can be a counterpart of many actual objects. In fact all actual objects that possess the same essences will have the exact same counterparts (with one exception, namely themselves). This means that objects with the same essences will have the same modal properties.[13] Whatever might be true of this electron, might also be true of that electron.

It is straightforward to observe that serious essentialism rejects all three variants of haecceitism (H1)–(H3). For the statement "*a* might not possess its actual property *P*" to be true, there has to exist a counterpart of *a* that does not possess *P* (requirement 5). But this counterpart must exist

[13] Strictly speaking, for this to be true we should assume that each actual object has a counterpart in some other world which is intrinsically indistinguishable from it. But this assumption seems uncontroversial: imagine a possible world which differs from the actual one only with respect to some unrelated fact while everything else remains the same.

in a possible world distinct from the actual world, as the only counterpart of *a* in its own world is *a* (requirement 4). However, according to desideratum 2, such a world must differ qualitatively from the actual world, which shows that (H1) cannot be satisfied. Haecceitistic claims (H2) and (H3) are made false simply by adopting requirement 2, which disallows the existence of distinct possible worlds that make the same qualitative sentences true.

In the next sections I will argue, using two study cases, that serious essentialism can have a non-trivial impact on the well-known debates in the philosophy of physics. The common theme present in the two considered cases is how to interpret certain symmetries characterizing models of particular theories (general relativity and the quantum theory of many particles). A symmetry is a transformation which preserves certain features of a model (usually it transforms one admissible solution of the equations of a theory into another one). Seen from a metaphysical perspective, two symmetry-related models can be conceived of as mathematical representations of two ways the world might be (two possible worlds, in short). The main question that will occupy us is whether and in what sense such pairs of possible worlds are distinct from one another. Answering this question on behalf of the serious essentialist will enable us to derive certain useful lessons about the nature of objects considered by each theory (spacetime points and quantum particles of the same type), and about the relations between physical objects and their mathematical representations.

3. Spacetime Points in General Relativity

3.1. *The Hole Argument*

The general theory of relativity (GR) can be seen as our best (to date) theory of spacetime and its interconnections with matter. Speaking loosely, this theory (understood as a class of differential field equations) delimits a set of mathematical models of the form $\langle \mathbf{M}, O_i \rangle$, where **M** is a set of points (a manifold) and O_i some geometric objects defined on **M** (including the metric field and the affine connection) which are solutions to the equations of the theory. General relativity possesses the feature known as general covariance, which basically means that there is a certain general class of mathematical transformations of the manifold **M** (called diffeomorphisms) that transform models into models. If d: $\mathbf{M} \rightarrow \mathbf{M}$ is a diffeomorphic mapping of **M**, then d applied to a model $\langle \mathbf{M}, O_i \rangle$ gives another admissible model $\langle \mathbf{M}, d^*O_i \rangle$ (objects d^*O_i are so-called carry-along geometric objects, whose values for image-points $d(p)$ equal the values of O_i for respective points p).

One particularly troublesome example of a diffeomorphic transformation is the infamous hole transformation h, which is identity outside a particular spatiotemporal region H (a 'hole') and then it smoothly changes into a non-identity inside H. Two models $\langle \mathbf{M}, O_i \rangle$ and $\langle \mathbf{M}, h^*O_i \rangle$ connected by a hole transformation are both solutions to the equations of GR, but they are identical outside H and differ inside it. The existence of such models threatens the fundamental concept of determinism, since it seems that our theory is incapable of fixing what is happening inside the region H even if the entire region outside H is fixed. However, the severity of this threat depends on some additional assumptions. The difference between the two models in the area H regards the assignment of the values of some geometric objects O_i to individual points, but this does not affect the observable or measureable properties of H, which are invariant under the diffeomorphism h. Thus it is commonly accepted that the hole example is a challenge only for those who are realists with respect to the existence of spatiotemporal points (this position is usually referred to as substantivalism). Leibnizian relationists, on the other hand, reject points as independent entities, arguing that they are only a means of representing relations between material objects (fields). Hence both diffeomorphically connected models are different representations of the same physical reality, which ultimately consists of material fields.[14]

In order to delve deeper into the metaphysical underpinnings of the hole argument, we have to make the crucial distinction between mathematical models and physical possibilities. Each model $\langle \mathbf{M}, O_i \rangle$ of our theory is a mathematical entity, but this entity is supposed to describe a certain physical situation, which we will refer to as a possible world. Now the central question pertaining to our problem is: what worlds are described by diffeomorphically-related models $\langle \mathbf{M}, O_i \rangle$ and $\langle \mathbf{M}, d^*O_i \rangle$? Jeremy Butterfield (1989) insists that even though the hole argument can be avoided if we assume that both models represent the same physical reality, for substantivalists this is not an option. As in these models the same spatiotemporal points are apparently assigned different values of geometric objects (e.g. different metric tensors), the substantivalist is forced to reject the supposition that diffeomorphically-related models are just different representations of the same reality. This leaves her with two options: either

[14] The philosophical literature on the hole argument is enormous. The argument in its modern version was originally formulated in (Earman and Norton 1987). A nice introductory essay on this topic, containing an up-to-date bibliography, is (Norton 2011).

only one of the models represents a physically admissible world, or each model describes its own possible world. The latter option seemingly clashes with determinism, while the former requires some further elucidations.

3.2. *Metrical Essentialism*

Tim Maudlin's metrical essentialism is a prime example of a well-developed position that gives good metaphysical reasons for the claim that at most one of the models ⟨**M**, O_i⟩ and ⟨**M**, d^*O_i⟩ can represent a physically possible reality (Maudlin 1988, 1990). Maudlin adopts an essentialist perspective on the identity of spatiotemporal points, arguing that they possess their metrical properties essentially. Thus it is metaphysically impossible for a given point to have metrical properties different from the ones it has in the actual world, same way as it is impossible for Socrates not to be human. Consequently, if model ⟨**M**, O_i⟩ represents the actual world, its diffeomorphically-transformed variant ⟨**M**, d^*O_i⟩ does not represent any metaphysically possible situation, since according to it some spatiotemporal points receive metrical properties different from the actual ones. Thus essentialist substantivalism survives the onslaught of the hole argument.

Maudlin is an essentialist but, as I will argue, not a serious one. That is, in his approach there are clearly visible remnants of haecceitism. When he compares the worlds described by diffeomorphically-related models ⟨**M**, O_i⟩ and ⟨**M**, d^*O_i⟩, he implicitly assumes that because those models are built out of the same base-set **M**, each point from **M** must represent the same unique physical point in either model. But this effectively means that the identification of objects in both possible worlds is done *before* any qualitative similarities between them are taken into account (being represented by one and the same mathematical entity is obviously not a qualitative feature). The only difference between full-blown haecceitism and Maudlin's essentialism is that, on the latter approach, if such a pre-identification does not preserve essential properties, it is to be discarded. But suppose that the diffeomorphism d happens also to be an isometry, i.e. a transformation which preserves all the metrical properties of points. In such a case Maudlin would have no qualms about accepting the existence of two distinct possible worlds corresponding to the models ⟨**M**, O_i⟩ and ⟨**M**, d^*O_i⟩, even though these worlds would clearly be qualitatively indiscernible.

Serious essentialism, on the other hand, prohibits any form of identification between objects in different possible worlds which is not supervenient on some qualitative facts. Essential properties are the only guide as to which objects in one possible world correspond to which actual objects. It is clear that for every point p in model ⟨**M**, O_i⟩ its image-point $d(p)$ in

$\langle \mathbf{M}, d^*O_i \rangle$ will possess the exact same metric properties, and hence will be its counterpart. And because d^* drags not only metric properties but all geometric objects O_i from p to $d(p)$, the resulting structure is isomorphic with the original one, and therefore qualitatively indiscernible from it. Hence models $\langle \mathbf{M}, O_i \rangle$ and $\langle \mathbf{M}, d^*O_i \rangle$ must refer to one and the same possible world. We were wrong in the supposition that the same points in both mathematical structures correspond to the same physical points.[15]

By bringing the metaphysical position of serious essentialism to bear on the debate we are able to argue that the substantivalist has yet another option at her disposal that has not been included in Butterfield's classification. According to Butterfield, only anti-realists with respect to spacetime can admit that both models $\langle \mathbf{M}, O_i \rangle$ and $\langle \mathbf{M}, d^*O_i \rangle$ refer to one and the same physical reality. But I argue that this option is open to the substantivalist too, if only she abandons the otiose metaphysical baggage of primitive, ungrounded identities. We can believe in the objective and independent existence of actual spatiotemporal points without assuming that they are equipped with some non-qualitative primitive thisnesses that enable us to keep track of their identities when moving from one mathematical description to another. As I already stressed, the premise that same elements of the mathematical base set **M** represent same spatiotemporal points in both models is based on the assumption that it makes sense to talk about the 'sameness' of points in various possible worlds regardless of their qualitative characteristics. But, as can be argued, this extra haecceitistic assumption need not be part of what we mean by substantivalism.[16]

---

[15] The serious essentialist solution to the hole problem is very similar to the modified variant of metrical essentialism advocated by Andreas Bartels (1996). Bartels observes that Maudlin does not fully respect his own assumption of essentialism, since he accepts that a particular manifold point p represents the same physical point regardless of what model p occurs in. But in different models p can be equipped with different values of the metric tensor, thus the identification of the physical referents of p in different possible worlds seems to be independent of the essential properties of these referents. Bartels himself assumes that the same mathematical points can represent different physical objects in different models, and consequently argues for the conclusion that diffeomorphically-related models refer to one and the same physical possibility.

[16] The variant of substantivalism which assumes that diffeomorphically-related models describe the same physical reality has already been seriously considered in the literature. The earliest anti-haecceitistic solutions of the hole problem were proposed in (Maiden 1993), (Brighouse 1994) and (Hoefer 1996); see also (Rickles 2008, pp. 89–125) for an excellent overview. Gordon Belot and John Earman (2000) refer to anti-haecceitistic substantivalism as "sophisticated substantivalism", but they raise some general objections to it. Oliver Pooley (2006) on the other hand is more sympathetic towards sophisticated substantivalism. He points out that one possible motivation for it should come from a structuralist metaphysics of individual substances that does not sanction haecceitistic differences. The discussion given

Substantivalism based on serious essentialism differs in some small but significant details from its main competitors: Maudlin's metrical essentialism and also Butterfield's counterpart theory. One interesting contrast of serious essentialism with metrical essentialism is provided by their different interpretations of the Leibniz shift. According to the standard analysis, substantivalists are committed to the claim that the actual world and the world in which all matter is shifted three feet in a given direction, are two distinct but indistinguishable worlds (given, of course, that spacetime is assumed to be classical, i.e. flat). Maudlin agrees with this assessment, adding however that the Leibniz shift is of a different type than the diffeomorphic transformations known from the hole argument. In the Leibniz case only matter-fields are dragged, while the metric properties remain unchanged. As each point retains its essential metrical property, the result of the Leibniz translation is another metaphysically admissible world in which numerically distinct points receive different values of particular matter fields. Quite predictably, serious essentialism disagrees with this assessment. Uniformly dragging matter-fields in a universe with classical spacetime produces a universe which is qualitatively indistinguishable from the original one, and therefore it must be seen as identical with it. Thus serious essentialism embraces Leibniz equivalence while still being committed to the existence of spatiotemporal points.

3.3. *Serious Essentialism vs. Butterfield's Counterpart Theory*

In the previous section we noted that serious essentialism is a variant of Lewis's counterpart theory. Butterfield's answer to the hole argument is also based on Lewis's denial of transworld identity, thus one may expect that these two approaches will turn out to be closely related. Indeed, there are some striking similarities between the two conceptions, but there are also glaring disagreements. Most notably, Butterfield insists that in his interpretation of the hole argument it still remains true that the diffeomorphically-related models $\langle \mathbf{M}, O_i \rangle$ and $\langle \mathbf{M}, d^*O_i \rangle$ cannot refer to the same physical reality. This clearly contradicts the lesson from serious essentialism, so it might be instructive to see where this difference comes from. To begin with, Butterfield assumes, following Lewis, that no individual can inhabit two different worlds, and thus *de re* modal sentences have to be interpreted with the help of some counterpart relation rather than identity. So far this looks similar to the seriously essentialist approach. But Butterfield defines his counterpart relation in a slightly different way. First

in section 2 of this article can be seen as a first step toward developing and justifying such a metaphysics.

of all, he relativizes it to a particular diffeomorphism *d*. According to his definition, point *d*(*p*) is a counterpart of *p* relative to the diffeomorphism *d* iff all the objects dragged by *d** at *d*(*p*) coincide with the objects at *p*. The counterpart relation can also be applied to regions: regions *S* and *d*(*S*) are counterparts relative to *d* iff *S* and *d*(*S*) are isomorphic under *d*.[17]

It is unclear to me what to make of this relativization of the counterpart relation to a particular diffeomorphism. Does this mean that *de re* modal sentences regarding actual objects (spatiotemporal points and regions) have to be similarly relativized? Interestingly, Butterfield admits that sometimes the relativization may be suppressed, if the class to which we relativize the counterpart relation is 'natural' (if, for instance, it contains all isomorphisms between two worlds). Be that as it may, one thing is certain: Butterfield's counterparts are meant to share all their geometric properties, and not only the essential ones (the metric ones). This clearly departs from the position developed in the current article. A common feature of both approaches is the claim that an object can have multiple counterparts in the same world. However, for Butterfield these counterparts will be images of the original point under various isomorphisms, while for the serious essentialist the set of all counterparts is extended to include all images under isometries.

The main advantage of Butterfield's counterpart theory is supposed to be the fact that it delivers the preferred answer to the question of the relation between the worlds described by models $\langle \mathbf{M}, O_i \rangle$ and $\langle \mathbf{M}, d^*O_i \rangle$. As Butterfield explains, because both worlds are assumed to consist of the same points, the world described by the second model is an impossibility, because no two distinct worlds can literally contain the same objects. Seen from the serious essentialist perspective, this reasoning is based not on one but two unjustified premises. It may be instructive to see in more detail why this is the case. Butterfield's argument can be reconstructed as follows (for brevity's sake I will symbolize model $\langle \mathbf{M}, O_i \rangle$ as $\mathcal{M}$ and model $\langle \mathbf{M}, d^*O_i \rangle$ as $d(\mathcal{M})$):[18]

(1) Model $\mathcal{M}$ is assumed to represent the actual world w_0.

---

[17] Both definitions can be alternatively presented as being relativized to a particular class of diffeomorphisms which agree in their values for *p* and *S*.

[18] Butterfield's argument is also reconstructed and analyzed informally in (Brighouse 1994, p. 120). The main difference between the two reconstructions is that in Brighouse's version the argument is based on the unreasonably strong premise that distinct mathematical models always represent distinct possible worlds. In contrast, I use the much weaker (and more plausible) assumption that if two models assign different values of a physical quantity to one and the same object, these models describe different realities.

(2) If model $d(\mathcal{M})$ represents a possible world w_1, then w_0 and w_1 consist of the same points.
(3) If model $d(\mathcal{M})$ represents a possible world w_1, then there is a point p (inside the hole) which is assigned one value of a geometric object O_i in w_0 and another value of O_i in w_1.
(4) If model $d(\mathcal{M})$ represents a possible world w_1, w_1 is distinct from w_0 (from premise 3).
(5) No two distinct possible worlds consist of the same points.

Therefore, model $d(\mathcal{M})$ does not represent any possible world (from 2, 4 and 5).

Both premises 2 and 3 can be legitimately questioned on the assumption of serious essentialism. The sameness of points in the worlds represented by models $\mathcal{M}$ and $d(\mathcal{M})$ is taken as a consequence of the fact that these models are built out of the same base-set **M**, but we have already emphasized that the sameness of some elements of mathematical representations does not guarantee that physical objects corresponding to these elements will be the same.[19] The same ungrounded transition from the identity of mathematical entities to the identity of the corresponding physical objects is responsible for accepting premise 3. The serious essentialist will reject any arbitrary identification of points in both models $\mathcal{M}$ and $d(\mathcal{M})$, arguing instead that we have to derive facts about identity (or counterparts) from the qualitative facts. In the end the serious essentialist accepts premise 2, but only because $\mathcal{M}$ and $d(\mathcal{M})$ turn out to be isomorphic. Premise 3, on the other hand, is rejected, because no isomorphism connects points which are equipped with different values of geometric objects in both models. Consequently, model $d(\mathcal{M})$ refers to a possible world which is identical with the world described by $\mathcal{M}$.

Let me end this brief analysis of Butterfield's metaphysically-motivated approach to the hole argument with a more general remark. While I generally believe that high-level metaphysical stipulations can have non-trivial consequences for specific problems that arise within philosophy of physics, the mere choice between transworld identity and the counterpart relation seems to me entirely neutral with respect to the substantivalism-relationism debate. That is, we can opt for the thesis that no individual literally occupies more than one world, and still hold on to the claim that diffeomorphically-related models faithfully represent distinct possible worlds.

[19] And some authors go even further, questioning the assumption of the identity between mathematical points figuring in two different, diffeomorphically-related models. They point out that in mathematics we should use the structuralist criteria of identity based on isomorphisms (see Pooley 2006).

We could for instance stipulate that one and the same point p in **M** represents in different models the counterparts of a given actual point rather than the point itself. As long as the counterpart relation possesses the same formal features as the transworld identity relation, it is inconsequential which one we decide to pick. Only when we couple counterpart theory with some additional assumptions (for instance by adding the principles of strong essentialism, or by defining the counterpart relation in such a way that it could not possess the formal features of identity) can we limit the available range of possible solutions to the hole argument.

3.4. *A Problem with Determinism*

Recall that the crux of the hole argument was that substantivalism leads to the conclusion that General Relativity is a radically indeterministic theory. Serious essentialism with its strong anti-haecceitist component solves the problem by implying that two diffeomorphically related models of GR describe one and the same possible world. Thus determinism is saved: there are no distinct possible worlds that would be identical in a certain spatio-temporal region and diverge in other regions. However, it may be argued that any theory which takes qualitative similarity as the only criterion relevant to the issue of determinism is bound to produce incorrect verdicts regarding some special cases of indeterministic systems. These cases involve highly symmetric systems which break the existing symmetry in a random way.[20] One widely discussed example of that sort is the tower collapse case. In this scenario a perfectly symmetrical cylindrical tower standing on an empty, featureless plain collapses under its own weight. The direction in which the tower collapses is not determined by the laws of physics, therefore we have a case of indeterministic behavior here. And yet all the alternative ways the tower might have collapsed are qualitatively identical to each other due to the underlying symmetry. Consequently, according to serious essentialism there is only one way the tower could collapse, and the situation is wrongly described as deterministic.

An even simpler example of the same kind involves a universe containing only two indistinguishable spheres. According to the laws of this universe, after a fixed period of time exactly one sphere turns pink, but which one remains undetermined. But both possible scenarios are qualitatively indistinguishable, hence serious essentialism does not classify this case as indeterministic. It may now be argued that cases like the ones described above support some form of haecceitism, because the alternative

[20] Detailed discussions of such cases of indeterministic systems can be found in (Wilson 1993, Belot 1995, Brighouse 1997, Melia 1999).

evolutions of the system (not fully determined by its past state) differ from each other only non-qualitatively. However, we should first note that admitting haecceitistic differences has equally unacceptable consequences concerning the classification of various systems as deterministic or indeterministic. Haecceitism creates too fine-grained distinctions between possible worlds that give rise to cases of spurious indeterminism. Consider the following modification of the indistinguishable spheres example: rather than randomly turning pink, each sphere emits a particle of a new kind. In addition, the new particles are indistinguishable from one another. The process of creating new particles is clearly deterministic, and yet according to haecceitism there are two distinct possible worlds which do not differ up to the point of the creation of the two new particles, and then differ only in that the newly created particles are swapped.

A solution to this problem has been proposed by Joseph Melia (1999). He notes that if a universe as a whole is governed by deterministic laws, then not only its future evolution, but also the evolution of all its parts, is fixed by its past states. Thus if there is a part of the universe such that there are two, qualitatively distinct ways this part may evolve, the universe is not deterministic. And it is not difficult to observe that in all cases of symmetric systems there are parts whose qualitative evolution is not fixed by the past states, in spite of the fact that the evolution of the whole universe is always qualitatively the same. In the tower example we can consider any sector of a circle with the tower at its center, and it will be true that there are two qualitatively discernible possible evolutions of such a section: one, in which the tower falls directly onto it, and the other in which the tower falls somewhere else. Similarly, the evolution of the two-sphere system when restricted to one of the spheres is qualitatively indeterministic, as each sphere has two distinguishable futures: it can either turn pink or remain the same. On the other hand, cases of spurious indeterminism do not satisfy the proposed criterion. In the example involving creations of new particles, in both scenarios each sphere gives rise to a qualitatively indistinguishable particle which can differ only with respect to its haecceity.

Melia's proposal can be adopted in order to enable the serious essentialist to make the proper distinction between cases of real and spurious indeterminism. The only technical complication is how to express the statement that a given part of the universe has two different possible futures without infringing upon any principle of serious essentialism. For instance, it would be inappropriate to point at the one of the two spheres that happened to turn pink and say "This sphere might not have turned pink while the entire history of the universe remained the same." For a universe with its past identical to that of the actual world will have its future qualitatively indistinguishable from the actual one, and by postulate 2 such a world is to be identified with the actual world. And the only counterpart of the selected sphere in the actual world is the very same sphere – this is non-negotiable. But it is a mistake to think that a proper expression of determinism must make reference to individual objects and their trans-world identities. On the contrary, determinism is supposed to be presented in a purely qualitative way. Thus we can use the following expression of indeterminism in the particular case being considered: "Things could be such that the world would have the exact same history, and yet an object qualitatively identical to this sphere up to moment *t* would not turn pink after *t*." And this last sentence is made true by the existence of the second sphere which in fact didn't turn pink. That way the serious essentialist can agree that the symmetric scenarios indeed violate some form of determinism, and the problem is solved.

4. Indistinguishable Quantum Particles

4.1. *The Theoretical and Metaphysical Roots of the Indiscernibility Thesis*

The quantum formalism represents states of a system of many particles as vectors in the tensor product of Hilbert spaces of individual particles: $\mathscr{H}_1 \otimes \mathscr{H}_2 \otimes \ldots \otimes \mathscr{H}_n$. If these particle are of the same type (in physical textbooks they are usually called identical, which is unfortunately confusing), then all the Hilbert spaces $\mathscr{H}_i$ are copies of one single-particle space. The main assumption made in the quantum theory of many particles is that the physical state of a system of particles of the same type should be invariant under permutations of particles. That is, if $\Psi(1, 2, \ldots, n)$ is a state of n indistinguishable particles, then the state $\Psi(\pi(1), \pi(2), \ldots, \pi(n))$ should be empirically indistinguishable from the original one for any permutation π. But clearly not all available states in the tensor product Hilbert space satisfy this requirement. Thus the theory introduces the well-known symmetrization postulate which limits the accessible states to either symmetric or antisymmetric ones. As a consequence, vectors that do not respect permutation invariance, such as direct products of the form $|u_1\rangle \otimes |u_2\rangle \otimes \ldots \otimes |u_n\rangle$, are not allowed as legitimate representations of the physical states of multipartite systems.[21]

[21] As Pooley (2006) observes, permutation invariance is the key feature that separates the case of identical quantum particles from the case of diffeomorphic models of GR considered in the previous section. For in the former case the result of applying a permutation to an admissible state is a state mathematically identical (or 'almost' identical, i.e. differing at most

The introduction of the symmetrization postulate has many interesting implications regarding the physics and metaphysics of quantum particles of the same type. Of primary importance to us is the fact expressed in the form of the Indiscernibility Thesis: distinct quantum particles of the same type turn out to possess exactly the same physical properties (both intrinsic and relational) and therefore are qualitatively indiscernible. The violation of the Principle of the Identity of Indiscernibles is often taken as a sign that quantum particles are not individuals in the full metaphysical sense of the word. The Indiscernibility Thesis has been proven formally in (French and Redhead 1988) for intrinsic properties, and in (Butterfield 1992) for relational ones. The central assumption of these proofs is that if O is a Hermitian operator acting in a single-particle Hilbert space, and O represents an admissible physical property of a given particle, then the tensor product $O_i = I^{(1)} \otimes I^{(2)} \otimes \ldots \otimes O^{(i)} \otimes \ldots \otimes I^{(n)}$ will describe the property O of the i-th particle out of an n-element system of particles of the same type. Now it can be easily verified that for all i, j the expectation values of operators O_i and O_j in symmetric and antisymmetric states are identical.

Operators O_i possess one interesting feature: namely, they distinguish between a state and its permuted variant in the case when the original state is neither symmetric nor antisymmetric. To see this, let us assume that $n = 2$, and let us consider the product state $\Psi(1, 2) = |u\rangle_1 \otimes |v\rangle_2$, where $|u\rangle$ and $|v\rangle$ are two orthonormal vectors in the single-particle Hilbert space. The expectation value of the operator $O_1 = O^{(1)} \otimes I^{(2)}$ in state $\Psi(1, 2)$ equals $\langle u|O|u\rangle$, whereas the same operator's expectation value in the permuted state $\pi_{12}\Psi(1, 2) = |v\rangle_1 \otimes |u\rangle_2$ is given by $\langle v|O|v\rangle$, which in general is different from $\langle u|O|u\rangle$. Of course this fact can be easily dismissed as having no deeper meaning, since vectors $\Psi(1, 2)$ and $\pi_{12}\Psi(1, 2)$ are disallowed as representations of physically possible states, because they violate the symmetrization postulate. However, I will argue that seen from a broader metaphysical perspective, the aforementioned fact has some significant consequences.

Let us first note that vector $\Psi(1, 2)$ represents a possible state of affairs which, although not physically possible, does not seem to be metaphysically or logically impossible. After all, the symmetrization postulate is at best a law of nature which may be violated in universes radically different from ours. If we agree with this supposition, then we have to admit next that the permuted vector $\pi_{12}\Psi(1, 2)$ describes a world which arises from the original one as a result of switching two particles of the same type. But now we

with respect to the sign) with the original one, while in the latter case the permuted models are distinct, and therefore in principle capable of describing different physical realities.

can apply the principles of serious essentialism which dictate that because the two worlds considered are qualitatively indistinguishable (since, by assumption, both particles have the exact same essential properties), they must be treated as numerically identical. Taking this into account, we can now return to the analysis of the above-stated fact that operator O_1 has different expectation values in transposed states $\Psi(1, 2)$ and $\pi_{12}\Psi(1, 2)$. If O_1 represented a genuine property of the two-particle system, this would imply that the system would possess different properties in numerically identical worlds, which is obviously impossible. Thus the only solution seems to be to dismiss O_1 as not representing any feature of the system whatsoever. Operators that 'see' the difference between a world and itself cannot be allowed as part of our description of physical possibilities.

It is not difficult to observe that the formulated argument eliminates all non-symmetric operators (i.e. operators that do not commute with permutation operators), since they are prone to the same problem of creating a haecceitistic difference between qualitatively indistinguishable worlds. This move is in agreement with the general practice of physicists who tend to limit physically meaningful operators to the symmetric ones in the case of systems of many particles of the same type.[22] However, if we followed their suggestion, we would have to reassess the main argument in favor of the Indiscernibility Thesis, as it clearly relies on the assumption that non-symmetric operators O_i represent physically meaningful properties of many-particle systems. French and Redhead readily agree that O_i does not represent any *observable* property, as there is no experimental procedure that could distinguish between particles whose only differentiating feature is the fact that they bear different labels. Yet they unquestionably accept that properties encoded by operators O_i are *real* (they are *beables*,

---

[22] Limiting meaningful operators to symmetric ones is also urged, albeit somewhat half-heartedly, in (Huggett, Norton 2013). An important exception to the symmetricity requirement that they seem to make concerns precisely the way the tensor product formalism is supposed to represent properties of individual particles. While Huggett and Norton (following Huggett 2003) propose a formalization of such properties that is slightly more general than French and Redhead's, still one crucial element of the standard approach remains: namely that properties of individual particles are labeled by the same labels that are used in the tensor product, and therefore that applying the permutation of two labeled particles transforms one particle's property into the other particle's property. Consequently, the only possibility of satisfying the postulate of symmetricity with respect to individual observables is to assume that for every property P its possession by any particle is represented by one and the same operator, which makes the Indiscernibility Thesis trivially true. In subsection 3.2 I will directly answer the challenge posed by Huggett and Norton in the following statement: "...it is hard to see in what other way we could define physical properties except by schema such as $P_t(a)$ iff $\langle\Psi|Q_a|\Psi\rangle = t$" (p. 4).

as opposed to *observables*). Thus, I claim, they commit themselves to some form of haecceitism: numerical distinctness between observationally indistinguishable states of affairs which therefore cannot be grounded in any qualitative facts.[23]

The proposed argument questioning the admissibility of the asymmetric operators O_i and therefore undermining the standard way to prove the Indiscernibility Thesis can be obviously turned around in order to support haecceitism against serious essentialism. If the only method of formalizing properties of subsystems of larger systems of particles was with the help of the Hermitian operators O_i, such an argument would have considerable force. We would be able to give a reason based on our best scientific theories for why we need haecceitistic differences between possible worlds. Fortunately for the essentialist, the quantum-mechanical formalism is capable of an alternative representation of the properties of individual particles in the form of fully symmetric operators. However, the proper symmetrization of the asymmetric Hermitian operators requires some care. It wouldn't be appropriate, for instance, to simply replace a single-particle operator O with the symmetric operator $O^{(1)} \otimes I^{(2)} + I^{(1)} \otimes O^{(2)}$ (in the case of two particles of the same type). It can be easily verified that the aforementioned symmetric operator acting in the product of two one-particle Hilbert spaces cannot represent a quantum-mechanical property O of one of the two particles, since generally the product $|\lambda\rangle \otimes |\varphi\rangle$, where $|\lambda\rangle$ is an eigenvector of O, will not be an eigenvector of the new symmetrized operator. We have to proceed more cautiously.

4.2. *Discernibility Redivivus*

As a first step we may want to represent O as the sum of one-dimensional projection operators: $O = \Sigma\, a_i P_i$ (for simplicity's sake we will assume that there is no degeneracy, and hence all values a_i are distinct). Next, we can introduce the following symmetric projection operators acting on the tensor product of two Hilbert spaces: $\Omega_i = P_i \otimes (I - P_i) + (I - P_i) \otimes P_i + P_i \otimes P_i$. We can easily verify that vectors $|\lambda_i\rangle \otimes |\varphi\rangle$, $|\varphi\rangle \otimes |\lambda_i\rangle$, and $|\lambda_i\rangle \otimes |\lambda_i\rangle$, where $|\lambda_i\rangle$ is an eigenvector of P_i, are eigenvectors of Ω_i. Thus it stands to

[23] One may complain that I am blurring here the distinction between observational differences and qualitative differences. In epistemology it is customary to treat the latter concept as being broader than the former: there may be qualitative differences that for some reason are not accessible to our perception. But in the context of the interpretation of quantum mechanics the separation of these two notions is essentially equivalent to embracing some form of the hidden variable hypothesis, and I do not wish to saddle French and Redhead with this controversial claim.

reason to interpret Ω_i as expressing the quantum-mechanical statement "At least one of the particles possesses the definitive value of O equal a_i."[24] Now, if we wish, we can build a new Hermitian operator out of projectors P_i which will represent the correct symmetrized version of O. But the crucial point to note is that projectors W_i can have their expectation values equal 1 in some (anti-)symmetric states of two particles. If the state of the two particles is represented by vectors of the type $|\lambda_i\rangle \otimes |\varphi\rangle \pm |\varphi\rangle \otimes |\lambda_i\rangle$, the expectation value of Ω_i is 1, and this means that in such states (at least) one particle possesses the precise value a_i of the observable O. Actually, an even stronger claim can be made if $|\varphi\rangle$ is assumed to be orthogonal to $|\lambda_i\rangle$,[25] since the symmetric component $P_i \otimes P_i$ of Ω_i in such a case gives zero, and therefore it can be argued that exactly one particle possesses the definitive property in question while the other one definitely does not possess it. But this implies that in some symmetric and antisymmetric states particles are discerned by their categorical properties.

Admittedly, the claim that particles of the same type can be sometimes absolutely discerned is bound to raise a few eyebrows. One possible objection to it can be spelled out as follows. It is easy to verify that in the fermionic states of the form $|\varphi\rangle \otimes |\psi\rangle - |\psi\rangle \otimes |\varphi\rangle$ there is an infinite number of symmetric two-particle operators Ω_i corresponding to incompatible projectors P_i for which this state is an eigenstate. For instance, if we consider a pair of electrons on the S-shell in an atom, it can be shown that no matter which direction $\overline{n}$ we choose, the sentence "Exactly one electron has spin in direction $\overline{n}$ equal +1/2" will come out true (given that we interpret this sentence with the help of the appropriate operators Ω_i). However, this seems to imply that individual electrons will be assigned more definite properties than it is allowed in the standard quantum-mechanical descriptions. It may be even argued that this situation can lead, via Bell's inequalities, to experimental predictions incompatible with quantum mechanics (see Pooley 2006 for a similar argument).

A thorough analysis of this problem would probably require a separate article. Nonetheless, we can sketch one possible strategy of defending the discernibility claim against the accusation that it runs afoul of Bell's theorem. It may be observed that in order to experimentally test whether Bell's inequalities are violated in the case of particles of the same type, we have

---

[24] This interpretation of operators Ω is adopted as self-evident in (Ghirardi et al. 2002). A more detailed analysis of this interpretation can be found in my (Bigaj 2015).

[25] In the case of fermions this assumption is always satisfied. That is, even if we start with two non-orthogonal vectors $|\varphi\rangle$ and $|\lambda\rangle$, we can always find a pair of orthogonal vectors such that their antisymmetric combination gives the initial state produced by antisymmetrizing $|\varphi\rangle \otimes |\lambda\rangle$.

to prepare the pairs of particles in a state whose spatial components for each particle are well separated. But this creates an altogether different situation from the one described above which involved two electrons occupying the same atomic shell. Now, as it can be verified, it is not the spin components in different directions that presumably receive well-defined values for both particles, but rather new observables represented by operators which are intricate combinations of positions and spins. Such observables are practically impossible to be measured in standard experimental settings, and therefore no violation of the quantum-mechanical statistics can be recorded in that case.

As we have argued, serious essentialism provides a strong motivation for the claim that bosons and fermions of the same type, whose joint states have a special symmetric (antisymmetric) form,[26] can be individuated by some selected physical properties. But we should also note that this does not mean that quantum particles behave as if they were fully classical corpuscles equipped with unique identities. The possibility of qualitatively differentiating between particles at a given moment does not imply that an analogous differentiation can be done in a unique way at different times (a similar point is forcefully made by Pooley 2006). To illustrate this, suppose that we start off with two particles of the same type whose initial state is such that one particle can be said to occupy location *L* while the other occupies location *R*. When we consider the same system at a later time, we cannot tell whether the particle now occupying location *L* is identical with or distinct from the one that was here a moment earlier. Such identification would be possible if the two particles differed with respect to their essential properties, but this is excluded. On the other hand, quantum mechanics excludes other means of identification over time, such as the continuity of trajectories.

This fact explains why we can't continue using direct products of states to describe joint states of indistinguishable particles. The representation $|L(t_1)\rangle \otimes |R(t_1)\rangle$ wrongly suggests that particle labeled 1 in the tensor product will be forever identified with the one that at time t_1 occupied location *L*, so that the two vectors $|L'(t_2)\rangle \otimes |R'(t_2)\rangle$ and $|R'(t_2)\rangle \otimes |L'(t_2)\rangle$ describe two distinct ways the entire system could evolve. But that's not the way systems of particles of the same type behave. As is well-known, interference effects observed in experiments with scattering identical particles

[26] Even though I don't have space to explain this in detail, I should note that for two bosons/fermions to be absolutely discerned by their properties (represented by the symmetric operators Ω_i) their joint state should arise as a result of symmetrization/antisymmetrization of a direct product of two orthogonal states. For more details see (Bigaj 2015).

strongly suggest that the identification of particles after and before the interaction does not even make sense (even though it is clear that the particles are distinct when taken at any given moment). Hence the proper way of describing the temporal evolution is to use the (anti-)symmetric forms of the joint states: $|L(t_1)\rangle \otimes |R(t_1)\rangle \pm |R(t_1)\rangle \otimes |L(t_1)\rangle$ and $|L'(t_2)\rangle \otimes |R'(t_2)\rangle \pm |R'(t_2)\rangle \otimes |L'(t_2)\rangle$. These formal representations do not determine whether the particle that at t_2 occupies location L' is identical with the particle that at t_1 was located at L.

5. Conclusion

We have seen how the principles of serious essentialism can suggest certain solutions to some long-lasting debates in the philosophy of physics. The obvious take-home lesson with respect to space-time theories is that realism regarding spatiotemporal points does not require the assumption that points are equipped with primitive identities, and that distinct mathematical models can nevertheless represent the same physical possibilities. The consequences of serious essentialism in the context of quantum mechanics are somewhat less banal. The doctrine of essentialism seems to undermine the standard proof of the indiscernibility of same-type quantum particles, and this result opens the door to alternative solutions, including the surprising claim that absolute discernibility by appropriately formalized properties of individual particles is possible. I am convinced that the metaphysical conception of serious essentialism can have more interesting consequences regarding the ontological interpretation of other physical theories, including QFT, but for now I have very little to back this conviction.

University of Warsaw
Institute of Philosophy
e-mail: t.f.bigaj@uw.edu.pl

ACKNOWLEDGMENTS

Earlier versions of this paper were presented in 2013 in Lausanne and San Diego. I would like to thank the audiences of my talks for their comments that exposed numerous holes in my initial arguments. I also wish to express my thanks to an anonymous referee for this volume whose critical remarks led to a substantial modification of this paper. Last but not least I would like to thank Ewa Bigaj for reading and commenting on the paper.

The work on this paper was supported by the Marie Curie International Outgoing Fellowship Grant No. 328285.

REFERENCES

Adams, R.M. (1979). Primitive Thisness and Primitive Identity. *The Journal of Philosophy* **76**, 5–26.

Bartels, A. (1996). Modern Essentialism and the Problem of Individuation of Spacetime Points. *Erkenntnis* **45**, 25–43.

Belot, G. (1995). New Work for Counterpart Theorists: Determinism. *British Journal for the Philosophy of Science* **46**, 185–195.

Belot, G., Earman, J. (2000). From Metaphysics to Physics. In: J. Butterfield and C. Pagonis (eds.), *From Physics to Philosophy*, pp. 166–186. Cambridge: Cambridge University Press.

Bigaj, T. (2015). Exchanging quantum particles. In: P.E. Bour, G. Heinzmann, W. Hodges and P. Schroeder-Heister (eds.), *14th CLMPS 2011 Proceedings, Philosophia Scientiae* (Vol. 19, issue 1), pp. 185–198.

Brighouse, C. (1994). Spacetime and Holes. In: D. Hull, M. Forbes and R.M. Burian (eds.), *PSA* 1994 (Vol. 1), pp. 117–125.

Brighouse, C. (1997). Determinism and Modality. *British Journal for the Philosophy of Science* **48**, 465–481.

Butterfield, J. (1989). The Hole Truth. *British Journal for the Philosophy of Science* **40**, 1–28.

Butterfield, J. (1993). Interpretation and Identity in Quantum Theory. *Studies in History and Philosophy of Science* **24**, 443–476.

Earman, J., Norton, J. (1987). What Price Spacetime Substantivalism? The Hole Story. *British Journal for the Philosophy of Science* **38**, 515–525.

Fara, D.G. (2009). Dear Haecceitism. *Erkenntnis* **70**, 285–297.

Forbes, G. (1985). *The Metaphysics of Modality.* Oxford: Oxford University Press.

Forbes, G. (1986). In Defense of Absolute Essentialism. In: P. French, T. Uehling and H. Wettstein (eds.), *Midwest Studies in Philosophy XI: Studies in Essentialism*, pp. 3–31. Minneapolis: University of Minnesota Press.

French, S., Redhead, M. (1988). Quantum Physics and the Identity of Indiscernibles. *British Journal for the Philosophy of Science* **39**, 233–246.

Ghirardi, G., Marinatto, L., Weber, T. (2002). Entanglement and Properties of Composite Quantum Systems: A Conceptual and Mathematical Analysis. *Journal of Statistical Physics* **108** (112), 49–122.

Hoefer, C. (1996). The Metaphysics of Space-Time Substantivalism. *The Journal of Philosophy* **93**, 5–27.

Huggett, N. (2003). Quarticles and the Identity of Indiscernibles. In: W.K. Brading, E. Castellani, *Symmetries in Physics*, pp. 239–249. Cambridge: Cambridge University Press.

Huggett, N., Norton, J. (2013). Weak Discernibility for Quanta, the Right Way. *British Journal for the Philosophy of Science*, doi:10.1093/bjps/axs038.

Kripke, S. (1980). *Naming and Necessity.* Oxford: Blackwell.

Lewis, D. (1968). Counterpart Theory and Quantified Modal Logic. *The Journal of Philosophy* **65**, 113–126.

Lewis, D. (1983). Postscript to Counterpart Theory and Quantified Modal Logic. In: *Philosophical Papers* vol. I, pp. 39–46. Oxford: Oxford University Press.

Lewis, D. (1986). *On the Plurality of Worlds.* Oxford: Blackwell.

Mackie, P. (2006). *How Things Might Have Been: Individuals, Kinds, and Essential Properties.* Oxford: Clarendon Press.

Maidens, A. (1993). *Substantivalism and the Hole Argument.* Cambridge: University of Cambridge.

Maudlin, T. (1988). The Essence of Space-Time. *Proceedings of the Biennial Meeting of the Philosophy of Science Association* **2**, 82–91.

Maudlin, T. (1990). Substances and Space-Time: What Aristotle Would Have Said to Einstein. *Studies in History and Philosophy of Science* **21**, 531–561.

Maudlin, T. (2007). *The Metaphysics within Physics.* Oxford: Oxford University Press.

Melia, J. (1999). Holes, Haecceitism and Two Conceptions of Determinism. *British Journal for the Philosophy of Science* **50**, 639–664.

Norton, J. (2011). The Hole Argument. In: E.N. Zalta (ed.), *The Stanford Encyclopedia of Philosophy* (Fall 2011 Edition), URL = <http://plato.stanford.edu/archives/fall2011/entries/spacetime-holearg/>.

Pooley, O. (2006). Points, Particles, and Structural Realism. In: D. Rickles, S. French and J. Saatsi (eds.), *The Structural Foundations of Quantum Gravity*, pp. 83–120. Oxford: Oxford University Press.

Rickles, D. (2008). *Symmetry, Structure and Spacetime.* Amsterdam: Elsevier.

Robertson, T., Atkins, P. (2013). Essential vs Accidental Properties. In: E.N. Zalta (ed.), *The Stanford Encyclopedia of Philosophy* (Winter 2013 Edition), URL = <http://plato.stanford.edu/archives/win2013/entries/essential-accidental/>.

Skow, B. (2008). Haecceitism, Anti-Haecceitism and Possible Worlds. *Philosophical Quarterly* **58**, 98–107.

Skow, B. (2011). More on Haecceitism and Possible Worlds. *Analytic Philosophy* **52**, 267–269.

Stalnaker, R. (2012). *Mere Possibilities: Metaphysical Foundations of Modal Semantics.* Princeton: Princeton University Press.

Swoyer, C., Orilia, F. (2014). Properties. In: E.N. Zalta (ed.), *The Stanford Encyclopedia of Philosophy* (Fall 2014 Edition), URL = <http://plato.stanford.edu/archives/fall2014/entries/properties/>.

Wilson, M. (1993). There is a Hole and a Bucket, Dear Leibniz. *Midwest Studies in Philosophy* **18**, 202–241.

Thomas Møller-Nielsen

SYMMETRY AND QUALITATIVITY

ABSTRACT. In this paper the relationship between the notion of symmetry and that of qualitativity is examined. In particular, it is argued that, on the standard construal of the notion of qualitativity, a widely-held view about the relationship between these two notions is mistaken. However, it is argued that on a nonstandard construal of the notion of qualitativity due to Ismael and van Fraassen (2003), the claimed relationship between the two notions holds much more promise. The paper ends with an attempt to build and improve upon Ismael and van Fraassen's own account of the notion of qualitativity relevant to the notion of symmetry.

1. Introduction

In this paper I wish to examine the relationship between two notions whose close connection has often been remarked upon in the philosophy of physics literature but which as yet has escaped any detailed analysis. The first notion – one of central concern to many areas of physical enquiry – is **symmetry**, where (to a first approximation) a symmetry is a transformation which preserves the space of solutions of a given theory. The second notion – one of comparably central concern to many areas of *meta*physical enquiry – is **qualitativity**, where (again to a first approximation) a qualitative property is one which is not essentially "individual-involving" in any respect. Defined in this rough-and-ready manner, the existence of any kind of interesting relationship between these notions might appear surprising. After all, if a symmetry is just a transformation which maps solutions of a theory to solutions, then little would seem to stand in the way of our labelling "symmetries" transformations which map solutions to solutions whose qualitative characters are distinct, and similarly not much would seem to stand in the way of our also labelling "symmetries" other transformations which map solutions to solutions whose qualitative

In: Tomasz Bigaj and Christian Wüthrich (eds.), *Metaphysics in Contemporary Physics* (*Poznań Studies in the Philosophy of the Sciences and the Humanities*, vol. 104), pp. 179-214. Amsterdam/New York, NY: Rodopi | Brill, 2015.

characters are identical.[1] Whence, then, the relevant connection between these two notions?

The connection – at least according to one influential school of thought – begins to come into much clearer focus when one attempts to construe symmetries in a more specific sort of way: namely, as *tools of metaphysical inference*. On this view, symmetries are no longer construed as arbitrary mappings of solutions to solutions of a given theory. Rather, symmetries are taken to map only those solutions to each other that are in agreement with regard to the genuine metaphysical structure they ascribe to the concrete world. In other words – and assuming for argument's sake that the theory in question is in fact true, or at least approximately true – whatever is left *invariant* under a given theory's symmetries is taken to represent structure that is genuinely real; conversely, whatever is *variant*, or varies, under a given theory's symmetries is not taken to represent anything genuinely real: rather, such variant structure is construed as being "surplus," "redundant," not real – a mere mathematical artefact of our theory, nothing more (cf. Saunders 2003).

It will be useful for our purposes to give this particular view concerning symmetries – *viz.* that only the structure left invariant under them is genuinely real – a name: adopting Saunders' (2007) useful coinage, then, let us label it the **Invariance Principle** (IP).[2] Now, one immediate difficulty with understanding symmetries in this way is that of specifying *which* particular transformations among all possible solution space-preserving ones in fact relate solutions which agree on all such "genuine structure."[3] It is here that the notion of symmetry (putatively) makes contact with the notion of qualitativity. Symmetries, it is claimed, should be taken to relate – and only relate – those solutions that are *qualitatively indiscernible*, the corollary being that any such solutions' variant *non*qualitative structure is thereby *not real*. So understood, then, symmetries become razors which may be wielded in order to excise the variant nonqualitative structure that our theories are reputed to ascribe to the concrete world.

---

[1] Here and and below, when I speak of solutions' "qualitative characters" what I mean more specifically are the qualitative characters solutions *represent* the (concrete) world as possibly having: we do not want to claim that two solutions are "qualitatively distinct" purely in virtue of the fact that they are distinct *mathematical* entities. Cf. Lewis (1986), pp. 224–225.

[2] Cf. Dasgupta's (forthcoming) "symmetry-to-reality"-type reasoning.

[3] Neither Saunders nor Dasgupta, of course, sign up to the view according to which symmetries are properly construed as arbitrary mappings among the space of solutions of a given theory. Rather, they each have a clearly restricted notion of "symmetry" in mind in their respective papers. (They seem to differ, however, on the issue of how it is exactly that the notion is to be properly restricted.)

Such an assessment of the metaphysical implications of physical symmetries – *viz*. that they only ever demonstrate the superfluousness of variant nonqualitative structure, and that we are justified in excising such structure from our general metaphysical framework – is seemingly so widely accepted that it would not, in my view, be much of an overstatement to label it the **Received View**.[4] As I shall argue in this paper, however, the Received View is not only straightforwardly false, but it is also seriously naïve. It is straightforwardly false because as a matter of plain fact symmetries do sometimes reveal the superfluousness of specific kinds of variant *qualitative* structure (see section 3); and it is seriously naïve because there are plausible reasons to suspect that symmetries can *never*, in fact, reveal the superfluousness of nonqualitative structure (see section 4). In addition to this discussion, I shall also consider (in section 5) a recent alternative (and slightly idiosyncratic) construal of the notion of "qualitativity" that is – it is claimed – relevant to drawing IP-based metaphysical inferences, namely that due to Ismael and van Fraassen (2003). In particular, I shall argue that although their account is far from perfect, it is nevertheless much closer to the truth than the Received View is, and moreover does a far better job in providing an adequate philosophical and justificatory account of the IP. The paper will end, then, with my attempt to sketch an outline of what I take such a correct account to be, one which I think improves on but nevertheless owes much to Ismael and van Fraassen's own account.

Before we begin all this, however, we would do well to examine the relevant notions of symmetry and qualitativity, as well as their supposed connection, in a bit more detail. This we do in the next section.

2. More On Symmetry and Qualitativity

2.1. *Symmetry*

In what follows it will be helpful – if only for illustrative purposes – to construe theories "semantically": that is, as specifying from a set $\mathscr{K}$ of kinematically or "metaphysically" possible models a subset $\mathscr{S}$ of dynamically or "physically" possible models. $\mathscr{K}$ should thus be taken to represent the set of all possibilities (or "possible worlds") consistent with the posited basic ontology of the theory, while $\mathscr{S}$ should be construed as a particular subset of these possibilities, namely those also consistent with the theory's *laws*.

[4] I should mention that one notable recent dissenter from the Received View is Pooley (2013), whose overall line of argument is very much in tune with this paper's.

One quite natural way of construing what a symmetry is on this view is to take it to be any transformation T which acts on $\mathcal{K}$ such that it maps $\mathcal{S}$ to itself: more formally, and where $\mathbf{m}$ is an arbitrary model in $\mathcal{K}$, a transformation T acting on $\mathcal{K}$ is said to be a *symmetry* just in case $\mathbf{m} \in \mathcal{S}$ iff $T(\mathbf{m}) \in \mathcal{S}$. (This is basically a precisified version of the notion of "symmetry" as approximately defined at the beginning of the previous section.) As several authors have noted, however, such a notion of symmetry, at least when combined with the IP, quickly leads to theoretical disaster.[5] For, on this view, *arbitrary* permutations of $\mathcal{S}$ will count as a symmetry of any given theory. Thus, for instance, in Newtonian Gravitation Theory ("NGT") – i.e. the theory comprising Newton's three laws, plus his inverse gravitational square law, governing the behaviour of point-particles in Newtonian spacetime – models putatively representing n number of particles moving in such-and-such a dynamically possible fashion will be mapped under some symmetry transformation to dynamically possible models of arbitrary cardinality. Particle cardinality, then, would not appear to constitute a genuinely real feature of a solution of NGT according to this notion of symmetry: a seemingly unacceptable conclusion, given that particle number is arguably a *canonical example* of a "genuine feature" of solutions of NGT. (Indeed, it is a conclusion particularly unpalatable for those Newtonians who would have liked to believe that *some* material entities actually exist!)

The question naturally arises, then, as to whether we can devise a *restricted* notion of "symmetry" such that arbitrary models in $\mathcal{S}$ are not mapped onto one another: a notion which, in addition, yields *correct* metaphysical inferences regarding the reality or otherwise of various putative physical quantities. As Belot (2013) has recently emphasised, however, attempting to work out what the correct, and suitably restricted, formal definition of "symmetry" is to combine with the IP is no easy task. As he points out, many of the formal notions of "symmetry" used by practising physicists are quite unsuited to the general task of providing IP-based metaphysical inferences. For instance, identifying the relevant symmetries of one's theory as its Hamiltonian symmetries[6] plausibly does not give us what we want: among other things, it would seem to rule out Galilean boosts as counting as genuine symmetries of

[5] See, e.g. Belot (2003, p. 402; 2013, section 3); Ismael and van Fraassen (2003, pp. 378–379).

[6] Very briefly, one determines the Hamiltonian symmetries of a given theory by (i) giving the theory the canonical Hamiltonian ("phase space") treatment and (ii) identifying the relevant symmetries as those diffeomorphisms on phase space that preserve both the symplectic form (which encodes the geometric structure of the phase space) and the Hamiltonian function (which assigns to each point of phase space the total energy of the relevant physical state).

NGT (boosting a system will normally fail to preserve the Hamiltonian in virtue of its altering the system's kinetic energy).[7] Our focus here, however, is on quite a different issue, one that arguably remains largely implicit in Belot's paper: namely, in virtue of what, precisely, do we decide that a particular formal notion of symmetry is, or is not, yielding "correct" metaphysical inferences? What, in other words, are the relevant *criteria* by which we decide whether or not a particular set of transformations on $\mathscr{S}$ is giving us reasonable IP-based metaphysical conclusions? Why, for example, do we think that the *purely formal* notion of a theory's Hamiltonian symmetries are giving us *philosophically* the wrong result, when combined with the IP? Why do we *want* to include boosts, as well as translations and rotations, among the relevant $\mathscr{S}$-preserving transformations on NGT's space of kinematically possible models?[8]

For many theorists, it would seem that the answer to this question is straightforward: a given formal definition of symmetry is adequate if it relates all and only those solutions to one another that are *qualitatively indiscernible*. Thus, for instance, Rickles (2008, p. 4) writes that

> It is a certain class of symmetry that I am interested in: those that preserve all *qualitative* structure of a model or world [...].

Similarly, Thébault (2012, p. 814) affirms that the antihaecceitist – that is, someone who "denies the possibility of non-qualitative determinants of cross-identification" of objects in different models or worlds, and who moreover denies that there are any worlds that differ "only with regard to which objects play which role[s]" – is invariably able to deny that "models related by symmetry transformations are [representationally] distinct": the obvious implication being that the relevant symmetry transformations

[7] See Belot (2013, section 6) for further discussion of the pitfalls of defining the relevant symmetries in this way, and for reasons why other such formal notions are similarly problematic.

[8] From what Belot (2013, pp. 329–333) writes in his paper it might seem that he would answer the question of how it is that we determine whether a particular formal definition is philosophically satisfactory or not by appealing to something like *intuition*, or perhaps the working intuition of the mathematician, physicist, or philosopher of physics: after all, it certainly does seem intuitive that in the case of NGT the relevant symmetries should include boosts, but not mappings of solutions of (say) different particle cardinality! However, unless we have good reason to believe that all theorists' intuitions will cohere in all cases, and unless we think that we are somehow justified in taking such intuitive judgements as being epistemically privileged or beyond suspicion, this would only appear to push the core issue one step further back: for on what grounds do these theorists make the intuitive judgements that they do – assuming, of course, that they are generally correct judgements?

therefore relate models that differ at most nonqualitatively. And Belot (2003, p. 394) himself claims that

> [o]bjects [including possible worlds] related by a symmetry occupy identical roles in the pattern of relations described by their structure [...] so objects related by symmetries will be qualitatively indistinguishable.[9]

According to these theorists, then, it would seem that it is the metaphysical notion of qualitative indiscernibility (or discernibility) which provides the basis or grounds for any claims to the effect that a particular notion of symmetry is yielding correct (or incorrect) IP-based inferences. That is, to the extent that any given formal notion of symmetry yields correct IP-based metaphysical inferences, then it relates and only relates qualitatively indiscernible solutions; but to the extent that any such formal notion does not yield correct IP-based metaphysical inferences, then it fails to relate and only relate qualitatively indiscernible solutions.

But are things really so simple? Is it really the case that the restriction to qualitative character-preserving transformations gives us the principled basis that we want, insofar as it would allow us invariably to draw legitimate IP-based inferences – specifically, inferences to the effect that only the variant nonqualitative structure of our (best) theories is not real? Before we turn to examine this question more closely, it will be helpful if we first attempt to explicate the notion of "qualitativity" in some more detail, if only to aid ourselves in getting a better handle on what the view just expressed properly amounts to.

2.2. *Qualitativity*

The idea that the world is in some sense fundamentally "qualitative" in character has fairly deep historical roots.[10] It is also a doctrine that is notoriously difficult to state to any great degree of precision. Ever since Adams' (1979) classic discussion of the distinction between the qualitative

[9] For further seeming endorsements of this view, see, e.g. Belot (2001), Saunders (2003; 2013, p. 356). (Baker 2010 seems to express some sympathy for the view, though officially his own view is different.) It should also perhaps be noted that in more recent work (Belot 2013) Belot has refrained from using the word "qualitative" at all in connection with the notion of symmetry. However, given that in this later paper Belot does not provide any *other* kind of principled basis on which to determine when and on what basis symmetries can be guides to superfluous structure (see fn. 8 above), this leaves us pretty much back where we started – after all, if it isn't the qualitative indiscernibility of solutions which allows us to identify them as representationally equivalent, then what is the relevant feature that does?

[10] Adams (1979, section 2) has argued that the view itself dates back at least as far as Leibniz, although O'Leary-Hawthorne and Cover (1996) have more recently disputed whether Leibniz was in fact a "generalist" in any straightforward sense.

and the nonqualitative, however, a "linguistic approach" to stating the view has become quite popular in the literature. Imagining for simplicity that standard predicate logic (with the identity predicate taken as primitive)[11] contains all of the expressive resources required in order for us to clearly and perspicuously describe the actual world (thus we are ignoring, e.g. plural, higher-order, and other more complicated linguistic resources), we can define a **general sentence** of PL to be any sentence that neither (i) contains any proper names nor (ii) is constructed using any essentially "individual-involving" predicates, such as *Pegasizes* or *being Nicolas Cage*. Correspondingly, a **singular sentence** of PL is simply any sentence that is not general, i.e. which does contain proper names and/or individual-involving predicates. Thus, for instance, "$\exists x\ Fx$" is a general sentence of PL – provided that F is a predicate that isn't in any way "individual-involving" – for it doesn't explicitly refer to any particular individual, while "Fa" is a singular sentence of PL, for it does explicitly refer to some particular individual, namely a.

The **generalist** will then go on to claim that only general (or "qualitative") sentences are required in order to correctly and perspicuously describe the fundamental structure of the world.[12] The **singularist**, on the other hand, will disagree: for him, singular sentences are also required. The notion of "perspicuous description" is again one that is notoriously difficult to make precise, but nevertheless it can plausibly be elucidated by example. Thus, for instance, just as the statement "There is a table" could never amount to a perspicuous description of a world in which mereological composition never occurred (i.e. in which, strictly speaking, there *aren't* any tables), so "Nicolas Cage is balding" would not amount – indeed, could not amount, given that it expresses a singular proposition – to a perspicuous description of a world which was ultimately purely qualitative in character.[13] Some propositions, to put the point more figuratively, "cor-

[11] Although note that on Saunders' (2003) version of generalism the identity predicate is taken to be a defined rather than an ideologically primitive entity.

[12] The view goes by a variety of names in the literature, although "generalism" seems to be the term favoured by most theorists. See, e.g. O'Leary-Hawthorne and Cover (1996), Saunders (2003), Maunu (2005), Dasgupta (2009), Pooley (forthcoming), and Russell (MS). Compare also van Fraassen's (1991) "semantic universalism," Fine's (2005) "metaphysical anti-haecceitism," Kment's (2012) "anti-individualism," and Dasgupta's (2014) "qualitativism." The view also bears obvious resemblances to various "ontic" forms of structural realism (see, e.g. Ladyman and Ross 2007), although I shall not pursue the connection between these views here.

[13] For more on the relevant notion of perspicuity, see O'Leary-Hawthorne and Cortens (1995, pp. 154–157). More physics-based examples might include claims regarding the ("absolute") simultaneity of two distinct events in a relativistic world exhibiting a Minkowskian

respond to" or "limn" the world's structure. The disagreement between the generalist and the singularist ultimately boils down to which propositions, general or singular, are required in order to properly "limn" such structure in just this way. The generalist will claim that only general sentences or propositions are required in order to carry out such "limning." The singularist will deny this.[14]

I shall be assuming in what follows that the qualitative/nonqualitative distinction is at least clear enough for us to work with. (Those doubtful of its coherence might want to get off the boat now.[15]) With the distinction at our disposal, we can now state the two propositions whose viability we shall be scrutinising for the majority of this paper, and whose conjunction I take to comprise the core of the Received View on symmetries:

(P1) Symmetries – insofar as they are guides to "surplus" theoretical structure – only ever relate models which differ at most nonqualitatively.

(P2) The world is fundamentally purely qualitative, or "general," in character.[16]

According to the Received View, then, symmetries are an ontological razor: a razor which, more specifically, may be wielded to excise the variant nonqualitative structure that our theories are (at least putatively) committed to.

Though much of our discussion in what follows will primarily focus on (P1), it might be useful at the outset to say just a few words on (P2), and its apparent relationship to (P1). For, on reflection, there would appear to be a rather curious tension that arises for those who subscribe to both

---

spatiotemporal structure, or claims regarding one's ("absolute") velocity in a world with (say) a Galilean spatiotemporal structure. Such claims might be true in a "contextual" or "loose" or "non-literal" sense; but they are not true – indeed, could not be true – *simpliciter*.

[14] Of course, it almost goes without saying that no generalist or singuralist would want to sign up to the view just stated in this precise form: after all, no on seriously believes that PL is a language anywhere near expressive enough to capture facts about our actual world! The hope, however, is that presenting the views in this "linguistic" way captures *something* about the views, if only their broad philosophical "flavour."

[15] But see Cowling (forthcoming), who notes the seemingly ineliminable role played by the qualitative/nonqualitative distinction in the formulation of a wealth of philosophical doctrines (the intrinsic/extrinsic distinction, physicalism, etc.). He goes on to argue that the qualitative/nonqualitative distinction should be taken as metaphysically primitive.

[16] It is worth noting the logical distinctness of the two propositions: one could coherently believe in both, neither, or one or the other of them. However, as the view expressed (albeit largely implicitly) by philosophers who work on symmetries generally seems to include a commitment to the truth of both, it is their conjunction that we shall focus on here.

propositions. That is, if we already believed that the world is fundamentally purely qualitative in character, of what precise *use* are symmetries in telling us about the nonreality of various nonqualitative structure that our theories purport to ascribe to the concrete world, given that – *ex hypothesi* – we would *already know* that the world fundamentally contains no nonqualitative elements? In other words, for the theorist who ascribes to both (P1) and (P2), it would seem that symmetries could at best only confirm the nonreality of structures that she antecedently took not to be real (cf. Dasgupta forthcoming, section 5.3).

There is also a more obvious question to consider, namely: What is our justification for believing (P2) in the first place? What, in other words, are our grounds for believing that the world is fundamentally general rather than singular: grounds which, in turn, are presumably meant to legitimise our excision of the nonqualitative structure that our theories are putatively committed to?

As far as I am aware, there have been three explicit, and variously interrelated, arguments in the literature which actually purport to argue independently for (P2). Thus, Saunders (2003) has attempted to derive[17] generalism from modern-day versions of "Leibniz's Principles," namely the Principle of Sufficient Reason and the Principle of the Identity of Indiscernibles; Morganti (2008) has claimed that "the best (if not the only)" way of arguing for generalism, one which he does not in fact ultimately endorse, is to appeal to a modern-day version of Bertrand Russell's Principle of Acquaintance; and Dasgupta (2009) has adopted a generalist metaphysics on the basis of a generalised symmetry argument involving the permutation of individuals: an argument which, at its core, was explicitly formulated by Clarke and whose conclusion was likewise explicitly endorsed by Leibniz in their *Correspondence*.[18] I shall not comment on the individual merits of each such argument here (although we shall touch on a version of Dasgupta's in section 4.1 below): each of them, in my view, has its own particular strengths, as well as its own particular deficiencies. All I wish to emphasise here is the rather trivial point that the claim that the world fundamentally possesses no nonqualitative structure cannot go without saying, but must be explicitly *argued for*. And if the response to this (and the previous) challenge is to claim that it is the (canonical) *symmetries* of our best physical theories that (in conjunction with the IP) ultimately justify our belief in the purely qualitative structure of the world

[17] The turn of phrase Saunders actually uses is "put in place."

[18] See Alexander (1956, ed.: see esp. C.II.1; C.III.2; L.IV.3–4). The argument found in Wilson (1959) is also plausibly made in much the same spirit.

(i.e. that (P1) in turn establishes (P2)), then the reason *why* symmetries might be legitimately wielded in this fashion – in other words, why it is that we should think that (P1) and the IP are true in the first place – will stand in similar, non-trivial need of justification.

Before we close this subsection one further comment regarding (P2) should be made. For while it is indeed true that there exists a body of related arguments in the literature which aims to establish the generalist picture, it is also important to note that there exists an important body of work (e.g. Adams 1979; Kment 2012) which pulls in the opposite direction – that is, which attempts to *refute* the generalist picture. The rough idea behind these arguments is that, should the generalist accept the possibility of worlds which exhibit a certain "symmetry-breaking" property – e.g. a Blackian world (cf. Black 1952) containing nothing except two spheres, two miles apart from one another, with one of them destined to be annihilated at (say) time t = 1 minute – then she will struggle to make sense of certain *de re* possibility claims (as in Adams' 1979 argument) and/or will struggle to provide a decent account of counterfactual and probabilistic discourse in such worlds (as in Kment's 2012 argument). I draw attention to these arguments not because I believe that they refute the generalist picture. (I don't believe they do.) Rather, I mention them because I believe it is crucial to stress the perhaps obvious point that, insofar as we believe that the world is qualitative and that symmetries drawn from physics are supposed to provide some support for this belief, we would also do well to examine just what exactly the belief in the fundamentally qualitative nature of the world ultimately commits us to, and indeed whether in fact such commitments are ones that, all things considered, we might prefer to do without.

3. Symmetries Do Not Only Relate Qualitatively Indiscernible Solutions

The view that surplus structure-indicating symmetries only relate solutions that are qualitatively indiscernible (*viz.* (P1) above) is not a *prima facie* implausible one. Below are three examples of physical symmetries – one drawn from quantum mechanics, one drawn from classical Newtonian physics, and the other from general relativity – all of which seem to provide support for the thesis.

Ex. 1: Permutation symmetry in quantum mechanics. In the labelled tensor product Hilbert space formalism of ordinary quantum theory, if we apply the unitary permutation operator **P** corresponding to the "exchange" of indistinguishable particles 1 and 2 in a two-particle quantum mechanical state Ψ_{12} such that $\mathbf{P}\Psi_{12} = (-)\Psi_{21}$ and $\mathbf{P}\Psi_{21} = (-)\Psi_{12}$, then it is true that the quantum state will be left qualitatively invariant under the action of the permutation group. Indeed, the fact that the quantum state is left qualitatively invariant under such a permutation follows almost trivially, for according to the quantum formalism itself, $\Psi_{12} = (-)\Psi_{21}$: the states themselves are equivalent according to the very formalism of the theory (up to a phase), and hence cannot be used to represent states that differ *in any way at all*, qualitatively or otherwise.[19]

Ex. 2: Translational/rotational symmetry in Newtonian gravity. Taking a generic ("coordinate-free") model of Newtonian gravitation theory set in Newtonian spacetime to be of the form $\mathcal{M} = \langle M, t_a, h^{ab}, \sigma^a, \Phi_i \rangle$ (where M is a differentiable 4-dimensional manifold, and t_a, h^{ab}, σ^a, and Φ_i encode the temporal structure, the Euclidean structure of instantaneous 3-space, the inertial "rigging" structure of absolute space, and the various mass contents of the spacetime respectively)[20] and applying the appropriate diffeomorphism[21] d to yield a new model $\mathcal{M}_{\text{static}} = \langle M, t_a, h^{ab}, \sigma^a, d^*\Phi_i \rangle$, then (on the stipulation that each point of M represents the very same point of physical spacetime as it occurs in both $\mathcal{M}$ and $\mathcal{M}_{\text{static}}$)[22] the two models will differ at most nonqualitatively. More specifically, they will differ at most with regard to which *particular* points of the spacetime manifold are underlying various parts of the matter fields.

[19] Another way to put this point is to note that permutations of the particle labels in quantum mechanics correspond not just to symmetries of the *theory* – in the sense that they map solutions of the theory to solutions – but are also symmetries of the *solutions themselves* – in the sense that they act as the *identity* on solutions of the theory (up to a phase factor of – 1 in the case of fermions, if the number of permutations is odd). For further discussion of this point, see Pooley (2006, section 4.7).

[20] See Pooley (2013) and references therein for further details on the model-theoretic treatment of NGT and other spacetime theories besides.

[21] That is, a diffeomorphism which is both a dynamical and a spacetime symmetry of the theory, as explicated by Earman (1989, section 3.4). To see what this means exactly, begin by taking a model of some generic spacetime theory T to be $\mathcal{M} = \langle M, A_1, A_2, ..., P_1, P_2, ... \rangle$, where M is a four-dimensional differentiable manifold, A_i are the geometric-object fields on M characterising the structure of spacetime, and P_j are geometric-object fields on M characterising the model's matter contents. A **spacetime symmetry**, Ψ, is defined as a diffeomorphism (i.e. a bijection from M to M such that both it and its inverse are differentiable) such that $\Psi^* A_i = A_i$ for all i. A **dynamical symmetry**, Y, of a given spacetime theory T, on the other hand, is defined as a diffeomorphism such that, if $\mathcal{M} = \langle M, A_1, A_2, ..., P_1, P_2, ... \rangle$ is a model of T, then $\mathcal{M}_Y = \langle M, A_1, A_2, ..., Y^* P_1, Y^* P_2, ... \rangle$ is also a model of T.

[22] This qualification cannot go without saying; see Pooley (2006, pp. 102–103).

Ex. 3: Diffeomorphism invariance in general relativity. Taking a generic model of general relativity (GR) to be of the form $\mathcal{M} = \langle M, g_{ab}, T_{ab}\rangle$ and applying an arbitrary diffeomorphism d to yield a new model $\mathcal{M}_{diff} = \langle M, d^*g_{ab}, d^*T_{ab}\rangle$ (where M is again a differentiable 4-dimensional manifold, g_{ab} is the metric tensor, and T_{ab} is the stress-energy tensor which, roughly speaking, represents the model's matter content[23]), then the two models will again (under the exact same representational stipulation as before) differ at most nonqualitatively. That is, they will differ at most with regard to which *particular* points of the spacetime manifold are underlying various parts of the metric and matter fields.

In addition to all of these examples of symmetries relating qualitatively indiscernible states of affairs and thus putatively revealing the "superfluousness" of various nonqualitative structures – "labelled" quantum particles; points of space; and points of spacetime respectively[24] – there also appear to be several important cases of symmetries relating models that are *not* qualitatively indiscernible, thereby revealing the "superfluousness" of various *qualitative* structures as well. I list three such counter-examples here.

C-Ex. 1: "Galilean boost" symmetry in Newtonian gravity. Taking again our model of Newtonian gravitation theory set in Newtonian spacetime to be $\mathcal{M} = \langle M, t_a, h^{ab}, \sigma^a, \Phi_i\rangle$, then applying an appropriate diffeomorphism[25] d (corresponding to a "velocity boost") to $\mathcal{M}$ will invariably yield a new model $\mathcal{M}_{\text{kin}} = \langle M, t_a, h^{ab}, \sigma^a, d^*\Phi_i\rangle$ such that the two models will straightforwardly represent worlds that *do* differ qualitatively. $\mathcal{M}$ might, for instance, represent a world in which the matter contents are entirely stationary with respect to absolute space, while $\mathcal{M}_{\text{kin}}$ might represent a world in which the matter contents are all moving uniformly with a constant absolute velocity. These are, I take it, models which in no conceivable way differ merely nonqualitatively, given how the qualitative-nonqualitative distinction is normally understood: the worlds represented by the models in question differ more than merely with regard to which particular objects are "playing which roles." Furthermore, this boost symmetry is in fact a *canonical* example of a symmetry that is "surplus structure-revealing" in the sense defined above, insofar as the existence of this symmetry is what is commonly taken to justify the reconceptualisation of Newtonian absolute space so as to excise the "absolute" rigging field σ^a and its

23 Strictly speaking, the matter content of GR is represented by other fields in terms of which T_{ab} is defined.

24 For recent endorsements of the stated metaphysical implications of each of these symmetries, see, e.g. French (2014), Russell (2014), and Dasgupta (2011), respectively.

25 That is, a dynamical symmetry which is *not* also a spacetime symmetry; see fn. 21 above.

replacement with an affine ("straightness") connection ∇ as a primitive piece of ideology, in the move from a Newtonian gravitation theory set in Newtonian spacetime to an ideologically sparser but empirically equivalent theory set in Galilean spacetime.

C-Ex. 2: "Dynamical boost"[26] symmetry in Newtonian gravity. Taking $\mathcal{M} = \langle M, t_a, h^{ab}, \nabla, \rho, \Phi \rangle$ to be a solution of Newtonian gravitation theory now set in Galilean spacetime (with ρ and Φ representing the matter density and the gravitational potential field respectively), then for a diffeomorphism d applied to $\mathcal{M}$ corresponding to an element of the so-called "Maxwell group" of transformations and for which the gravitational potential field is appropriately transformed, one is able to yield a new dynamically possible model of Newtonian gravitation theory set in Galilean spacetime, $\mathcal{M}_{\text{dyn}} = \langle M, t_a, h^{ab}, \nabla, d^*\rho, \Phi' \rangle$. By wide agreement, what this symmetry is supposed to indicate is that the laws of the theory plus the global matter distribution ρ underdetermine the exact combination of the inertial structure and the gravitational force, rendering a given observer incapable in principle of determining whether she is moving in a force-free inertial manner or whether she is being accelerated under a gravitational force. Moreover, this difference is qualitatively specifiable: the difference between a model which represents an observer moving in a force-free inertial manner and a model which represents an observer being accelerated under a gravitational force is one that can be perfectly well articulated in general or non-individual-involving terms. Furthermore, it is this so-called "gauge-redundancy" of Newtonian gravitation theory which is widely thought to motivate the "geometrisation" of Newtonian gravity and consequent reconceptualisation of the theory's posited qualitative structure by moving to a theory of Newtonian gravity set in *Newton-Cartan* spacetime. Here, in brief, the flat inertial connection ∇ is replaced by a new kind of *dynamical* inertial structure ∇^{NC}, with models of the form $\mathcal{M}_{NC} = \langle M, t_a, h^{ab}, \nabla^{NC}, \rho \rangle$. Up to isomorphism, the two models $\mathcal{M}$ and $\mathcal{M}_{\text{dyn}}$ described in this paragraph correspond to a unique model of a Newtonian gravity geometrised in this way, thus (it is said) removing the undesirable "gauge-redundancy" inherent in all non-geometrised versions of Newtonian gravitation theory.[27]

[26] I draw the term "dynamical boost" loosely from Huggett (1999, pp. 166–167), who labels it the "dynamic shift."

[27] For further details and discussion, see e.g. Knox (2014).

C-Ex. 3: "Gauge" symmetry in electromagnetism. For our final counter-example, consider the (source-free) theory of electromagnetism set in a fixed Minkowskian spatiotemporal background, but with the matter fields, rather than being represented in terms of the Faraday tensor F_{ab} (satisfying the equations $\nabla_{[a}F_{bc]} = 0$ and $\nabla_a F^{ab} = 0$), instead being represented in terms of the electromagnetic 4-potential A^a (i.e. such that $F_{ab} = \nabla_{[a}A_{b]}$, and with the corresponding field equation of the theory being written as $\nabla_a \nabla^a A^b = \nabla^b \nabla_a A^a$). Picking a model of this theory to be given by the triple $\mathcal{M} = \langle M, \eta_{ab}, A^a \rangle$, and considering the transformation which maps models of this form to models of the form $\mathcal{M}_{\text{gauge}} = \langle M, \eta_{ab}, A'^a \rangle$ where $A'^a = A^a + \nabla^a \Psi$ (and where Ψ is some smooth scalar field), it will be the case that, if $\mathcal{M}$ is a dynamically possible model of this theory, then $\mathcal{M}_{\text{gauge}}$ will also be a dynamically possible model of this theory (cf. Weatherall MS, section 4). Read literally, $\mathcal{M}$ and $\mathcal{M}_{\text{gauge}}$ once again assign qualitatively *distinct* material distributions over the manifold: the distributions do not differ merely with regard to which particular manifold points are underlying the various parts of the vector potential field. *Modulo* various concerns which arise as a result of the Aharanov-Bohm effect, the conclusion invariably drawn is that this "gauge invariance" indicates the superfluousness of the vector potential *qua* physical quantity: that it is merely a mathematically convenient "shorthanded" way of describing and determining the values of the Faraday tensor, which in turn *is* taken to represent the genuine material contents of the theory.[28]

The moral in each of the above three examples is the same: the fact that the theory allows for these putatively unpalatable qualitative distinctions between models is supposed to motivate the search for and adoption of an alternative physical theory which collapses the qualitative distinctions between the symmetry-related models in question. (Why these distinctions *are* considered unpalatable is a question that we shall address in section 5.) Symmetries, therefore, sometimes *do* relate qualitatively discernible models; moreover, symmetries do not (merely) motivate us to excise nonqualitative structure, but rather can and do relate qualitatively distinct solutions,

[28] Furthermore, trying to defuse this argument by regarding "absolute" values of the vector potential as nonqualitative will not be of much use here, for in general $\mathcal{M}$ and $\mathcal{M}_{\text{gauge}}$ will differ "comparatively" as well; that is, the *variation* in field values from spacetime point to spacetime point will nearly always be different across gauge-related models. (This is due to such gauge symmetry's being suitably "local," or being variable from spacetime point to spacetime point.) To see this, merely consider the gauge transformation which maps the vector potential to itself at all spacetime points bar one: the model thereby yielded will differ from the original in more than purely "absolute" terms. (For more on the issue of "absolutism" versus "comparativism" about quantity, see Dasgupta 2013.)

and are capable of motivating the excision or reconstrual of the structure which "varies" in the appropriate way.

All of this might seem utterly transparent. Why, then, do so many theorists appear to insist on symmetries relating – and only relating – qualitatively indiscernible solutions? Unfortunately, I do not have any particularly convincing answer to this question. To close this section, then, I will consider what I take to be three of the most promising candidate responses.

The first, very natural response would be that all of these theorists are using the word "qualitative" in a somewhat different manner to the way in which I have used it here. I do not think this is at all the case,[29] though I admit that this claim is difficult to substantiate without engaging in a rather long (and tiresome!) exegetical discussion. I shall therefore leave it as an exercise for the committed reader to try and work out what such authors *could* possibly mean by "qualitative," at least on any charitable interpretation, unless they mean what I am taking them to mean here.[30]

A second possible answer to the above question might be that these authors are implicitly claiming that symmetries only motivate the excision of nonqualitative structure in "correctly formulated" theories.[31] Thus, for instance, they might follow Earman (1989, section 3.4) in claiming that counter-examples 1 and 2 above are not in fact counterexamples to proposition (P1) – which, to recall, was the claim that symmetries, insofar as they are guides to "surplus" theoretical structure, only ever relate models which differ at most nonqualitatively – because the relevant theories violate a "condition of adequacy" on theories that the symmetries of the dynamics should not outstrip the symmetries of the relevant spacetime (Earman's criterion "SP1"), and are therefore not "correctly formulated" in the appropriate sense.

But although I find this line of reasoning intriguing, I do not, for various reasons, ultimately find it convincing. One immediate worry is that it is difficult to see how one might flesh out what is meant by a "correctly formulated" theory in a such a way that the notion is both (a) broadly

[29] One important exception to this however is Ismael and van Fraassen (2003), who are explicit in their distinct construal of the notion of "qualitative." We discuss their paper in section 5.

[30] Perhaps I should add in my initial defence that the fact that (i) these authors usually contrast their notion of "qualitative" with the doctrine of haecceitism (i.e. the view according to which worlds might differ solely with regard to which objects are playing which roles), (ii) they often cite metaphysicians (e.g. Lewis 1983) who use the word in the same (standard) sense used here, and (iii) the canonical examples of symmetries that they consider (e.g. Leibniz shifts) *are* transformations that relate qualitatively indiscernible solutions, very strongly suggests (to my mind) the interpretation according to which they are simply mistaken.

[31] Thanks to Nick Huggett for suggesting this response to me.

applicable to all theories and (b) non-*ad hoc*, insofar as the notion of a theory's being "correctly formulated" isn't merely concocted on a case-by-case basis to preserve the truth of proposition (P1). (Why, for instance, isn't ordinary quantum mechanics as formulated in the labelled tensor product Hilbert space formalism also not a "correctly formulated" theory, given the availability of the Fock space formalism?) A second worry is that often what we *mean* by a particular theory's being "correctly formulated" is only evident in retrospect, *after* the relevant surplus structure has been identified and theoretically dispensed with. After all, it took quite some mathematical ingenuity and innovation (invention of the notion of an affine connection, etc.) before we could even meaningfully speak of there being such things as "Galilean" and "Newton-Cartan" spacetimes: indeed, it is plausible to think that for Newton, who had no alternative, "Newtonian" spacetime *was* the "correct" spacetime setting for his theory at the time he wrote the *Principia*! And a third and final worry about this kind of response is that, even assuming that a plausible, broadly applicable and non-*ad hoc* notion of a theory's being "correctly formulated" can be found, it is at best not obvious why we should expect such a notion to line up with the qualitative/nonqualitative distinction in any uniform way. *A priori*, what reason do we have to think that such a notion will invariably affirm that those theories involving "excess" qualitative structure are not "correctly formulated," but that those theories that involve "excess" non-qualitative structure are?[32]

Finally, a third, slightly more speculative answer – my preferred answer – to the question of why the view that symmetries only ever relate qualitatively indiscernible solutions is so prevalent is that since the explosion of literature on structural realism over the last two decades or so, discussion of theories' symmetries has (perhaps overly-) focussed on two specific symmetries – namely, the permutation invariance of quantum mechanics, and the diffeomorphism invariance of general relativity – and their connection to various structuralist theses concerning the metaphysics of individuality, objecthood, and relations.[33] These symmetries, of course, and as we noted above, *do* invariably relate solutions to qualitatively indiscernible solutions. It is, perhaps, not entirely implausible to think that many theorists have on this basis illegitimately extrapolated to the general

[32] Recall that Earman (1989, section 9) himself was led to conclude that the diffeomorphism invariance of GR motivated the search for a theory of gravitation that did not quantify over manifold points. For Earman, then, a theory's violation of SP1 plausibly constitutes only a sufficient condition of its being "incorrectly formulated," but not a necessary one.

[33] For broadly "structuralist" discussions of these specific symmetries, see e.g. Stachel (2002), Ladyman and Ross (2007), and Rickles (2008).

conclusion that symmetries thereby *always* relate qualitatively indiscernible solutions, when in fact a moment's reflection would have revealed that, in truth, this is not so.

But whatever the reasons might be for the predominance of the view that symmetries only ever relate qualitatively indiscernible solutions, the important point to note is that, as a matter of plain fact, it is false. Symmetries do in fact map solutions to qualitatively distinct solutions, and moreover can on occasion motivate the excision of theoretical structure that can only be plausibly construed as qualitative. In the next section, we change tack. Having shown in this section that proposition (P1) is straightforwardly false, in the next section I will attempt to push the claim that (P1) is also, quite plausibly, philosophically naïve. More specifically, I first want to press a "*reductio*" point originally made by Clarke in his *Correspondence* with Leibniz – namely, that if one is motivated to excise points of space on the basis of "shift-style" arguments, then one should likewise be motivated to excise intrinsically indiscernible atoms as well – and secondly, I wish to develop and offer some defence of a separate line of argument according to which symmetries can, in fact, *never* serve as guides to the nonreality of the nonqualitative.[34]

4. Symmetry and the Nonqualitative

In this section I shall (to repeat) consider two distinct problems associated with the view that symmetries can act as guides to redundant nonqualitative structure. The first problem begins to emerge when one considers what proposition (P1) above might plausibly commit one to: namely, the possible non-existence of individuals *tout court*. The second problem is that there appears to be a plausible way of *resisting* symmetry arguments directed against the reality of nonqualitative structure: a way of resisting which, it seems, is not available in the case of symmetry arguments directed against the nonreality of qualitative structure. We shall consider each of these problems in turn below.

[34] To fix ideas in what follows I shall primarily be focussing on spacetime symmetries in the classical Newtonian case, although I see no reason why the points made in this setting will not generalise to the general relativistic or indeed other theoretical contexts as well. (However, it should be noted that interpreting the permutation invariance of quantum mechanics as indicating the non-individuality of quantum particles faces various other, more specific difficulties, one of which is the fact that *classical* statistical mechanics is plausibly interpreted as being a permutation invariant theory as well; see Saunders 2013.)

4.1. *The First Problem*

In getting to grips with this problem it might be helpful to quickly remind ourselves of (one of) the core issues discussed in the famous *Correspondence* between Clarke and Leibniz, which took place almost exactly three centuries ago in 1715–1716. Here, recall, Leibniz argued against the existence of substantival space on the grounds that, were space a real substance, then "'tis impossible there should be a reason" why God would choose to place the world's material contents in one region of space rather than another, the two scenarios' being "absolutely indiscernible" from one another (L.III.5). Clarke, on the other hand, admonished Leibniz for such a principle's entailing that God "has neither created, nor can possibly create any matter at all" (C.IV.3), for a structurally identical argument could easily be levelled against the existence of atoms as well (cf. C.II.1; C.III.2) – there would equally be no reason for God to have arranged two (intrinsically indiscernible) bodies one way rather than another, "permuted" way – and on this basis Clarke concluded that Leibniz's principles (namely, the Principle of Sufficient Reason and the Principle of the Identity of Indiscernibles) were illegitimate tools of metaphysical theorising. Leibniz, however, was perfectly happy to bite the bullet both when it came both to arguments against the existence of substantival space *and* when it came to arguments against the existence of intrinsically indiscernible atoms: thus for him (and to Clarke's astonishment) the actual existence of "atoms [...] are confuted [...] by the principles of true metaphysics" (L.IV.4).

It is important to note two things about this (very) brief summary of the Leibniz-Clarke *Correspondence*. First, one should note that both Clarke and Leibniz were in dialectical agreement insofar as they both believed that, should our actual – or indeed any possible – world contain such intrinsically indiscernible entities, then there is also a possible world in which such entities have swapped roles (i.e. a merely "haecceitistically distinct" possible world). Second – and more importantly – both Leibniz and Clarke agreed (in modern terminology) that *if* symmetry considerations ultimately motivated the excision of absolute space from our fundamental metaphysics, then it *also* motivated to a precisely equal extent the excision of atoms as well.

"Anti-haecceitist" or "sophisticated" substantivalists, of one form or another, deny the possibility of such haecceitistically distinct worlds. I take it as given that one cannot simply declare that one denies the possibility of such worlds by *fiat*. One wants to know *why* – on what metaphysical grounds, or basis – one is in fact able to deny the possibility of such individuals' switching roles. (Similarly, in a Newtonian context one cannot simply deny that there are no possible worlds differing solely over an absolute velocity boost: one must, I take it, demonstrate the metaphysical coherence of such a denial by developing an alternative metaphysical framework – e.g. the notion of a Galilean spacetime – on which basis one is able to make such claims; cf. Dasgupta 2011.) One fairly straightforward way to accomplish this would be to claim that only one world – namely, our actual one – is metaphysically possible. Such a rigidly essentialist response, however, seems blatantly *ad hoc*. We would not, after all, have accepted an analogous response to the boost invariance of NGT: merely stipulating that whatever absolute velocity I have I have essentially does nothing to rid us of our unease about accepting absolute velocity as a genuine physical quantity. A second, better way of responding to the possibility of such shifted worlds, then, would be to try to develop a particular metaphysical framework from which the supposed representational equivalence of shifted models might follow in a more-or-less natural way. One *prima facie* plausible way to do this might thus be to construe individuals (and in particular, spacetime points) in "modestly structural" terms:[35] that is, as constituting nothing more than "nodes" in the relational physical structures in which they are embedded, or as "contextually individuated" entities (cf. Ladyman 2007). This is not meant to be equivalent to an eliminativism about individuals: one is perfectly free to take facts about the numerical identity and distinctness of such individuals as metaphysically primitive. Nor is it necessarily equivalent to the bundle theory, where individuals are ontologically secondary to properties and various "compresent" collections of them. Rather, the claim is that concrete, metaphysically robust individuals fundamentally exist, but that nevertheless there is no substratum or haecceity underlying them which grounds facts about their identity and distinctness. A shifted scenario, then, in virtue of the exact same relational structures being instantiated as in the original, amounts to a mere redescription of the original scenario. Crucially, this spatiotemporal metaphysics still purports to be a version of substantivalism – construed as the view that spacetime points constitute fundamental, basic elements of the actual world – very much worthy of the name.[36]

Recently, however, theorists have questioned the viability of this kind of "modestly structuralist" metaphysics (Dasgupta 2011, section 5; Russell 2014, section 4-section 5). In particular, these theorists have claimed not to be able to understand what kind of metaphysics "modest structuralism" truly amounts to without it collapsing into either (a) a version of strong

[35] I draw the term from Pooley (2006), p. 102.

[36] For further defence of this sort of view, see e.g. Hoefer (1996), Saunders (2003), Esfeld and Lam (2008), and Pooley (2013), section 7.

essentialism about which spacetime points are occupied or (b) a version of "straightforward" or non-structural substantivalism about spacetime points. At the very least, these authors have argued, the "modestly structuralist" or sophisticated substantivalist position needs to be more fully worked out before it can truly be said to constitute a response to the "Leibniz shift" argument against absolute space (or, in the context of GR, the "hole" argument against the existence of spacetime points) at least roughly on a par with the way in which the move to Galilean spacetime is taken to constitute an adequate response to the boost invariance of NGT.

I do not wish to take a position here on whether these authors are right on the issue of whether more needs to be said if one wants successfully to respond to these sorts of shift arguments. What I wish to point out here, rather, is merely the Clarkean point that *should* these theorists be right that some kind of modest structuralism about spacetime points is not a coherent position, then *exactly* parallel considerations should militate against the existence of atoms as well. That is to say, *if* one thinks that the static shift argument is a good argument against absolute space, *then* – and as Clarke pointed out to Leibniz in the *Correspondence* – one should also believe that "permutation" arguments against intrinsically indiscernible particles should mitigate against the existence of them as well: what goes for one should go for the other. Thus, if symmetry considerations can plausibly be used as a guide to the nonreality of nonqualitative variant structure (and assuming that modest structuralism is not in fact a tenable or coherent position), then they should be taken to indicate not just the nonreality of points of space(time), but the nonreality of all individuals.

One might, of course (*à la* Leibniz), still find shift and permutation-style arguments appealing: indeed, it is for precisely these reasons that Dasgupta (2009) has attempted to formulate a metaphysical position ("algebraic generalism") that purports to contain no fundamental individuals at all. Whether such an individual-less metaphysics is truly viable is an open question.[37] But what I wish to emphasise here is that, to the extent that one finds the notion of a modest structuralism about spacetime points or individuals in general untenable, then attempting to draw IP-based metaphysical consequences on the basis of symmetries' relating qualitatively indiscernible solutions is a far from straightforward task: a

[37] Though I must say that the early signs of developing such an "individual-less" metaphysics do not look promising. See, e.g. Rynasiewicz (1992) for a famous criticism of the the "Einstein Algebra" approach to GR advocated by (among others) Earman (1989, section 9); Lam and Wüthrich (forthcoming) for some trenchant criticisms of the "category-theoretic" approach advocated by Bain (2013); and Turner (forthcoming) for a critical discussion of Dasgupta's "algebraic" approach to cashing out an individual-less metaphysics.

wholly revisionary, fundamentally "individual-less" metaphysics will be necessitated; a metaphysics which tells against not just the existence of points of space (as in Leibniz's "shift" argument) or points of spacetime (as in the "hole" argument), but against the existence of all individuals: a consequence that many theorists believe, not entirely implausibly, constitutes a *reductio* of Leibniz-style arguments directed against the nonreality of the nonqualitative.[38]

4.2. *The Second Problem*

Having emphasised in the previous subsection that the claim that symmetries motivate the excision of nonqualitative structure from our theoretical framework would plausibly excise too much (i.e. not just spacetime points, but individuals *tout court*), we focus in this subsection on a possible way of *resisting* any symmetry argument to the effect that some particular nonqualitative aspect of the world is not real – a way of resisting which is apparently not available in the case of symmetries' being putative guides to the nonreality of variant qualitative structure, and which arguably must be responded to by those who think that symmetries can indeed be taken to reveal the redundancy of the variant nonqualitative aspects of our theories.[39]

It is helpful to start by considering a pertinent difference between what Maudlin (1993) labels the **static shift** – which involves a global, time-independent repositioning of all matter fields in the space (see Ex. 2 above) – and the **kinematic shift** – which involves "boosting" the absolute velocity of all matter contents by a constant value in the same direction against the background of Newtonian spacetime (see C-Ex. 1 above). As Maudlin makes explicit, NGT's commitment to kinematically-shifted possibilities would appear to indicate the existence of a genuine, contentful *question* that an observer living in such a Newtonian world would be in principle unable to answer: namely, the question as to what her absolute

---

[38] See, e.g. Horwich (1978, p. 409), who calls the consequence "unacceptable"; Field (1984, p. 77), who claims it renders Leibniz's argument "obvious[ly] unsound"; and most recently Arntzenius (2012, p. 178), who thinks it entails that Leibniz's shift argument "is just not a good argument."

[39] Although in what follows I shall focus almost exclusively on Maudlin's (1993) discussion, it should nevertheless be noted that Maudlin's point was independently anticipated in more-or-less identical form earlier in the literature by Horwich (1978) and Teller (1987). (Although, as Pooley forthcoming, p. 81, notes, Horwich and Maudlin draw very different conclusions from their shared observation.) What sets Maudlin's discussion apart, however, is the fact that he was the first person to note the important *difference* between the two sorts (static and kinematic) of shift argument, and it is for this reason that his paper will be the primary focus of our discussion here.

velocity actually *is*. In the case of the static shift, on the other hand, no such analogous question expressing our ignorance appears to be available. Indeed, as Maudlin (1993, p. 190) writes, the only question one might sensibly ask oneself – namely, "Where am I in absolute space?" – deserves the pithy answer: "I am *here*, not three meters north or anywhere else." In other words, no *directly analogous* epistemological problem to the kinematic shift appears to be generated in the case of the static shift: for the only way in which one can even *state* the static shift argument is in such a way as to antecedently determine that all such statically shifted scenarios are *in fact* counterfactual, requiring as they do essential reference to the actual world. Another way to put essentially the same point is to note that in a Newtonian world you will always be certain that you are *not* (say) 5 metres to the left of where you *actually are* right now; but you will be never certain whether you are moving at 0 m/s, or 5 m/s, or 10 m/s with respect to absolute space.

Now, it is crucial to see that the disanalogy Maudlin notes between the static and kinematic shift arguments is not one that arises due to some inherent difference in the nature of the static and the kinematic shifts *per se*.[40] Rather, it arises as a consequence of the fact that, while the kinematic shift involves *qualitative* differences between distinct models, the *static* shift involves differences that are purely haecceitistic: that is, which involve differences only with regard to the scenarios' respective (putative) *singular* facts. Indeed, depending on the nature of the spacetime in question, a "static" shift can be used to generate qualitatively distinct models, while a "kinematic" shift can correspondingly be employed in certain spacetimes to generate distinctions between models that are merely haecceitistic. Thus, in an "Aristotelian" spacetime[41] setting – where models are of the form $\mathcal{M} = \langle M, t_a, h^{ab}, \sigma^a, \gamma, \Phi_i \rangle$, where γ is a privileged integral curve of σ^a, intuitively representing a "special" persisting point of absolute space (perhaps the "centre of the universe") – a "static" shift can be wielded to yield a distinction between models that *do* differ qualitatively: for such shifted models would typically represent all physical systems' differing with respect to their (qualitatively specifiable) spatial distance relative to the privileged point. Conversely, in a Galilean spacetime setting, a "kinematic" shift would introduce merely haecceitistic distinctions between

[40] Though Maudlin does not explicitly mention in his paper that his point generalises in this way, he has confirmed (in personal correspondence) that it does. Dasgupta (MS) also notes that Maudlin's objection similarly applies to cases of property-switching as well, so long as properties are construed "quiddistically," such that it makes sense to speak of (say) mass and charge "swapping roles" in some possible world.

[41] I draw the term from Earman (1989, section 2.6).

models – for the boosted models would differ only with regard to *which particular* spacetime points were successively (un)occupied by the relevant physical systems – without differing qualitatively in any way at all (cf. Pooley 2002, p. 59).[42] Maudlin's point can thus be put more broadly as follows: models which differ qualitatively present an epistemological problem which is not present in the case of models which differ merely haecceitistically, for qualitative distinctions entail that a *relevant question* can be asked regarding what it is that I, *qua* agent "embedded" in the model, am ignorant of, and which cannot be asked in the case of models or worlds that differ purely haecceitistically.

Maudlin (1993, p. 191) is entirely explicit concerning the upshot of all of this:

> [T]he static shift does not result in an indistinguishable state of affairs, nor does it imply that there are any real but empirically undeterminable spatiotemporal facts about the world. The world described by the shift may be *qualitatively* indistinguishable from the actual world in the sense that no purely qualitative predicate is true of the one which is false of the other. But we have more than purely qualitative vocabulary to describe the actual world; we have, for example, the indexicals, without which the Leibniz shift cannot be described.

Thus, for Maudlin, statically shifted solutions are distinguishable from one another: we distinguish between them *indexically*, i.e. by noting that *I* am *here*. Conversely, kinematically shifted solutions are not distinguishable from one another: for merely stating that my absolute velocity is in fact *whatever it actually is right now* does not get one anywhere nearer to answering the pertinent question of *how fast* I am actually moving with respect to absolute space. (Is it 5 m/s? 10 m/s? 20 m/s?) And, Maudlin argues, it is in virtue of this very distinguishability that one might legitimately resist drawing any relevant metaphysical inferences when the symmetry-related models in question are qualitatively indiscernible: that is, when they represent at most haecceitistically distinct possible worlds.[43]

Theorists who are resistant to Maudlin's point, and who believe that symmetries can in fact motivate the excision of nonqualitative facts from

[42] This is because in Newtonian gravitation theory as set in Galilean spacetime "Galilean boosts" now count both as a spacetime and as a dynamical symmetry.

[43] It is perhaps worth noting that uniformly altering the absolute velocities of each material system in a given world will not *always* give rise to a qualitatively distinct world: for instance, changing the absolute velocity of each such material system by merely changing the *direction* of their absolute motion will give rise to a world that differs merely haecceitistically from the original, for such a transformation will only alter which *particular* points of space are occupied by material particles or fields at any given time. It would therefore be more correct to say that only alterations in absolute *speeds* invariably give rise to qualitatively distinct solutions. (Many thanks to Nick Huggett for help with this point.)

our fundamental metaphysics, have attempted to counter Maudlin's point in a multitude of ways. Thus, Dasgupta (2011, p. 146) argues that Maudlin's construal of the relevant notion of distinguishability is flawed because it "implies that whether something is detectable depends on factors that are, intuitively, entirely irrelevant to the matter [of detectability]": for in the case of the static shift in NGT, he writes, the issue as to whether we are able to "detect" our absolute position would then seemingly depend on whether or not we lived in (say) a Newtonian or Aristotelian spacetime setting – or whether "God had a favourite point of space," as Dasgupta himself puts it – something which (Dasgupta claims) seems intuitively wrong: the thought being that in either such spacetime our absolute position would presumably be as "undetectable" as in the other.

I am, however, inclined to think that Dasgupta simply misses the point of Maudlin's objection here: indeed, he appears to me to simply be *re-stating* Maudlin's core claim that merely haecceitistically distinct scenarios are never indistinguishable. For as we saw above, in an Aristotelian spacetime setting "shifting" the entire material universe *will* invariably yield a world qualitatively distinct (but nevertheless empirically indistinguishable) from the original. But now, in virtue of the qualitative distinctness of the two worlds, there is a fact which I can perfectly coherently express in qualitative vocabulary but which I am in principle unable to know, namely my distance relative to the privileged point of space. Moreover, it is for this very reason – *viz.* that there is a stateable fact that I am in principle unable to determine or detect – that (Maudlin would presumably argue) one should excise the privileged point of space from one's fundamental metaphysics, and move to (e.g.) Newtonian spacetime. Rather than posing a problem for Maudlin, then, Dasgupta appears merely to have provided a classic instance of Maudlin-type reasoning at work.

A second way of responding to Maudlin's point – one which Dasgupta also appears to endorse in the same paper – might be to claim that, just because one cannot *state* what it is that one is ignorant of in the case of merely haecceitistically distinct scenarios, it doesn't automatically follow that there is thus *no fact* of which one is ignorant. There might, in other words, be genuine facts that I am in principle unable to detect, and which, moreover, I am in principle unable to express; all Maudlin has in fact shown us is that any such fact is inexpressible – not that there isn't any such fact. Indeed, as Dasgupta (2011, p. 146) notes, if this is all that Maudlin has actually shown us, then "[O]ur epistemic situation *vis à vis* our location in space is much *worse* than our epistemic position *vis à vis* our velocity through space, since in the former case we cannot even formulate questions about what it is we cannot detect!"

But Maudlin might respond: whoever said anything about facts being inexpressible? I *can* in fact detect what my position in absolute space actually is: specifically, I am right *here*, and/or the object before me is *that one*. In other words, Maudlin might respond by claiming that *singular facts are in fact detectable after all*, and that they are stateable using spatial indexicals and/or demonstrative terms; the fact that symmetries cannot be used as legitimate guides to nonqualitative structure is because this very nonqualitative structure is structure that we can in fact unequivocally detect: (P1) above is undermined because we know ("indexically") that proposition (P2), or the claim that the world is fundamentally purely qualitative, is false. There is no genuine fact of which I am ignorant, and there is nothing of which I am ignorant that I cannot express: our metaphysics, our epistemology, and our semantics – they all line up.

If this interpretation of Maudlin is correct (and I think it is), how might the anti-Maudlinian – that is, someone who believes that symmetries can in fact be used as a guide to the nonreality of nonqualitative structure – respond? One very natural way to do it would be to offer a *positive* account of what detection "truly" is such that merely haecceitistically distinct scenarios turn out to be indistinguishable (i.e. "undetectably distinct") after all: indeed, this is the strategy that Dasgupta (MS) and Russell (2011, section 3.5) have both independently tried to pursue in recent work. I am, however, quite sceptical of the general viability of this kind of anti-Maudlinian strategy. This is because I suspect that our pre-theoretical notion of what "detection" is much too ambiguous for any account of what it "truly" or "genuinely" amounts to not to simply beg the question against Maudlin: that is, by simply ruling out by *fiat* cases of "indexical detection" at some point in one's positive account of detection.[44] Rather, what I think the generalist must do if his response to Maudlin is to be dialectically effective is to try to *account for* Maudlin's (apparently false) intuition that he is in fact able to "indexically detect" which singular facts are true at his world; that is, the generalist must try to provide a way of accommodating Maudlin's intuition that the possibility of such indexical or "singular" detection decisively (indeed, almost trivially) refutes generalism *from within* the generalist framework.

---

[44] This is especially evident in Russell's (2011) account, in which he first distinguishes between the "semantic" and the "functional" contents of belief (in which the former, but not the latter, such contents are sensitive to the absolute position of a given subject), and then simply asserts on p. 149 that "what should count as *observed* is an observational belief's functional content," and that this conjecture is permissible because "I'm not worried about capturing intuitive judgements about the ordinary use of the word 'observe'." This, I take it, will hardly go very far in convincing the committed Maudlinian.

I do not wish to advocate any particular way of doing this, but here is at least one plausible way that the generalist might try.[45] The trick that the generalist might be able to exploit here is that not all "knowledge" is plausibly *propositional* knowledge; that, in other words, some knowledge which one acquires is merely "locational," in the sense that by acquiring it one does not thereby acquire another *objective fact*, but rather merely (to paraphrase Pooley forthcoming, p. 97) "locates oneself in the objective order." Thus, this account will run, by "detecting where he is" at any given time Maudlin is not thereby acquiring knowledge of a genuine singular fact, but is instead only acquiring knowledge of "where he is" in the full and complete generalist description of the world in which he is located: in Lewis' (1979) terminology, it is knowledge *de se*, not *de dicto*, which Maudlin lacks and subsequently acquires when he "detects" where he is in absolute space/which individual is in front of him at any given time. The subject knows *de se* that he is *here* and not *there*; but he does not "detect" this difference, for only propositional knowledge is knowledge that one can properly-speaking "detect."

To repeat, this is only one plausible way for this line of argument to go. All that I wish to emphasise is here is that Maudlin's intuition that he can directly detect specific singular facts (i.e. his position in absolute space and, more broadly, facts concerning particular individuals) must one way or another be accounted for by any generalist metaphysics which attempts to collapse any number of putatively haecceitistically distinct possibilities to one. Maudlin's objection would thus appear to constitute yet another significant, though perhaps on this occasion not insurmountable, problem for the Received View.

Let us now take stock. Recall the two propositions first stated in section 2 above:

- (P1) Symmetries – insofar as they are guides to "surplus" theoretical structure – only ever relate models which differ at most non-qualitatively.
- (P2) The world is fundamentally purely qualitative, or "general," in character.

In section 3 we showed that proposition (P1) is straightforwardly false, for the reason that symmetries do, on occasion, reveal the superfluousness of variant qualitative structure. In this section (section 4), we have seen that (P1) is not only straightforwardly false, but also plausibly naïve as

[45] This approach is advocated by Pooley (forthcoming, pp. 96–97); see also van Fraassen (1991, pp. 465–466).

well: for not only is it plausibly the case that granting that symmetries are a guide to the nonqualitative grants too much (insofar as such a view might well also indicate the nonreality of, e.g. electrons and other material entities besides), but there also seems to be a generic (Maudlinian) way of resisting arguments to the effect that some variant nonqualitative aspect of our theoretical formalism is not real (which would seem to appeal to the fact that (P2) can be proved trivially false by cases of "indexical" detection) – a way of resisting which, it seems, is not available in the case of symmetries' revealing the superfluousness of qualitative structure.

In the next and final section, our philosophical focus will shift slightly. Rather than addressing problems specific to the Received View *per se*, our attention will now be directed on the issue of how and why symmetries can be used as legitimate guides to superfluous structure in the first place (to the extent, that is, that they can in fact be so used). It is in addressing this question that Ismael and van Fraassen (2003) have recently proposed a notion of "qualitativity" distinct from its usual metaphysical connotation. Thus, they claim, it is the fact that symmetries relate "qualitatively indiscernible" solutions *in their sense* which, they argue, ultimately justifies the use of symmetries in picking out "surplus" theoretical structure. The next section will assess their proposal, and examine to what extent they are correct.

5. A New Notion of "Qualitative"?

Ismael and van Fraassen's (2003) main concern in their paper is, in brief, to provide a philosophical account of how and why (and when) it is the case that drawing symmetry-based metaphysical inferences is justifiable. Similarly to those theorists who would appear to ascribe to the Received View, they too argue that the "qualitative indiscernibility" of a theory's solutions plays a crucial role in such an account. However, their construal of what it means for two solutions to be "qualitatively indiscernible" is crucially different from the notion as construed by adherents of the Received View. For rather than construing qualitative properties as being those which are not "individual-involving" in some intuitive sense, Ismael and van Fraassen (2003, p. 376) construe qualitative properties as being those that are "directly observationally accessible to the observer," and are "distinguishable by [...] a gross discrimination of colour, texture, smell, and so on." Furthermore, they are careful in their paper to contrast qualitative properties with those that are merely "measurable" in their sense: such measurable properties are those that are able to "make some discernible impact on gross discrimination of colour, texture, smell and so on [... no] matter how attenuated the connection is, how esoteric the impact, or how special the conditions under which it can be discerned."

Ismael and van Fraassen (2003, p. 380) go on to summarise their proposal as follows:

> [*W*]*e submit that it is precisely the qualitative-structure-preserving symmetries of the laws that are indicative of the presence of superfluous theoretical structure and should always be interpreted as trivial.* [Italics in original.]

The thought would seem to be this. Take the space of solutions $\mathscr{S}$ of your favourite theory; determine which solutions are qualitatively indiscernible in the relevant sense of containing the exact same distribution of "directly observable" quantities; take the relevant set of $\mathscr{S}$-preserving transformations on the space of kinematically possible models to be just those that map qualitatively indiscernible solutions to one another; and then go on to take solutions related to one another in this way to represent the same physical state of affairs.

In assessing this account, the first obvious question to ask is: Why are Ismael and van Fraassen so careful to stress the distinction between quantities or properties that are "qualitative" or directly observable, and those that are merely "measurable"? Moreover, why do they think that this is a distinction that is *relevant* to the IP? Unfortunately, the answer to both of these questions is rather difficult to discern from what they write: indeed, they often write as if it *is* the measurable/unmeasurable distinction, rather than the qualitative/nonqualitative distinction, that is relevant to the IP. (E.g. "[...O]ur main topic: superfluous structure will align with the presence of *unmeasurable* quantities in the theory's world picture. [...] To sum up: we are going to connect superfluous structure with the presence of *unmeasurable* quantities." pp. 376, 378; italics mine.) They do on occasion (e.g. pp. 378; see also p. 376), however, seem to suggest that the distinction is crucial because "what is measurable/unmeasurable cannot be read off directly from the theory," and that "[w]e need to make use of what is observable [i.e. qualitatively discernible] in order to make this distinction." But, even assuming the correctness of this claim (a big assumption), the question arises: Why should the fact that certain quantities can be "read off" more-or-less "directly," rather than "indirectly," from a given theory's formalism, have any bearing on what kind of quantities superfluous structure-indicating transformations should necessarily preserve? To be sure, what quantities are directly observable might be *easier to determine* than what quantities are merely measurable: but why should this have any bearing on the IP, or the issue of which symmetry transformations on $\mathscr{K}$ among the $\mathscr{S}$-preserving ones are the *right* ones? Ismael and van Fraassen do not appear to provide an answer to this question in their paper; nor is it easy to imagine what a satisfactory answer to it could even be.

A second, related, problem also confronts their account. Given Ismael and van Fraassen's (2003, p. 376) explicit admission that "many measurable quantities will be non-qualitative," their account would seem to imply that $\mathscr{S}$-preserving $\mathscr{K}$ transformations which map *measurably distinct*, but nevertheless "qualitatively indiscernible," solutions to one another could plausibly be interpreted as revealing the presence of superfluous structure. But such a consequence is patently absurd. Now, of course, the obvious way to remedy this defect in their account would be to connect the presence of superfluous structure to the existence of unmeasurable quantities – and, indeed, we noted above that this is how Ismael and van Fraassen often (informally) appear to frame their view – but then it would seem that the qualitative/nonqualitative distinction has lost its primary relevance for their account: it would be symmetries that preserve all *measurable* structure that would be indicative of surplus structure, with the qualitative/nonqualitative distinction, quite simply, having no substantive role to play in their account at all (other than, perhaps, insofar as qualitative quantities help us "get a fix on" what is measurable according to a given theory).[46]

A third and final problem facing Ismael and van Fraassen's account is also perhaps the most interesting.[47] That is, if we agree with Ismael and van Fraassen's proposal that one can "directly observe" certain quantities, it would appear to follow almost ineluctably that *once* "directly observed," such quantities *must* be left invariant under the symmetries of any future theory. The history of physics, however, would seem to suggest a very different lesson. Thus, for instance, in the transition from Newtonian to relativity theory, it is plausible to think that symmetry considerations ultimately led us to accept that what we "directly observe" in terms of the distance between two physical events is not their spatial distance *simpliciter*,

---

[46] One might, of course, also be suspicious of the the very coherence of the distinction between the qualitative and the measurable that Ismael and van Fraassen feel is important to draw in their paper: a distinction which, I suspect, ultimately derives from the importance that van Fraassen (1980) places upon this distinction in his constructive empiricist philosophy of science. (Where, roughly speaking, one should believe what one's theory says about the directly observable or "qualitative" to be true, but remain agnostic about what it says about the unobservable or merely "measurable.") Indeed, it is interesting to note that in earlier work Ismael (2001, pp. 131–132) herself appeared to raise just such a concern about the relevance and indeed even coherence of such a distinction: "[T]here is not an epistemologically interesting difference between unimplemented sight and sight augmented by imaging instruments. [...] Seeing, whether with our bare eyes or through a microscope, is just measuring [...]."

[47] This point has been explicitly noted by Dasgupta (forthcoming, pp. 30–31), while Saunders (2003, p. 300) also appears to make an essentially identical observation.

but rather only their "spatial distance" *relative to a particular frame of reference*, the latter quantity being left invariant under relativistic (Lorentz) symmetry transformations. In other words, what we take ourselves to "directly observe" is *itself* seemingly a function or product of our theory: even supposedly "directly" observable quantities might turn out not to be real in the end. The move to a new theory, and a new associated set of transformations, might well yield the conclusion that quantities that we previously thought we could straightforwardly and unproblematically detect are, in fact, not detectable or indeed not real after all. The notion that we "directly observe" certain quantities *simpliciter*, then, *pace* Ismael and van Fraassen, is one that is extremely difficult, if not impossible, to square with the history of physics.

Despite these faults, however, I think that Ismael and van Fraassen are latching onto something relatively important in their attempted explanation and justification of IP-based inferences. For I, too, think that issues to do with observation and, more specifically, detection and measurability will turn out to be essential in justifying IP-based metaphysical inferences.[48] I will end, then, by sketching the barest outlines of how I think the correct account of the IP goes. (One which I take to be broadly van Fraassen-esque in spirit.)

To keep things relatively concrete, consider again the case of NGT and absolute velocity. This theory has an associated set of $\mathscr{S}$ of dynamically possible models. Now, we seem to have very good reason to think that, for instance, models related by a boost transformation are empirically indistinguishable: there is no possible measurement that any observer "embedded" in such a Newtonian world could perform in order to determine what her absolute velocity actually is. Absolute velocity, in other words, is a quantity that no Newtonian observer could ever empirically *measure*. But why do we think this? It is often claimed (by, e.g. Friedman 1983) that it is a simple consequence of Newton's *laws*: that it follows merely from the fact that accelerations are left invariant under the Galilean transformations that absolute velocity is an undetectable physical quantity. As several theorists have pointed out (e.g. Barbour 1989; Brown 1993), however, this is simply incorrect: in order to derive the conclusion that absolute velocity is a truly undetectable dynamical quantity one needs an additional assumption – one which in fact Newton implicitly made in his derivation of Corollary V in

[48] Indeed, in recent work Ismael (2014), Caulton (forthcoming), and Dasgupta (forthcoming) have all – apparently independently – pressed this same point, albeit in different ways. Working out the relevant differences and similarities between these (and my) accounts is something that I hope to accomplish in future work.

the *Principia* – namely, that both the inertial and gravitational mass of bodies (and the corresponding forces which act upon them) are *independent* of the bodies' absolute state of motion.[49] But this assumption is plausibly still only a necessary, and not a sufficient, condition to guarantee absolute velocity's undetectability. One additional (extremely obvious, but nevertheless non-trivial) assumption that needs to be made in order to ensure undetectability concerns subjects' internal mental states: namely, that they do not covary in a systematic way with their absolute motion. That is (and Wittgensteinian "private language" arguments to the contrary notwithstanding), one must assume that subjects do not (e.g.) possess "absolute velocity recorders" in the corners of their visual fields, or indeed any other devices which would allow them to have direct knowledge of what their own particular absolute velocity is at any given time.[50] And, finally, one plausible assumption that would appear to be necessary to guarantee absolute velocity's undetectability is that no concurrent or future theory (e.g. classical electrodynamical theory) will eventually render one's absolute velocity detectable according to some other method.[51]

All of these assumptions are, of course, very natural ones to make. Indeed, most theorists writing on these topics appear to make them implicitly. What I wish to emphasise here, however, is simply their *non-triviality*; none of them, after all, follow purely from "the truth" of Newtonian theory alone; if one of them were not to obtain, absolute velocity might well have turned out not to have been an undetectable quantity after all.

But why do these assumptions feel natural? Well, one thing worth pointing out is that all of them seem to fit in with our "scientific knowledge" in the broadest sense. We are, indeed, reasonably confident that forces do

[49] Indeed, this assumption is required in order to ensure that NGT counts as a Galilean (or "boost") invariant theory in the first place.

[50] This point in particular is emphasised by Roberts (2008, section 6). As he notes, if such knowledge did exist it would be a very strange kind of knowledge indeed: in particular, it would not be communicable through standard physical channels (for instance, writing a letter to someone to inform him or her of what your absolute velocity actually is wouldn't work, as the relative positions of the ink particles on the written paper would be preserved in the boosted scenario as well).

[51] It could of course be debated whether such a proper subclass of frames in classical electrodynamical theory should be identified with the absolute space rest frame in NGT (for why can't the "ether" be moving with respect to absolute space?). I follow Friedman (1983, p. 105) (and, apparently, Earman (1989, pp. 51–55)), however, in believing that such an identification is a perfectly natural and obvious one to make. (Perhaps Occamist reasoning could also be mobilised here: for Occam's razor would plausibly dictate that we should, other things being equal, posit only one privileged subclass of inertial frames common to both such theories (NGT and classical electrodynamics), rather than two.)

not systematically covary with systems' "absolute" velocities (as our experiences on trains and aeroplanes, along with more detailed experiments, will testify); that humans do not, for instance, possess "absolute velocity recorders" in the corners of their visual field; and that no other theory will eventually allow one to determine such a thing as one's "true" motion with respect to some privileged inertial frame. It would thus seem to be the case that it is to *science itself*, in some very general sense, that we must appeal to if we are to determine whether or not some specific quantity is detectable or not, and to whom we must ultimately defer if we are to justify IP-based metaphysical inferences. In a certain sense, then, we should embrace van Fraassen's (1980, section 3.7) "hermeneutic circle": we should let science itself be our guide to what we (think we) can detect.

This general proposal that I am suggesting (or, rather, sketching) here might therefore be summarised as follows. First, one determines what the superfluous structure-indicating transformations on the solution space of one's theory actually are by engaging in hermeneutic circle-type reasoning: this allows us to determine which models of our theory represent empirically indistinguishable, but nevertheless putatively physically distinct, solutions. Second, one justifies the excision of the superfluous "variant" structure in question by appealing to a mixture of both scientific realism (i.e. the view that the models of our theory should be construed more-or-less literally) and by an appeal to Occam's razor (i.e. the assumption that, other things equal, it is better for one's theory *not* to allow for physically distinct but empirically indistinguishable solutions).

Now, of course, in the Newtonian case things are relatively straightforward: working out which solutions are empirically indistinguishable is relatively easy once we have determined (by hermeneutic circle-type reasoning) that all that we in fact have empirical access to are the relative positions and velocities that material systems instantiate with respect to one another. In more complex or heavily mathematicised theories, working out the empirical content of one's theory will almost certainly constitute a far less straightforward task. Nevertheless, I think that the general process of reasoning that applies in these more complex cases will be the same in all essential respects to the one I am here suggesting applies in the simpler case of NGT. That is, in the first instance we engage in hermeneutic circle-type reasoning to work out which solutions in $\mathscr{S}$ represent empirically indistinguishable but physically distinct ways for the world to be: in the Newtonian case, this will involve identifying those solutions in which the same histories of relative distances and velocities of material systems are instantiated. And in the second instance, we seek an interpretative schema or novel theory according to which these two solutions are not only *empirically* indistinguishable, but represent *physically* identical states of affairs: in the case of the "variant" quantity of absolute velocity in NGT, for instance, this will involve moving to Galilean (or, better, Newton-Cartan) spacetime.

In closing: I would of course admit that much of what I have just claimed could be disputed. Nevertheless, I believe that, at the very least, this account provides a modest improvement over Ismael and van Fraassen's own account, insofar as it appears to solve what are arguably its most pressing difficulties. For in my account, there is no dubious distinction between the "qualitative" and the "measurable" drawn; my account does not have the (unwanted) consequence that measurably distinct solutions can sometimes be indicative of the presence of superfluous structure; and furthermore my account would seem to be fully capable of accommodating the transition from classical Newtonian physics – in which we "directly observe" spatial distances *simpliciter* – to relativity theory, in which we only "directly observe" space*time* intervals. This is because on my account the hermeneutic circle serves (as it appears to have served for van Fraassen (1980, section 3.7)) an essential *dual* role: it is not only the means by which we determine *whether* two solutions are empirically indistinguishable; but it also, in a crucial and important sense, informs us of *what* it is that we think we can empirically detect. Of course, there is no guarantee that what we take ourselves to detect now will be preserved under theory change. But this is just the Humean predicament that we, as scientists and philosophers, find ourselves in: a predicament which, as Quine (1969, p. 72) noted several decades ago, needn't trouble us, for it would appear to be essentially equivalent to the human one.

University of Oxford
Balliol College
e-mail: thomas.moller-nielsen@balliol.ox.ac.uk

ACKNOWLEDGEMENTS

Thanks to the editors for affording me the opportunity to contribute to this volume, and to Jeff Russell, Neil Dewar, Teru Thomas, Hugo Maxwell, Simon Saunders, and Oliver Pooley for discussion. Special thanks must go to Nick Huggett, who provided numerous helpful comments on a previous version of this paper.

REFERENCES

Adams, R.M. (1979). Primitive Thisness and Primitive Identity. *The Journal of Philosophy* **76** (1), 5–26.

Alexander, H.G. (1956). *The Leibniz-Clarke Correspondence.* Manchester: University of Manchester Press.

Arntzenius, F. (2012). *Space, Time, and Stuff.* Oxford: Oxford University Press.

Bain, J. (2013). Category-Theoretic Structure and Radical Ontic Structural Realism. *Synthese* **190** (9), 1621–1635.

Baker, D. (2010). Symmetry and the Metaphysics of Physics. *Philosophy Compass* **512,** 1157–1166.

Barbour, J. (1989). *Absolute or Relative Motion?* Vol. 1: *The Discovery of Dynamics.* Cambridge: Cambridge University Press.

Belot, G. (2001). The Principle of Sufficient Reason. *The Journal of Philosophy* **98** (2), 55–74.

Belot, G. (2003). Notes on symmetries. In: K. Brading and E. Castellani (eds.), *Symmetries in Physics: Philosophical Reflections*, pp. 393–412. Cambridge: Cambridge University Press.

Belot, G. (2013). Symmetry and Equivalence. In: R. Batterman (ed.), *The Oxford Handbook of Philosophy of Physics*, pp. 318–339. Oxford: Oxford University Press.

Black, M. (1952). The Identity of Indiscernibles. *Mind* **61** (242), 153–164.

Brown, H.R. (1993). Correspondence, Invariance, and Heuristics in the Emergence of Special Relativity. In: S. French and H. Kamminga (eds.), *Correspondence, Invariance and Heuristics: Essays in Honour of Heinz Post*, pp. 227–260. Dordrecht: Springer Press.

Caulton, A. (forthcoming). The Role of Symmetry in the Interpretation of Physical Theories. *Studies in the History and Philosophy of Modern Science.*

Cowling, S. (forthcoming). Nonqualitative Properties. *Erkenntnis.*

Dasgupta, S. (2009). Individuals: An Essay in Revisionary Metaphysics. *Philosophical Studies* **145** (1), 35–67.

Dasgupta, S. (2011). The Bare Necessities. *Philosophical Perspectives* **25** (1), 115–160.

Dasgupta, S. (2013). Absolutism vs Comparativism about Quantity. In: K. Bennett and D. Zimmerman (eds.), *Oxford Studies in Metaphysics*, Volume 8. Oxford: Oxford University Press.

Dasgupta, S. (2014). On the Plurality of Grounds. *Philosophers' Imprint* **14** (20), 1–28.

Dasgupta, S. (forthcoming). Symmetry as an Epistemic Notion (Twice Over). *The British Journal for the Philosophy of Science.*

Dasgupta, S. (unpublished). *Inexpressible Ignorance.* Available online at www.shamik.net.

Earman, J. (1989). *World-Enough and Space-Time.* Cambridge, MA: MIT Press.

Esfeld, M., Lam, V. (2008). Moderate Structural Realism About Space-Time. *Synthese* **160** (1), 27–46.

Field, H. (1984). Can We Dispense with Space-Time? *PSA: Proceedings of the Biennial Meeting of the Philosophy of Science Association*, 38–90.

Fine, K. (2005). *Modality and Tense.* Oxford: Oxford University Press.

French, S. (2014). *The Structure of the World: Metaphysics and Representation.* Oxford: Oxford University Press.

Friedman, M. (1983). *Foundations of Space-Time Theories.* Princeton, NJ: Princeton University Press.

Hoefer, C. (1996). The Metaphysics of Space-Time Substantivalism. *The Journal of Philosophy* **93** (1), 5–27.

Horwich (1978). On the Existence of Time, Space, and Space-Time. *Noûs* **12** (4), 397–419.

Huggett, N. (1999). *Space from Zeno to Einstein: Classic readings with a contemporary commentary*. Cambridge, MA: MIT Press.

Ismael, J., van Fraassen, B. (2003). Symmetry as a guide to superfluous theoretical structure. In: K. Brading and E. Castellani (eds.), *Symmetries in Physics: Philosophical Reflections*. Cambridge: Cambridge University Press.

Ismael, J. (2001). *Essays on Symmetry*. New York: Garland.

Ismael, J. (2014). *Symmetry as a Guide to Superfluous Structure*. Talk given at the UC Irvine Conference on Gauge Symmetries, March 2014.

Kment, B. (2012). Haecceitism, Chance, and Counterfactuals. *Philosophical Review* **121** (4), 573–609.

Knox, E. (2014). Newtonian Spacetime Structure In Light of the Equivalence Principle. *The British Journal for the Philosophy of Science* **65** (4), 863–880.

Ladyman, J. (2007). Scientific structuralism: On the identity and diversity of objects in a structure. *Aristotelian Society Supplementary Volume* **81** (1), 23–43.

Ladyman, J., Ross, D. (2007). *Every Thing Must Go: Metaphysics Naturalized*. Oxford: Oxford University Press.

Lam, V., Wüthrich, C. (forthcoming). No categorial support for radical ontic structural realism. *The British Journal for the Philosophy of Science*.

Lewis, D.K. (1979). Attitudes de Dicto and de Se. *Philosophical Review* **88** (4), 513–543.

Lewis, D.K. (1983). Individuation by Acquaintance and by Stipulation. *Philosophical Review* **92** (1), 3–32.

Lewis, D.K. (1986). *On the Plurality of Worlds*. Oxford: Blackwell.

Maudlin, T. (1993). Buckets of Water and Waves of Space: Why Spacetime Is Probably a Substance. *Philosophy of Science* **68** (2), 183–203.

Maunu, A. (2005). Generalist Transworld Identitism (or, Identity Through Possible Worlds Without Nonqualitative Thisnesses). *Logique et Analyse* **48** (189–192), 151–158.

Morganti, M. (2008). Weak Discernibility, Quantum Mechanics and the Generalist Picture. *Facta Philosophica* **10** (1–2), 155–183.

O'Leary-Hawthorne, J., Cortens, A. (1995). Towards Ontological Nihilism. *Philosophical Studies* **79** (2), 143–165.

O'Leary-Hawthorne, J., Cover, J.A. (1996). Haecceitism and anti-haecceitism in Leibniz's philosophy. *Noûs* **30** (1), 1–30.

Pooley, O. (2002). *The Reality of Spacetime*. D. Phil Thesis, University of Oxford.

Pooley, O. (2006). Points, Particles, and Structural Realism. In: D. Rickles (ed.), *The Structural Foundations of Quantum Gravity*, pp. 83–120. Oxford: Oxford University Press.

Pooley, O. (2013). Substantivalist and Relationalist Approaches to Spacetime. In: R. Batterman (ed.), *The Oxford Handbook of Philosophy of Physics*, pp. 522–586. Oxford: Oxford University Press.

Pooley, O. (forthcoming). The Reality of Spacetime. Oxford: Oxford University Press.

Quine, W.V. (1969). *Ontological Relativity and Other Essays*. New York: Columbia University Press.

Rickles, D. (2008). *Symmetry, Structure, and Spacetime*. Amsterdam: Elsevier Press.

Roberts, J. (2008). A Puzzle about Laws, Symmetries, and Measurable Quantities. *The British Journal for the Philosophy of Science* **59** (2), 143–168.

Russell, J.S. (2011). *Possible Worlds and the Objective World*. Ph.D. Thesis, New York University.

Russell, J.S. (2014). On Where Things Could Be. *Philosophy of Science* **81** (1), 60–80.

Russell, J.S. (unpublished). *Quality and Quantifiers*. Available online.

Rynasiewicz, R. (1992). Rings, Holes, and Substantivalism: On the Program of Leibniz Algebras. *Philosophy of Science* **59** (4), 572–589.

Saunders, S. (2003). Physics and Leibniz's Principles. In: K. Brading and E. Castellani (eds.), *Symmetries in Physics: Philosophical Reflections*, pp. 289–307. Cambridge: Cambridge University Press.

Saunders, S. (2007). Mirroring as an A Priori Symmetry. *Philosophy of Science* **74** (4), 452–480.

Saunders, S. (2013). Indistinguishability. In: R. Batterman (ed.), *The Oxford Handbook of Philosophy of Physics*. Oxford: Oxford University Press.

Stachel, J. (2002). The Relations Between Things versus the Things Between Relations: The Deeper Meaning of the Hole Argument. In: D. Malament (ed.), *Reading Natural Philosophy: Essays in the History and Philosophy of Science and Mathematics*, pp. 231–266. Chicago, IL: Open Court.

Teller, P. (1987). Space-Time as a Physical Quantity. In: P. Achinstein and R. Kagon (eds.), *Kelvin's Baltimore Lectures and Modern Theoretical Physics*, pp. 425–448. Cambridge, MA: MIT Press.

Thébault, K. (2012). Symplectic Reduction and the Problem of Time in Nonrelativistic Mechanics. *British Journal for the Philosophy of Science* **63**, 789–824.

Turner, J. (forthcoming). Can We Do Without Fundamental Individuals? In: E. Barnes (ed.), *Current Controversies in Metaphysics*. London: Routledge.

van Fraassen, B. (1980). *The Scientific Image*. Oxford: Clarendon Press.

van Fraassen, B. (1991). *Quantum Mechanics: An Empiricist View*. Oxford: Clarendon Press.

Weatherall, J. (unpublished). *Are Newtonian Gravitation and Geometrized Newtonian Gravitation Theoretically Equivalent?* Available online at http://arxiv.org/pdf/1411.5757.pdf.

Wilson, N.L. (1959). Substances Without Substrata. *The Review of Metaphysics* **12** (4), 521–539.

Matteo Morganti

RELATIONAL TIME

ABSTRACT. The present paper looks at a particular point of intersection between contemporary physics and metaphysics, and claims that a relatively neglected metaphysical theory could become of interest (again) thanks to the interaction between the two. In particular, the paper discusses the relational view of time, whereby time (but not, at least not necessarily, space) is reduced to a structure of relations between events. The argument takes its main cue from Barbour's recent 'Machian' perspective on physical theory. It is contended that it is possible to endorse Barbour's basic insights while not following him in his outright rejection of time as a non-existent entity. As a matter of fact, doing this makes it possible to circumvent some problems that Barbour's theory has to face, especially with respect to the explanation of temporal experience. At the same time, a scientifically credible background to the relational view of time is provided.

1. Introduction

According to substantivalists about space and/or time, space and/or time exist independently of physical objects and processes and are prior to them, as they constitute the 'stage' in which objects exist and physical processes take place. According to relationists, instead, space and/or time depend on physical objects and events: they are derivative on, and even reducible to, relations between things. Slightly differently, according to the substantivalist an ontological catalogue of what exists as a fundamental entity includes portions of space and/or time, according to the relationist it doesn't.

Why 'space *and/or* time'? On the one hand, from the purely logical point of view, one's metaphysical theory of space is independent of one's metaphysical theory of time: for instance, a philosopher could be a substantivalist about time while believing that there are good reasons for adopting a relational theory of space. On the other hand, historically relationism and substantivalism have been regarded as 'package deals', whereby space and

In: Tomasz Bigaj and Christian Wüthrich (eds.), *Metaphysics in Contemporary Physics* (*Poznań Studies in the Philosophy of the Sciences and the Humanities*, vol. 104), pp. 215-236. Amsterdam/New York, NY: Rodopi | Brill, 2015.

time are to be subjected to the same metaphysical treatment. And this has happened, it seems, essentially for two reasons. A first, generally methodological, motivation is that one's basic philosophical stance should be implemented as uniformly as possible. And the arguments for either substantivalism or relationism have often seemed to apply equally to time and to space – two obviously related concepts/entities. Leibniz, for instance, based on his ideas concerning God's creation, the Principle of Economy and the Principle of Sufficient Reason, opposed substantivalism both with respect to space and with respect to time: in both cases, he argued, God would have had no reason for creating a 'container' over and above all the things existing in the universe, nor grounds for locating the universe exactly where it is in fact located within either the spatial or the temporal container.[1] There is also an important, more science-related, reason for what one may label the 'unitary' attitude with respect to the ultimate nature of space and time. It is the fact that 20th century physics – in particular, Einstein's Theory of Relativity – seems to have decreed that space and time are not two separate things but rather aspects of a unique, fundamental, four-dimensional entity: space-time.

It is not surprising, then, that the more or less recent philosophical discussion has been concerned with the opposition between relationism and substantivalism in general. That is, it has by and large taken for granted that, whatever the winner of the contention, the corresponding metaphysical perspective would apply to both space and time (or, better, to space-time and nothing else). Indeed, a winner might seem to have emerged clearly quite some time ago. In the 1960s and 1970s, it became a widespread opinion that the space-time manifold could, and should, be regarded as an autonomous entity, fundamental for our physics (see, e.g., Earman 1970). In the 1980s, difficulties were raised in the form of the 'Hole Argument' (for an illustration, see Norton 2011). But this 'only' led to the formulation of a 'sophisticated substantivalism', according to which certain space-time models should be regarded as representations of the same state of affairs (see, for example, Hoefer 1996) – the general opinion with respect to relationism remaining generally unfavourable.

More recently, however, there have been interesting developments. Some have suggested that the metaphysical debate concerning the nature of space and time has become outdated, as it doesn't properly transfer from the classical context Newton and Leibniz worked in to contemporary physics (Rynasiewicz 1996). Others have argued instead that not only does contemporary physics justify the continuing interest in the metaphysical

[1] It goes without saying that this is a very rough reconstruction of Leibniz's views.

dispute concerning the nature of space and time, but it also offers important new insights. Here, obviously enough, we will side with the supporters of the relevance of the metaphysical debate. In this context, we will be concerned with two main elements of (relative) novelty: first, the ongoing re-evaluation of the relationist programme in physics (for examples, see Pooley and Brown 2002 and Pooley 2001); secondly, the progressively more popular idea that time is only an illusion (for an overview, see Callender 2010), which appears particularly compelling exactly in the context of the 'new relationism'. The leading thought will be, in a nutshell, that it is in fact possible to decouple the revival of the relationist programme and the closely related eliminativist stance with respect to time. And that this leads to a particular answer to the question concerning the nature of time, one affirming:

(a) The mutual independence of space and time from each other (and, consequently, between one's metaphysical views about the former and about the latter);
(b) That there might be physics-based reasons for endorsing a relational theory of time *only* (leaving it open whether space is a substance or it too should be reduced to relations between physical entities).

I believe the resulting conception, i.e., relationism about time only, deserves serious attention for two reasons. First of all, because, as already mentioned, it is a rather underdeveloped philosophical view. Secondly, because, remarkably, a lot of the arguments that have been recently provided in favour of the idea that physics tells us that time is an illusion – most notably, the arguments developed in the last 15–20 years by Julian Barbour – crucially rely upon a) relationist ideas and, as already pointed out in the past (for example, by Healey 2002 and Rickles 2006), b) on an important ambiguity between reductionism and eliminativism. Consequently, the possibility appears worth exploring (especially from the perspective of a broadly naturalistic methodology) of providing a scientific basis to temporal relationism as an explicitly non-eliminative stance.

In what follows, in particular, I will argue that the approach to physics delineated by Barbour constitutes a plausible basis for a respectable form of 'selective' relationism, about time but not (necessarily) about space, and make some suggestions as to how to articulate such a relationist stance. To be clear, my aim here is not so much to discuss the merits and/or limits of Barbour's approach to physics; nor of relationism as a view on the metaphysical nature of time. Rather, the paper only aims to make two *conditional* claims. First, that *if* one has independent, a priori reasons for being a relationist about time, then s/he should also try to make his/her views credible from a naturalistic, scientifically informed perspective; and the best way to do so is to endorse an elaboration of Barbour's views. Secondly, that *if* one has independent, scientific reasons for taking Barbour's views seriously, then one should consider interpreting those views in terms of (a particular ontologically committed version of) relationism, rather than antirealism, about time. In this context, what will be said specifically about space can and should be intended as a relatively independent philosophical addition. Another thing that must be made explicit is that this paper does not aim to do anything like providing a detailed solution to the so-called 'problem of time' in quantum gravity, which is at the basis of the idea that time is an illusion. Rather, it will exploit some ways of dealing with the problem that have been proposed in, or at least can be reconstructed from, the recent literature with a view to lending support to the specific metaphysical view under scrutiny.

The structure of the paper is as follows. The next section provides a – very brief! – reconstruction of the relationist view of time (and space) – and of its opposition to substantivalism – from Aristotle to modern and contemporary physics. This will be instrumental to providing some important distinctions, definitions and qualifications. Section 3 explores in some detail recent arguments against the reality of time – in particular, those based on the Machian perspective developed by Barbour – and employs them for sustaining instead a realist, relational metaphysics of time (and not necessarily space too). Section 4 deals with some potential problems for the proposed metaphysical account, and Section 5 contains a concluding summary.

2. A Brief History of Relationism (About Time)

Intuitively, events take place at specific times which pre-exist as potential 'containers' of physical happenings: there might be a moment, the layman is likely to think, in which nothing happens, but that moment would still be real, *hence* temporal instants must be ontologically basic. Relationism opposes this idea, starting from the observation that talk about time is (or at any rate seems to be) exclusively talk about what happens (or may happen), and about relations between certain events and other events. As a matter of fact, relationists argue, quantification over times should not be read literally, as it is simply a convenient instrument.

At least according to its supporters, relationism is made clearly preferable by epistemological considerations. In the present case, it is unquestionable that events are less mysterious entities than times – if only because we can make direct experience of (some) events but not of any instant of time *per se*. And, of course, this epistemological aspect goes hand in hand with considerations of economy. In the present case, if I can meaningfully talk about time and things that happen in some sort of sequence while only positing the latter sequence in my ontology, there are good reasons not to postulate the existence of time as a fundamental entity as well. Here, we will not discuss this in detail but, rather, take the above considerations as *prima facie* compelling reasons for carefully exploring the prospects for relationism – primarily, of course, in the form of relationism about time.

The best known supporter of relationism is indubitably Leibniz. However, a relational view of time can be traced back to Aristotle. In the *Physics*, and in particular in Sections 10–14 of Book IV, Aristotle explicitly argues that time is distinct from, but existentially dependent on, change, that is, on relations between *numerically and qualitatively* distinct events. In particular, for Aristotle time is "a number of change with respect to the before and after" (219b 1–2): instants of time, that is, can be distinguished, counted and ordered in a series only on the basis of modifications in what exists, and of the properties that things exemplify in different 'sections' of their existence.[2] Leibniz makes a similar, but more radical, point when he says that "instants, consider'd without the things, are nothing at all; [...] they consist *only* in the successive order of things" ([1704] 1956, third paper, Section 6, emphasis added).

That Leibniz is a relationist also about space is well-known. For example, he states: "I hold space to be something merely relative, as time is; [...] I hold it to be the order of coexistences, as time is an order of successions" (Ibid.; 25–26). As already mentioned, the motivation for this particular sort of relationism, applied to both space and time, was for Leibniz – very roughly – that both space and time as containers for physical objects and events are superfluous; and that, so understood, they would entail too many unrealised possibilities, a choice among which couldn't be grounded in any way (these are the well-known observationally indistinguishable shifted universes that Leibniz used to refute Newton and his followers).

Leibniz's relationism didn't remain unchallenged. In 1689, Newton had already formulated his notorious bucket experiment: if we picture a bucket suspended by a rope which starts rotating around the rope's axis, first bucket and water are at rest with respect to each other, then the bucket rotates but the water is at rest (bucket and water are in motion relative to each other), but ultimately bucket and water are mutually at rest (the water moving with the same velocity as the bucket). Since in the last phase the water's surface becomes concave but water and bucket are at rest in relation

[2] It is arguable that Aristotle's definition indicates that he takes time to be essentially the by-product of conscious experience. We will set this aside here.

to one another – which was not the case in the initial stage of mutual rest – there's something that must be explained but cannot be explained by having recourse to the relativity of the water's motion with respect to the bucket.[3] Strictly speaking, the bucket argument was intended by Newton to prove the weakness of Descartes' views of space as identical with matter and of the 'proper motion' of bodies as relative to other, *neighbouring* bodies – *not* that motion *requires* an absolute background. Exegetical matters aside, however, Newton can be legitimately regarded as the paradigmatic enemy of relationism; historically, his arguments have indeed been primarily interpreted as direct arguments in favour of substantivalism and absolute space and time. As is well known, Leibniz debated at length the force of Newton's considerations with Newton's follower Clarke. Independently of the details of this philosophical dispute, it is unquestionable that the Newtonian reflections[4] turned out to constitute formidable obstacles to relationism. In more detail, the dispute remained open for a long time, and many thinkers sided with Leibniz, trying to strengthen the relationist stance. Among these, Huygens, Berkeley and, most importantly, Mach between the second half of the 19th and the beginning of the 20th century. Mach's response to Newton's bucket thought-experiment, in particular, was that absolute acceleration is relative to the fixed stars, that is, to the universe as a whole. More generally, absolute space was systematically replaced by Mach with a fully relationist analogue, and the same goes for time. From the point of view of philosophical motivation, Mach (especially in his *Science of Mechanics* 1883) started from the observation that Newton departed in a striking way from his own key methodological tenet, according to which one should not go beyond observational facts. Based on this, Mach tried instead to truly and fully implement Newton's principle, which more or less immediately led him to dispense with absolute space and time altogether. In particular, Mach tried to make do with *relative distances* only, the key idea behind the Machian relationist approach to physics. However, Mach's project was not a success. Even though people such as Reichenbach went as far as to argue that Einstein's Theory of Relativity fully vindicated the Machian perspective, hence relationism,

[3] Newton also imagined a pair of globes connected by a rope and revolving about their centre of gravity. This he took to show that, despite the fact that absolute space is invisible to the senses, it is nonetheless possible to infer the quantity of absolute motion of individual bodies.

[4] Kant reinforced the substantivalist case by bringing considerations about 'chirality' to bear on the issue, and presenting them as providing further indisputable support to substantivalism. Using a famous example, Kant argued that in a space containing only a single hand, its being a right or a left hand could not be established on the basis of relational facts, but is an objective matter nonetheless.

it seems rather the case that the workability of a completely relational physics remained doubtful to say the least. First, as mentioned, Einstein's relativity was almost unanimously taken to show that space and time form a unitary whole and cannot be considered as separate entities. And such a unitary space-time progressively came to be regarded as a basic element in the context of the theory, *to be understood as a fundamental substance.*[5] More generally, modern-day relationists – exactly like Leibniz – seem to remain in any case unable to produce a 'relationally pure' physics, where everything we need to know about physical systems can be expressed without making any reference to an objective spatio-temporal background. In other words, substantivalism about space and time came to be regarded as correct, as an absolute background, i.e., an inertial structure and a temporal metric, appear necessary for our physics to work (see, e.g., Pooley and Brown 2002, Section 4).

However, in spite of the foregoing facts, which unquestionably give substantivalism about (space-)time some advantage, there are other elements to be taken into account, having to do with some of the most recent developments in physical theorising.

3. Contemporary Physics and Barbour's Machian Perspective

One of the fundamental tasks of contemporary physicists is that of putting quantum mechanics and General Relativity (GR) together in a theory of quantum gravity, so solving the obvious problem represented by the mutual incompatibility of the two theories when taken as they are currently formulated. Roughly speaking, there are two main approaches: some physicists give priority to quantum mechanics (superstring theories, for example, go in this direction), others regard relativity as more important, thus essentially attempting to quantize Einstein's GR. The focus here will be on the latter approach. Let us, therefore, look at it in more detail.

Canonical quantum gravity emerged in the 1950s and 1960s as a first attempt to employ the same techniques used to give a quantum formulation of electromagnetism in the case of gravity. In particular, in the late 1960s DeWitt formulated the basics of the theory by making use of previous work of Bergmann and, even earlier, Dirac. This kind of research developed in

[5] Although, for reasons already mentioned in the introduction, not one constituted by points provided with primitive, irreducible identities that could give rise to distinct, yet indiscernible, spatio-temporal arrangements in scenarios such as those contemplated in the hole argument.

a number of different directions. In so-called 'loop quantum gravity', for instance, space becomes discrete, a network of finite entities (the loops). Other options include the use of so-called 'Regge calculus', or of cosmological ideas developed by Hartle and Hawking. All these approaches share the 'problem of time': very briefly, time is not *presupposed* as a background 'container' but rather *described* by GR, and it turns out to disappear in the context of quantum gravity, as the wave-functional that supposedly describes the evolution of the universe does not change (more on this in a moment). This has led to the idea that physics, and the physical world itself, should be regarded as *timeless*. According to Rovelli, for example, as one moves to a deeper and deeper level of reality, time as a fundamental quantity disappears: mechanics turns out to be the theory of relations between variables, not a description of how things evolve in time, the latter being regarded as a somehow 'special' variable. Time only emerges at a later stage, more or less as a variable singled out as preferred given the specific state of the system.[6]

But it is Julian Barbour's views that deserve the most attention here. In work carried out in the last three decades or so, including his book (1999), Barbour translated the Machian idea of grounding the whole of physics on relations between (observable) quantities – in particular, as we have already seen, instantaneous relative distances – into the view that the only things that exist are configurations of physical systems – that is, of interrelated objects and properties; and that reference to absolute space and time can, and in fact should, be systematically interpreted in terms of relationships between different configurations. To support this view, Barbour noticed first of all that, although a bunch of particles obeying Newton's laws and their relative distances are not enough for reconstructing the entire sequence that we would identify with a 'history' of the relevant physical system – that is, with a physically possible evolution of that system – only a little more is required. In particular, while only referring to the space of configurations makes one unable to describe how physical evolution takes place (i.e., to specify the total kinetic energy and the 'orientation' of the dynamical behavior of the relevant system), there is a family of dynamical principles formulated on configuration space alone that allows one to predict a unique curve for each point in configuration space and direction of evolution from that point. These principles correspond to the

[6] More precisely, Rovelli puts forward a 'thermal time hypothesis', according to which when we identify a certain physical variable as 'time' we only make a statement about the statistical distribution that we use to describe a physical system at the macroscopic level, ignoring the full microstate. Nothing like an 'objective' temporal magnitude exists.

Jacobi principle formulations of dynamics. In a nutshell, these formulations implement a generalised principle of least action and identify possible histories with paths in configuration space that are of extremal length with respect to a metric defined on the basis of the metric structure of three-dimensional physical space. In this way, the geodesics of the newly defined metric correspond to inertial trajectories of particles. In this context, the configurations that can follow each other in the relevant sequences of (descriptions of) physical systems are determined by what Barbour calls 'best matching' – basically, a generalisation of Pythagoras' theorem applies, and the overall differences among distinct configurations with respect to all the quantities appearing in them turn out to be systematically minimised. Starting from this, to obtain a fully Machian physics one 'just' needs one additional element: namely, an assumption concerning the angular momentum of the universe. Indeed, there is a subset of the solutions to any Newtonian theory that indicates that, *provided that* the total angular momentum of the physical system under consideration (measured with respect to its centre-of-mass inertial frame) is zero, then *relative quantities are sufficient for a complete physical theory*.[7] In view of this, Barbour concludes, relationism is eventually fully vindicated by postulating that the total angular momentum of the universe is actually zero.[8]

Relationism seems to become even more natural when one moves beyond the classical domain. Together with Bruno Bertotti, Barbour showed that, once GR is formulated as the dynamical theory of the geometrical features of space coupled with that of matter fields, it turns out to have a purely Machian nature, and thus not to require any further treatment to satisfy the relationist desiderata. More specifically, when intended in the sense of geometrodynamics on superspace (the latter being the configuration space

[7] It also follows that isolated subsystems are correctly described by the full Newtonian theory, which straightforwardly accounts for the seeming greater naturalness of the non-relationist approach in the more realistic scenarios available to us, which clearly involve less than the entire universe.

[8] Or, maybe better, is not a well-defined quantity, as there is nothing with respect to which it can be measured. One might complain that what was just described is a very relevant assumption, and that the fact that it is necessary to make it shows that relationism is untenable, as it crucially depends on contingent features of the universe. However, it can also be argued (Belot 1999) that what may look like a very strong, and possibly even ad hoc, assumption is in fact a decisive prediction that provides a fundamental bit of empirical support to the theory. For, granted that relationism doesn't get off the ground if the universe as a whole is not 'static' in the above sense, one can regard the latter state of affairs as a consequence of the theory rather than an assumption, and proceed to see whether or not the ensuing prediction is correct, so effectively 'testing' the Machian approach. And Belot points out that there is evidence that our universe is in fact non-rotating.

constituted by the set of 'acceptable' geometries of space[9]), GR can be regarded as the theory of relationships between 3-geometries, and the relevant configuration space is an entirely relative configuration space. In a sense, then, GR seems to contain right from the start exactly the sort of action principle that the relationist had, as it were, to actively 'plug into' classical mechanics.[10]

Lastly, and most importantly, Barbour puts quantum considerations into the picture, and argues that quantum gravity too can be formulated in terms of relative configurations. This, he argues, requires one to give priority to the time-*independent* Schrödinger equation. That is, in the case of the universe as a whole the Wheeler-DeWitt equation

$$\hat{H}|\psi\rangle = 0$$

must be used, which is naturally interpreted as conveying the information that the wave-function of the universe is constant. Thus, it looks as though one has to do with a 'frozen formalism' that mirrors the fact that *the universe does not change* – not, of course, in the relatively uninteresting sense that it should be considered as a four-dimensional 'block', already familiar from Relativity alone – but rather in the sense that, given the Hamiltonian constraint, the relevant transformations in phase space do not affect the state, the physical observables having to be invariant with respect to such transformations.

The upshot is, then, that all (physically) possible states of the universe are 'ontologically on a par', and what seems to be a sequence of states in time is instead a completely different path *in a timeless space including all the possible 'ways the universe could be'*. In particular, Barbour endorses a 'many instants' interpretation that turns the Everettian many-worlds line of thought into the idea that all 'worlds' – he calls them 'Nows' – exist together – a sort of physically-motivated version of Lewisian modal realism. Is this view compelling? What else can be said about time, and relationism, in this framework? Can realism about time still be regarded as a viable conception? To answer these questions, we now turn to the more constructive part of the paper.

[9] In particular, the quotient space of all Riemannian metrics defined on the 3-manifold under the action of spatial diffeomorphisms that map different spaces into one another.

[10] More precisely, under certain conditions the orthodox four-dimensional action of GR can be put in a form (the BSW form firstly formulated by Baierlein, Sharp and Wheeler (see, for instance, their 1962) which is a particular case of the general form defined by Barbour.

4. Problems for Barbour, Definition (and Limited Defence) of an Alternative

The biggest problem for the supporter of the view that the universe is fundamentally timeless is to explain why we perceive it as evolving in time. What is Barbour's solution to this problem?

Barbour's many-instants view relies on the idea that the path connecting the various Nows is determined by best matching together with the probability distribution described by quantum mechanics. Crucially, Barbour *conjectures* that the quantum probability distribution is such that the most probable configurations, i.e., among other things, those which are most likely to be part of a trajectory *which involves human observers making experiences*, are those highly structured ones that contain 'time capsules'. The latter are physical subsystems that encode highly structured information, that our brains process as if it were information about Newtonian trajectories across canonical time.

According to Barbour, then, one should be an 'error theorist' about time: whenever we have the perception that something shifted from being possible to being actual, and from being present to having been present, we are in fact elaborating in our brains peculiar kinds of information that are, in actual fact, non-temporally 'written' in the physical configuration(s) in which we happen to find ourselves. To use well-known labels, Barbour follows McTaggart in accepting the existence of an objective but non-temporal structure in the physical domain (something like McTaggart's C-series) which grounds the relations of being 'earlier than', 'simultaneous with' and 'later than' holding between events (McTaggart's B-series) in virtue of a decisive contribution coming from the human mind, which is the only place where a fundamental distinction between future, present and past (the A-series of McTaggart) can be found. Crucially, though, while McTaggart's C-series is an objective ordering, a sequence of elements that follow one another in some sense, Barbour's so-called 'Platonia' is instead like a completed puzzle, all the pieces 'given together at once'. In light of this, it should appear clear that the idea of a time capsule and the assumption that quantum probabilities are peaked around appropriately structured Nows are essential in Barbour's framework.

While of course naturalists should be ready to endorse (in fact, should actively seek) an error theory of this sort whenever they find that one of our common sense and/or philosophical presuppositions is put into question by our best science, the problem here is that it is not entirely clear that the error theory provided in the present case is compelling. There are two reasons for this claim, one having to do with explanatory strength and the other with the physical basis of the proposal. Starting from the first point, it could be argued that the sort of Leibnizian 'temporal monadism' endorsed by Barbour – whereby time reduces entirely to the internal structure of the individual configurations – "does a fair job of capturing the central features of the *experience* of time" (Ismael 2002, p. 326). Ismael grounds this assertion on the fact that a lot of our everyday experience involving the past is indeed based on 'mementos', i.e., physical signs in the present that get the mechanisms of temporality started in our heads. However, Ismael herself acknowledges the problem that something being a record of something else presupposes the sort of causal relations that Barbour rules out, and thus it remains unclear exactly what makes an instantaneous configuration of the universe a time capsule, i.e., what mechanism is responsible for there being mementos in our experience. Developing a suggestion made in passing by Baron, Evans and Miller (2010, p. 53), one may add to this the following consideration: our temporal experience is primarily experience of change, but to be experience of change, such experience must itself change, as it cannot but consist of a 'diachronic process of interpretation', as it were, of the available data by our brains. Even more generally, we perceive our experience as a process and as itself changing, and this again seems to *presuppose* something like a temporal dimension and, possibly, to consequently open the way for a vicious regress – analogous to that employed by McTaggart to demonstrate the unreality of time, but now directed to Barbour's error theory! If the foregoing is correct, it follows that Barbour's view is unworkable: our temporal experience might well be illusory, but to be so it has to have certain 'dynamic' features that Barbour's picture doesn't seem to be able to reconstruct – exactly because it denies that physical reality is objectively ordered along *some* dimension.[11]

As for the second problem with Barbour's error-theoretic reconstruction of temporal experience, it can be stated quite briefly: Barbour never formulates a truly convincing argument in support of his claim that (some) solutions to the Wheeler-DeWitt equation give high probability to configurations containing time capsules – which is clearly crucial for his peculiar

---

[11] Analogous worries are raised by Healey (2002), who emphasises the fact that physical theorising essentially relies on observations and experiments that, it would seem, necessarily occur in time. Related to this, Baron, Evans and Miller also argue that Barbour's view threatens to lead to temporal solipsism as, according to it, we are confined to single points in configuration space and can never access other configurations. Time capsules, that is, only give us the (wrong) impression of being able to collect information about several distinct physical configurations. This criticism, however, seems to be off the mark, as it is exactly Barbour's intention to replace our traditional, commonsense understanding of our perceptual (in particular, temporal) experience in this way. On the other hand, it is true that the acceptance of solipsism represents a non-negligible cost for those accepting Barbour's views.

explanation of the 'psycho-physical parallelism' holding with respect to time and temporal experience. It is at least unclear whether this is compensated by the overall strength of the theory or, to the contrary, makes the proposed reconstruction of the psychological side of the issue irredeemably ad hoc because based on a mere conjecture.

In light of the above, one may start to think that, if at all possible, one should be a realist relationist (in particular, about time) rather than accepting Barbour's antirealism about time. That is, try to preserve the Machian perspective construed by Barbour without also accepting Barbour's eliminativist views, so avoiding the need for a consistent error-theory of temporal experience altogether. After all, it is an obvious, yet important, philosophical fact that relationism is a form of reductionism, but not all forms of reductionism amount to full-blown eliminativism.[12] At this point, then, it is necessary, first, to say more about what time could look like from a non-eliminativist viewpoint in the context of a Barbour-like quantum gravity. In connection to this, secondly, more should be said about the differentiated treatment of space and time that we have indicated as at least a conceptual possibility. There is a lot to be said of course, and, as mentioned earlier, we will not pretend to have a final solution (this would not only be overambitious, but also beyond the aims of this paper, which is only intended as a general overview of a particular area of interplay between traditional metaphysics and contemporary physics, with the addition of one or two more specific suggestions for further research).

Let us begin by looking again at the origin of the timelessness of quantum gravity, i.e., the Wheeler-DeWitt equation. While it is true that the route followed by Wheeler and DeWitt for quantizing GR with specific initial 'constraints' are normally taken to lead to a manifestly time-independent equation, this 'only' entails that the t variable does not refer to a fundamental physical magnitude. But of course this leads to eliminativism about time in the metaphysical sense only if one assumes that, if it is real, time must be a fundamental entity – which is exactly what the non-eliminativist relationist denies. Thus, one should explore the possibility of either

[12] Compare with Healey's (2002, p. 303) idea of replacing the 'Parmenidean' timeless view with the 'Lockean' view that time is a secondary quality. To be absolutely clear, the point being made is not that one should buy Barbour's views as a package and add the label 'real!' to them. Rather, the ideal aim is i) to re-establish the objectivity of the sequence of 'stages' that Barbour's 'temporal monadism' had eliminated (so replacing, as it were, the completed puzzle whose pieces do not and cannot communicate with each other with an ordered deck of cards which get uncovered one by one, the figure of each one of which depending on that on the previous one); ii) say more about the relevant relational structure, so recovering at least a minimal notion of physical time.

recovering time in some way from the available degrees of freedom (the 'internal time' approach) or 'imposing' it from outside as an additional element (the 'external time' approach). Either way, what one obtains is, essentially, the view that i) the intrinsic sequential ordering typical, say, of McTaggart C-series should be also attributed to the Machian relationist universe; and, additionally and crucially, ii) contrary to what McTaggart and Barbour would say, that objective structure is itself intrinsically temporal.[13] How can this be done?

The view I would like to put forward (once again, only as a suggestion for future work) is a sort of 'hybrid approach' between the internal time option and the external time alternative. The idea is that the observables that act as the relata of temporal relations are all the fundamental properties of physical systems; and that the relations that connect these relata with each other are nothing but Barbour's 'best matching' relations – which consequently end up coinciding with the canonical 'earlier than' and 'later than' relations.[14] An obvious objection is that this only amounts to naïve relationism, as a physically-respectable formulation of the view requires the related quantities to be gauge invariant, which they haven't been shown to be yet. A response to this objection can be given in the terms recently suggested by Barbour himself together with Koslowski and Mercati (2013). According to these authors, in the context of so-called 'Shape Dynamics' (a theory of gravity that implements Mach's principle, and in which the spacetime picture is replaced by a picture of evolving spatial conformal geometry) it is possible to define a variable τ with the desired features in terms of the overall expansion of the universe D. In particular, $\tau = D/D_0$

[13] Notice that the crucial assumption that the Machian has to make concerning the 'staticity' of the universe as whole can also be made in a non-eliminativist setting, regarding it as a constraint on the possible evolution of physical systems ordered in a linear series rather than a description of something that is true of them when they are all considered together (perhaps cosmology can provide a non-teleological elucidation of such a constraint). For one, Butterfield (2002) explicitly agrees with this separation of Barbour's relationism from his antirealism about time: "Barbour's views are by no means a package deal [... and his] denial of time, and speculations about quantum theory and quantum gravity [...can be] left on the shelf" (Ib.; 291). In particular, Butterfield explicitly notices that that based on eliminativism with respect to time is only one possible interpretation of the Wheeler-DeWitt equation.

[14] For each and every configuration in the overall relevant space there will exist either only one other configuration (or, at most, only one type of configurations) that 'follows' it as its best match, which naturally corresponds to a deterministic temporal evolution; or more (types of) configurations that, in different ways, match the initial configuration equally well, which is naturally interpreted in terms of non-deterministic, or at any rate 'chancy', temporal evolution. In any case, best matching intended as a relation holding between different physical configurations ordered in an objective sequences (rather than coexisting in a timeless 'Platonia') seems sufficient to reconstruct a canonical temporal sequence.

(with D_0 being the value of D at the point chosen to begin evolution), and $\partial f / \partial \tau = D_0\{H, f\}$ (with f being the relevant shape variable); and τ turns out to be a monotonic and dimensionless independent variable whose growth is determined 'by all physical degrees of freedom working together'. Importantly, in the words of Barbour, Koslowski and Mercati, the "variable τ appears as an internal time, but it is an external evolution parameter in the scale-invariant dimensionless description." Of course much more can and should be said, but this seems a promising avenue for the non-eliminativist relationist about time who intends to follow Barbour's suggestions for the development of a Machian physics and a working theory of quantum gravity.[15]

Having said this, let us move on to our second crucial question. Why draw on Barbour's relationism in order to be relationists about time only? Well, first of all let me remind the reader that the fundamental claim of this paper is that one can be a Machian relationist without accepting the idea that time is a mere illusion, while leaving it completely open what the metaphysical status of space is. Thus, the position being put forward is best intended as neutral with respect to space, and equally liable to be developed in the sense of a) a full-blown relationism about both time and space, and of b) a 'mixed' view whereby spatial substantivalism goes together with temporal relationism.

However, there are reasons for not regarding (a) above as the obvious choice, and indeed for regarding (b) as a serious alternative. As Pooley (2002) argues, and Rickles (2006) approvingly reports, Barbour's claims with respect to GR might be best interpreted in terms of substantivalism about space. In particular, Barbour's treatment of GR might be said to demand that the relative configurations taken to be fundamental be relational

---

[15] The ground for the work just mentioned was at least partly prepared by James W. York. In David Brown and York (1989), for instance, it is suggested that the usual action principle of General Relativity is analogous to Jacobi's form of the principle of stationary action, and that elaborating on this analogy one can arrive at a time-dependent Wheeler-DeWitt equation in which the role of physical time is played by the four-volume of space-time. On this note, at least two other recent works must be mentioned. Gryb and Thebault have recently elaborated upon the idea of a 'York ontology' and a 'York time variable' which is the canonical conjugate to the spatial volume. In their (2012), they argue more extensively that, once General Relativity is translated into the terms of shape dynamics as suggested by Barbour, modulo certain formal extensions of the theory a non-standard procedure of quantization can be implemented that leads to a dynamical theory of quantum gravity which retains a canonical temporal structure (while, however, not reintroducing an absolute, non-relational notion of duration). Lastly, on a different note, Okon and Sudarsky (2014) suggest that objective collapse models of quantum mechanics could usefully contribute to the solution of the problem of time by selecting a privileged frame of reference.

specifications of the properties of space (matter fields need not be present in all cases); and that this entails that such a Machian rendering of the theory in fact *requires* a substantival treatment of space.[16] Without entering a detailed discussion of whether this claim is correct, we can certainly say here that, if Pooley and Rickles are at least remotely on the right track, it immediately follows that the Barbour-inspired relationalist about time *may have good reasons* to a substantivalist about space (and, obviously enough, also reasons to refrain from saying anything about space). Whatever one decides, the fundamental point is that it is *physics itself* that shows that space and time may not be on a par, in terms of conceptual, formal and ontological standing, within Barbour's perspective. In particular, not only does this follow from the general fact that Barbour is an antirealist about time but not about space, together with the considerations just reported concerning substrantivalism about space possibly being the best interpretation of what Barbour says about GR. It also follows from the fact that ADM formulation of general relativity[17] which Barbour employs is based on the idea that the dynamics of the theory does not concern four-dimensional distances in a 4D block, but rather distances in three-dimensional space-like surfaces. This makes it possible (although, of course, not necessary!) to treat the diffeomorphism constraints and the Hamiltonian constraints of the theory differently, following a suggestion made, for instance, by Kuchar (1993).

In view of the above, it seems that we are now in a position to make at least the conditional claim that *if* one regards relationism about time (and perhaps *only* about time) as compelling, then such a view can be shown to be perfectly compatible with (some parts of) contemporary physics; and to even be supported by (a specific interpretation of) the latter, developed along the lines suggested by Barbour in his work in support of the Machian approach to physical theory. Obviously enough, it should not be forgotten that Barbour's vision is not mainstream, and indeed his proposal is incompatible with both standard GR and other, perhaps better developed,

[16] Of course, for reasons that we have already mentioned at the beginning of the paper, if this were the case substantivalism would then have to be made 'sophisticated' enough not to attribute haecceities to space(-time) points. This opens the way for the criticism, raised for instance by Belot and Earman (2001, pp. 248–249), that there is no third way between traditional substantivalism and relationism, and thus one should be a relationist about space too. Without entering into the details, this is exactly the position that Pooley attacks in his abovementioned (2002), and I agree that 'mild' substantivalism about space is in fact a viable option.

[17] That is, the Hamiltonian formulation of GR proposed and developed from 1959 onwards by American physicists Arnowitt, Deser and Misner.

approaches to quantum gravity. Nonetheless, that something counts as a conceptual possibility which is not straightforwardly ruled out by science is already an important fact from the philosophical viewpoint, at least in a broadly naturalistic context. Moreover, it should be acknowledged that, when it comes to assessing and choosing theories and hypotheses at the intersection between science and philosophy, no straightforward algorithm is present – let alone a direct derivation from scientific 'data' – and a careful consideration of different factors should instead be carried out. In this sense, one's independent, and more philosophical, reasons for being a relationist in general, and about time in particular, might outbalance the disadvantages of the chosen approach from the purely physical viewpoint. For sure, whatever one's personal views on these matters may be, only a cautious assessment based on a preliminary identification of the various options can lead to progress.

Having said this, let us now conclude by considering three other potential objection to Barbour's Machian perspective and/or its proposed reinterpretation in terms of realist relationism about time.

As we have seen, the geometrodynamic formulation of GR turned out to be the Machian version of the theory that Barbour was looking for, and in it the metric based on best matching between configurations makes it possible to ground the dynamics exclusively on relative dimensional configurations. This, however, has been taken (e.g., by Pooley 2001; Sec. 3.2) to lead to a problem of indeterminism. The idea is, essentially, that, contrary to what Barbour contends, there are many different sequences of configurations of the desired type (i.e., satisfying the least action principle in geometrodynamics) that can be 'extracted' from the canonical four-dimensional relativistic space. But such sequences constitute observationally indistinguishable 'histories'. Moreover, two sequences can be identical up to a point and radically differ afterwards and, consequently, the specification of an initial sequence appears insufficient for predicting the rest of the evolution of that sequence. A form of indeterminism thus emerges, analogous to that emerging in the context of the hole argument. This suggests that the traditional four-dimensional formulation of GR should, after all, be preferred to its Machian reformulation as the only framework that guarantees the needed uniqueness of configuration-sequences. However, the surplus degree of freedom might be disposed of by formulating the theory on so-called conformal superspace. This allows one to identify families of sequences in such a way that, as a matter of fact, a unique curve in relative configuration space is individuated in the relevant cases in any general relativistic space-time. This suggestion (see, for instance, the work by Barbour and O'Murchadha 2010) is still under study but it does seem to have the potential to eliminate the difficulty just pointed at. One might point out that, since the above amounts to adopting Leibniz-equivalence for spatial configurations, it naturally suggests relationism about space too, not only time. This might be the case but, first, it does not affect our claim that relationism about time only is at least a philosophical possibility which is in no way explicitly refuted by our best physics. Moreover, the sophisticated substantivalist, especially if s/he agrees with Pooley's above-mentioned claims about the need to interpret space in substantival terms in the context of Barbour's reconstruction of GR, can point out an analogy with the debate concerning the hole argument, and insist that the claimed equivalence has no obvious metaphysical consequences, as a consistent version of substantivalism can in any case be formulated.[18]

On a more general note, it could be objected that both geometrodynamics and the Wheeler-DeWitt equation are very questionable: the former was popular from the mid 1960s to the early 1980s, but has now been replaced by physicists with more sophisticated programmes such as, for example, loop quantum gravity; the latter has always been difficult to interpret and make sense of, even at the purely mathematical level. Couldn't these facts be sufficient for keeping away from approaches to the interpretation of physics that require one to put one's emphasis exactly on geometrodynamics and on the Wheeler-DeWitt equation? As may be expected, the reply is that this objection misses the key point: namely, that certain approaches to quantum gravity are being pursued because they have turned out to lead to some progress, and that since progress was based in this case on the use of geometrodynamics and the Wheeler-DeWitt equation, this is sufficient for re-establishing the respectability of the latter two elements. Indeed, as pointed out, for example, by Brown (1996, p. 197), quantum gravity is problematic in itself, and *every* attempt and approach in that context has its own difficulties. Given this, one could contend that – even admitting the difficulties he has to face – Barbour provides a new motivation and a new interpretation for the geometrodynamical programme (although, arguably, this does not yet put geometrodynamics on an equal footing with respect to the extant alternatives) and a new way of dealing with the Wheeler-DeWitt equation.[19]

[18] After all, if this were not the case, relationism would have been nearly-unanimously preferred to substantivalism already in the case of the traditional hole argument, which clearly not been the case.

[19] It shouldn't be forgotten, in this connection, that progress with respect to the Wheeler-DeWitt equation has been made: some (e.g., Smolin 2001) claim to have found solutions to it (the existence of such solutions is central, for example, in both loop quantum gravity and the path-integral approach to quantum gravity).

Another relevant issue, although of a rather different, more purely metaphysical nature, has to do with temporal vacua. While, strictly speaking, this is not a problem a philosopher of physics needs to be worried about, it becomes relevant when one attempts, at a more general level, to put together the strongest possible combination of a coherent metaphysical view and a workable interpretation of the relevant physics. Since this is exactly the level of discourse that the present paper is intended to be at, we will discuss the temporal vacua objection to relationism about time, albeit briefly, before closing. This will, at the very least, be instrumental to illustrating the 'other side' of the metaphysics-physics interplay as well.

With the aim of putting the originally Aristotelian conviction that time implies change into question, Shoemaker (1969) (see also LePoidevin 1991, pp. 94–98) considered a world consisting of three disjoint regions each one of which completely 'freezes' and remains changeless for a precise period of time at regular intervals. These intervals are different for the three regions, so as to entail a) that freezes in each region can be observed (or, better, indirectly reconstructed based on the available evidence) by the inhabitants of the other regions, who can then inform the inhabitants of the relevant region of what happened to their part of the world; and b) that some freezes can occur simultaneously in the three regions. Shoemaker argues that a) lends support to the idea that the global freezes suggested by b) are not only possible but also something that is reasonable to expect and regard as real given the available data. Of course, this means that in Shoemaker's imaginary world a global freeze does not imply that time stops (for, local freezes do not entail this, and global freezes are entirely analogous to local ones), and this seems to count against relationism. For, clearly, in such a world there are no changes, no relations between different events in terms of which the passing of time can be analysed, and yet – we just concluded – time does pass. Now, the mere conceptual possibility of temporal vacua can certainly be taken to add to the force of the substantivalist perspective on time. But there are answers available to the relationist. Against the temporal vacua objection, first of all, some (among others, Newton-Smith 1980, pp. 42–47, and Butterfield 1984) suggested a reformulation of relationism in modal terms, so that time is said to pass between two distinct instants t_a and t_b if and only if there is either an actual *or a possible* event occurring at an instant t_n in between t_a and t_b. Others argued instead that, upon scrutiny, it can be maintained that in fact "we are unable to conceive of a world about which it is clearly reasonable to claim that time passes but no events occur" and, thus, despite appearances to the contrary Shoemaker fails to prove his point (Warmbrõd 2004, p. 282). Last but not least, I take it that it is also possible to claim that change is in fact not required for real passage of time in the relationist framework, as sequences of *merely numerically distinct* events are sufficient. After all, the identity conditions of events do not analytically entail qualitative novelty and/or qualitative uniqueness, and if time is reducible to relations, why should it matter whether or not the relata of such relations are qualitatively distinct? If this is correct, the relationist can contend, contra Shoemaker, that time can in fact pass without change.[20] When this happens, s/he will add, for every thing that appears to persist 'frozen' in the relevant interval there in fact is a sequence of several events that are exactly similar qualitatively and yet non-identical.[21] This appears to be compatible, quite importantly, with the idea of grounding temporal succession on best matching and the evolution of physical observables along the lines suggested above. For on Barbour's construal it seems to be an open possibility that an instantaneous configurations best matches an absolutely indiscernible one (which, on the present proposal, plays the role of its temporal successor).

5. Conclusions

There seems to be some relatively unexplored space for manoeuvre at a particular point of the thin, and sometimes blurred, boundary between science and metaphysics. In particular, at the intersection between the related but distinct debates between relationism vs. substantivalism and '3+1' space and time vs. four-dimensional space-time. The present essay has attempted to identify this area of potential philosophical research, illustrate some of its features and put forward some suggestions. In particular, some recent work on the implementation of a Leibnizian-Machian relationist perspective to contemporary physics has been exploited with a view to recommending both a separation – at the metaphysical and physical level – of space and time, and a defence of relationism with respect to the latter only (remaining instead neutral with respect to space, if not suggesting substantivalism about it). Julian Barbour's recent work has been analysed in special detail, and his Machian perspective preserved while rejecting his anti-realism with respect to time. Besides being instrumental to a defence of a specific, intuitively more plausible, metaphysical perspective where

[20] An obvious consequence of this is, clearly, that relationism should *not* be formulated as the view that time is reducible to change, but rather as the view that temporal relations are reducible to physical relations between objects and properties, independently of whether or not the latter, as relata of the relevant relations, are qualitatively different from each other. This suggests a Leibnizian, rather than Aristotelian, conception of relational time.

[21] Interestingly, independent metaphysical arguments in favour of the primitive identity of events have been provided: see, in particular, Diekemper (2009).

time is not eliminated altogether, this has been presented as advisable on independent grounds, having to do with the strength of Barbour's account when it comes to making sense of our perception of time. Turning the 'flat and frozen' relational structure identified by Barbour into a genuine sequence and enriching it with further features, it has been argued, allows one to avoid antirealism about time while also augmenting the theory's explanatory power – most notably, with respect to temporal experience. The discussion aimed to nothing more than a first, quite general illustration of certain philosophically relevant facts and possibilities. Nonetheless, its outcomes appear sufficient for confirming the fruitfulness of the interplay between contemporary science, especially physics, and metaphysics, and to suggest that the relatively neglected option of relationism about time deserves more careful examination in the future (perhaps in connection with the other fundamental opposition in the metaphysics of time, namely, that between presentism and eternalism – or, more generally, A-theories and B-theories of time).

University of Rome III
Department of Philosophy, Communication and Visual Arts
e-mail: matteo.morganti@uniroma3.it

REFERENCES

Baierlein, R.F., Sharp, D.H., Wheeler, J.A. (1962). Three-dimensional Geometry as Carrier of Information about Time. *Physical Review* **126** (5), 1864–1865.

Barbour, J., (1999). *The End of Time. The Next Revolution in Our Understanding of the Universe.* London: Weidenfeld and Nicholson.

Barbour, J., Bertotti, B. (1982). Mach's Principle and the Structure of Dynamical Theories. *Proceedings of the Royal Society*, London, A **382**, 295–306.

Barbour, J., Koslowski, T., Mercati, F. (2013). *The Solution to the Problem of Time in Shape Dynamics.* Online at: http://arXiv:1302.6264v1.

Barbour, J., O'Murchada, N. (2010). *Conformal Superspace: The Configuration Space of General Relativity.* http://arxiv.org/abs/1009.3559.

Baron, S., Evans, P., Miller, K. (2010). From Timeless Physical Theory to Timelessness. *HumanaMente* **13**, 35–60.

Belot, G. (1999). Rehabilitating Relationism. *International Studies in the Philosophy of Science* **13** (1), 35–52.

Belot, G., Earman, J. (2001). Pre-Socratic Quantum Gravity. In: C. Callender and Huggett, N. (eds.), *Physics Meets Philosophy at the Planck Scale*, pp. 213–255. Cambridge: Cambridge University Press.

Brown, H.R. (1996). Mindful of Quantum Possibilities. *British Journal for the Philosophy of Science* **47**, 189–200.

Butterfield, J. (1984). Relationism and Possible Worlds. *British Journal for the Philosophy of Science* **35**, 101–113.

Butterfield, J. (2002). The End of Time? *British Journal for the Philosophy of Science* **53**, 289–330.

Callender, C. (2010). Is Time an Illusion? *Scientific American* **302** (6), 58–65.

David Brown, J., York Jr., J.W., (1989). Jacobi's Action and the Recovery of Time in General Relativity. *Physical Review D* **40** (10), 3312–3318.

Diekemper, J. (2009). Thisness and Events. *Journal of Philosophy* **106**, 255–276.

Earman, J. (1970). Who's Afraid of Absolute Space? *Australasian Journal of Philosophy* **48** (3), 287–319.

Gryb, S., Thebault, K. (2012). The Role of Time in Relational Quantum Theories. *Foundations of Physics* **42**, 1210–1238.

Healey, R. (2002). Can Physics Coherently Deny the Reality of Time. In: C. Callender (ed.), *Time, Reality and Experience*, pp. 293–326. Cambridge: Cambridge University Press.

Hoefer, C. (1996). The Metaphysics of Space-Time Substantivalism. *Journal of Philosophy* **93** (1), 5–27.

Ismael, J. (2002). Remembrances, Mementos, and Time-Capsules. In: C. Callender (ed.), *Time, Reality and Experience*, pp. 317–328. Cambridge: Cambridge University Press.

Kuchar, K. (1993). *Canonical Quantum Gravity.* arXiv:gr-qc/9304012.

Leibniz, G.W. ([1704] 1956). *The Leibniz-Clarke Correspondence*, edited by H.G. Alexander, Manchester: Manchester University Press.

LePoidevin, R. (1991). *Change, Cause and Contradiction.* London: MacMillan.

Norton, J.D. (2011). The Hole Argument. In: E.N. Zalta (ed.), *The Stanford Encyclopedia of Philosophy* (Fall 2011 Edition). URL = <http://plato.stanford.edu/archives/fall2011/entries/spacetime-holearg/>.

Newton-Smith, W.H. (1980). *The Structure of Time.* London: Routledge and Kegan Paul.

Okon, E., Sudarsky, D. (2014). Benefits of Objective Collapse Models for Cosmology and Quantum Gravity. *Foundations of Physics* **44** (2), 114–143.

Pooley, O. (2001). *Relationalism Rehabilitated? II: Relativity.* http://philsci-archive.pitt.edu/221/1/rehab2ps.pdf.

Pooley, O., Brown, H.R. (2002). Relationalism Rehabilitated? I: Classical Mechanics. *British Journal for the Philosophy of Science* **53** (2), 183–204.

Rickles, D. (2006). Time and Structure in Canonical Gravity. In: S. French, D. Rickles, and J. Saatsi (eds.), *Structural Foundations of Quantum Gravity*, pp. 152–195. Oxford: Oxford University Press.

Rynasiewicz, R. (1996). Absolute Versus Relational Space-Time. An Outmoded Debate? *Journal of Philosophy* **93** (6), 279–306.

Shoemaker, S. (1969). Time without Change. *Journal of Philosophy* **66**, 363–381.

Smolin, L. (2001). *Three Roads to Quantum Gravity.* New York: Basic Books.

Warmbrõd, K. (2004). Temporal Vacua. *Philosophical Quarterly* **54**, 266–286.

Antonio Vassallo

GENERAL COVARIANCE, DIFFEOMORPHISM INVARIANCE, AND BACKGROUND INDEPENDENCE IN 5 DIMENSIONS

ABSTRACT. The paper considers the "GR-desideratum," that is, the way general relativity implements general covariance, diffeomorphism invariance, and background independence. Two cases are discussed where 5-dimensional generalizations of general relativity run into interpretational troubles when the GR-desideratum is forced upon them. It is shown how the conceptual problems dissolve when such a desideratum is relaxed. In the end, it is suggested that a similar strategy might mitigate some major issues such as the problem of time or the embedding of quantum non-locality into relativistic spacetimes.

Notation: In the following, Einstein's convention will apply. Lower-case indices will be taken to range from 0 to 3, while upper-case ones will range from 0 to 4. A semicolon before one or more indices will indicate covariant differentiation with respect to those indices, a comma will instead indicate ordinary differentiation. Moreover, all equations will be written in natural units such that $c = G = 1$.

1. Introduction

General relativity (GR) is the most corroborated theory of gravitation we have so far. It is also considered a spacetime theory because it is taken to unify gravitational phenomena and physical geometry.[1] The dynamics of GR is encoded in the Einstein's field equations:

[1] I will not even try to address here questions regarding the meaning (if any) of the unification involved. The interested reader can take a look, for example, at Lehmkuhl (2008, 2014).

In: Tomasz Bigaj and Christian Wüthrich (eds.), *Metaphysics in Contemporary Physics* (*Poznań Studies in the Philosophy of the Sciences and the Humanities*, vol. 104), pp. 237-258. Amsterdam/New York, NY: Rodopi | Brill, 2015.

(1) $\mathbf{G}[\mathbf{g}] = 8\pi\mathbf{T}[\mathbf{g}, \boldsymbol{\varphi}]$.[2]

The left-hand side of the equation is the Einstein tensor defined over a 4-dimensional differentiable semi-Riemannian manifold M_4: since it depends on the 4-metric tensor **g**, it is taken to encode information on the 4-geometry of physical spacetime. The right-hand side comprises the stress-energy tensor, which encodes information such as the energy-momentum density of some matter field **φ** distributed over (a region of) spacetime. A solution (or model) of (1) is then a triple $\langle M_4, \mathbf{g}, \mathbf{T}\rangle$ that represents a physical scenario where the geometry of spacetime (aka the gravitational field) interacts with matter fields distributed over it. In short: "Space acts on matter, telling it how to move. In turn, matter reacts back on space, telling it how to curve" (Misner et al. 1973, p. 5).

Equations (1) are formulated in terms of geometrical field-theoretic objects, more precisely tensor fields, defined over a 4-manifold M_4. In order to make them simpler to handle for the sake of calculation, it is possible to introduce some coordinate system $\{x^i\}$ over a neighborhood U of M_4, and then rewrite them in terms of components of the geometrical objects in that coordinate system:

(2) $G_{ij}[g_{ij}] = 8\pi T_{ij}$.

Unlike (1), the equations compactly written as (2) involve derivatives and symmetric matrices, which simplify (but not *over*-simplify) the work of the physicists dealing with calculations. The important fact is that, by construction, the physical information encoded in the relations between field-theoretic objects as described by (1) remain unchanged when switching to (2), independently of the particular coordinatization chosen. For example, for two coordinate systems $\{x^i\}$ and $\{\bar{x}^i\}$ and two associated bases $\{e^i\}$and $\{\bar{e}^i\}$ on a neighborhood of a point $P \in M_4$, we have:

(3) $\mathbf{g} = g_{ij}|_P e^i \otimes e^j = \bar{g}_{ij}|_P \bar{e}^i \otimes \bar{e}^j$,

the (invertible) transformation law relating g_{ij} and $\bar{g}_{ij}$, being:

(4) $g_{ij} \rightarrow \bar{g}_{ij} = (\partial x^h / \partial\bar{x}^i)(\partial x^k / \partial\bar{x}^j) g_{hk}$.

Since we know from elementary tensor calculus that (3) holds for whatever tensorial object, it follows that it is possible to recover (1) from (2) for whatever coordinate system $\{x^i\}$ defined on a neighborhood of a point in the manifold.

[2] I choose to disregard the cosmological constant, since it does not affect the analysis carried out in this paper.

Further, it can be shown that to any (sufficiently smooth) coordinate transformation $\{x^i\}\rightarrow\{\bar{x}^i\}$ defined in (a neighborhood of) M_4 corresponds a self-diffeomorphism[3] f such that, for each point P in (a neighborhood of) M_4, it is the case that $x^i(f(P)) = \bar{x}^i(P)$.[4]

In the following, we will take advantage of the duality between the geometrical or intrinsic formulation of GR (involving geometrical objects and diffeomorphisms) and its local formulation (involving components of geometrical objects and change of coordinate systems), and we will switch back and forth from the intrinsic to the coordinates-related language for the sake of simplicity, according to the specific circumstances considered.

Nowadays, GR is so widely and firmly accepted by the scientific community to be considered one of the pillars of modern physics. At the root of such an overwhelming agreement lies not only the empirical success of the theory, but also its numerous applications to, just to mention two major fields, astrophysics and cosmology. In the words of Misner, Thorne, and Wheeler:

> Einstein's theory attracts the interest of many today because it is rich in applications. No longer is the attention confined to three famous but meager tests: the gravitational redshift, the bending of light by the sun, and the precession of the perihelion of Mercury around the sun. The combination of radar ranging and general relativity is, step by step, transforming the solar-system celestial mechanics of an older generation to a new subject, with a new level of precision, new kinds of effects, and a new outlook. Pulsars, discovered in 1968, find no acceptable explanation except as the neutron stars predicted in 1934, objects with a central density so high ($\sim 10^{14}$ g/cm^3)that the Einstein predictions of mass differ from the Newtonian predictions by 10 to 100 per cent. About further density increase and a final continued gravitational collapse, Newtonian theory is silent. In contrast, [GR] predicted [...] the properties of a completely collapsed object, a "frozen star" or "black hole." (Misner et al. 1973, pp. viii–śix)

However, the success of GR is not only due to its empirical adequacy and its richness in applications. Many physicists (and mathematicians) would in fact add that GR is an extremely elegant and mathematically beautiful theory. This further virtue of the theory should not be regarded as

---

[3] A self-diffeomorphism (from now on called simply diffeomorphism), is a mapping $f\colon M_4 \rightarrow M_4$ that is bijective, continuous, and differentiable together with its inverse f^{1}.

[4] Metaphorically speaking, while in the "coordinate" case we "keep P fixed" and we evaluate it (in the sense of associating a 4-tuple of real numbers to it) under a coordinate system *different* from the starting one, in the "diffeomorphic" case we "move P around the manifold" and then we evaluate it under the *same* coordinate system. For the reader unsatisfied by this, indeed, very rough characterization, Norton (2011) has a section that nicely explains the difference between extrinsic (or passive) transformations expressed in terms of coordinate systems, and intrinsic (or active) transformations expressed in terms of diffeomorphisms.

merely aesthetic, but it should be taken as a hint of the fact that GR points at some deep physical "truth" about the world. Dirac addresses this view as follows:

> Let us now face the question, suppose a discrepancy had appeared, well confirmed and substantiated, between the theory and observations. [...] Should one then consider the theory to be basically wrong? [...] I would say that the answer to the [q]uestion is emphatically NO. The Einstein theory of gravitation has a character of excellence of its own. Anyone who appreciates the fundamental harmony connecting the way Nature runs and general mathematical principles must feel that a theory with the beauty and elegance of Einstein's theory *has* to be substantially correct. If a discrepancy should appear in some applications of the theory, it must be caused by some secondary feature relating to this application which has not been adequately taken into account, and not by a failure of the general principles of the theory. (Dirac 1978, pp. 6, 7, Dirac's emphasis)

The questions about how, in general, a physical theory becomes accepted by the community, and what features should be privileged over others when assessing the merits of such a theory (e.g. empirical over formal), are beyond the scopes of this essay (see, for example, Brush, 1999 for a historical reconstruction and a philosophical discussion of the reasons that led to the acceptance of Einstein's theory). Here we will just make the working hypothesis that there is "something" physical *and* formal about GR (what Dirac calls the "general principles of the theory") that, according to mainstream physicists, places a constraint on future generalizations of the theory (to include, for example, quantum effects). In the remainder of the section, we will briefly consider some candidates for this "something," and we will finally propose a tentative definition that captures the "character of excellence" of GR that should therefore be preserved in any decent generalization of the theory.

To start our inquiry, let us consider the most straightforward candidate to be such "something": general covariance. With the designation "general covariance" here we will intend the formal invariance of the dynamical equations (2) under an arbitrary coordinate transformation. Einstein originally thought that general covariance so intended was some sort of generalization of the principle of relativity that holds in Newtonian mechanics and special relativity. Very simply speaking, while in these two latter theories the description of the dynamics of a physical system remained unchanged under coordinate transformations between inertial reference frames, so that these frames where considered physically equivalent,[5] in GR such

[5] The question of the equivalence between coordinate systems and frames of reference is another extremely delicate matter I will not touch upon here. See, e.g., Norton (1993), section 6.3.

a description is unchanged under whatever coordinate transformation, so one might claim that *all* reference frames are physically equivalent. Can the "virtue" of GR reside in its general covariance simpliciter?

The answer, as firstly pointed out by Kretschmann (1917), is no since it is possible to render generally covariant also a theory that patently accords a privileged status to inertial reference systems. Hence, general covariance cannot be taken as the implementation of the physical requirement that all reference frames are physically equivalent. To understand why it is so, consider a theory of a massless Klein-Gordon field φ over a manifold endowed with a Minkowski metric $\boldsymbol{\eta}$. We can write the field equations of the theory as:

(5) $\eta^{ij}\varphi_{,ij} = 0.$

Equation (5) is not generally covariant because it holds only in inertial coordinate systems.

In order to make it generally covariant we just have to make the purely formal move of rewriting it in a way such that it preserves its form in whatever coordinate system *and* reduces to (5) in those coordinate systems that are inertial.

This means that we have to write $\boldsymbol{\eta}$ in a generalized form $\mathbf{g}$ and introduce the (unique) compatible covariant derivative operator $\boldsymbol{\nabla}$,[6] so that (5) finally reads:

(6) $g^{ij}\varphi_{;ij} = 0.$

It is easy to understand that (6) is invariant in form under whatever coordinate transformation but, still, the theory – being specially relativistic – accords to inertial frames of reference a privileged status. In the language of geometrical objects, (6) is written

(7) $\Box_{\mathbf{g}}\varphi = 0,$

and, as expected, it is invariant in form (that is, covariant) under diffeomorphic transformations. Since such a move can be performed for whatever spacetime theory – even Newtonian mechanics – then (i) general covariance per se has no physical import and (ii) the generally covariant nature of GR cannot be linked to a peculiar feature such as a generalized principle of relativity. If Kretschmann's point is correct, then what is it that renders GR such a peculiar and successful theory?

The simplest reaction would be to claim that the peculiarity of GR does not reside just in the mere formal invariance of its field equations, but

[6] "Compatible" means that $\nabla \mathbf{g} = 0$.

also in the physical import of the solutions space of such equations. If, in fact, we consider two solutions of (6)/(7) related by a diffeomorphism f, they will represent one and the same physical situation if and only if f is an isometry[7] of the Minkowski metric. Not surprisingly, the group of isometries $iso(\boldsymbol{\eta})$ of the Minkowski metric is just the Poincaré group. In the case of equation (1), instead, since the metric g is not fixed a priori but it is subject to the dynamical evolution, the group $diff(M_4)$ is unrestricted and, as expected, whatever two solutions of (1) related by a diffeomorphism are physically indistinguishable.[8]

The above reply points at the fact that what makes the fortune of GR is not just its general covariance, but its diffeomorphism invariance: in virtue of having a dynamical geometry, GR has no fixed isometries different from the whole $diff(M_4)$ and, hence, diffeomorphisms are physical symmetries of the theory. Indeed, $diff(M_4)$ can be considered the *gauge group* of GR.

The question whether GR can be really considered a gauge theory in the sense commonly intended in particle physics is complex and highly debated (to have a substantive example of such debate, see Earman 2006; Pooley 2010); here we will just sketch a counter-reply to the above argument.

If the real difference between GR and previous spacetime theories is the dynamical status of spacetime's geometry, then we can just straightforwardly rewrite whatever spacetime theory by adding dynamical constraints on the spatiotemporal structures, so that such a difference can be obliterated. To see this, let us take the above theory of a Klein-Gordon field over a Minkowski spacetime and rewrite the dynamical equations as:

(8a) $\mathbf{Riem}[\mathbf{g}] = 0,$

(8b) $\Box_{\mathbf{g}}\varphi = 0,$

[7] Roughly speaking, $f \in diff(M_4)$ is an isometry for a geometrical object $\boldsymbol{\theta}$ if the application of such a transformation to the object does not change it: $f^*\boldsymbol{\theta} = \boldsymbol{\theta}$.

[8] Just to have a more concrete idea of what this means, consider a solution $\mathfrak{M}$ of (1) where we can construct a physical observable interpreted as a 4-distance function $d_{\mathbf{g}}: M_4 \times M_4 \to \mathbb{R}$ defined in terms of the metric $\mathbf{g}$. Let us now take a diffeomorphically transformed model $f^*\mathfrak{M}$, where $f \in diff(M_4)$; in such a model we can construct an observable diffeomorphically related to $d_{\mathbf{g}}$, namely a function $d_{f^*\mathbf{g}}$ defined by the diffeomorphic metric $f^*\mathbf{g}$. It is now easy to see that, taken two arbitrary points $P, Q \in M_4$, we have $d_{\mathbf{g}}(P, Q) = d_{f^*\mathbf{g}}(f(P), f(Q))$. It follows that no 4-distance measurement can discriminate $\mathfrak{M}$ from $f^*\mathfrak{M}$. By repeating the same reasoning for all the observables constructible in $\mathfrak{M}$ and $f^*\mathfrak{M}$, we can conclude either that the two models represent different physical possibilities which are empirically indistinguishable – and then we would immediately fall pray to arguments from indeterminism, such as the infamous "hole argument", cf. Earman and Norton (1987) – or that the two models represent one and the same physical scenario.

Riem[**g**] being the Riemann curvature tensor. The Minkowski metric, now, is a dynamical object picked up as a solution of (8a). The theory whose dynamics is encoded in (8) has the feature that two models related by whatever diffeomorphism represent in fact one and the same physical situation.

The immediate reaction to such an example is that it exploits a mere mathematical trick to render the transformations in $diff(M_4) \setminus iso(\boldsymbol{\eta})$ just *mathematical* symmetries of the theory and then mix them up to the *physical* symmetries in $iso(\boldsymbol{\eta})$ in order to patch them together into a fake gauge group.

In this particular case, a possible implementation of this counter-argument is, following Anderson (1967) and Friedman (1983), to point out that some geometrical objects have been artificially rendered dynamical and, thus, the theory can be deparametrized by quotienting out the redundant structures in the solutions space introduced by adding the fake dynamical sector.[9] Such a strategy makes sense in the present case: if we deparametrize (8) by quotienting out the solution space by $diff(\mathrm{M}_4) \setminus iso(\boldsymbol{\eta})$, we end up again with (7). If this reasoning could be carried out for all spacetime theories prior to GR, we would then have a reasonable candidate to be the peculiar feature of this latter theory, namely, the absence of spacetime structures that are not genuinely dynamical. Some authors would dub such a feature "background independence."

However, the above reasoning cannot be the final word in the debate. The present case per se is, in fact, too artificial to represent a really difficult challenge. Here the Minkowski metric represents an elephant in the room in the sense that we are perfectly able to see that such a structure "persists" unaltered in all models of the theory modulo a diffeomorphism transformation.[10] In general, however, making the distinction between genuine dynamical and pseudo-dynamical structures varies from difficult to impossible so that defining background independence in an accurate manner is a hard task (Pitts 2006; Giulini,2007, discuss at length the problem; see also Belot 2011 for an alternative account of background independence).

So, in the end, what is it that confers on GR a "character of excellence of its own"? Let us try to filter from the above discussion the salient uncontroversial claims regarding the features of GR, and come up with the following tentative characterization: Unless its predecessors, GR is a $diff(M_4)$-gauge theory whose only non-dynamical constraint is to be

[9] Just to have a rough idea of what adding redundant structure means when passing from (7) to (8), it is sufficient to point out that to each solution $\langle M_4, \mathbf{g}, \varphi\rangle$ of (7) corresponds a set of solutions $\{\langle M_4, \mathbf{g}, \varphi\rangle, \langle M_4, f^*\mathbf{g}, f^*\varphi\rangle, \langle M_4, h^*\mathbf{g}, h^*\varphi\rangle, \ldots\}$ of (8) with $f, h, \ldots \in diff(M_4) \setminus iso(\boldsymbol{\eta})$.

[10] Which means that, for whatever two models $\langle M_4, \mathbf{g}, \varphi\rangle$ and $\langle M_4, \mathbf{g}', \varphi'\rangle$, there is *always* some $f \in diff(M_4)$ such that $\mathbf{g}' = f^*\mathbf{g}$.

formulated over some 4-dimensional semi-Riemannian manifold, and that does not postulate more spatiotemporal structure than just that encoded in a Lorentzian 4-metric. If this is (or gets close to) what is at the root of the major success of GR, then we can consider it as a (minimal) constraint on any future (Quantum) spacetime theory that seek to supersede GR:

GR-desideratum. Any theory which is a valid candidate for superseding GR must be [become in the classical limit] a *diff*(M_d)-gauge theory whose only non-dynamical constraint is to be formulated over some *d*-dimensional semi-Riemannian manifold and that does not postulate more spatiotemporal structure than just that encoded in a *d*-metric.[11]

Note that, in general, besides arguments from physics, the GR-desideratum can be backed up by metaphysical arguments. One might for example argue that (1) is preferable to (8) because, in this latter theory, there is a "lack of reciprocity" between the Minkowski metric and the scalar field: in (8) spacetime tells matter how to move, but matter does not tell spacetime how to curve.

In the following, I will show how a too strict pursuit of the GR-desideratum might negatively affect the interpretation of prima facie genuine generalizations of GR.

2. 5-Dimensional Extensions of General Relativity

The Kaluza-Klein approach, so called because, historically, the first attempts in this sense were made by Theodor Kaluza (1921) and, slightly later, by Oskar Klein (1926), started as a theoretical program that sought to unify gravitational and electromagnetic forces as curvature effects of a 5-dimensional semi-Riemannian manifold. This is achieved by considering the 5-dimensional vacuum Einstein's equations[12]:

(9) $R_{AB}[g_{AB}] = 0,$

with the 5-Ricci tensor R_{AB} depending on a metric of the form:

(10) $g_{AB} = \begin{pmatrix} g_{ij} + \kappa^2\varphi^2 A_i A_j & \kappa^2\varphi^2 A_i \\ \kappa^2\varphi^2 A_j & \varphi^2 \end{pmatrix},$

[11] The g_{ij}-part of such a metric having Lorentzian signature.

[12] In the 4-dimensional case, if we take (1) and we set the stress energy tensor equal to zero (that is, no matter is present), we obtain that the equations reduce to **Ric[g]** = 0, **Ric** being the so-called Ricci tensor. This is why they are in general said to be "vacuum" field equations.

where **A** is the 4-vector potential, φ is a scalar field, and κ is a free parameter. The signature of the 5-metric will be assumed to be Lorentzian, thus identifying the extra fifth dimension as space-like.

The original Kaluza-Klein theory placed two very strong constraints on the fifth dimension, namely, (i) that all partial derivatives with respect to the fifth coordinate are zero (cylinder condition), and (ii) that the fifth dimension has a closed short-scale topology (compactification condition).[13] The most important consequences of these conditions are that there is no change in 4-dimensional physical quantities that can be ascribed to the presence of an extra spatial dimension and that such a fifth dimension is unobservable at low energies. Condition (ii) was also a vital ingredient in the attempt to explain the quantization of electric charge.[14] En passant, it is interesting to note that condition (ii) obviously prevents whatever macroscopic object from spanning the fifth dimension.[15]

If we set $\varphi = 1$ and $\kappa = 4\sqrt{\pi}$ in (10), and we substitute it in (9), the ij-components of the field equations become:

(11) $G_{ij} = 8\pi T_{ij}^{EM}$,

with G_{ij} the Einstein 4-tensor and $T_{ij}^{EM} = (1/4)g_{ij}F_{kl}F^{kl} - F_i^{\ k}F_{jk}$ the electromagnetic stress-energy 4-tensor of standard GR. The i4-components, instead, become:

(12) $F_{ij}^{\ \ ;i} = 0$,

with $F_{ij} = A_{j,i} - A_{i,j}$ the Faraday 4-tensor. In short, (9) collapse into Einstein's field equations of GR and the (source free) Maxwell's equations of electromagnetism coupled together. This is the so called "Kaluza-Klein miracle" and represents a remarkable result in that it "geometrizes away" the electromagnetic field as GR does with gravity, thus suggesting that both gravity and electromagnetism are just a manifestation of the geometry of a 5-dimensional spacetime. Is this a right suggestion? Let us consider a reason why it does not seem so.

---

[13] To be fair to historians of physics, (i) was proposed by Kaluza, while (ii) was added later by Klein, who further showed that, in fact, (ii) implies (i).

[14] Attempt which experiments proved unsuccessful: According to the Kaluza-Klein theory, the electron had a mass of twenty-two orders of magnitude higher then the measured one.

[15] The reader is invited to think about the strange consequences for spatial 3-dimensional beings like us of living in an uncompactified 4-space. For example, if we were allowed to move in an extra spatial dimension, we could "switch" the left and right sides of our body without altering the up-down and front-back orientations. Moreover, from the perspective of the extra dimension, the interior of our bodies would be exposed (our skin represents just a 3-dimensional boundary).

In order to see what is fishy about the interpretation of the Kaluza-Klein theory as a genuine 5-dimensional theory, we just have to notice that conditions (i) and (ii) restrict the possible topologies of the 5-manifold to $M_4 \times S^1$, with M_4 whatever topology compatible with the *ij*-part of the metric tensor (10) and S^1 the topology of the circle. This hints at the fact that, as long as we restrict our attention to the *ij*-part of the theory, we are dealing with standard GR and, hence, whatever coordinate transformation $x^i \rightarrow \bar{x}^i = f(x^i)$ we apply to the metric (10), we always end up with the same equations (11) *and* we do not change the physical information originally encoded in the 4-metric before the transformation. What about the components of the metric depending on the fifth coordinate? In this case, the freedom in choosing the coordinate transformations that leave (12) invariant in form and do not alter the physical information encoded in (10) is severely limited. In fact, the only possible choice that we can make is $x^4 \rightarrow \bar{x}^4 = x^4 + f(x^i)$. To realize why it is so, we have just to plug this transformation in the 5-dimensional extension of the law (4) for the transformation of the metric tensor: it is easy to see that such a transformation induces just one change in the metric (10), namely, $A_i \rightarrow \bar{A}_i = A_i + (\partial f(x^i)/\partial x^i)$, which obviously does not alter the electromagnetic part of the theory. In short, the gauge group of the theory is not $diff(M_5)$ but $diff(M_4) \times U(1)$.

If, now, we claim that a genuine 5-dimensional extension of GR should admit $diff(M_5)$ as gauge group, we cannot but reach the conclusion that the Kaluza-Klein theory is a fake 5-dimensional theory in that it is a version of 4-dimensional GR that camouflages the gauge group of electromagnetism by "spatializing" the symmetries of the theory using an additional space-like dimension: there is no substantial unification here, just a mere algebraic play.[16]

Note that the above reasoning is based on the simple fact that the conditions of cilindricity and compactification are fixed a priori, i.e., they are not part of the dynamics of the theory. Simply speaking, the theory is not really dealing with solutions of (9) but with a small subset of it, namely Ricci-flat spacetimes with topology $M_4 \times S^1$ where, additionally, S^1 has to "appear" at short scales. This situation is nothing but a more complex instance of that occurring with the theory (6)/(7). It is then clear that we can in principle modify the field equations (9) so that they can encode

[16] This line of reasoning can be found, for example, in Maudlin (1989, p. 87).

cilindricity and compactification conditions[17] in order to have a fully *diff*(M_5)-invariant theory, exactly as (8) is a *diff*(M_4)-invariant extension of (6)/(7).

Such a move is exactly as controversial as the one discussed in the previous section, but can it be charged of hiding an elephant in the room as it was the case with the Minkowski metric in (8)? Well, if we hold the firm commitment to the GR-desideratum, then we can clearly hear the $M_4 \times S^1$ structure trumpeting beneath the dynamics. Notice, however, that in this case the GR-desideratum cannot be backed up with compelling metaphysical arguments as, for example, the "reciprocity" one mentioned at the end of the previous section. Here we are dealing with an empty spacetime, so there is nothing acting on anything without being affected in return. Hence it would not be that weird to claim that the dynamically extended version of the Kaluza-Klein theory (9) just depicts spacetimes that are, *as a bare matter of fact*, more structured then a generic (empty) general relativistic spacetime, in that it adds to such generic spacetime an extra-dimensional tiny ring for each point. The counter-reply would be that, in this case, such a theory would be unable to explain *why* spacetime has to have such a structure. This is fair enough, but any theory has to start from some assumptions in order to be developed: even standard GR does not explain why spacetime has to be 4-dimensional and Lorentzian; it just takes as a bare fact that special relativity holds at quasi-point-sized regions. The whole discussion, then, seems to turn into metaphysical tastes about what has to be counted as a bare fact and what has not. We could, for example, endorse some variant of the regularity account of laws where the dynamical laws of the Kaluza-Klein theory just supervene on a pure spatiotemporal Humean mosaic of local matters of particular fact. In such a context, asking why the Kaluza-Klein 5-spacetime has this structure is a misleading question: it just happen to show a pattern of regularities that is best described by the equations of the theory.

Let us recap: if we insist on the GR-desideratum, then we are driven to judge the Kaluza-Klein theory as a mathematical trompe-lœil that spatializes the symmetries of the electromagnetic theory, so it cannot count as a genuine unification of gravity and electromagnetism. However, such a judgment seems too harsh and does not involve in any way the empirical adequacy of the theory (whose failure is the only uncontroversial argument that historically led to the dismissal of the Kaluza-Klein theory). It

[17] For example, by adding some conditions that pick up at each point of 4-spacetime a fifth-dimensional-like vector field whose integral curves are circles with small radii.

is in fact possible to make perfect physical and metaphysical sense of the Kaluza-Klein theory as a unified theory by relaxing the GR-desideratum.

The moral to draw from the discussion so far is that, all things considered, perhaps the GR-desideratum is not that strict a desideratum: we can relax such a requirement without necessarily falling into patently artificial theories such as (6)/(7). But let us go ahead in the analysis, and show a case where the GR-desideratum is not only too rigid, but has also metaphysically odd consequences.

Since the strongest source of resistance to the idea that the Kaluza-Klein theory represents a genuine 5-dimensional theory lies in the conditions of cilindricity and compactification, let us now consider an implementation of the Kaluza-Klein program[18] that dispenses with such constraints. This is the case of induced matter theory, especially known in its "space-time-matter" version as firstly put forward by Wesson (1984). This theory (or, better, class of theories) rests on the mathematical result that any analytic N-dimensional [semi-] Riemannian manifold can be locally embedded in a $(N + 1)$-dimensional Ricci-flat [semi-] Riemannian manifold (Campbell-Magaard theorem, see for example Romero et al. 1996); hence, the field equations of the theory are, again, (9).

If we now write a 5-metric in its diagonal form:

$$(13)\quad g_{AB} = \begin{pmatrix} & & & & 0 \\ & & & & 0 \\ & & g_{ij} & & 0 \\ & & & & 0 \\ 0 & 0 & 0 & 0 & g_{44} \end{pmatrix}$$

and we substitute it in (9), we obtain again the 4-dimensional Einstein's field equations plus an expression for the 4-stress-energy tensor of the form (see Wesson and Ponce de Leon 1992, for the detailed calculations):

$$(14)\quad T_{ij} = \frac{(\phi_{,i})_{;j}}{\phi} - \frac{1}{2\phi^2}\left\{\frac{\phi_{,4}g_{ij,4}}{\phi} - g_{ij,44} + g^{kl}g_{ik,4}g_{jl,4} - \frac{g^{kl}g_{kl,4}g_{ij,4}}{2} + \right.$$
$$\left. + \frac{g_{ij}}{4}\left[g^{kl}{}_{,4}g_{kl,4} + (g^{kl}g_{kl,4})^2\right]\right\}$$

where, for notational simplicity, it is assumed $g_{44} = \varphi^2$.

[18] See Overduin and Wesson (1997) for an extensive survey of Kaluza-Klein gravity.

From these constructions it follows that, whenever we consider a 4-hypersurface Σ_4 by fixing $x^4 = const.$, we obtain a 4-metric g_{ij} and a stress-energy tensor T_{ij} both well-defined on Σ_4.

Just to have a rough idea of how this works, let us consider an example taken from Ponce de Leon (1988) and, in order to simplify the notation, let us follow the author in naming the coordinates as follows: $x^0 = t$, $x^1 = x$, $x^2 = y$, $x^3 = z$, $x^4 = \psi$. In case of a homogeneous and isotropic 5-dimensional universe, a class of 5-line elements (Parametrized by $\alpha \in \mathbb{R}_{\neq 0}$) correspondent to 5-metrics which are solutions to (9) is:

(15) $ds^2 = \psi^2 dt^2 - t^{2/\alpha}\psi^{(2/1-\alpha)}(dx^2 + dy^2 + dz^2) - (\alpha^2/(1-\alpha)^2)t^2 d\psi^2.$

It is easy to see that, if we restrict to 4-hypersurfaces ($\psi = const.$, $d\psi = 0$), (15) is reduced to:

(16) $ds^2 = dt^2 - R^2_{\alpha}(t)(dx^2 + dy^2 + dz^2),$

which is just the family of Friedman-Lemaître-Robertson-Walker metrics in 4-dimensions corresponding to flat 3-geometries. Moreover, if we substitute (15) in (14), we recognize that the stress-energy tensor resembles that of a perfect fluid with density ρ and pressure p given by:

(17a) $\rho = 3/(8\pi(\alpha\psi t)^2),$
(17b) $p = (2\alpha/3 - 1)\rho.$

Hence, by fixing α and ψ we obtain different possible FLRW 4-models, with state equation for matter given by (17). For example, for $\alpha = (3/2)$, we would have a 4-dimensional universe filled with dust, while for $\alpha = 2$ we would have one filled with radiation.

The moral to be drawn from the above facts is that matter and energy are 4-dimensional physical properties of spacetime "induced" by the 5-metric tensor. Therefore, the cosmological solutions of (9) all agree in giving us a picture of a 5-dimensional empty universe whose restriction to a 4-surface is a spacetime curved by matter.

In conclusion, it seems that now we have a truly 5-dimensional generalization of GR: the theory is *diff*(M_5)-invariant and there are neither background geometrical objects camouflaged as dynamical, nor bizarre restrictions on the topology of 5-spacetime. However, such a complete fulfillment of the GR-desideratum does not come for free.

To see what is the price to be paid, consider, for example (Wesson 1995), a 5-dimensional Minkowski spacetime. The correspondent line element in polar coordinates reads:

(18) $ds^2 = dt^2 - dr^2 - r^2 d\Omega^2 - d\psi^2,$

whose sections $\psi = const.$ are 4-dimensional Minkowski spacetimes.

However, if we consider the transformation:

(19) $t' = t,\ r' = (r/\psi)(1 + (r^2/\psi^2))^{-1/2},\ \psi' = \psi(1 + (r^2/\psi^2))^{1/2}$,

then (18) becomes

(20) $ds^2 = dt'^2 - \psi'^2((dr'^2/1 - r'^2) + r'^2 d\Omega^2) - d\psi'^2$,

whose sections $\psi' = const.$ are FLRW spacetimes.

Using the intrinsic language, we can think that (18) belongs to a 5-model $\mathfrak{M} = \langle M_5, g_{AB}\rangle$, and that the transformation (19) is dual to a 5-diffeomorphism f, such that the metric (20) belongs to the diffeomorphic model $f^*\mathfrak{M} = \langle M_5, f^*g_{AB}\rangle$. We therefore have two 5-models which are diffeomorphically equivalent, but that induce 4-models, which are empty in the former case and non-empty in the latter. What is the physical meaning of this fact? None, and that is exactly the problem.

If, in fact, we claim that f is a gauge transformation, then we cannot but accept that $\mathfrak{M}$ and $f^*\mathfrak{M}$ are physically indistinguishable, which means that (18) and (20) carry exactly the same physical information, namely, that related to the 5-geometry of spacetime. The fact that we are free to convey such an information by writing the line element either in the form (18) or (20) means that such freedom has no physical consequences: this is exactly what physicists refer to as "gauge freedom." The puzzling consequence is that it makes no physical difference whether we take the 5-spacetime under consideration as a "pile" of empty Minkowski 4-spacetimes or matter-filled FLRW 4-spacetimes, since, in the end, the only thing that counts for the theory is the 5-geometry of spacetime. Crudely speaking, the theory is "blind" to whatever change in 4-spacetimes (as long as their piling up leads to the same 5-spacetime), and if we insist that $\mathfrak{M}$ and $f^*\mathfrak{M}$ are ontologically distinct albeit physically indistinguishable, we commit the theory to indeterminism, as already mentioned in footnote on page 242. For let us assume for the sake of argument that the dynamics of the theory under scrutiny can be cast in a (4 + 1) fashion, that is, that a 5-geometry can be recovered by considering a generic foliation of 4-hypersurfaces, specifying some initial data on one of such surfaces, and then evolving these data on the subsequent leaves of the foliation. In this case, the gauge freedom of the theory – as hinted above – would be translated into the freedom of choosing such a foliation without altering the dynamical evolution. Now let us consider a foliation $\mathcal{F}$ whose folios are Minkowski 4-spacetimes and another one $\mathcal{F}'$ which is identical to $\mathcal{F}$ except for the fact that some leaves in-between are FLRW 4-spacetimes. If we claim that $\mathcal{F}$ and $\mathcal{F}'$ are ontologically distinct, then so will be the corresponding dynamical evolutions and, hence, we are forced to accept that, even by specifying with arbitrary precision the data on an initial Minkowski 4-surface (which is common to

$\mathcal{F}$ and $\mathcal{F}'$), the theory is unable to single out a unique dynamical evolution between $\mathcal{F}$ and $\mathcal{F}'$, which means that the theory is indeterministic. Such a situation is of course avoided if we take $\mathcal{F}$ and $\mathcal{F}'$ to represent one and the same physical situation.

In conclusion, since the theory is $diff(M_5)$-invariant, all the quantities that are not invariant under 5-diffeomorphisms are just "gauge fluff" that can be fixed at will without changing the physical information conveyed by the theory. This is the case for all the 4-dimensional quantities extracted by 5-dimensional ones by fixing the gauge (for example by setting $\psi = const$): under a different gauge fixing, in fact, we would obtain different 4-quantities from the same 5-ones. Of course, the 4-dimensional stress-energy tensor (14) is among the non-gauge invariant quantities.

If we take seriously this picture, then we cannot but claim that, in this theory, matter is just an unphysical illusion which merely depends on the 4-dimensional "perspective" from which 5-dimensional spacetime is seen, just like a holographic sticker that shows a different image depending on how it is inclined with respect to the light source. One could say that the unreality of matter in this theory already stems from the fact that, at the fundamental level, there is nothing but 5-spacetime, but that would be a half-truth. While in fact we can agree that according to equations (9) *at the fundamental ontological level* there is nothing but 5-dimensional geometry, this does not rule out the possibility that matter configurations *supervene* on such a geometry, and hence are real albeit ontologically non-fundamental.

Naively, we could solve such a "problem of matter" just by declaring our firm pre-theoretical commitment to the existence of matter as a substance and hence arguing for the dismissal of induced matter theory. This line of reasoning, however, would be weird to say the least, since it pushes us to reject a well-defined and consistent physical theory *based only* on metaphysical considerations. The other option would be to grant the theory physical dignity, in the sense that we should judge its validity based on its empirical adequacy. This second choice is potentially even worse, because it would drive us into something that smells like a case of empirical incoherence[19]: the truth of this theory would undermine our empirical justification for believing it to be true. This is because, if induced matter theory fulfills the GR-desideratum, there is no way to account for whatever measurement in a 5-diffeomorphic invariant way.

[19] The definition of empirical incoherence given here is taken from Huggett and Wüthrich (2013, section 1), who follow in turn Barrett (1999, section 4.5.2).

Such a situation is all the more strange for the following reason. Since induced matter theory rests on the Campbell-Magaard theorem, GR is embedded in such a theory already at the mathematical level, which means that the former is a formal generalization of the latter and hence, if we forget for a moment the GR-desideratum, all the empirical tests that corroborate GR, would corroborate induced matter theory. Moreover, the theory would give novel testable predictions, mainly related to deviations in the geodesic motion of 4-dimensional objects due to the presence of the uncompactified fifth dimension (see, e.g., Wesson 2006).

In order to find a solution to the problem of matter compatible with the GR-desideratum, some coherent account could in principle be put forward that saves the reality of 4-dimensional material entities (including measuring devices), perhaps by appealing to some distinction between "partial" and "complete" observables *à la* Rovelli (Rovelli 2002), but that would look more like a patch rather than a real solution. Of course, we can prevent the problem from happening by putting in the theory 5-dimensional matter, thus retrieving a 5-dimensional analog of Einstein's field equations (1)/(2). However this would not count as a genuine solution because (i) it would just amount to dismissing induced matter theory in favor of another one, and (ii) it would not straightforwardly wash away any "perspectival" character from 4-dimensional matter: such a "full" 5-dimensional generalization of GR would for example fire up a metaphysical debate between 5- and 4-dimensionalists that would be very similar to that between 4- and 3-dimensionalists in standard relativistic physics (for an example of the literature, see Balashov 2000; Gilmore 2002).

Be it as it may, the moral to draw from the above analysis is that, in the induced matter theory, the GR-desideratum has rather undesired consequences. Hence, if we are willing to solve the problem of matter without dismissing induced matter theory, we can do it in a straightforward way, i.e., by relaxing the GR-desideratum.

A possible strategy in this direction would be simply to add more structure to 5-spacetime, for example a 5-flow in spacetime. This would amount to selecting a privileged fifth-dimension-like vector field $\boldsymbol{\psi}$ with components $\psi_A \equiv (0, 0, 0, 0, \psi)$[20] by adding to the field equations (9), the following ones:

(21a) $g^{AB}\psi_A\psi_B = -1,$
(21b) $\psi_{A;B} = 0.$

[20] Or, more precisely, such that the ij-part of $g^{AB}\psi_A$ is identically null.

It is easy to see that ψ induces a privileged decomposition of the 5-manifold M_5 into 4-hypersurfaces that are normal to the vector field and, hence, spacetime-like: each 4-slice of the foliation will have equation $\psi = const.$ For this reason, equations (21) restrict the possible topologies compatible with (9) to $M_5 = M_4 \times \mathbb{R}$. In a sense, we are back to the case of the compactified the Kaluza-Klein theory: we are using (21) as a dynamical constraint on the possible topologies. But there is more to that: (21) tells us that not only M_5 has to be foliable by spacetime-like hypersurfaces, but that there is, for each model, a *distinguished* way to foliate it. In short, according to (21), M_5 is a "pile" of 4-spacetimes.[21]

Under this new reading, a 5-model $\mathfrak{M}$ of the theory will not be just a couple $\langle M_5, g_{AB}\rangle$ but a *triple* $\langle M_5, g_{AB}, \psi_A\rangle$. Let us return to the two models whose line elements are (18) and (20), which are related by the intrinsic transformation $f \in diff(M_5)$ (dual to the extrinsic transformation (19)): we see that they correspond to the new situation where the first model is, say, $\langle M_5, g_{AB}, \psi_A\rangle$ and the second model is $\langle M_5, f^*g_{AB}, \psi_A\rangle$. We immediately notice that, due to the introduction of the structure ψ_A, the two models represent *distinct* physical situations unless $f^*\psi_A = \psi_A$.

Also in this case, we can be accused of cheating – not once but twice! Firstly, because we are not considering the starting theory but a brand new one that admits a structure (namely, a privileged foliation). Secondly because we are hiding a metaphysically suspicious background structure behind (21).

As regards the first charge, the plea is: guilty. It is undeniable that the starting theory dealt just with a generic 5-metric (13) solution to the field equations (9) and nothing more (e.g. no cilindricity or compactification conditions), but this is exactly at the root of the problem of matter: a 5-dimensional theory just resting on (13) and (9) has simply not enough structure to ground claims about the existence of 4-matter. Can a physical theory that places dynamical constraints so weak to risk to be empirically incoherent be considered really a theory in any useful physical sense? Under this reading, adding further structure to such a theory is more like *completing* it, rather than merely changing it.

As regards the second charge, the plea is: not guilty. Although ψ technically is a background object as the Minkowski metric in (8), still it cannot be charged of being metaphysically suspicious. While, in fact, it is correct to say that ψ affects the 4-matter distributions in each model without being

[21] In this case, 4-matter would still supervene on 5-dimensional state of affairs, such as the rigging of the pile (e.g., a certain matter configuration would not have existed having the 5-flow been different).

affected in return, still the "influence" we are talking about is not a physical interaction (ψ does not really push 4-matter in any physical sense), but an ontological one: in each model 4-matter supervenes on the 5-dimensional structure comprising a 5-metric **g** *and* a vector field ψ.

To sum up, it is true that ψ represents an additional and "rigid" structure besides the 5-metric but, as in the Kaluza-Klein case, there is nothing particularly wrong in that as long as we do not mind relaxing the GR-desideratum. Indeed, the payoff for this relaxation is huge: the problem of matter just vanishes.

3. Conclusion

The starting point of this paper was considering the possibility that, although general covariance, diffeomorphism invariance, and background independence are not features uniquely ascribable to GR, there is something in the way GR implements such features that renders this theory better than its predecessors. We tentatively identified this "something" in the GR-desideratum, and we then discussed two cases in which forcing such a desideratum on theories that seek to generalize GR leads to conceptual difficulties. We finally suggested to overcome these difficulties just by relaxing the GR-desideratum, in particular by introducing further (dynamical) spatiotemporal structures besides the metric. From what has been discussed in section 1, it appears clear that questioning the GR-desideratum is not questioning general covariance, diffeomorphism invariance, or background independence by themselves, but the necessity of implementing these features in the way GR does.

The analysis developed in this paper is far from being just an otiose conceptual exercise, since fully understanding what the GR-desideratum really is might shed light on two huge open problems in the philosophy and physics of spacetime theories. Namely, (i) the problem of time in Hamiltonian GR and (ii) the problem of constructing a theory that combines the main tenets of relativistic physics with the empirically proven non-locality of quantum theories.

The first problem (debated for example in Earman 2002; Maudlin 2002), roughly speaking, arises when we decompose the 4-dimensional spacetime of GR by means of space-like 3-surfaces in the so-called ADM formulation of Hamiltonian GR (Arnowitt et al., 1962). The idea behind this (3 + 1) decomposition is that we can render the general relativistic dynamics simpler for calculational purposes by using the machinery of Hamiltonian dynamics. In such a setting, we specify a set of initial conditions on a starting 3-surface and we evolve it using Hamilton's equations of motion. Of course, such a decomposition is merely formal: it is just a convenient way to recast the standard dynamics of GR. However, there is a conceptual problem lurking beneath the formalism.

To see this, let us consider some arbitrary model of GR; since the (3+1) decomposition is merely formal, we can imagine two different ways to foliate the model by means of the foliations $\mathcal{F}$ and $\mathcal{F}'$. Let us further assume that the foliations agree only on the initial surface Σ_0.

We are now in the strange situation already experienced in the previous section: even by specifying with arbitrary precision the initial data on Σ_0, the dynamics is unable to single out one evolution among $\mathcal{F}$ and $\mathcal{F}'$. Does this imply indeterminism? Of course no, because both evolutions represent one and the same physical situation. However, if we take the freedom to foliate the model as a gauge freedom, then we are forced to say that $\mathcal{F}$ and $\mathcal{F}$' are physically indistinguishable, which means that the only physically meaningful observables definable in this context are those that do not change whatever foliation we choose. This excludes that whatever physically significant quantity changes in time: taking this picture seriously seems to imply that the universe is a frozen block. Note how this problem stems from forcing a strict requirement of gauge invariance on the theory: from this point of view, the problem of time in Hamiltonian GR is strikingly similar to the problem of matter in induced matter theory. Also in this case, then, a way out of the conundrum would be to loosen or modify the gauge requirement so that other observables besides the unchanging ones could be defined (a solution of the problem has been in fact proposed by Pitts 2014).

The second problem (discussed at length in Maudlin 1996, 2011), can be roughly summarized as follows. Let us imagine a quantum system made up of space-like separated parts *A* and *B* in an entangled state, and let us consider an event *E*(*A*) such as a certain outcome of a measurement performed on *A*. It has been experimentally shown (Aspect et al. 1981) that *E*(*A*) is not determined solely by the events in its past light-cone. This is what Bell called non-locality (Bell 2004, ch. 2) and that might mean that *E*(*A*) can be affected either by events in its future light-cone or by space-like separated events such as a measurement taking place on *B*. In any case, quantum non-locality seems to be at odds with the physical interpretation of the light-cone structure of relativistic spacetimes, so that some modification of such a structure might be called for.

If we consider that one of the most promising theoretical programs towards the construction of a quantum theory of gravitational phenomena (viz. canonical quantum gravity) seeks to quantize GR starting from its Hamiltonian formulation, we immediately realize that this program faces a conceptual tangle that is basically the combination of the above mentioned problems. This is a further reason for reflecting on the GR-desideratum and finding ways to relax it without giving up general covariance, diffeomorphism invariance, and background independence.

University of Lausanne
Department of Philosophy
CH-1015 Lausanne
e-mail: antonio.vassallo@unil.ch

ACKNOWLEDGEMENTS

I would like to thank J. Brian Pitts and an anonymous referee for the useful comments on an earlier draft of this paper. Research contributing to this paper was funded by the Swiss National Science Foundation through the research grant no. 105212_149650.

REFERENCES

Anderson, J. (1967). *Principles of Relativity in Physics*. New York: Academic Press.

Arnowitt, R., Deser, S., Misner, C. (1962). The Dynamics of General Relativity. In: L. Witten (ed.), *Gravitation: An Introduction to Current Research*, pp. 227–265. New York: John Wiley & Sons. http://arxiv.org/abs/grqc/0405109.

Aspect, A., Grangier, P., Roger, G. (1981). Experimental Tests of Realistic Local Theories via Bell's Theorem. *Physical Review Letters* **47** (7), 460–463.

Balashov, Y. (2000). Enduring and Perduring Objects in Minkowski Space-Time. *Philosophical Studies* **99**, 129–166.

Barrett, J. (1999). *The Quantum Mechanics of Minds and Worlds*. Oxford: Oxford University Press.

Bell, J. (2004). *Speakable and Unspeakable in Quantum Mechanics*. Cambridge: Cambridge University Press.

Belot, G. (2011). Background-Independence. *General Relativity and Gravitation* **43**, 2865–2884. http://arxiv.org/abs/1106.0920.

Brush, S. (1999). Why Was Relativity Accepted? *Physics in Perspective* **1**, 184–214.

Dirac, P. (1978). *The Excellence of Einstein's Theory of Gravitation*. Paper given at the symposium on the impact of modern scientific ideas on society held in Munich, 18–20 September 1978. Available at http://unesdoc.unesco.org/images/0003/000311/031102eb.pdf.

Earman, J. (2002). Thoroughly Modern McTaggart or What McTaggart Would have Said If He had Read the General Theory of Relativity. *Philosopher's Imprint* **2** (3), 1–28.

Earman, J. (2006). Two Challenges to the Requirement of Substantive General Covariance. *Synthese* **148** (2), 443–468.

Earman, J., Norton, J (1987). What Price Spacetime Substantivalism: The Hole Story. *British Journal for the Philosophy of Science* **38**, 515–525

Friedman, M. (1983). *Foundations of Space-Time Theories: Relativistic Physics and Philosophy of Science.* Princeton: Princeton University Press.

Gilmore, C. (2002). Balashov on Special Relativity, Coexistence, and Temporal Parts. *Philosophical Studies* **109**, 241–263.

Giulini, D. (2007). Remarks on the Notions of General Covariance and Background Independence. *Lecture Notes in Physics* **721**, 105–120.

Huggett, N., Wüthrich, C. (2013). Emergent Spacetime and Empirical (In)Coherence. *Studies in History and Philosophy of Modern Physics* **44**, 276–285.

Kaluza, T. (1921). Zum Unitätsproblem in der Physik. *Sitzungsber. Preuss. Akad. Wiss. Berlin. Math. Phys.* **33**, 966–972.

Klein, O. (1926). Quantentheorie und fünfdimensionale Relativitätstheorie. *Zeitschrift für physik A* **37**, 895–906.

Kretschmann, E. (1917). Über den physikalischen Sinn der Relativitätspostulate: A. Einsteins neue und seine ursprüngliche Relativitätstheorie. *Annalen der Physik* **53**, 575–614.

Lehmkuhl, D. (2008). Is spacetime a gravitational field?. In: D. Dieks (ed.), *The ontology of Spacetime*, (*Philosophy and Foundations of Physics*, vol. 2) pp. 83–110. Amsterdam: Elsevier.

Lehmkuhl, D. (2014). Why Einstein did not believe that general relativity geometrizes gravity. *Studies in History and Philosophy of Modern Physics* **46**, 316–326.

Maudlin, T. (1989). The essence of space-time. *PSA 1988* **2**, 82–91.

Maudlin, T. (1996). Space-time in the quantum world. In: J. Cushing, A. Fine, and S. Goldstein (eds.), *Bohmian mechanics and quantum theory: An appraisal* (*Boston studies in the Philosophy of Science*, vol. 184), pp. 285–307. Dordrecht: Kluwer.

Maudlin, T. (2002). Thoroughly muddled McTaggart or how to abuse gauge freedom to generate metaphysical monstrosities. *Philosopher's Imprint* **2** (4), 1–23.

Maudlin, T. (2011). *Quantum non-locality and relativity. Metaphysical intimations of modern physics.* New York: Wiley-Blackwell.

Misner, C., Thorne, K., Wheeler, J. (1973). *Gravitation.* San Francisco: W.H. Freeman and Company.

Norton, J. (1993). General covariance and the foundations of general relativity: Eight decades of dispute. *Reports on Progress in Physics* **56**, 791–858.

Norton, J. (2011). The hole argument. In: E.N. Zalta (ed.), *The Stanford Encyclopedia of Philosophy.* http://plato.stanford.edu/archives/win2008/entries/spacetime-holearg/. See http://www.science.uva.nl/~seop/entries/spacetime-holearg/Active_passive.html for a toy model explaining the difference between active and passive transformations.

Overduin, J., Wesson, P (1997). Kaluza-Klein Gravity. *Physics Reports* **283**, 303–378.

Pitts, J. (2006). Absolute objects and counterexamples: Jones-Geroch dust, Torretti constant curvature, tetrad-spinor, and scalar density. *Studies in History and Philosophy of Modern Physics* **37**, 347–351. http://arxiv.org/abs/gr-qc/0506102v4.

Pitts, J. (2014). Change in Hamiltonian general relativity from the lack of a time-like Killing vector field. *Studies in History and Philosophy of Modern Physics* **47**, 68–89. http://arxiv.org/abs/1406.2665v1.

Ponce de Leon, J. (1988). Cosmological models in a Kaluza-Klein theory with variable rest mass. *General Relativity and Gravitation* **20**, 539–550.

Pooley, O. (2010). Substantive general covariance: Another decade of dispute. In: M. Suàrez, M. Dorato, and M. Rèdei (eds.), *EPSA Philosophical Issues in the Sciences: Launch of the European Philosophy of Science Association*, Volume 2, pp. 197–209. Dordrecht: Springer. http://philsci-archive.pitt.edu/9056/1/subgencov.pdf.

Romero, C., Tavakol, R., Zalaletdinov, R. (1996). The Embedding of General Relativity in Five Dimensions. *General Relativity and Gravitation* **28**, 365–376.

Rovelli, C. (2002). Partial Observables. *Physical Review D* **65** (12), 124013. http://arxiv.org/abs/grqc/0110035.

Wesson, P. (1984). An Embedding for General Relativity with Variable Rest Mass. *General Relativity and Gravitation* **16**, 193–203.

Wesson, P. (1995). Consequences of Covariance in Kaluza-Klein Theory. *Modern Physics Letters A* **10** (1), 15–24.

Wesson, P. (2006). The Equivalence Principle as a Probe for Higher Dimensions. *International Journal of Modern Physics D* **14**, 2315–2318.

Wesson, P., J. Ponce de Leon (1992). Kaluza-Klein Equations, Einstein's Equations, and an Effective Energy-Momentum Tensor. *Journal of Mathematical Physics* **33**, 3883–3887.

Ioan Muntean

A METAPHYSICS FROM STRING DUALITIES: PLURALISM, FUNDAMENTALISM, MODALITY

ABSTRACT. Some philosophers of science have suggested that contemporary science should be the source of inspiration to the new analytic metaphysics (A. Chakravartty, C. Callender, S. French, J. Ladyman, T. Maudlin, etc.). This paper explores the prospect of a "string metaphysics": a research program in analytic metaphysics based on string theory. Different forms of fundamentalism and pluralism are discussed in this context. The paper focuses on string metaphysics with S-dualities (a relation between models of string theory at different coupling regimes) and argues that fundamentality and compositionality have to be reconceptualized. String metaphysics with dualities is better couched in terms of metaphysical pluralism. Grounding, as well as a sketch of a string modality, are briefly discussed. The paper concludes with a suggestion for future work: the metaphysician may find a productive ground for research in discussing other dualities (especially T-dualities, or AdS/CFT duality), the emergence of spacetime, the concept of time in string theory, the multiverse etc.

1. A New Work for Physics in Metaphysics

What is the relation between science and philosophy, especially metaphysics? Revisiting this perennial question is more intriguing nowadays. We are several decades after philosophy of science had progressed beyond its positivistic roots, and after developments in contemporary physics (Quantum Gravity, cosmology, the Standard Model, black hole thermodynamics, etc., or what happened in physics roughly after the 1960s) have settled enough to be scrutinized by philosophers. Is "analytic metaphysics" in a

In: Tomasz Bigaj and Christian Wüthrich (eds.), *Metaphysics in Contemporary Physics* (*Poznań Studies in the Philosophy of the Sciences and the Humanities*, vol. 104), pp. 259-292. Amsterdam/New York, NY: Rodopi | Brill, 2015.

better relation with science than metaphysics was in the decade of the Carnap-Quine debate?[1]

One can witness some radical proposals: the (Post-)Carnapian "metaphysics is dead" argument that completely dismisses analytic metaphysics (Price 2009); a need for a fundamental reformation of it in the light of contemporary physics (Ladyman et al. 2007; Maudlin 2007), or a need to acknowledge the "division of labor" between science and metaphysics (Lowe 2011; Ross et al. 2014). There are also some conciliatory stances: expand existing views in metaphysics in the light of scientific results; analyze the "metaphysical foundations of physics," or the more traditional "metaphysics-in-physics" view (Friedman 2011). Last but not least, some emphasize the *similarities* between the method, the subject-matter or the ideals in metaphysics and science (Godfrey-Smith 2006, 2012; Paul 2012) or, stronger, a continuum scale between them (Callender 2011; Chakravartty 2013). What is needed in the present argument is just a minimal base of shared *ideals* between metaphysics and science. Science and metaphysics may have different methods, different subject-matters, or different paths to success.

1.1. *Different Is Better: Metaphysics, Physics and Pluralism*

This paper aligns with the reconciliatory stances and starts with two questions: (i) Is contemporary physics a legitimate source of inspiration, innovation or creativity in analytic metaphysics?, and (ii) How does a metaphysician build a viable program inspired by contemporary physics?

Science is considered a source of inspiration and a fact-checker in analytic metaphysics, at least in some quarters of the new analytic metaphysics. As Callender puts it, metaphysics is "best when informed by good science, and science is best when informed by good metaphysics" (Callender 2011, p. 48). Callender sketches a "symmetric picture" on a continuum scale between science and metaphysics, such that both are part of the 'Best representation system'. Both metaphysical and scientific endeavors might or might not be worth pursuing epistemically. Scientific theories used in metaphysics should possess certain marks of success: empirical adequacy, simplicity, novel predictions, novel explanations, unification, consilience,

[1] "Analytic metaphysics" became preeminent long after the Quine-Carnap debates of the 1950s; conventionally, the landmark is Kripke's *Naming and Necessity* (1973) and subsequent work by D. Armstrong, D. Lewis, R. Chisholm. New 'analytic metaphysics' covers probably the last two decades or so. 'Analytic metaphysics' is as a whole a wide enterprise that includes or not naturalism. For a critical review of the scope and aim of analytic metaphysics, see Ladyman et al. (2007).

etc. (Callender 2011). In both camps there are proponents and dissenters of the affirmative answer to (i).

Building on arguments from C. Callender, T. Maudlin, J. Ladyman and D. Ross, this paper proposes a preliminary example of (ii) based on string theory, a program in quantum gravity.[2] The "metaphysics-from-physics" approach includes programs in metaphysics inspired from virtually all areas of theoretical physics, including quantum gravity, cosmology, etc.

What is a *good* metaphysics-from-physics program? The present approach lowers the expectations about science: our existing theories (or better, models) in quantum gravity are put at work in metaphysics without imposing too much on their empirical adequacy. These models are farther away from empirical work than other areas in physics, and "riskier" in confronting the world. Science must sometimes address issues where there is a dearth of empirical data or when *direct* confirmation is almost impossible: early cosmology, the fundamental nature of space and time at very small scale or very large scale (multiverse?), the distant future of the universe, the cosmological process before Big-Bang, etc. Qualifying Callender's suggestion, this paper assumes that metaphysics is enriched by *good enough* science. Interpreting a scientific theory is a part of this. Hence, it makes sense to talk about quantum metaphysics based on different interpretations (Everettian, Bohmian, the "Copenhagen" interpretation, GRW, etc.), spacetime metaphysics, LQG (loop quantum gravity) metaphysics, or "string metaphysics." Second, this approach does not aim to answer *all* 'mortal' questions in metaphysics, nor does it claim to reform existing views. Many questions cannot be addressed by a string metaphysics, or are heavily undermined by its current epistemological or theoretical problems. The current proposal pleads with the metaphysician to be tolerant enough and accept science-inspired programs with their fresh and unexpected perspectives, and with their drawbacks.

A plurality of metaphysical projects inspired from quantum gravity programs, alongside with projects from quantum mechanics, relativity, would populate this metaphysical landscape. Some fare better than others on different dimensions, on Callender's marks of success, some are deficient on others. The metaphysician cannot simply eliminate one candidate because it does not satisfy all dimensions. Each program comes with a new metaphysical signature and cries for philosophical analysis. This paper

[2] String theory is discussed in the philosophy of physics literature quite extensively. The first philosophical discussions are Weingard (1988) and the relevant papers in Callender and Huggett (2001a). More recent literature is: Cappelli et al. (2012); Dawid (2007, 2009, 2013); Dawid et al. (2014); Matsubara (2011); Rickles (2011, 2013, 2014); Teh(2013).

endorses a more inclusive view about fundamentality and, to a less extent, modality, to accommodate some 'string-inspired' programs. The main task is to emphasize the novelty and the difference that a string metaphysics makes.

Is metaphysics an *a priori* endeavor? Is metaphysics the only discipline to study modality? Glossing on the idea of a 'division of labor' between science and metaphysics, S. French and K. McKenzie (2012) have urged metaphysicians to engage with physics by continuing to focus on their subject-matter: the study of the possible. The metaphysical results may offer the resources that philosophers of physics can use in physics. Even the scientifically disinterested metaphysicians cannot ignore those parts of physics that do not fit their metaphysical system or more brazenly to claim that metaphysical systems replace physics.

Typically, those willing to relate metaphysics to science use laws of nature, causality, space and time, etc. as common grounds. In this paper the starting point is different: the scientific model and the model-building operation as practiced in quantum gravity. Given this model-centered metaphysics, the questions have now a counterfactual form:

(1) **Metaphysics from models**: If one has some minimal commitments to a model *M*, what kind of program in metaphysics can be inferred from *M*?

(2) **Model-centered metaphysics**:
- If there is enough evidence that *something like* model *M* happens in a world, what would that world look like, metaphysically?
- Suppose one were serious about the model *M* in one world: how is one to think of what world is like in that case?
- Assume that model *M* is possible: then what is possible and what is not possible in other worlds, including our own?

In short, by starting with (1) and some of (2), one urges the metaphysician to ask *not* whether a theory is true, well confirmed by data, etc., but ask what it implies metaphysically about a world where it holds and about other worlds, if one takes some or all of its models (*M*) seriously.

2. Prolegomena to a String Metaphysics

Should we take string models seriously in metaphysics? The typical metaphysician sees unwarranted assumptions, mathematical and logical compromises, and ultimately falsity, in contemporary physics. One can always run a variant of the pessimistic meta-induction argument against current physics, or the 'new induction' against scientific realism by Stanford

(2006). Not a very long time ago, D. Lewis was unenthusiastic to take lessons from quantum mechanics because its instrumentalist frivolity, doublethink logic and "supernatural tales about the power of the observant mind to make things jump" (Lewis 1986, p. ix). The same line of reasoning applies implicitly and *a fortiori* to string metaphysics or to quantum-gravity metaphysics in general. The philosopher of physics would immediately note that quantum mechanics is open to different interpretations; ditto about quantum gravity, where more interpretative work that goes beyond its popularizing 'coat' is needed. In the case of string theory, a handful of philosophers of science have addressed its consequences for metaphysics: Atkinson (2005); Baker (2014); Callender (2011); Hudson (2005); Monton (2011).

2.1. *Singular Truth, or 'Anything Goes'?*

Probably the majority of metaphysicians willing to use a scientific theory in metaphysics would require from it to be true. As Monton (2011) notes, neither quantum mechanics, nor relativity are true. One simple reason is that they are together inconsistent and their principles clash: their respective counterparts in metaphysics would also be inconsistent. Hence, there are limits of the pluralism entertained in this paper. What theory the metaphysician is supposed to use then? There is no singular true theory in physics (call it the "final theory of physics": true, unique, universal, referring to everything, etc.). Therefore, arguments against a metaphysical doctrine (*e.g.* presentism, essentialism, modality, causation) need to be "more sophisticated" than those presented in classical physics, relativity, quantum mechanics, quantum field theory, etc. because they are strictly speaking false. For Monton, without a final quantum gravity program, any "physics-based" metaphysics relies on shaky foundations.

A "truth condition" is a bridge too far; it acts more like an ideal than a necessary constraint. Not *any* program in metaphysics has to originate from a true theory in physics: the requirement, vaguely put, is to start with *good enough* theory, at least *better than* previous theories, such as relativity and quantum theory. Such a "requirement for improvement" promoted in this paper does not exclude concurrent or precedent programs in metaphysics. String metaphysics does not deliver *definitive* metaphysical lessons, and does not blow off existing doctrines in metaphysics. On the contrary, some new programs fare well compared to mainstream metaphysics in some respects, but not in others. The impact of string theory is more limited than the popularizing literature depicts it. The plurality of string models allows a variety of new answers to *some*, not all, metaphysical questions: some are corrections to old solutions, some are new and unexpected. Instead of metaphysician's 'intuition condition', one can require from "good enough" theory used in metaphysics: unification, universality, axiomatization – when compared to previous theories.

2.2. *Unification and Universality Without Laws and Principles?*

As ideals, unification and universality are common to metaphysics and physics. The new analytic metaphysics of D. Armstrong, K. Fine, D. Lewis, E. Lowe, T. Sider *i.a.* is no exception. They have striven to find its simplest and the most unificatory doctrine: the unique set of natural kinds, the ultimate structure of the world, the ultimate language that "cuts nature at its joints," etc. Similarly, the recent history of theoretical physics can be read, charitably, as a toil for a unification and universality.[3] It is widely acknowledged that, trivially or not, the core principles of relativity clash with the core principles of quantum theory and inconsistencies loom large. Hence the request of unification is a consequence of the natural abhorrence a metaphysician may have for inconsistencies between theories in physics. Philosophers usually acknowledge this advantage of string theory: its ability to provide a more coherent picture than everything before it is deemed by R. Dawid (2013, p. 33 sq.) as the "argument of its unexpected explanatory coherence." A theory with explanatory coherence may, or may not, evolve into an "final theory." String theory alleviates, at least, the conflict between quantum theory and relativity: it is able to explain the theory of relativity as a low energy limit of string dynamics. It is premised on the idea that gravitation needs to be quantized,[4] and that the "beable" of the quantized gravitation is then a particle, *i.e.* "the graviton," and not the classical gravitational field.[5] But one also witness that gravitation cannot be quantized as "yet another quantum field theory."[6] In string theory, unexpectedly, the features of the gravitons are inferred from vibrations of strings: the "beable" of string theory is not one entity but a collection of strings, branes or other high-dimensional objects.

[3] Some other philosophers of physics argue for the autonomy of different areas of physics based on emergence. See the details of the reduction-emergence debate in (Butterfield 2010).

[4] See philosophical discussions on the pros and cons of quantizing gravity in (Callender and Huggett 2001a; Wüthrich 2005).

[5] "Beable" carries the distinction between what is physical and what is non-physical (Bell 1987).

[6] To be fair, one approach, developed by St. Weinberg in the 1970s, the "asymptotic safety" theory, does start from a quantum field theory of gravity (Niedermaier 2007).

2.3. *Laws, Principles, Axioms – Or Simply Models?*

There is another reason to look with a jaundiced eye at string theory: it has been not yet lucky enough to benefit from a Newton, von Neumann, or a J.S. Bell figure to axiomatize the discipline, or to provide its principles. There are few, if any, high-level generalizations used to constrain the applicability of a theory, no laws of nature to rely on, no "methodological maxims" or principles.[7] B. Greene (1999, p. 171) asked a decade and a half ago: "Is string theory itself an inevitable consequence of some broader principle – possibly but not necessarily a symmetry principle – in much the same way that the equivalence principle inexorably leads to general relativity or that gauge symmetries lead to the nongravitational forces?"[8] Greene concedes that formulating the theory without its core principle is possible, but it would be extremely difficult: string theory is "in a position analogous to an Einstein bereft of the equivalence principle" (Greene 1999). In the same line of thought, L. Smolin wrote:

> ... string theory in its present form most likely has the same relationship to its ultimate form as Kepler's astronomy had to Newton's physics." (Smolin 2002, p. 149)

M. Douglas (2007, p. 138) proposed recently an interesting analogy. A mathematical discipline starts from simple axioms and large symmetry groups, and ultimately the complexity of the discipline is the result of the symmetry breaking. In chemistry one postulates some building blocks and the complexity is the result of the combination of these blocks. String theory is closer to chemistry than to mathematics, concludes Douglas. One starts with the blocks (branes, strings, gauge sectors, charges, symmetries, and probably even spacetime) and builds from them the complexity of the real world.

The lack of laws of nature, guiding principles, or axiomatical structure, can touch a metaphysical nerve. Can one build a metaphysics that includes modality from string theory? In the present interpretation, instead of laws and principles, or structures, a string metaphysics is grounded in the very operation of model-building, postulating and working with the symmetry group needed, with the type of spacetime, etc. The "string metaphysics" ilk sketched here is multifarious: as string theory went through several revolutions and paradigm changes, each of its incarnations containing new

---

[7] It is interesting to compare quantum field theory and string theory in respect of axiomatics. Although they are similar in some respects, an algebraic formulation of string theory does not exist yet.

[8] Since then, the holographic principle gained the status of a a relatively reliable candidate for a principle of string theory; see the debate in (Crowther and Rickles 2014; Sieroka and Mielke 2014).

models, new commitment to realism – or lack of thereof. At different stages during its history, string theory has been able to address some metaphysical questions: Is spacetime fundamental, or, on the contrary, emergent?; Is spacetime multidimensional? Do we live on a subspace of a higher-dimensional space? Is our world a hologram? How robust is the difference between the quantum and the classical world? The claim here is that at least in the 1990s string theory has addressed questions about fundamentality and modality: What entity is fundamental at different levels, or scales, at different energies or in different time intervals – even those not accessible to experiment? Ditto about what can be predicated at different levels, scales, etc. Can the physics of our world be inferred or explained by string models from other worlds?

3. Five Models...

String theory is both revolutionary and conservative: it attempts to change the ontology of physics (both the fundamental entity and the spacetime where fundamental entities live), although it preserves the formalism of quantum field theory. This section is centered on the more conservative facet of string theory: its connection to the formalism of quantum theory.[9] At its core, its mathematics follows quantum field theory: many of strings' features, including super-symmetry and dualities are present in other theories. Nonetheless, symmetries and dualities play a special role in string theory beyond mere formalism, well into its metaphysical foundations.

String theory describes the dynamics of strings propagating and vibrating in a flat, D-dimensional spacetime. Strings are more attractive than pointlike particles because they do not display singularities. Strings sweep out a two-dimensional "worldsheet" in spacetime that can be parameterized by a function $X^\mu(\sigma, \tau)$, where σ is typically associated to the proper space coordinate of the worldsheet and τ to the one time coordinate of the world-sheet. A string, a one-dimensional object, it is just a special case of p-branes which are objects with p dimensions. The world-sheet of a p-brane is a $p + 1$ world volume parametrized by one timelike coordinate $\sigma^0 = \tau$ and p spacelike coordinates σ^i with $i = 1 \ldots p$.

The simplest method used in the 1980s was to write down the action of a string on the worldsheet (the Polyakov action), and after studying its invariants and symmetries (Poincaré transformations of the X^μ coordinates, a

[9] See resources on the string theory formalism in Becker et al. (2007); Green et al. (1987); Polchinski (1998b).

local symmetry of the reparametrization of the worldsheet and a conformal transformation of the worldsheet metric), one can quantize the string and apply Feynman diagrams to infer its dynamics. The states are the result of applying operators on the Fock vacuum. The result is the Relativistic Quantum String Theory with a field theory in two dimensions (the "worldsheet theory"), which is a perturbative model having Feynman diagrams with a smooth topology.

There are two fundamental types of string: closed and open. The closed string is parametrized such that X is periodic in $X^{\mu}(\sigma, \tau) = X^{\mu}(\sigma + \pi, \tau)$. Closed strings are central to string theory as the graviton appears as a massless mode of the closed-string spectrum. Open strings can have two types of boundary conditions at the ends: the Dirichlet condition specifies a spacetime hypersurface associated to a physical particle; the Neumann condition: $\partial_{\sigma} X^{\mu} = 0$ means that the string ends move freely, unexpectedly at the speed of light! And, finally, strings can be oriented or not.

For quantization purposes, the most suitable action is the string sigma-model (or the Polyakov action):

$$S_{\sigma} = S_{P}\,[X, \gamma] = -\,T/2 \int \mathrm{d}\tau \mathrm{d}\sigma \gamma^{ab} \sqrt{-\gamma}\partial_{a} X^{\mu} \partial_{b} X_{\mu}$$

3.1. *Supersymmetry* (*SUSY*)

The first string model, the bosonic model, did not include any fermionic excitation. In order to account for fermions, some unobserved symmetries were needed. The claim: "the maximal symmetries of the real world is that observed in normal standard experiments with elementary particles," assumed in particle physics, is denied in string theory and replaced with Supersymmetry (SUSY). There are two options to integrate fermions: assuming SUSY on the worldsheet by the Ramond-Neveu-Schwarz formalism, or assuming SUSY on the spacetime itself, by the Green-Schwarz formalism.[10] Theorists use SUSY in string theory to represent the standard model by the means of the gauge fields with the gauge group $SU(3) \otimes SU(2) \otimes U(1)$. The argument pertains to show that the low energy effective limit of string theory is nothing more than a Yang-Mills gauge theory.

A new term is added to the Polyakov action, describing D free Dirac fermionic fields:

$$S = -\,T/2 \int d^2\sigma\,(\partial_{\alpha} X^{\mu}\, \partial^{\alpha} X_{\mu} - i\bar{\Psi}^{\mu}\, \rho^{\alpha}\, \partial_{\alpha} \Psi^{\mu})$$

[10] For $D = 10$ these two assumptions are in fact equivalent.

where both ρ and Ψ are defined on the worldsheet. ρ are Dirac matrices in $1 + 1$ dimensions and $\Psi^{\mu} = \Psi^{\mu}(\sigma, \tau)$ are Majorana spinors.

3.2. *Three Interpretations of String Models*

Depending on the types of strings (open or closed), on boundary conditions (Dirichlet or von Neumann), on various fields added to the spacetime or to the worldsheet (scalar fields or vector fields; for example the Majorana fermions are postulated in all models), on the topology of spacetime and its symmetries, one can create an impressive number of models, although only five of them are actually relevant to current developments. "What is string theory?" is a challenging foundational question: a model of hadrons, a theory of everything, a quantum gravity program?[11] The theory may well lack a definition, a subject-matter, principles or laws, but it is nevertheless open to some interpretative work. Right after the "second string revolution" (around 1995), one could envisage string theory as:

(3) **Strings *per se***: A collection of mathematical models of strings vibrating in various types of spacetimes, on which different symmetries and different fields are postulated. These models may include branes or other objects, and they may or may not represent aspects of known theories in physics: the Standard Model, gravitation, various gauge theories, black hole thermodynamics, information theory, condensed matter physics, etc.

(4) **Strings to strings**: A collection of conjectures about the relations among the string models, as in (3);

(5) **Strings to physics**: A collection of conjectures about the relations among string models, as in (3), and theories of known physics. String models are judged based on their similarity with the phenomenological world of experimental physics.

Witness that according to (3) and (4), string theory is not strictly speaking a theory of physics, but a collection of mathematical conjectures and mathematical models.[12] The present string metaphysics is modest enough to stay closer to (3) and (4) (which integrates well with string dualities). Adopting (5) paves our way to a stronger interpretation of string theory as a theory in physics, about our world.

---

[11] D. Gross asked in a conference in 1985: "how many more string revolutions will be required before we know what string theory is?," quoted in (Rickles 2014, p. 234).

[12] From now on, the difference between string models and string theories is not honored anymore. String theories are mathematical models, but the name "string theory" forces us to talk about them as theories.

There are five consistent models: four string models and the *SUGRA* (*i.e.* "supergravity," which has no strings, but it is useful to understanding low limit energy of string models):

Model	Space-time	Modes	Types of Objects	SUSY	Gauge Group
Open Bosonic String	$D = 26$	bosonic	open string	none ($\mathcal{N} = 0$)	$U(1)$
Closed Bosonic String	$D = 26$	bosonic	closed string	none	none
Type I superstring	$D = 10$	fermionic & bosonic	not oriented open strings (closed by interaction)	$\mathcal{N} = 1$	$SO(32)$
Type IIA superstring	$D = 10$	fermionic (non-chiral) & bosonic	closed oriented strings & D-branes	$\mathcal{N} = 2$	$U(1)$?
Type IIB superstring	$D = 10$	fermionic (chiral) & bosonic	closed strings	$\mathcal{N} = 2$	None ?
SO(32) Heterotic	$D = 10$ and $D = 26$	bosonic & fermionic	closed strings	$\mathcal{N} = 1$	$SO(32)$
E8 × E8 Heterotic	$D = 10$ and $D = 26$	bosonic & fermionic	closed strings	$\mathcal{N} = 1$	E8 × E8
Supergravity (SUGRA)	$D = 11$	bosonic & fermionic	No strings	$\mathcal{N} = 2$	

4. ...and their S-dualities

Dualities are transformations or symmetries of theories in a special space called the moduli space: some of these transformations are trivial, some are unexpected. Different types of dualities have existed long before string theory: the classical electromagnetic (E-M) duality, the Olive-Montonen duality, the Seiberg-Witten dualities in quantum theories etc. Nevertheless, from a metaphysical point of view, S-dualities in string theories are more prolific: although legitimate in themselves, a "field duality metaphysics" inspired by non-string dualities would invariably collapse to some string dualities.

Dualities constitute a paradigm shift in the history of string theory that happened in the mid 1990s, sometimes referred to as the "second string revolution": anybody would agree that they play a central role in string theory, whereas in all other cases (including field theories) they are probably just curious features, if not vicissitudes, of theories.[13]

In the mid 1990s, two questions were asked about the five models in Table 1:

(6) How are these models related one to the other?
(7) What is the strong coupling limit of each model?

Before the discovery of dualities, (6) was appraised separately from (7). The relation between models works between sectors of models, and not between the whole model. The strong coupling regime of any of these models can be mapped to the weak coupling regime of another, or the same, existing model. The proliferation of dualities inspired E. Witten (1995) to conjecture that *all* string models are different weak coupling limits of a unique, fundamental theory, called M-theory. Witten's hunch is the first attempt to consider string theory, a "theory of everything," now such that here dualities played the central role.

To discuss dualities, two parameters of string models are relevant: the string coupling constant g_S and the string length ℓ_S. The coupling g_S separates the perturbative models with $g_S \ll 1$ from the non-perturbative models with $g_S \geq 1$. Its values is related also related to the topology of spacetime. The length ℓ_s is related to the "size" of the spacetime.[14] In each string model, the perturbation formulation is part of the weak coupling sector, premised on the assumptions that strings are weakly coupled, and somehow sparsely distributed in spacetime. Strings vibrate against a flat, fixed background spacetime. At high coupling $g_S \geq 1$, the models are non-perturbative as

[13] For non-string S-dualities, see Harvey (1996); Intriligator and Seiberg (1996). S-dualities are discussed in almost any resource on string theory, but an excellent starting point is Sen (1998); foundational issues are discussed in: Polchinski (1996); Rickles (2011, 2013) and (Polchinski, 1998a, esp Ch. 14). As an interesting case of a non-string duality, N. Seiberg and E. Witten (1994) conjectured dualities about low-energy effective action, $\mathcal{N} = 2$ or $\mathcal{N} = 1$ Yang-Mills theories. The natural home of the Seiberg-Witten duality, has been said, is string theory, more precisely the self-dual string model. Some would even insist that one can deduce logically the Seiberg-Witten duality from string theory (Vafa, 1997).

[14] The string length: $\ell_S = \sqrt{\alpha'}$ where α' is the Regge slope, which at its turn is the inverse of the string "tension" $T = 1/2\pi\alpha'$. See Section 5.3 for a short discussion on string tension.

they assume that two strings can join, that single strings can split, and that strings interact with spacetime by acting on spacetime and getting back a reaction from spacetime.[15]

S-duality is a transformation that puts in correspondence the strong coupling sector to a weak coupling sector.[16] One can infer from the results of the weakly coupled model some (or all, depending on the relative strength of the duality) results in the strongly coupled model.

It is useful to have a working definition of S-dualities:

(8) **String S-Duality**: A string S-duality is a conjecture about a map between (i) the sector of a string model in the weak regime M_{weak} with a set of fundamental entities (possibly one) F_{weak}, a set of non-fundamental, derivative entities d_{weak} some fields φ_{weak} and a coupling constant g_{weak}, and (ii) the sector of its "dual" model in the strong regime M_{strong} with $g_{\text{strong}} = 1/g_{\text{weak}}$, another set of fundamental entities F_{strong}, non-fundamental, derivative entities d_{strong} and fields φ_{strong}. S-duality conjectures that there is an isomorphism between the consequences of M_{weak} and M_{strong}. Moreover, dualities map most of F_{weak} to d_{strong} and most of d_{weak} to F_{strong}.

The sector M_{weak} is represented by a perturbative formalism and sector M_{strong} by a non-perturbative formalism. S-dualities postulate the same "observational" consequences of M_{weak} and M_{strong} and map only *some* of their entities: the reason for the latter is the lack of knowledge about the perturbative formalism. The strong coupling sector of a model is asymptotic, not convergent, and the weak coupling formalism cannot be extended to describe it. The conjectured S-dualities do not have rigorous proofs: even if they are unwarranted epistemically, S-dualities are novel and unexpected mathematically, as there should be no correspondence between M_{weak} and M_{strong}. Discovering new S-dualities is always marked by surprise and very often by serendipity.

Some S-dualities, the so-called "self-dualities," relate two sectors (weak and strong coupling) of the same model: that is the case with the Type IIB model. Other S-dualities relate different models: the strong coupling limit of Heterotic $SO(32)$ model is the Type I model and *vice versa* (see The Appendix). In other cases, the simpler physics lives in a spacetime with fewer dimensions than its dual strong coupling model: it is now

[15] At strong coupling, there is enough energy in their vibrations to make the spacetime dynamic, and therefore to obtain "background independence."

[16] S-dualities of the heterotic string model were discussed for the first time in (Font et al. 1990; Hull and Townsend 1994; Sen 1994).

known that the type IIA model ($D = 10$) is dual to the *SUGRA* model in $D = 11$. In this case, the dual of a string model is *SUGRA*, a theory with no strings! The strong coupling limits of Type IIA and heterotic $E8 \times E8$ models are interesting examples. Both these models are ten-dimensional at weak coupling. In their strong coupling limit, an additional eleventh dimension appears! The extra eleventh dimension is compact and its size is related to the ten dimensional coupling constant. At weak coupling, the eleventh dimension is small and invisible, but as the coupling increases, this extra dimension unfolds.

How does one infer a duality conjecture if there is no knowledge of the M_{strong}? In the presence of SUSY one has to identify some invariants of the dual transformation, special states in the spectrum of the strong model, called the BPS (Bogomol'nyi-Prasad-Sommerfield) states. Based on the invariants, the theorist expects to encounter completely new, non-perturbative entities in the strong coupling regime: the D-branes are the typical example. In the perturbation formalism at weak coupling $g_s \ll 1$, Dp-branes do not play a crucial role in the M_{weak} and they can be deemed as non-fundamental, geometrical objects that are not directly represented in the model, as they are too heavy and too complicated to be captured by the formalism. Crucially for the present discussion is to realize that at $g_s \geq 1$, some p-branes become more fundamental than fundamental strings. For example, in some types of string models, strings end on Dp-branes (a special class of p-branes with Dirichlet conditions). Interestingly, Yang-Mills gauge theories may live on the world volumes of Dp-branes in p dimensions, while gravity extends beyond the branes in bulk space with $D > p$ dimensions.

5. Fundamentalism and Existence without Dualities

As the formalisms of string theory and field theory are similar, it is tempting to infer a string metaphysics from this or that interpretation of field theory. This is not the line of thought followed here: this section assumes that it is counterproductive and misleading to look at string ontology as "yet another quantum field ontology." There is one reason why string theory is not "yet another quantum field theory": it does not start with a "string field" and generate strings and their dynamics. This claim has some caveats though: there is a "string *field*" formalism, a *competitor* to string theory, which may have a completely different ontology. M-theory may (or may not) end up being an ordinary quantum field theory. The return to a field perspective in string theory is more than just a possibility, but it is outside the scope of the present endeavor.

The metaphysician has learned a lesson from past inquiries into theories in physics: read not the ontology of a theory straight from its formalism! The importance of strings is quite different than the role of particles in quantum field ontology, especially in the presence of dualities. Particles are emergent from fields: strings are not. String metaphysics *without* dualities follows much closer the Quinean existential agenda for metaphysics: *to say what exists*.[17] The metaphysician extracts existential commitments from string models based on the nature of the fundamental string postulated in (11). To anticipate the conclusion of the following sections: *with dualities*, string ontology looks even more different than field theory. The main aim is not to unveil the metaphysics within physics, but to evaluate the novelty of string metaphysics. Such new programs, when effective, point to new pathways of thinking where most metaphysicians see few alternatives.

In the line of the main question (1), Sections 5 and 6 address two related questions about string metaphysics[18]:

(9) **String ontology**: If one believes in a given model of string theory, what does one learn about fundamentality?

(10) **String ideology**: If one believes in a given model of string theory, what is the relation between properties and entities? What do theoretical terms mean in that model?

Some words of caution are in order: a relatively narrow episode in the development of the theory is under scrutiny here; only S-dualities are analyzed, and not T-dualities or the AdS/CFT duality; it does not cover the "string landscape" or string theory's progress towards M-theory, which are all later developments. The analysis is split between string theory as a collection of models with no dualities (the stage of the theory in the early 1990s, after the "second string revolution"), and a later stage when S-dualities have played a central role – roughly after 1995.[19] There is no problem with focusing on earlier stages in the history of a scientific theory when trying to devise a metaphysical program from it. The metaphysician needs

[17] See Schaffer's (2009) division of the analytic metaphysics between the Quinean one centered on *existence*, and neo-Aristotelian focused on *grounding*.

[18] The ontology-ideology difference was used by Quine (1951) and it is relatively popular with the new analytic metaphysics.

[19] The "second string revolution" is marked by an increasing interest in string dualities. Conventionally, it happened after E. Witten's talk at the conference *String '95*, and with the publication of two landmark articles that followed (Polchinski 1996; Witten 1995). As precursor ideas on dualities, see (Font et al. 1990). Historical details are in (Rickles 2014).

to adopt sometimes a more opportunistic "wait-and-see" attitude towards the latest development of a scientific theory.

There are some reasons to think of strings as fundamental entities: they have ontological *priority* over all other entities and everything else is *grounded* in them, and they have some individuality because they are countable. Historically, strings were the only fundamental entities in the "fundamental base" of the perturbative models of the early 1990s: the fundamental string, which is a 1-brane, is *extended* in spacetime. Some particles of known physics (graviton, photon, some gauge bosons, interesting properties of black holes, etc.) were derived from strings, and from their solitonic excitations. *Without* dualities, the fundamentalist derives many aspects of the reality from the features of the fundamental string. Hence, the string is *fundamental*. Because a string has no fundamental parts or components, it a "simple." When asked metaphysical questions such as: what are strings made of? do they have parts? in virtue of what is a string a string? the string theorist replies: parthood, compositionality, as well as other concepts in mereology, and the predication of intrinsic properties, do not apply literally to strings. The theorist B. Greene uses an argument to dismiss such concepts:

> paragraphs are made of sentences, sentences are made of words, and words are made of letters [...] from a linguistic standpoint, that's the end of the line [...] letters are the fundamental building blocks of written language; there is no further structure. Questioning about composition has no meaning [...] A string is simply a string – as there is nothing more fundamental, it can't be described as being composed of any other substance. (Greene 1999, pp. 141–142).

In the "good old fashioned string theory," *i.e.* prior to the discovery of dualities, the fundamental string is the entity quantized in the theory, not its "geometric parts." The fields, symmetries, boundary conditions, etc. are all reducible to states of the string. Strings have a different structure than pointlike particles, although they are still "simples." A fundamental string has pointlike charges at its ends, modes of vibration, waves, a tension, etc. By filling a region of spacetime, it does not have parts that are strings, or other entities. Parts of strings are not other entities or other strings, but "geometric parts." Unlike parts of fields in quantum field theory, strings cannot be decomposed on arbitrary subsets of the domain of the whole field and then glued back together. Non-fundamental entities are complicated and hard to capture by perturbative models and they are non-simple, i.e. composite. Ditto about solitonic objects.

To put the puzzle together, let us assume that non-dual string models entail this ontology:

(11) **Strong fundamentalism**: The string is the only fundamental entity of string models: all strings are extended, simple entities, that retain individuality, and are discrete (countable). Only strings are real.

As a prospective answer to (9), this claim (11) may satisfy the fundamentalist. Still, is an extended simple a desirable fundamental entity in metaphysics?

5.1. *Extended, Fundamental, and "Simple"?*

For the mainstream metaphysician, "simples" have to be non-extended or pointlike entities. Strings and *p*-branes are definitely extended: can they be "simple"? For the metaphysician, it is necessary that if an object occupies a region *R* and *r* is a sub-region of *R* (such that $r \supset R$), then there is a material object that occupies *r*. This can be called the doctrine of arbitrary undetached parts (DAUP).[20] In respect of the ideology, the metaphysician demands a sort of qualitative invariance of intrinsic properties of an object: if the object is distributed over spacetime, its intrinsic properties are distributed to its parts. It looks like an extended simple cannot meet these requirements. Some analytic metaphysicians such as K. Hawley, K. McDaniel, N. Markosian, D. Zimmerman, *i.a.* have entertained, or even endorsed, the conception of an "extended simples." For example McDaniel (2007) suggests that an extrinsic theory about fundamental objects can save "extended simples."[21]

Can strings be simple, extended, and fundamental entities? The mainstream metaphysician could have argued that superstrings cannot be fundamental, extended and simple, and could have opted to keep the official "zero-dimensional building block" theory. For example, H. Hudson claims it is physicist's job to argue whether strings are fundamental entities, but "it is the job of the metaphysician to tell us whether those uncuttable things are composite" (2005, p. 107). When it comes to the composition relation, Hudson believes that string fundamentalists have misused it. In the line of a naturalized metaphysics, Callender (2011) argues that superstring theory posits "extended simples," although they do not fit the metaphysician's Procrustean bed. The metaphysics should then change its agenda.

[20] As stated, and criticized, in (McDaniel 2007).

[21] Although McDaniel's argument based on 'shapes' in spacetime does not apply properties of string, it would be interesting to explore the extrinsicality of some properties of strings, especially as relations to spacetime.

5.2. *Non-Local but Discrete Strings*

The convoluted concept of "extended simple" is not the only conundrum of a prospective string metaphysics. While other problems can be more or less dodged by the metaphysician ready to take a look as string metaphysics, the issue of non-local fundamental entities unveils the role of spacetime in metaphysics. At a first take, spacetime is the same in quantum field theory and string theory. A deeper analysis shows some conceptual differences.

There are some reasons to think of string theory as non-local. In string models locality at certain levels can be an emergent property, in the same way in which spacetime may emerge. Unlike quantum field theory which is local and causal, in string theory one gives up locality *as fundamental*, at least at short scales.[22] One problem is that perturbative models do not incorporate well string interactions, especially because of their non-locality, which is in sharp contrast with particle physics, where interactions are local. The situation became even more complex after the "second string revolution," when other more sophisticated models include fundamental p-branes (with $p > 1$), which is essentially an object with geometrical properties – even less local than the string.

The interaction of two strings to create a third is essentially non-local. The reverse process of a string splitting in two other strings is non-local, too, because string theory is time-reversal. When these processes occur, the number of strings is not conserved in a region, although for any point of spacetime there is always only one string present, or none. In field theories a virtually infinite number of fields coexist at one spacetime point. Then interactions of strings are different. Think of a two interacting strings: the worldsheet of the interaction is always smooth. In quantum field theory, the information codifying interaction is inserted at vertices of the Feynman diagram. String diagrams have no points: locally, every section of the diagram is a free propagating string. Globally the diagram reveals the interaction.

Some authors believe that this non-locality is "harmless," but quite a few see a problem with the interpretation of non-locality. N. Seiberg writes:

> We have already asked whether we expect locality in a space, or in its dual space. It is hard to imagine that the theory can be simultaneously local in both of them. Then, perhaps it is local in neither. Of course, when a macroscopic weakly coupled natural description exists, we expect the theory to be at least approximately local in that description. It is important to stress that although intuitively the notion of locality is obvious, this is not the case in string theory [...] The theory has no local observables.

[22] At large scales string theory is supposed to behave similar to quantum field theory. For example, the perturbative string 8-matrix theory is analytic. See Giddings (2006).

> Most of the observables are related to the S-matrix or other objects at infinity. (Gross et al. 2007, p. 168)

In an "entity-oriented" interpretation of string theory, strings are simple, extended, non-local and fundamental. As extended entities, they do not have intrinsic properties the same way pointlike particles have: mass, energy, spin, etc. Even without dualities, string metaphysics does not look like a Bohmian quantum metaphysics in which fundamental entities are localized simples with intrinsic properties.

Strings are discrete, because they can be counted, even if they are extended, non-local, interacting entities. There are "number operators" in string theory similar to the $\hat{N}_k$ in field theory, that count strings of a variety k. The expectation value of such an operator quantify the average number of strings of variety k in that state. This is probably one preeminent similarity with field theories, that runs against the previous argument. Going back to its formalism, one can see why "superstring theory fits squarely within the fundamentalist tradition in particle physics, in which a Lagrangian field theory is combined with perturbative techniques and renormalization to yield phenomenological models" (Wayne 2006). As countable entities, strings retain individuality, both at weak and strong couplings. Strings are then countable, similar to excitations of fields in field theory (which are identified with particles) and fundamental.

Strings are discrete entities, with individuality: are countable, discrete individual entities necessarily fundamental? Does existence need individuality? For some, in quantum field theory, objects have "withered away," while fields or structures (group-theoretical structures) are more fundamental French (1998). Probably string theory is a backlash of an object oriented ontology, but one suspect that a fundamentality based on individuality is not enough in this case.

To conclude, barring non-locality and an unusual mereology, string metaphysics without dualities can accommodate an entity-oriented fundamentalism. Albeit the fundamentalist needs to do some legwork and to enrich the theory of individuals to include extended simple, non-localized fundamental entities, the "fundamental" nature of string is deeply rooted in the theory. Tables, chairs, photons, gravitons, etc. exist, insofar as derived from strings. Non-locality, interactions among strings, the splitting of strings, all complicate the fundamentality of strings, but cannot completely dismiss it: strings are individual, discrete and countable, grounding any other entities of the model.

5.3. *Ideologies without Dualities*

What is the ideology of string theory with no dualities? A nominalist can argue that theoretical terms are nothing more than mathematical 'names'. Although it is hard to find properties predicated by the model, there are some plausible candidates: the tension T predicated about strings, the fields postulated on the worldsheet or on the spacetime and finally, the geometry and topology of spacetime.

First, the string tension T can be predicated about strings or branes (P-branes have a tension T_p). The tension represents the energy density along the string. In the non-relativistic string, the tension is never constant on the string, as it has "tension waves." It turns out that the 'slope parameter' $\alpha' = 1 / (2\pi T\hbar c)$, which is the ratio between the angular momentum of a string and the square of its energy, replaces T. From α', one can calculate the string length $\ell_s = \hbar c\sqrt{\alpha'}$. It is more of a global parameter of the model, than an intrinsic property of one string, and it has only a limited impact on the economy of the string model. The fundamental energy scale of one string is huge (about 10^{19} higher than the energy of a photon), such that the fundamental vibrations of the string are too heavy and too energetic to account for the mass of elementary particles. It is true that the string tension generates the massive modes at the Planck scale, but the interesting physics belongs to the massless modes, due not to T, but to the quantum effects of the string vibrations. These low-energy modes are relevant to the deduction of the graviton and other gauge particles of the Standard Model. One can in fact suspect that T is explanatorily less powerful when it comes to the relation between string theory and known physics.

Second, the ideology of string theory may include fields and symmetries: either those postulated on the spacetime, or on the worldsheet, as needed. They are extremely useful mathematically, even indispensable, but nevertheless not fundamental. For the fundamentalist, most symmetries of fields can be inferred naturally from string theory: "in the context of string theory these symmetries [supersymmetry, gauge symmetries, the symmetries of spacetime], are *consequences*; although their importance is in no way diminished, they are part of the end product of a much larger theoretical structure" (Greene 2011, p. 333). On each string there is an infinite number of oscillations and its states are obtained by acting on their Fock *vacua*. Fields are not intrinsic properties of strings as they are inherited from the spacetime structure. The modes, on the other hand, are features of strings, but do not characterize points of spacetime.

Third, what role does spacetime play in string theory? Is it an object of the theory, or part of its ideology? Theorists see it as a computational device, or probably a necessary mathematical fiction, not as a fundamental entity. In the textbook form of string theory, spacetime is represented as a *necessary* background manifold against which strings vibrate. Dualities are a game changer here, as they may entail that spacetime is emergent. The fundamentalist is free to interpret spacetime as an emergent feature of strings, given its dynamical nature.[23] In different models, spacetime had $D = 26$, $D = 10$ or $D = 11$. It may well be the case that dimensionality of spacetime is a illusion or a non-fundamental feature of the world. At strong coupling, in some models, the dynamics of strings alone produces the eleventh dimension of spacetime. For each model, the topology of the spacetime, its symmetry, different geometrical features including dimensionality are inferred or corrected, if needed. Spacetime most likely qualifies as a necessary part of the ideology of the model, as a meaningful theoretical term, separated from its ontology, with some of its properties being inferred from strings (see *e.g.* the mechanism of "anomaly cancellation").

In the light of string metaphysics, the metaphysician may want to adopt a compositional pluralism, by admitting that parthood and compositionality are said in "many ways," and accept a fundamentalism able to incorporate non-locality and "extended simples." These are already counterintuitive enough. Nevertheless, the fundamentalist can rest assured that strings are discrete, individual and countable entities. Probably a monist would impose here some 'intuition conditions' upon fundamentalism – to satisfy whatever intuition we have about it. That would run against the idea of a science-inspired metaphysics: our intuitions do not work squarely in science. The metaphysician ends up accepting a very peculiar, stringy "way" of fundamentalism, hard to replicate or infer from analytic metaphysics. Last but not least, the way fundamental properties occupy (and ultimately interact with) spacetime is radically different than the passive "occupation relation" of an object with a place in metaphysics.[24] These are all aspects of pluralism at work in science-inspired metaphysics.

6. Fundamentalism, Modality and S-dualities

Section 5 capitalized on the parallel between the 'Quinean metaphysics of existence' and forms of standard fundamentalism with no dualities.

[23] See Huggett and Wüthrich (2013) and their blog at http://beyondspacetime.net/. There are arguments for and against the emergence of spacetime especially in the context of AdS/CFT duality and holographic principle (Rickles 2013; Teh 2013). The present discussion on the spacetime as part of the ideology is radically changed even in the presence of T-dualities, not covered here.

[24] This latter lesson is in fact now new, as it is the main moral of general relativity.

Schaffer (2009) contrasts this task with the Aristotelian 'grounding task': *to say what grounds what.* In this latter, neo-Aristotelian/Schafferian perspective that gains momentum in the new analytic metaphysics, fundamentality is relative to a hierarchical view of reality.[25] Fundamentalism asserts that all priority relations need to terminate in one minimal set of entities, the "fundamental base." One alternative to fundamentalism is the "infinite hierarchy" (Cameron 2008). Another option is to take fundamentality relative to a hierarchy of *levels* of reality (Wilson 2012). A set of entities at a level L is "a minimal relatively fundamental base" for all entities at L and for all entities at compositional levels "higher" than L. The relative fundamentality can also terminate at some fundamental level L_0.

Non-perturbative models add two crucial ideas to string fundamentality. First, fundamental entities F at strong coupling can split or join, can disintegrate and rejoin. Second, by S-dualities, they are related to some non-fundamental, "dual" entities d at a different coupling. The equivalence classes of grounding, the fundamental objects are all consequences of dualities and of belonging to a weak or strong coupling sectors.

This suggests a relation of "pairing through duality" among models, which is not hierarchical and not reductive, but terminates at the string level. Fundamentality ends in a web of dualities, rather than at the bottom of the ontological chain. One can read the present argument as yet another attempt to reject reductive fundamentalism in the vein of McKenzie (2011), or as an attempt to show string fundamentalism is a new, different type than the metaphysical ilk. This does not entail that string metaphysics gives up the ontological hierarchy view: in such a view, the strong coupling model M_{strong} is more fundamental than the weak coupling one M_{weak}, albeit the latter is much more tractable mathematically and computationally than the former. If reduction is also obtained, the entities of M_{weak} are reduced to entities of M_{strong}.

The present paper aims to appraise in what sense such a reductive hierarchical view is ruined by S-dualities: when dualities hold, properties of fundamental entities can be inferred from properties of non-fundamental entities of a different model. To argue against the hierarchical view of weak-strong models, the example of the $SO(32)$ heterotic (model H) – Type I (model I) duality is relevant (see Appendix). The dual transformation here is the reversal of the dilaton field: $\Phi^I \rightarrow -\Phi^H$ and the coupling constants are inverse one to the other: $g_S{}^I g_S{}^H = 1$. The H string is dual to an object in the I model called "D-string" (the d_{strong}): a D1-brane of one dimension,

[25] K. Fine, G. Rosen and J. Schaffer *i.a.* discuss 'grounding' and adopt some form of 'grounding realism'. See: (Correia and Schnieder 2012; Schaffer 2003).

"heavy," composite and complicated, which is fixed in spacetime such that fundamental open strings end on it. It is also a supersymmetric object that saturates a BPS bound.

There is no counterpart of the I string in the H model. Type I string does not carry enough conserved charges and is not supersymmetric; hence nothing can prevent it from breaking. As Becker, Becker and Schwarz (2007, p. 326) put it, the Type I string is "good enough" at low coupling: because the probability of breaking a Type I string is proportional to g_S, it is long-lived at weak coupling. But it *can* break, so it is not properly speaking fundamental at any coupling in the sense of (11). At strong coupling, this I string has no dual counterpart in the weakly-coupled H model. This example undermines the standard concept of "fundamentality as grounding" in several aspects. First, fundamentality is not only relative to a model, but also indexicalized by g_S and other coupling factors within the same model. A fundamental entity F_{strong} is put in a 1-to-1 correspondence with a non-fundamental entity d_{weak} of its "dual world." Such correspondence depends on the fields and the coupling factors. The coupling factors are not constants (this is a misnomer) and not fundamental: they are merely changes of the scale at which a model works, are dynamical and ultimately integrated in the field Φ (see Appendix).

This suggests than fundamentality is more of a *functional* concept in which the properties of entities, *i.e.* parts of its ideology, play the crucial role, and not entities themselves. Fundamentality is crucially determined by the values of couplings, and by the conserved charges or symmetries. What is mapped in S-dualities is not the entity with its fundamentality, but a bundle of properties. The non-fundamental entity has the ideology of the fundamental entity (from the same model of from a different one). For other dualities than self-dualities, each model lives in a different world so to speak, being subject to different symmetries, different coupling constant, different spacetimes. One may experience a shift from question (9) to (10) in which questions of 'grounding' still make sense, whereas 'existence' is harder to fathom.

Third, properties of one entity are related to the properties of its dual in unexpected ways. This is a very extravagant ideology going well beyond what the ontology allows: properties can be imported from one model to another and applied to fundamental and non-fundamental entities. The string metaphysician needs to be more tolerant about fundamentalism and concede to a new type of fundamentality in the line of ontological pluralism (Turner 2010). The above is a working definition of the new fundamentality envisaged here, definitely weaker than (11):

(12) **Relative and "dual" fundamentalism**: A class of entity (F) of a string model M_1 is more fundamental than (or, alternatively, they are in a 'priority' and 'grounding' relation with) all other entities (Particles, fields, gauge symmetries, black holes, soliton excitations of F, etc.). Class F is also (necessarily?) "dual" to a class d of non-fundamental entities from the dual model M_2 (a different model, or M_1 itself). Entity F can split in two other fundamental entities with a probability related to the coupling factor g_S. Two entities F can join, forming a new F.

Because F shares some properties with d, most likely properties of F are inferred (epistemically) from d. There is no grounding and priority relation between F and d. All these classes are relative to a set of coupling constants, most notably to g_S.

The reductive hierarchical doctrine, cherished by some metaphysicians, is inconsistent with (12). S-dualities do not reduce the number of fundamental entities to a more restrictive class, but show us in what sense fundamentality with duality is much weaker concept than the standard string fundamentalism (11). Here, ideology dominates the ontology.

6.1. *Stability of Fundamental Strings*

S-dualities can be used to clarify another question about strings as fundamental, "extended simples." Baker (2014) rejects the idea that strings are extended simples: the better option is to take string as "gunky," in its metaphysical meaning. He sees problems with both 'strings are simple' and with 'strings are extended'. The argument against (11) is that any string (closed or open) is able to split and to join. Strings will split eventually, sometimes, but they never split into a pointlike particle (or a zero-brane). The process of splitting is a probabilistic process and its amplitude is given by the coupling constant. For Baker, strings after splitting are strings, as the parent string before it. Hence, the "gunky" nature of strings. Strings are *not* like lines: divide them, and you always get strings, and never points. Further, Baker suggests that as the concept of distance is not well defined in string theory, the very idea of extended strings is problematic.

To save fundamentality as stated in (11) and the strings as simples against Baker's argument, the string metaphysician can make another move: use the conceptual distinction between perturbative and non-perturbative models through the lenses of S-dualities. In *some* perturbative models, at small scale and at strong coupling, splitting of strings may become relevant. The model-dependent metaphysics acknowledges the central role of abstraction and idealization in models. Sometimes, but not always, the string metaphysician can abstract away the splitting of strings. Past several string revolutions, string ontology is more complex: other fundamental entities are added and the fundamentality has to be reconceptualized. There is a class of "stable enough" entities that are approximately fundamental or "fundamental enough." In non-perturbative models then, "fundamental enough entities" are "simple enough," as it were. In perturbative models, other "fundamental enough" objects (D-branes, typically) will have the approximate role of "extended simples." There is always a class of heavy, composed and irrelevant objects, because are hard to model, which are not fundamental, but may become fundamental at another coupling regime, and a class of fundamental enough, stable enough entities which will be the "base." Put differently, the major heavylifting effort here is to accept that perturbative models of string theory with interactions are essentially probabilistic. The metaphysician may be very unhappy with this outcome and would resist to relinquish absolute stability for fundamental entities.[26]

6.2. *M-Fundamentalism and SUSY Skepticism*

It is germane to revisit the ideal of universality and unification which is obscured by the type of pluralism vindicated here. What if dualities are only mathematical redundancies paving the road to a unique and unified string model? Faced with string pluralism, the monist would hold on to reduction, to something more unified and more universal than the web of string models and their dualities. The best candidate is the M-theory (*aka* the "master theory") that arguably will come with its own fundamental entity and its ontology. All string models at all couplings will become low energy limits of a "master" theory with one fundamental entity – most likely it will not be a string, but some combination of non-commutative geometrical objects. There is then a duality between elementary entities and composite entities which may be "the structural realists' best 'physics-motivated case'" (Rickles 2011, p. 66). Rickles chooses to be a structural realist about M-theory. In a slogan, the ontic structural realist says: *dualities, because of structure*. A metaphysical program inspired by M-theory may become more committed to the ideal of a pure fundamentalism, unification and universality and deeply intertwined with ontic structural realism. The emphasis in this paper is that the jury is still out: *at this stage* of its development, a string metaphysics program entails pluralism: about parthood,

[26] The other parts of Baker's argument, about the concept of distance in string theory cannot be addressed with the resources of the present paper. String metaphysics, as devised here, places spacetime within its ideology.

about fundamentality, about grounding and priority and, about modality (subsection 6.3).

Finally, there is a skeptical argument against a string metaphysics: most of the aforementioned pluralist stances could be due to a "reification" of the symmetry put in all string models and in their dualities. Except the "exact" E-M duality, all dualities discussed need SUSY, which is a necessary constraint of all string models. The skeptic, opting for a more austere ideology, can retort: "you get as much metaphysics as you like if you assume rich structures (symmetries) in your models." SUSY is a maximal group, although it may not apply to our world, at least not at low energy. The skeptic insists that string theorists do not need to postulate anything above and beyond SUSY; it alone "dictates" which entities are fundamental, stable enough, simple enough etc., and which are not. Where SUSY is present, everything must go. The slogan of the skeptic is: *ex* SUSY *quodlibet sequitur*: "from SUSY, anything follows."

6.3. *Modality from Dualities: A First Stab*

Is there a new type of modality in string metaphysics? The naturalist claims that metaphysical modality is nothing above scientific modality, which is "murky" (Callender 2011, p. 44). Some types of modality (normative, scientific, epistemic, etc.) are not restricted forms of a generic modality. Nevertheless, they may be (or may be not) analogous one to the other, or related by a shared set of ideals. While only an informed speculation, this section argues that analytic modalities do not generalize well to string metaphysics. The proposal here is again to accept a modal pluralism that accommodates this type of semantics.

Other scientific modalities stemming from theoretical physics are deeply dependent on how laws of nature are conceptualized. This is already a problem for string metaphysics: as emphasized in subsection 2.3, the lack of laws in string theory shifts the emphasis on model-building and properties in the models. How much modality do string models "present"?[27] In the semantic view of science of P. Suppes, B. van Fraassen, etc., modality is generated by the model. This is probably a starting point for string models with no dualities. On the other hand, in the recent literature on structural realism, modality belongs to a collection of models connected by relations, and not to one model in particular (Brading 2011; French 2014).

[27] The distinction 'presentation-representation' of Brading and Landry (2006) illuminates the present argument.

This hints to a string modality in which models are collection of possible worlds and relations among models constraint what can happen in each world. Within one model, the structure of spacetime and the symmetries are the same, but the coupling constant g_s can have any value. The "minimal base of fundamental entities" also changes from strong coupling to weak coupling. The two sectors live in different worlds, as it were. Strong coupling brings in more quantum mechanical effects, less stable fundamental entities, and a hardly known formalism. We have more epistemic access to perturbative models and they fewer and simpler fundamental entities (strings, not branes for example). A model is too rich a structure, because different sectors of the same model do not have the same ontology and the same ideology. Here is a 'toy semantics' for possible worlds based on string models:

(13) **"Bare" string modality**: For a given model M, with a symmetry group G (*e.g.* SUSY), a set of fields Φ, a set of fundamental entities F and a set of non-fundamental, derivative entities d, and a spacetime manifold $\mathcal{M}$, a possible world is defined as the n-tuple: $w = \langle G, \Phi, \mathcal{M}, g_S, F, d\rangle$.

A world w is fixed by all the assumptions of the model and the value of g_S. Models then include sets of possible worlds, separated conventionally in two types: weak coupling worlds and strong coupling worlds, separated by the world with $g_S = 1$, call it $w_{g=1}$. While $w_{g=1}$ may look ordinary, it is more than a conventional borderline: here fundamental objects are equally composite and composite objects are equally fundamental. Metaphorically, in $w_{g=1}$ composition is fundamentality and vice-versa.

A metric relation is given foremost by g_S and then by G and $\mathcal{M}$: very weak coupling worlds ($w_{M, g \ll 1}$) of the same model M are similar to other weak coupling worlds of $w_{M, g < 1}$ and behave differently than any strong coupling worlds with ($w_{M, g \gg 1}$). As one moves to another model, worlds are less and less similar. Each world has a "fundamental enough" base with entities F, some entities derivable from F, but also other entities d, which are heavy, composite, or too complex. They are typically composed of Fs but too complicated to be derived in that model from F. Through duality, one can infer properties of the d from their weak duals.

Dualities create equivalence classes of models within (13): a class of possible worlds has the same BPS states, which are invariants of the duality transformations. Dualities relate worlds which share the BPS states, which is their shared ideology, but their spacetimes $\mathcal{M}$ and symmetry group G can be radically different. Call these worlds "dual worlds":

(14) "Dual" possible worlds: Two worlds:
$w_{strong} = \langle G_{strong}, \Phi_{strong}, \mathcal{M}_{strong}, g_{strong}, F_{strong}, d_{strong}\rangle$
and
$w_{weak} = \langle G_{weak}, \Phi_{weak}, \mathcal{M}_{weak}, g_{weak}, F_{weak}, d_{weak}\rangle$
are called "dual" if a relation of correspondence can be established between g_{weak} and g_{strong} (e.g. $g_{weak}\ g_{strong} = 1$) and some *BPS* states common to w_{strong} and w_{weak}. Two dual worlds share some properties of some of their objects, most notably: F_{strong} shares properties of d_{weak} and F_{weak} with d_{strong}.

Fundamental objects in the "dual worlds" are *always* different. A self-duality relates worlds of the same model, but not with the same *F*s. The invariant part of the ideology is shared by the 'dual worlds'.

Two final warnings to string modality enthusiasts. First, a M-fundamentalist may dissent by claiming that once we take M-theory seriously, all possible "dual worlds" are sub-worlds, or regions, of a larger, singular M-world built up from M-theory. Second, a SUSY skeptic may remind us that BPS states are all super-symmetric, such that the similarity of "dual" worlds is due to a common structure inserted by *fiat* in all models.

7. Conclusion

As a version of peaceful cohabitation between metaphysics and physics, this paper proposes a "metaphysics from physics" based on the operation of building string models. This project is permissive: models and theories in physics generate distinct metaphysical programs which do not aim primarily at replacing or reforming metaphysics. The focus is on the ontology and the ideology of string models, with and without dualities (only S-dualities are reviewed here). First, the metaphysician can opt for a strong string fundamentalism with no dualities, relatively similar to the one used in metaphysics. Second, if the metaphysician adopts a form of realism about S-dualities, this fundamentality looks different: it is relative to the "coupling constants" of the model; now fundamental entities split and join; and they are connected to their dual, not-fundamental entities. With duality, this paper suggests, ideology dominates the ontology and this entails a pluralism about fundamentality. There are still two pathways to retort this pluralism. First, a more optimistic argument that credits M-theory as the ultimate fundamental structure of string metaphysics. Second, a skeptical stance claiming that the whole string metaphysics is based on a reification of the supersymmetry. Based on fundamentality with duality, a string modality is sketched: possible worlds are identified with sectors of models and modality is "coded" by S-dualities. All in all, string models unveil enticing new ideas: the analytic metaphysician should take (a first) look at string theory!

University of Notre Dame
The Reilly Center for Science, Technology, and Values
e-mail: imuntean@nd.edu

ACKNOWLEDGMENTS

I want to thank Jeremy Butterfield, Craig Callender, Richard Dawid, Steven French, Jeffrey Harvey, Nick Huggett, Ken Intriligator, Dean Rickles, and Christian Wüthrich, for having the patience and the grace to discuss the relevance of quantum gravity to philosophy. Bits of my argument for a string-inspired metaphysics were presented between 2010 and 2013, sometimes in a radically different form, at the meetings of the *U.K. and European Association for the Foundations of Physics*, the *Philosophy of Science Association*, the *British Society for the Philosophy of Science*, the *Bucharest Colloquium in Analytic Philosophy* and *the Society for Exact Philosophy* and two conferences at the University of Bristol, U.K.: *Structure and Identity* and the *Structuralism in Physics and Mathematics*. Thanks to the audience, the referees and the organizers of these events. Last but not least, special thanks go to the editors and the referee of this volume for their patience and invaluable suggestions!

APPENDIX: TYPE *I* – *SO*(32) HETEROTIC DUALITY

The Type I – SO(32) duality was inferred from (a) the identification of the actions of the two models and (b) through the analysis of the spectrum of SUSY states.[28] Method (a) is described below.

The *SO*(32) model has bosonic massless states that come from the closed heterotic string, a metric $g^{(H)}_{\mu\nu}$, the dilaton field $\Phi^{(H)}$, the string tension and its inverse: $T_H = 1 / (2\pi\alpha'_H)$ and the coupling constant g_H by changing the metric. Its low energy action is:

[28] De Wit and Louis (1998) and Sen (1998) are used here.

(1)
$$S^{(H)} = \frac{1}{(2\pi)^7(\alpha'_H)^4 g_H^2}\int d^{10}x\sqrt{-g^{(H)}}\big[R^{(H)} - \frac{1}{8}g^{(H)\mu\nu}\Phi_\mu\Phi^{(H)}\Phi_\nu\Phi^{(H)}$$
$$-\frac{1}{4}g^{(H)\mu\mu'}g^{(H)\nu\nu'}e^{-\Phi^{(H)}/4}Tr(F^{(H)}_{\mu\nu}F^{(H)}_{\mu'\nu'})$$
$$-1\frac{1}{2}g^{(H)\mu\mu'}g^{(H)\nu\nu'}g^{(H)\rho\rho'}e^{-\Phi^{(H)}/2}H^{(H)}_{\mu\nu\rho}H^{(H)}_{\mu'\nu'\rho'}\big]$$

where $R^{(H)}$ is a Ricci scalar, $F^{(H)}_{\mu\nu}$ is a gauge field, and $H^{(H)}_{\mu\nu\rho}$ is the field strength of $B^{(H)}_{\mu\nu}$. For calculation purposes, one can rescale the whole action and some of the fields this way:

(2)
$$\begin{aligned}
g_H &\rightarrow e^C g_H\\
\Phi^{(H)} &\rightarrow \Phi^{(H)} - 2C\\
g^{(H)}_{\mu\nu} &\rightarrow e^{C/2} g^{(H)}_{\mu\nu}\\
B^{(H)}_{\mu\nu} &\rightarrow B^{(H)}_{\mu\nu}\\
A^{(H)\alpha}_{\mu} &\rightarrow A^{(H)\alpha}_{\mu}\\
\alpha'_H &\rightarrow \lambda\alpha'_H\\
\Phi^{(H)} &\rightarrow \Phi^{(H)}\\
g^{(H)}_{\mu\nu} &\rightarrow \lambda g^{(H)}_{\mu\nu}\\
B^{(H)}_{\mu\nu} &\rightarrow \lambda B^{(H)}_{\mu\nu}\\
A^{(H)\alpha}_{\mu} &\rightarrow \lambda^{1/2} A^{(H)\alpha}_{\mu}
\end{aligned}$$

By analyzing the invariance of this rescaling, one can show that g_H and α'_H are changed, so they are not universal parameters of the model. One can change the variables to: $e^{-C} = g_H$; $\lambda = (\alpha'_H)^{-1}$ which is the same as setting $g_H = \alpha'_H = 1$ by a choice of units. Here the coupling constant is given by the vacuum expectation of $e^{\Phi(H)/2}$. By changing the $\langle\Phi^{(H)}\rangle$, this string model covers all the values of coupling g_H and all values of the string tension T_H.

After rescaling, for $\alpha'_H = 1$; $g_H = 1$ from (1) one infers:

(3)
$$S^{(H)} = \frac{1}{(2\pi)^7}\int d^{10}x\sqrt{-g^{(H)}}\big[R^{(H)} - \frac{1}{8}g^{(H)\mu\nu}\Phi_\mu\Phi^{(H)}\Phi_\nu\Phi^{(H)}$$
$$-\frac{1}{4}g^{(H)\mu\mu'}g^{(H)\nu\nu'}e^{-\Phi^{(H)}/4}Tr(F^{(H)}_{\mu\nu}F^{(H)}_{\mu'\nu'})$$
$$-\frac{1}{12}g^{(H)\mu\mu'}g^{(H)\nu\nu'}g^{(H)\rho\rho'}e^{-\Phi^{(H)}/2}H^{(H)}_{\mu\nu\rho}H^{(H)}_{\mu'\nu'\rho'}\big]$$

and some corresponding transformations for the field $F_{\mu\nu}$.

The dual model of the $SO(32)$ theory is the type I string model theory in which the massless bosonic states come from different sectors: the closed string Neveu-Schwarz (NS) sector contains $g^{(I)}_{\mu\nu}$ and $\Phi^{(I)}$, and the closed string Ramond-Ramond (RR) sector has the tensor field $B^{(I)}_{\mu\nu}$. The low energy action needs the $\mathcal{N} = 1$ supergravity theory coupled to a super Yang-Mills theory:

(4) $$S^{(I)} = \frac{1}{(2\pi)^7}\int d^{10}x\sqrt{-g^{(I)}}\Big[R^{(I)} - \frac{1}{8}g^{(I)\mu\nu}\Phi_\mu\Phi^{(I)}\Phi_\nu\Phi^{(I)} - \frac{1}{4}g^{(I)\mu\mu'}g^{(I)\nu\nu'}e^{\Phi^{(I)}/4}Tr(F^{(I)}_{\mu\nu}F^{(I)}_{\mu'\nu'}) - \frac{1}{12}g^{(I)\mu\mu'}g^{(I)\nu\nu'}g^{(I)\rho\rho'}e^{\Phi^{(I)}/2}H^{(I)}_{\mu\nu\rho}H^{(I)}_{\mu'\nu'\rho'}\Big]$$

where $F^{(I)}_{\mu\nu}$ is the strength of a non-Abelian gauge field, and $H^{(I)}_{\mu\nu\rho}$ is the field strength associated of $B^{(I)}_{\mu\nu}$. The main result in Witten (1995) is the identification of different variables in (2) and (4):

(5) $$\Phi^{(H)} = -\Phi^{(I)}$$
$$g^{(H)}_{\mu\nu} = g^{(I)}_{\mu\nu}$$
$$B^{(H)}_{\mu\nu} = B^{(I)}_{\mu\nu}$$
$$A^{(H)\alpha}_{\mu} = A^{(I)\alpha}_{\mu}$$

As $e^{\langle\Phi\rangle/2}$, it follows that the strong coupling limit of $SO(32)$ is related to the weak coupling limit of the Type I model and *vice versa*.

REFERENCES

Atkinson, D. (2005). A New Metaphysics: Finding a Niche for String Theory. In: R. Festa, A. Aliseda, and J. Peijnenburg (eds.), *Cognitive Structures in Scientific Inquiry: Essays in Debate with Theo Kuipers* (*Poznań Studies in the Philosophy of the Sciences and the Humanities*, vol. 2). Amsterdam: Rodopi.

Baker, D. J. (2014). *Does string theory posit extended simples?* http://philsci-archive.pitt.edu/11053/.

Becker, K., Becker, M., Schwarz, J.H. (2007). *String Theory and M-theory.* New York: Cambridge University Press.

Bell, J.S. (1987). The Theory of Local Beables. In: *Speakable and Unspeakable in Quantum Mechanics: Collected Papers on Quantum Philosophy* (2 ed.), pp. 52–63. Cambridge: Cambridge University Press.

Brading, K. (2011). Structuralist approaches to physics: Objects, models and modality. In: A. Bokulich and P. Bokulich (eds.), *Scientific Structuralism* (*Boston Studies in the Philosophy of Science*, vol. 281), pp. 43–65. Dordrecht, New York: Springer.

Brading, K., Landry, E. (2006). Scientific structuralism: Presentation and representation. *Philosophy of Science* **73** (5), 571–581.

Butterfield, J. (2010). Less is Different: Emergence and Reduction Reconciled. *Foundations of Physics* **41**, 1065–1135.

Callender, C. (2011). Philosophy of Science and Metaphysics. In: S. French, J. Saatsi (eds.), *Continuum Companion to the Philosophy of Science*, pp. 33–54. London: Continuum.

Callender, C., Huggett, N., eds. (2001a). *Physics Meets Philosophy at the Planck Scale: Contemporary Theories in Quantum Gravity*. Cambridge: Cambridge University Press.

Callender, C., Huggett, N. (2001b). Why Quantize Gravity (or any Other Field for that Matter)? *Philosophy Of Science* **68** (3), S382-S394.

Cameron, R.P. (2008). Turtles all the Way Down: Regress, Priority and Fundamentality. *The Philosophical Quarterly* **58** (230), 1–14.

Cappelli, A., Colomo, F., Di Vecchia, P., Castellani, E. eds. (2012). *The Birth of String Theory.* Cambridge: Cambridge University Press.

Chakravartty, A. (2013). On the Prospects of Naturalized Metaphysics. In: J. Ladyman, H. Kincaid and D. Ross (eds.), *Scientific Metaphysics*, pp. 27–50. Oxford: Oxford University Press.

Correia, P.F., Schnieder, P.B. (2012). *Metaphysical Grounding: Understanding the Structure of Reality.* Cambridge: Cambridge University Press.

Crowther, K., Rickles, D (2014). Introduction: Principles of Quantum Gravity. *Studies in History and Philosophy of Science Part B, Studies in History and Philosophy of Modern Physics* **46 B**, 135–141.

Dawid, R. (2007). Scientific Realism in the Age of String Theory. *Physics & Philosophy* **11**.

Dawid, R. (2009). On the Conflicting Assessments of the Current Status of String Theory. *Philosophy of Science* **76** (5), 984–996.

Dawid, R. (2013). *String Theory and the Scientific Method.* Cambridge: Cambridge University Press.

Dawid, R., Hartmann, S., Sprenger, J.. The No Alternatives Argument. *The British Journal for the Philosophy of Science* **66**, 213–34. doi:10.1093/bjps/axt045.

de Wit, B., Louis, J. (1998). *Supersymmetry and Dualities in Various Dimensions.* arXiv:hep-th/9801132 .

Font, A., Ibáñez, L.E., Lüst, D., Quevedo, F. (1990). Strong-Weak Coupling Duality and Non-Perturbative Effects in String Theory. *Physics Letters B* **249** (1), 35–43.

French, S. (1998). On the Withering away of Physical Objects. In: E. Castellani (ed.), *Interpreting Bodies*, pp. 93–113. Princeton: Princeton University Press.

French, S., McKenzie, K. (2012). Thinking outside the Toolbox: Towards a more Productive Engagement between Metaphysics and Philosophy of Physics. *European Journal of Analytic Philosophy* **8** (1), 42–59.

French, S. (2014). *The Structure of the World. From Representation to Reality.* Oxford: Oxford University Press.

Friedman, M. (2001). *Dynamics of Reason.* Stanford, CA: Center for the Study of Language and Inf.

Giddings, S.B. (2006). Locality in Quantum Gravity and String Theory. *Physical Review D* **74** (10), 106006.

Godfrey-Smith, P. (2006). Theories and Models in Metaphysics. *Harvard Review of Philosophy* **14**, 4–19.

Godfrey-Smith, P. (2012). Metaphysics and the Philosophical Imagination. *Philosophical Studies* **160** (1), 97–113.

Green, M.B., Schwarz, J.H., Witten, E. (1987). *Superstring Theory.* Cambridge: Cambridge University Press.

Greene, B. (1999). *The Elegant Universe: Superstrings, Hidden Dimensions, and the Quest for the Ultimate Theory.* New York: W.W. Norton.

Greene, B. (2011). *The Hidden Reality: Parallel Universes and the Deep Laws of the Cosmos.* New York: Knopf.

Gross, D., Henneaux, M., Sevrin, A., eds. (2007). *The Quantum Structure of Space and Time (Proceedings of the 23rd Solvay Conference on Physics Brussels*, Belgium, 1–3 December 2005). World Scientific Pub Co Inc.

Harvey, J.A. (1996). *Magnetic Monopoles, Duality, and Supersymmetry.* arXiv:hep-th/9603086.

Hudson, H. (2005). *The Metaphysics of Hyperspace*. Oxford, New York: Oxford University Press.

Huggett, N., Wüthrich, C. (2013). The Emergence of Spacetime in Quantum Theories of Gravity. *Studies in History and Philosophy of Science Part B, Studies in History and Philosophy of Modern Physics* **44** (3), 273–275.

Hull, C.M., Townsend, P.K. (1994). Unity of Superstring Dualities. *Nucl. Phys. B* **438**, 109–137.

Intriligator, K., Seiberg, N. (1996). *Lectures on Supersymmetric Gauge Theories and Electric-magnetic Duality. Nuclear Physics B – Proceedings Supplements* **45** (2–3), 1–28. arXiv: hep-th/9509066.

Ladyman, J. (2012). Science, Metaphysics and Method. *Philosophical Studies* **160** (1), 31–51.

Ladyman, J., Ross, D., Spurrett, D., Collier, J.G. (2007). *Every Thing Must Go: Metaphysics Naturalized.* Oxford, New York: Oxford University Press.

Lerche, W. (1997). Introduction to Seiberg-Witten Theory and Its Stringy Origin. *Fortschritte der Physik/Progress of Physics* **45** (3–4), 293–340. arXiv:hep-th/9611190.

Lewis, D.K. (1986). *On the Plurality of Worlds*. Oxford, New York: Blackwell.

Lowe, E.J. (2011). The Rationality of Metaphysics. *Synthese* **178** (1), 99–109.

Maudlin, T. (2007). *The Metaphysics within Physics.* Oxford: Oxford University Press.

McDaniel, K. (2007). Extended simples. *Philosophical Studies* **133** (1), 131–141.

McKenzie, K. (2011). Arguing against Fundamentality. *Studies In History and Philosophy of Science Part B, Studies In History and Philosophy of Modern Physics* **42** (4), 244–255.

Monton, B. (2011). Prolegomena to any Future Physics-Based Metaphysics. In: J.L. Kvanvig (ed.), *Oxford Studies in Philosophy of Religion*, Volume 3, pp. 142–165. Oxford: Oxford University Press.

Niedermaier, M. (2007). The Asymptotic Safety Scenario in Quantum Gravity: An Introduction. *Classical and Quantum Gravity* **24** (18), R171.

Paul, L. (2012). Metaphysics as Modeling: The Handmaiden Tale. *Philosophical Studies* **160** (1), 1–29.

Polchinski, J.G. (1996). String Duality. *Reviews of Modern Physics* **68** (4), 1245–1258.

Polchinski, J.G. (1998b). *String Theory: An Introduction to the Bosonic String*, Volume 1. Cambridge, New York: Cambridge University Press.

Polchinski, J.G. (1998a). *String Theory: Superstring Theory and Beyond*, Volume 2. Cambridge, New York: Cambridge University Press.

Price, H. (2009). *Metaphysics after Carnap: The Ghost who Walks?* In: D. Manley, R. Wasserman and D.J. Chalmers (eds.), *Metametaphysics: New Essays on the Foundations of Ontology*, pp. 260–289. Oxford: Oxford University Press.

Quine, W.V. (1951). Ontology and Ideology. *Philosophical Studies* **2** (1), 11–15.

Rickles, D. (2011). A Philosopher Looks at String Dualities. *Studies In History and Philosophy of Science Part B, Studies in History and Philosophy of Modern Physics* **42** (1), 54–67.

Rickles, D. (2013). AdS/CFT duality and the emergence of spacetime. *Studies in History and Philosophy of Science Part B: Studies in History and Philosophy of Modern Physics* **44** (3), 312–320.

Rickles, D.P. (2014). *A Brief History of String Theory. From Dual Models to M-Theory*. Berlin-Heidelberg: Springer.

Ross, D., Ladyman, J., Kincaid, H. (2013). *Scientific Metaphysics.* Oxford: Oxford University Press.

Schaffer, J. (2003). Is There a Fundamental Level? *Noûs* **37** (3), 498–517.

Schaffer, J. (2009). On what grounds what. In: R. Wasserman, D.J. Chalmers, and D. Manley (eds.), *Metametaphysics: New essays on the foundations of ontology*, pp. 347–383. Oxford: Oxford University Press.

Seiberg, N., Witten, E. (1994). Electric-Magnetic Duality, Monopole Condensation, and Confinement in N=2 Supersymmetric Yang-Mills Theory. *Nuclear Physics B* **426** (1), 19–52.

Sen, A. (1994). Strong-Weak Coupling Duality in Four Dimensional String Theory. *Int. J. Mod. Phys. A* **9**, 3707–3750.

Sen, A. (1998). *An Introduction to Non-Perturbative String Theory.* arXiv:hep-th/9802051.

Sider, T. (2009). Ontological Realism. In: Chalmers, D.J., D. Manley, and R. Wasserman (eds.), *Metametaphysics: New Essays on the Foundations of Ontology*, pp. 77–129. Oxford, New York: Oxford University Press.

Sieroka, N., Mielke, E.W. (2014). Holography as a Principle in Quantum Gravity? Some Historical and Systematic Observations. *Studies in History and Philosophy of Science Part B, Studies in History and Philosophy of Modern Physics* **46**, 170–178.

Smolin, L. (2002). *Three Roads to Quantum Gravity.* New York: Basic Books.

Stanford, P.K. (2006). *Exceeding Our Grasp: Science, History, and the Problem of Unconceived Alternatives.* New York: Oxford University Press.

Teh, N.J. (2013) Holography and emergence. *Studies in History and Philosophy of Science Part B, Studies in History and Philosophy of Modern Physics* **44** (3), 300–311

Turner, J. (2010). Ontological Pluralism. *Journal of Philosophy* **107** (1), 5–34.

Vafa, C. (1997). Geometric Origin of Montonen-Olive Duality. *Adv. Theor. Math. Phys.* **1**, 158–166.

Wayne, A. (2006). Particle Physics. In: S. Sarkar, J. Pfeifer (eds.), *The Philosophy of Science. An Encyclopedia*, Volume 2 (N-Z). London: Routledge.

Weingard, R. (1988). *A Philosopher Looks at String Theory. PSA: Proceedings of the Biennial Meeting of the Philosophy of Science Association* **2**, 95–106.

Wilson, J. (2012). Fundamental Determinables. *Philosophers' Imprint* **12** (4), 1–17.

Witten, E. (1995). String Theory Dynamics in Various Dimensions. *Nuclear Physics B* **443** (1–2), 85–126.

Wüthrich, C. (2005). To Quantize or not to Quantize: Fact and Folklore in Quantum Gravity. *Philosophy of Science* **72**, 777–788.

Adam Caulton

IS MEREOLOGY EMPIRICAL? COMPOSITION FOR FERMIONS

ABSTRACT. How best to think about quantum systems under permutation invariance is a question that has received a great deal of attention in the literature. But very little attention has been paid to taking seriously the proposal that permutation invariance reflects a representational redundancy in the formalism. Under such a proposal, it is far from obvious how a constituent quantum system is represented. Consequently, it is also far from obvious how quantum systems compose to form assemblies, i.e. what is the formal structure of their relations of parthood, overlap and fusion. In this paper, I explore one proposal for the case of fermions and their assemblies. According to this proposal, fermionic assemblies which are not entangled – in some heterodox, but natural sense of 'entangled' – provide a *prima facie* counterexample to classical mereology. This result is puzzling; but, I argue, no more unpalatable than any other available interpretative option.

1. Introduction

The fact that quantum mechanics forces a revision of many of our dearly held metaphysical beliefs is by now familiar. In this article, I aim to provide one more example of such a metaphysical belief; namely that classical mereology – by which I mean the formal theory of parts and wholes developed by, amongst others, Leśniewski (1916), Tarski (1929) and Leonard and Goodman (1940) – gives a true account of the structure of composition for physical objects.

One might suppose that the culprit responsible for the failure of mereology in the quantum domain is entanglement. As is well known, any entangled state fails to supervene on (i.e. be determined by) the states of the joint system's constituents – supposing that we are happy attributing

In: Tomasz Bigaj and Christian Wüthrich (eds.), *Metaphysics in Contemporary Physics* (*Poznań Studies in the Philosophy of the Sciences and the Humanities*, vol. 104), pp. 293-322. Amsterdam/New York, NY: Rodopi | Brill, 2015.

the constituents with states at all.[1] This has prompted some (e.g. Maudlin 1998) to claim that a certain strong version of reductionism fails for quantum systems.

Could this failure of reductionism lead to a failure of mereology? Surely not. For the failure of reductionism is merely a failure of the joint system's properties (as encapsulated by its state) to supervene on the constituents' *properties* (as encapsulated by *their* states). And the possibility that there may be irreducible relations was countenanced at least as far back as Russell (1918). Thus, entanglement merely demands that we accept irreducible relations into our ontology of quantum mechanics. (A point also argued by Ladyman and Ross 2007, pp. 149–150.)

A more pressing problem for quantum mereologists is raised for those who take quantum particles to be indiscernible in a sense stronger than can be articulated in the framework of classical logic and set theory. This stronger sense is explicated by 'quasi-set' theory, developed by da Costa (1980), Krause (1990, 1996), Dalla Chiara, Giuntini and Krause (1998) and pursued in further detail by French and Krause (2006, Chs. 7 and 8). According to quasi-set theory, it may be said of certain individuals – or rather "non-individuals", called '*m*-atoms' – that they are either discernible or indiscernible, but not that they are identical, nor that they are distinct. As French and Krause (2006, pp. 278–281) note, although quasi-set theory defines the containment relation between quasi-sets, no theory yet exists which describes how their ur-elements, the *m*-atoms, compose.[2]

While it is far from unanimous that quasi-set theory can offer the right account of quantum particles, the conviction that motivates the theory's application to quantum particles enjoys near universal assent: namely, that elementary quantum particles of the same species are indiscernible in some important sense. The precise sense of this indiscernibility has been honed in the recent literature (see French and Krause 2006, §4.2.1; Muller and Saunders 2008; Muller and Seevinck 2009). But these indiscernibility results are underpinned by an interpretative assumption that I wish here to deny – or, at least, entertain the cogency of denying.[3]

The interpretative assumption regards an important constraint in quantum mechanics, known as *permutation invariance*. The assumption is that, while states of elementary particles obey permutation invariance, still sense can be given to the permutations: specifically, they represent a literal

[1] I am picking up here on the fact that, in any entangled state, the constituents' states, as calculated by performing a partial trace on the joint state, are improper, i.e. not ignorance-interpretable, mixtures.

[2] I am grateful to George Darby for correspondence on these matters.

[3] I return to the issue of indiscernibility in Section 4.1, following Corollary 4.7.

swapping of the particles. On the view I wish to pursue here, permutation invariance is not a feature of states that has anything to do with the physical swapping of particles, but rather reflects a representational redundancy in the standard formalism of quantum mechanics. This redundancy is somewhat analogous to the representational redundancy that counterpart theorists see in standard Tarskian semantics, in which two distinct models can differ purely as to which object plays which role in the pattern of instantiation of properties and relations.

The problem posed for mereology that I wish to present here arises just for those, like me, who wish to take the hard line that permutation invariance reflects representational redundancy in the standard tensor product Hilbert space formalism. The problem arises specifically for fermions, which include the particles which "make up" all stable matter – in *some* sense of "make up"! We shall see that the hard line interpretation of permutation invariance prompts a revision to some common quantum concepts, most importantly for us here, what counts as an entangled state. On this revision, some joint states of fermions count as not entangled. The contradiction with mereology will be proven for these states alone: thus the failure of mereology has nothing do with entanglement. (Whether the contradiction between mereology and the hard line on permutation invariance speaks more against the latter than the former is a question I shall address in due course.)

The argument will run as follows. Upon interpreting the permutation invariance of fermionic joint states in the correct way, we shall see that these states may be represented naturally by subspaces of the single-system Hilbert space. More specifically, a non-entangled joint state of N fermions will correspond to an N-dimensional subspace. A natural definition of parthood will emerge, which is represented by the relation of subspacehood. The associated notion of mereological fusion between fermionic assemblies – which can be defined using parthood alone – will then be shown *not* to match the structure that smaller fermionic assemblies (i.e. assemblies containing fewer fermions) bear to larger ones (i.e. assemblies containing more fermions).

Some strategies to save mereology in the face of this result will be considered. As I will argue, one successful strategy exists, in the sense that the truth of the axioms of mereology may be preserved. However, it will remain the case that the operation which unifies many fermions into one fermionic assembly is *not* mereological fusion.

The title of this article is intended to reference the famous discussion of quantum logic, inspired by the work of Birkhoff and von Neumann (1936), in Putnam (1968, 1974) and Dummett (1976), and continued by Maudlin (2005) and Baciagaluppi (2009). Indeed we shall see that the analogy here is exact: fermionic composition is to quantum logic as mereological composition is to classical logic. That is, while it is well-known that the mathematical structures used to describe the relationship between objects in mereology or propositions in classical logic are the same (namely, Boolean algebras), so too the mathematical structures used to describe the relationship between *fermions* or propositions in *quantum* logic are the same (namely, Hilbert lattices).

The structure of the article is as follows. In Section 2, I discuss briefly the issue of permutation invariance in quantum mechanics. In Section 3, I outline classical mereology and offer a means to "translate" the states of quantum mechanics into Tarskian models, so that the question of whether mereology holds of them can meaningfully be put. Section 4 contains the main results, and considers one saving strategy for mereology.

2. Permutation-Invariant Quantum Mechanics

In this Section, I will introduce the key formal and interpretative basics for our discussion of fermionic composition. The main motivation is to define (or redefine) *entanglement* in a permutation-invariant setting in a physically salient way, particularly for fermions. I will argue that, under the most appropriate definitions of these terms, there are fermionic states that are not entangled; these states will be the focus of the results of Section 4.

2.1. *Permutation Invariance and Its Interpretation*

Permutation-invariant quantum mechanics is standard quantum mechanics with the additional condition of permutation invariance. We begin with the single-system Hilbert space $\mathcal{H}$. From this we define the N-fold tensor product $\otimes^N\mathcal{H}$, the *prima facie* state space for N "indistinguishable" systems (their indistinguishability is expressed by the fact that any two factor Hilbert spaces are unitarily equivalent). The joint Hilbert space $\otimes^N\mathcal{H}$ carries a natural unitary representation U: $S_N \rightarrow \mathcal{U}(\otimes^N\mathcal{H})$ of the group S_N of permutations on N symbols. For example, the permutation (ij), which swaps systems i and j, is represented by the unitary operator $U(ij)$ defined on basis states (having chosen an orthonormal basis $\{|\varphi_k\rangle\}$ on $\mathcal{H}$) by

(1) $$U(ij)\,|\varphi_{k1}\rangle \otimes \ldots \otimes |\varphi_{ki}\rangle \otimes \ldots \otimes |\varphi_{kj}\rangle \otimes \ldots \otimes |\varphi_{kN}\rangle = |\varphi_{k1}\rangle \otimes \ldots \otimes |\varphi_{kj}\rangle \otimes \ldots \otimes |\varphi_{ki}\rangle \otimes \ldots \otimes |\varphi_{kN}\rangle$$

and then extended by linearity. Permutation invariance, otherwise known as the *Indistinguishability Postulate* (Messiah and Greenberg 1964, French and Krause 2006), is the condition on any bounded operator $Q \in \mathcal{B}(\otimes^N\mathcal{H})$, that it be *symmetric*[4]; i.e. for all permutations $\pi \in S_N$ and all states $|\psi\rangle \in \otimes^N \mathcal{H}$,

(2) $\langle\psi|U^\dagger(\pi)QU(\pi)|\psi\rangle = \langle\psi|Q|\psi\rangle$

The representation U is reducible, and decomposes into several copies of inequivalent irreducible representations, each irreducible representation corresponding to a different *symmetry type*; namely bosonic states, fermionic states and (if $N \geq 3$) a variety of paraparticle states (see e.g. Tung 1985, Ch. 5). If we consider only the information provided by the symmetric operators, we treat permutation invariance as a superselection rule, and each superselection sector corresponds to one of these symmetry types.

What does it mean to "impose" permutation invariance? Isn't it rather that permutation invariance holds of some operators and not others? I propose that imposing permutation invariance means to lay it down as a necessary condition on any operator's receiving a physical interpretation. This justifies, and is justified by, treating the factor Hilbert space labels – i.e. the order in which single-system operators and states lie in the tensor product – as nothing but an artefact of the mathematical formalism of quantum mechanics. That is a heterodox position and, as we shall see (Section 4.1), it leads to the overturning of many commonly held beliefs in the quantum philosophy literature.

So what is the justification for interpreting the factor Hilbert space labels in this heterodox way? Of course, the ultimate justification is that it leads to an empirically adequate theory. And while it is an empirical fact that elementary particles exhibit statistics consistent with their being either bosons or fermions, this fact is logically weaker than the claim that factor Hilbert space labels represent nothing. It *could* be, as is commonly assumed, that factor Hilbert space labels represent (or name) the elementary constituent systems, and that the joint state of any assembly of elementary particles remains in the fermionic or bosonic sector under all actual physical evolutions due only to the fact that the corresponding Hamiltonian happens to be permutation-invariant. Indeed, this interpretative gloss is offered by many authors (e.g. French and Redhead 1988; Butterfield 1993; Huggett 1999, 2003; French and Krause 2006; Muller and Saunders 2008; Muller and Seevnick 2009; Caulton 2013).

However, I wish to suggest that the physical emptiness of the factor Hilbert space labels offers the *best explanation* of the empirical fact that permutation invariance always holds true. This suggestion is in line with a

---

[4] This use of 'symmetric' is not to be confused with the condition that $\langle\psi|Q\varphi\rangle = \langle Q\psi|\varphi\rangle$ for all $|\psi\rangle, |\varphi\rangle \in \mathrm{dom}(Q)$.

more general interpretative stance in physics: that any exact symmetry is a symptom of representational redundancy in the corresponding theory's formalism.

My claim to offering the 'best explanation' of permutation invariance is in need of some elaboration.[5] Of course, I am not claiming that the interpretative line that permutation-invariance reflects representational redundancy is empirically better supported than its rivals: the relevant empirical predictions (namely, Bose-Einstein and Fermi-Dirac statistics) are identical under any interpretation and not at issue. Moreover, it must be admitted that any proposed explanation is to be judged according to the commitments it entails. Given the results of Section 4.1, if mereology is held sacred, this (modulo Section 4.2's considerations) will be an overriding consideration against the interpretation of permutation invariance proposed here. With this in mind, I will now articulate two theoretical considerations which, all else being equal, favour the representational redundancy interpretation over the orthodox interpretation. I will then address quasi-set theory separately.

The first consideration is Ockham's razor, according to which, all else being equal, simpler explanations are to be preferred over less parsimonious ones. In this case, Ockham's razor is to be applied to the ontology required for each explanation of permutation invariance; specifically, the state-independent properties one takes quantum particles to possess. According to the orthodox line, aside from the familiar properties (mass, charge, spin) and the state-dependent properties (represented by projectors on the single-particle Hilbert space) each particle must have one other thing: whatever is represented by the factor Hilbert space labels. According to the heterodox line, each particle has only mass, charge, spin, and whatever state-dependent properties. It will not help to clarify, in defence of the orthodox line, that a factor Hilbert space label does not represents any *property* (except perhaps '$x = a$' for each particle a), but rather the particle itself, and so no extra ontological commitments are incurred. For it still follows, under the orthodox line, that particles may be individuated independently of their state-dependent properties. There is no call for such 'transcendental individuality' (to use French and Redhead's 1988 phrase), as the heterodox line shows.

The second consideration is the injunction that one's explanations should minimize, as much as possible, the facts one needs to take as brute. According to the orthodox line, permutation invariance is just a contingent feature of the assembly's Hamiltonian, and so must be posited as a brute

[5] Thanks are due to an anonymous referee for pressing me on this point.

fact alongside the existence of the particles themselves. According to the heterodox line, this feature of the Hamiltonian is mandated at the outset: since factor Hilbert space labels represent nothing at all, any physical quantity (such as the Hamiltonian) must be invariant under their arbitrary permutation.

These two considerations are related, since a more parsimonious ontology allows for fewer possibilities, and consequently fewer *unrealised* possibilities that one must rule out by positing brute facts. Similar considerations have motivated a variety of existence-denying moves in the history of physics: e.g. Ptolemy's epicycles (why do they always lead to elliptical orbits?); the luminiferous ether (why does length contraction and time dilation conspire so as to prevent us from identifying the ether rest frame?); and gravity, as separate from geometrical-inertial structure (why are inertial mass and gravitational charge always equal?).

So much for the orthodox interpretation of permutation invariance. But there is a significant rival heterodox interpretation that I have so far failed to mention: the "non-individuals" approach, as explicated by quasi-set theory. If quantum particles are quasi-set-theoretical *m*-atoms, then they do not possess transcendental individuality, and they presumably cannot be described in anything other than a permutation-invariant theory. So the heterodox interpretation I wish to urge here cannot claim the edge over the "non-individuals" interpretation on the basis of the above considerations. However, there is an additional consideration. My heterodox line preserves classical logic – specifically, the law that every object is self-identical ($\forall x\, x = x$) – while quasi-set theory suspends this law for *m*-atoms (French and Krause 2006, p. 5). And while I wish to suggest that empirical considerations may bring the sanctity of mereology into question, I demur from suggesting the same of logic.

2.2. *Fermionic States and GMW-Entanglement*

The focus of this paper is fermionic states and their compositional structure. Picking some orthonormal basis $\{|\varphi_i\rangle\}$ in $\mathscr{H}$, these states are spanned by states of the form

$$(3) \qquad (1/\sqrt{N!})\Sigma_{\pi \in S(N)} (-1)^{\deg \pi} |\varphi_{i(\pi(1))}\rangle \otimes |\varphi_{i(\pi(2))}\rangle \otimes \ldots \otimes |\varphi_{i(\pi(N))}\rangle$$

and carry the alternating irreducible representation of S_N; i.e. any permutation π is represented by multiplication by $(-1)^{\deg \pi}$, where deg π is the *degree* of the permutation π (i.e. the number of pairwise swaps into which π may be decomposed).

Following Ladyman, Linnebo and Bigaj (2013), we may use the mathematical apparatus of *Grassmann* or *exterior algebras* to represent fermionic states. The exterior algebra $\Lambda(V)$ over the vector space V (over the field of complex numbers $\mathbb{C}$) is obtained by quotienting the tensor algebra $T(V) := \oplus_{k=0}^{\infty} T^k(V) = \mathbb{C} \oplus V \oplus (V \otimes V) \oplus (V \otimes V \otimes V) \oplus \ldots$ with the equivalence relation ~ defined so that $\alpha \sim \beta$ iff α and β have the same anti-symmetrization;[6] i.e.

(4) $\Lambda(V) := T(V) /\!\sim$

For example, $[x \otimes y] = [-y \otimes x]$ and $[x \otimes x] = [\mathbf{0}]$. We may set $V = \mathcal{H}$, then there is a natural isomorphism ι from the elements of $\Lambda(\mathcal{H})$ onto the vectors of the fermionic Fock space $\mathcal{F}(\mathcal{H}) := \oplus_{N=0}^{\dim H} \mathcal{A}(\otimes^N\mathcal{H})$. ι simply takes any ~-equivalence class of degree-r vectors of $T^r(\mathcal{H})$ to the anti-symmetric degree-r vector in $\mathcal{A}(\otimes^r\mathcal{H})$ that is their common anti-symmetrization. Therefore we may pick out any N-fermion state in $\mathcal{A}(\otimes^N\mathcal{H})$ by specifying its pre-image under ι in $\Lambda^N(\mathcal{H})$ (i.e. the subalgebra of $\Lambda(\mathcal{H})$ containing only degree-N vectors).

Elements of $\Lambda(V)$ are called *decomposable* iff they are equivalence classes $[x_{i(1)} \otimes x_{i(2)} \otimes \ldots \otimes x_{i(r)}]$ containing product vectors. Not all elements are decomposable; an example is given at the end of Section 3.3.

The product on the exterior algebra is the *exterior* or *wedge product* $\wedge$, defined by its action on decomposable elements as follows:

(5) $[x_{i(1)} \otimes x_{i(2)} \otimes \ldots \otimes x_{i(r)}] \wedge [x_{i(r+1)} \otimes x_{i(r+2)} \otimes \ldots \otimes x_{i(r+s)}] = [x_{i(1)} \otimes x_{i(2)} \otimes \ldots \otimes x_{i(r+s)}]$

where $\{x_1, x_2, \ldots, x_{\dim V}\}$ is an orthonormal basis for V and each $i_k \in \{1, 2, \ldots, \dim V\}$. We then extend the definition of $\wedge$ to non-decomposable elements by bilinearity. (Note that if there is a pair $i_j = i_k$ for $j \neq k$, then the righthand side of (5) is $[\mathbf{0}]$.) For any $\alpha \in \Lambda^r(V)$ and any $\beta \in \Lambda^s(V)$, $\alpha \wedge \beta = (-1)^{rs} \beta \wedge \alpha \in \Lambda^{r+s}(V)$.

In the following, I will, like Ladyman, Linnebo and Bigaj (2013), make use of a harmless abuse of notation by referring to anti-symmetric states by their corresponding wedge product. In particular, given an orthonormal basis $\{|\varphi_i\rangle\}$ on $\mathcal{H}$,

(6) $|\varphi_{i(1)}\rangle \wedge |\varphi_{i(2)}\rangle \wedge \ldots \wedge |\varphi_{i(N)}\rangle$

will be used as a shorthand for

(7) $(1/\sqrt{N!})\Sigma_{\pi \in SN} (-1)^{deg\,\pi} |\varphi_{i\pi(1)}\rangle \otimes |\varphi_{i\pi(2)}\rangle \otimes \ldots \otimes |\varphi_{i\pi(N)}\rangle$

[6] Equivalently, $\Lambda(V)$ is the quotient algebra $T(V)/D(V^2)$ of $T(V)$ by the two-sided ideal $D(V^2)$ generated by all 2-vectors of the form $x \otimes x$. See e.g. Mac Lane and Birkoff (1991, §XVI.6).

The distinction between decomposable and non-decomposable fermionic states has a clear analogy with the distinction between product and non-product states. However, decomposable fermionic states have the property, unlike product states, that there are (up to a possible factor of -1) invariant under arbitrary permutations in their factor space indices. Therefore the wedge product offers a permutation-invariant way of constructing joint states of, say N fermions, out of N fermion states, much as the tensor product offers a permutation *non-invariant* way of constructing joint states for "distinguishable" systems.

The analogies between the tensor product in the permutation-non-invariant case and the wedge product in the fermionic case suggest a redefinition of entanglement for fermionic states. The standard definition, with which we do not take issue in the permutation-non-invariant case, is that an assembly's state is entangled iff it is non-separable, i.e. it cannot be written as a product state (see e.g. Nielsen and Chuang 2010, p. 96). This suggests redefining entanglement for fermions so that *a fermionic joint state is entangled iff it is not decomposable*, in the sense given above.

In fact this redefinition has been suggested already, by Ghirardi, Marinatto and Weber in a series of papers (Ghirardi, Marinatto and Weber 2002, Ghirardi and Marinatto 2003, 2004, 2005), and endorsed by Ladyman, Linnebo and Bigaj (2013).[7] Therefore I call the proposed notion *GMW-entanglement*. Further discussion of the physical salience of this notion is taken up in Caulton (2015).

The important fact for Section 4 is that decomposable fermionic states have a feature that is not shared by non-entangled states under the standard definition. That is that decomposable fermionic states, corresponding as they do to decomposable elements of the exterior algebra on $\mathscr{H}$, correspond to *subspaces* of $\mathscr{H}$. More specifically the state

(8) $|\varphi_1\rangle \wedge |\varphi_2\rangle \wedge \ldots \wedge |\varphi_N\rangle$

where $\langle\varphi_i|\varphi_j\rangle = \delta_{ij}$, corresponds to the N-dimensional subspace spanned by the degree-1 vectors $|\varphi_1\rangle$, $|\varphi_2\rangle$, ... $|\varphi_N\rangle$. This offers a glimpse of two of our main results in Section 4, namely: (i) parthood between fermionic assemblies is represented by subspacehood; and (ii) the state of a larger assembly is given by the *span* of the states of its constituents. The tension between the two notions of fusion implicit in (i) and (ii) embodies the tension between classical mereology and the quantum mechanics of fermions.

[7] The definition that Ghirardi, Marinatto and Weber actually offer is equivalent to the one above.

3. Setting Up Quantum Mechanics for Mereology

3.1. *Classical Mereology*

There are several axiomatizations of classical mereology available (see Hovda 2009 and Varzi 2014 for a discussion); for the purposes of this paper, I have chosen the one that allows the most perspicuous discussion of its troubles for fermionic systems.

Classical mereology requires only one primitive term, $\sqsubseteq$ (parthood). From this we define *proper parthood*:

(9) $\forall x\, \forall y\, (x \sqsubset y \leftrightarrow (x \sqsubseteq y\ \&\ x \neq y))$

the *overlap* relation $x \circ y$ ('x overlaps y') in terms of common parthood:

(10) $\forall x\, \forall y\, (x \circ y \leftrightarrow \exists z\, (z \sqsubseteq x\ \&\ z \sqsubseteq y))$

and the *disjointness* relation $x \ltimes y$ as the contrary of overlap:

(11) $\forall x\, \forall y\, (x \ltimes y \leftrightarrow \neg x \circ y)$

Finally, given any 1-place formula φ, something is a *fusion* of the φs iff all and only its overlappers overlap some φ. So we define $\mathscr{F}_\varphi(x)$ ('x is a fusion of the φs') as follows:

(12) $\forall x\, (\mathscr{F}_\varphi(x) \leftrightarrow \forall y\, (y \circ x \leftrightarrow \exists z\, (\varphi(z)\ \&\ z \circ y)))$

With these definitions, we may now present the two axioms and one axiom schema. I also include a third axiom, *Axiomicity*, which is not essential to classical mereology, but which will hold in all of the theories discussed here.

- *Partial Order*. $\sqsubseteq$ is a partial order (i.e. it is reflexive, anti-symmetric and transitive).
- *Strong Supplementation*. If something is not a part of a second thing, then some part of the first thing is disjoint from the second thing:

(13) $\forall x\, \forall y\, (x \not\sqsubseteq y \rightarrow \exists z\, (z \sqsubseteq x\ \&\ z \ltimes y))$

Or, equivalently (and perhaps more elegantly), if every part of some thing overlaps a second thing, then the first thing is a part of the second thing:

(14) $\forall x\, \forall y\, (\forall z\, (z \sqsubseteq x \rightarrow z \circ y) \rightarrow x \sqsubseteq y)$

- *Atomicity*. Everything has a part that has no proper parts.

(15) $\forall x\, \exists y\, (y \sqsubseteq x\ \&\ \neg \exists z\, z \sqsubset y)$

- *Unrestricted Fusion.* If there are some φs, then there is a fusion of the φs:

(16) $(\exists x\, \varphi(x) \rightarrow \exists x\, \mathscr{F}_{\varphi}(x))$

This is imposed for all substitution instances of φ.

3.2. *Finding the Subsystems in the Quantum Formalism*

It will be key to proving the results in Section 4 that we have some way of identifying in the quantum formalism when the joint system has subsystems in particular states. This requires giving some physical interpretation to that formalism. In this we are constrained by the requirements of permutation-invariance to give a physical interpretation *only* to those quantities which are permutation invariant.

We assume that we are dealing with an N-fermion assembly, so the joint state lies in $\mathscr{A}(\otimes^N \mathscr{H})$, where $\mathscr{H}$ is the single-system Hilbert space. We expect any subsystem's state to lie in $\mathscr{A}(\otimes^r \mathscr{H})$, where $1 \leq r \leq N$. I will categorise projectors according to the Hilbert space they act on. A projector is of degree-r iff it acts on $\mathscr{A}(\otimes^r \mathscr{H})$ (where $r = 1$ corresponds to the single-system Hilbert space $\mathscr{H}$).

Choose any degree-1 projector P. Its orthocomplement is $P\perp := \mathbb{1}-P$. From P we may define a family of projectors $\{\sigma^s_r(P) \mid 1 \leq r \leq s \leq N\}$:

(17)
$$\sigma^s_0(P) := \underbrace{P_\perp \otimes \ldots \otimes P_\perp}_{s}$$
$$\sigma^s_1(P) := P \otimes \underbrace{P_\perp \otimes \ldots \otimes P_\perp}_{s-1} + P_\perp \otimes P \otimes \underbrace{\ldots \otimes P_\perp}_{s-2} + \ldots + \underbrace{P_\perp \otimes \ldots \otimes P_\perp}_{s-1} \otimes P$$
$$\sigma^s_2(P) := P \otimes P \otimes \underbrace{P_\perp \otimes \ldots \otimes P_\perp}_{s-2} + P \otimes P_\perp \otimes P \otimes \underbrace{\ldots \otimes P_\perp}_{s-3} + \ldots + \underbrace{P_\perp \otimes \ldots \otimes P_\perp}_{s-2} \otimes P \otimes P$$
$$\vdots$$
$$\sigma^s_s(P) := \underbrace{P \otimes \ldots \otimes P}_{s}$$

These projectors will be the most important ones below. Each $\sigma^s_r(P)$ is a symmetric projector, and so may be interpreted as corresponding to a physical property. I propose the following interpretation:

> $\sigma^s_r(P)$ corresponds to the property 'Exactly r of s degree-1 constituents have property P'.

This interpretation can obviously be justified in the context in which permutation invariance is *not* imposed. In that case, each summand of $\sigma^s_r(P)$, which acts on exactly r of s factor Hilbert spaces with P and on the remaining $s - r$ with $P_\perp$, can itself be given a physical interpretation, according to which some selection of r named degree-1 systems have property P and the remaining degree-1 systems do not. The sum over all summands can then be interpreted as a (quantum) disjunction over all possible selections of r named degree-1 systems.

However, when permutation invariance *is* imposed, this justificatory story is not available to us. For the individual summands of $\sigma^s_r(P)$ are typically not themselves permutation-invariant, and so, as per our discussion in Section 2, cannot receive a physical interpretation. Instead, the physical interpretation offered above must be taken as primitive.

It is worth pointing out some formal properties of $\sigma^s_r(P)$, which are consistent with the interpretation offered. First, it must be emphasised that the domain of $\sigma^s_r(P)$ is restricted to $\mathcal{A}(\otimes^s\mathcal{H})$: so, in particular, if $d := \dim(P) \geq s$, then $\dim(\sigma^s_s(P)) = \binom{d}{s}$; otherwise $\sigma^s_s(P) = 0$, due to Pauli exclusion. Second, we have that $\sigma^s_r(P) = \sigma^s_{s\text{-}r}(P_\perp)$, so exactly r of s degree-1 constituents have the property P iff exactly $s\text{-}r$ degree-1 constituents have the property $P_\perp$, which is the quantum negation of P. Third, due to Pauli exclusion, $\sigma^s_r(P) = 0$ if $\dim(P) < r$, or, since $\sigma^s_r(P) = \sigma^s_{s\text{-}r}(P\perp)$, if $\dim(P_\perp) = d - \dim(P) < s - r$, where $d := \dim(\mathcal{H})$. So a non-vanishing $\sigma^s_r(P)$ requires $r \leq \dim(P) \leq d - s + r$.

An important result for later will be

Proposition 3.1. For any degree-1 projectors P, Q: $P \leq Q$ iff $\sigma^s_s(Q) \leq \sigma^s_r(P)$, where $r := \dim(P)$ and $s := \dim(Q)$.

Proof.

(*Left to Right.*) Since $\Sigma_{i=0}{}^s\sigma^s_i(P) = \sigma^s_s(\mathbb{1})$, which is the identity on $\mathcal{A}(\otimes{}^s\mathcal{H})$, we can multiply $\sigma^s_s(Q)$ on the right with the identity to obtain

$$(18)\quad \sigma^s_s(Q) = \sigma^s_s(Q)(\Sigma_{i=0}{}^s\sigma^s_i(P)) = \Sigma_{i=0}{}^s\sigma^s_s(Q)\sigma^s_i(P)\,.$$

Now $\dim(P) = r$, so $\sigma^s_i(P) = 0$ for $i > r$; so at most the first r terms of this sum are non-vanishing. Now decompose Q into $Q = P + R$, where $R := P_\perp Q = QP_\perp$. Then $\sigma^s_s(Q)\sigma^s_i(P) = \sigma^s_s(Q)\sigma^s_{s\text{-}i}(P_\perp) = \sigma^s_s(Q)\sigma^s_{s\text{-}i}(R)$. But $\dim(R) = s - r$, so $\sigma^s_{s\text{-}i}(R) = 0$ for $s - i > s - r$, i.e., $i < r$; so at most the last $s - r$ terms of the sum in (18) are non-vanishing. It follows that the only non-vanishing term in (18) is for $i = r$; so $\sigma^s_s(Q) = \sigma^s_s(Q)\sigma^s_r(P)$. By multiplying $\sigma^s_s(Q)$ on the left with the identity, we can similarly show that $\sigma^s_s(Q) = \sigma^s_r(P)\sigma^s_s(Q)$. It follows that $\sigma^s_s(Q) \leq \sigma^s_r(P)$.

(*Right to Left.*) $\sigma^s_s(Q) \leq \sigma^s_r(P)$ means that $\sigma^s_r(P)\sigma^s_s(Q) = \sigma^s_s(Q)\sigma^s_r(P) = \sigma^s_s(Q)$. It follows that P and Q commute, and that either $P > Q$ or $P \leq Q$. Define $S := QP = PQ$. Assume for *reductio* that $P >$ Q; then, since $\dim(P) = r$, it must be that $\dim(S) < r$. In that case $\sigma^s_r(S) = 0$, due to Pauli exclusion. But

$\sigma^s_s(Q)\sigma^s_r(S) = \sigma^s_s(Q)\sigma^s_r(P) = \sigma^s_s(Q)$ and $\dim(\sigma^s_s(Q)) = 1$; so $\dim(\sigma^s_r(S)) \geq 1$. Contradiction. So we must have $P \leq Q$. □

The physical interpretation of this result is as follows: for any two degree-1 projectors P and Q, $P \leq Q$ is equivalent to the condition that, if a number $\dim(Q)$ of elementary constituents satisfy Q, then exactly $\dim(P)$ of them satisfy P. We can understand this as a result of Pauli exclusion.

3.3. *Translation Rules*

The question whether mereology holds true or not for quantum mechanics is *prima facie* ill-formed: mereology is a theory axiomatised in a first-order formal language, while quantum mechanics has no first-order axiomatisation and is instead presented in the mathematics of linear operators on Hilbert spaces. Therefore we require some way to "translate" the claims of one theory into the framework of the other. It will be simplest to run the direction of translation from quantum mechanics to mereology.

More specifically, we will set up a correspondence between the states of fermionic assemblies, which are normalised vectors in some Hilbert space, and Tarskian models. This correspondence will be constrained by what I call "translation rules". I hasten to add that the goal is *not* to do quantum mechanics in first-order logic! The goal is simply to represent the states of quantum mechanics in a form appropriate for comparison with mereology.

First some general remarks regarding the objects and properties of the "translated" quantum states:

Objects. The domain of any model will contain only two kinds of objects: quantum systems and projectors. In the following, I will use lower case variables to range over quantum systems and upper case variables to range over projectors. (This is just a notational convenience: our models are first-order, and both kinds of object are objects in the Frege-Quine sense of belonging to the first-order domain.) Any model will have the total system in its domain.

Predicates. There will only be two primitive predicates. The first is $\sqsubseteq$, denoting the mereological parthood relation, already discussed. For our purposes, we may stipulate that this relation holds (if at all) only between quantum systems.

The second primitive predicate is $E(x, P)$, which denotes a certain dyadic relation between a quantum system x and projector P. We stipulate that E never holds between two quantum systems (e.g. $E(x, y)$) or two projectors (e.g. $E(P, Q)$), or between a projector and a system in the wrong order (e.g. $E(P, x)$). $E(x, P)$ has the following interpretation: *x has the property associated with P.*

The translation rules are now presented as follows:

(1) (*Total System*). The total system Ω, whose state is represented by a ray in $\mathcal{A}(\otimes^N \mathcal{H})$, exists.

(2) (*Existence and Completeness of Projectors*). All and only *symmetric* projectors of rank r, where $r \in \{1, 2, ..., N\}$, exist.

(3) (*Eigenstate-Eigenvalue Link*). For any system x and any projector P: $E(x, P)$ iff x's state is an eigenstate of P with eigenvalue 1.

(4) (*Existence of Subsystems*). For any degree-1 projector P and all $r = 1, 2, ..., N$: $E(\Omega, \Sigma^N_{i=r}\sigma^N_i(P))$ iff there is some system x such that $E(x, \sigma^r_r(P))$.

(5) (*Uniqueness of Subsystems*). For any degree-1 projector P and all $r = 1, 2, ..., N$: $E(\Omega, \sigma^N_r(P))$ iff there is some unique system x such that $E(x, \sigma^r_r(P))$.

(6) (*Non-GMW-Entangled Systems*). For all systems x, there is some $r \in \{1, 2, ..., N\}$ and some degree-1 projector P with $\dim(P) = r$ such that $E(x, \sigma^r_r(P))$.

(7) (*Definition of Parthood*). For any quantum systems x and y, $x \sqsubseteq y$ iff: for any degree-1 projector P and all $s \in \{1, 2, ..., N\}$, if $E(y, \sigma^s_s(P))$, then there is some $r \leq s$ such that $E(x, \sigma^r_r(P))$.

Some discussion about these rules is in order. I take each one in turn.

(1) (*Total System*). This rule ensures that the total system Ω belongs to the domain.

(2) (*Existence and Completeness of Projectors*). This rule expresses two essential interpretative assumptions. The first is that the quantum formalism is *complete*, so that no physical facts are left unrepresented by the quantum state. The second assumption is none other than the interpretation of permutation invariance as underpinned by representational redundancy, as discussed in Section 2.

(3) (*Eigenstate-Eigenvalue Link*). This rule also expresses the completeness of the quantum formalism. It has a controversial element, which is that it applies not only to the total system Ω, but also all subsystems; see below.

(4) (*Existence of Subsystems*). This is the only rule which introduces systems other than Ω into the domain. The guiding idea, following from the discussion in Section 3.2, is that *at least* r of the total system's N degree-1 constituents have some property P iff there is at least one system all r of whose degree-1 constituents have property P. However, we will see later that this interpretation cannot quite be correct, if by 'constituent' we mean atomic part.

(5) (*Uniqueness of Subsystems*). The guiding idea here is that exactly r of N degree-1 constituents of the total system have property P iff there is a unique system all r of whose degree-1 constituents have property P. The existence-entailing component is redundant, given (*Existence of Subsystems*), but is included here for expedience.
This rule permits us to extend (*Eigenstate-Eigenvalue Link*) to subsystems, in the following way. Given a unique x such that $E(x, \sigma^r_r(P))$, we may say that x has a state whose corresponding density operator has its domain and range in the range of $\sigma^r_r(P)$ (which is a projector). For any (degree-r) projector Q such that $Q\sigma^r_r(P)Q = \sigma^r_r(P)$, we may infer $E(x, Q)$.

(6) (*Non-GMW-Entangled Systems*). Since $\dim(P) = r$, $\dim(\sigma^r_r(P)) = 1$. In fact the range of $\sigma^r_r(P)$ is the ray spanned by the degree-r non-GMW-entangled state $|\varphi_1\rangle \wedge |\varphi_2\rangle \wedge \ldots \wedge |\varphi_r\rangle$, where $\{|\varphi_i\rangle\}$ is any family of orthonormal degree-1 states which span the range of P. So this rule entails that *all* systems occupy decomposable, i.e. non-GMW-entangled, states.
This rule is problematic if $N \geq 4$. For, in that case, it is not true that any non-GMW-entangled state can be decomposed only into states that are themselves decomposable. This corresponds to the well known result for exterior algebras that the non-decomposable degree-2 vector $\xi := (1/\sqrt{2})(a \wedge b + c \wedge d)$ satisfies $\xi \wedge \xi = a \wedge b \wedge c \wedge d$. However, we may take this rule as a *restriction of the domain* to those systems which *are* non-GMW-entangled. All future reference to systems is then to be taken as implicitly concerning only non-GMW-entangled systems.

(7) (*Definition of Parthood*). This connecting principle can only be justified for non-GMW-entangled fermionic systems. The idea is that x is a part of y iff all the degree-1 constituents of x are also constituents of y, so if all of y's degree-1 constituents have some property P, then *a fortiori* all of x's degree-1 constituents have that same property.

4. Composition for Fermions

We are now in a position to establish the main results of this paper. They are presented in Section 4.1. Section 4.2 contains a concluding discussion.

4.1. *Main Results*

Proposition 4.1 (Unique Degree) *Every system has a unique degree in {1, 2, ..., N}; i.e. if E(x,P) and E(x, Q), then deg(P) = deg(Q).*

Proof.
Given (*Non-GMW-Entangled of Systems*), for system x there is some $r \in \{1, 2, ..., N\}$ and some degree-r projector P such that $\dim(P) = r$ and $E(x, \sigma^r_r(P))$. By (*Eigenvector-Eigenvalue Link*), we can therefore attribute to x the state $|\varphi_1\rangle \wedge |\varphi_2\rangle \wedge \ldots \wedge |\varphi_r\rangle$, where $\text{span}(\{|\varphi_i\rangle \mid i \in \{1, 2, ..., r\}\}) = \text{ran}(P)$. This state is an eigenstate only of projectors of degree-r; so by (*Eigenvector-Eigenvalue Link*) again, if $E(x, Q)$ for any projector Q, then Q has degree r. □

Definition 4.1 (Degree of Systems) *For any system x,* deg(x) *is the unique degree of any projector P such that E(x, P).*

Propostion 4.2 (Reflexivity of ⊑) *For any system x, x ⊑ x.*

Proof. This follows straightforwardly from (*Definition of Parthood*). □

Proposition 4.3 (Transitivity of ⊑) *For any systems x, y, z, if x ⊑ y and y ⊑ z, then x ⊑ z.*

Proof. This follows straightforwardly from (*Definition of Parthood*). □

Proposition 4.4 (State-System Uniqueness 1) *For any degree-1 projector P, if there is some system x such that* $E(x,\sigma_r^r(P))$, *where* $r = \dim(P)$, *then x is unique.*

Proof. Let P be any degree-1 projector with $\dim(P) = r$. Assume that there is an x such that $E(x, \sigma^r_r(P))$. By (*Existence of Subsystems*), $E(\Omega, \sigma^N_r(P))$. Since $\dim(P) = r$, $\Sigma_{i=r}^{N} \sigma^N_i(P) = \sigma^N_r(P)$; so $E(\Omega, \Sigma_{i=r}^{N} \sigma^N_i(P))$. By (*Uniqueness of Subsystems*), there is a unique system y such that $E(y, \sigma^r_r(P))$; so $x = y$ and x is unique. □

Proposition 4.5 (State-System Uniqueness 2) For any system x, the degree-1 projector P such that $E(x,\sigma^r_r(P))$, where $r = \dim(P)$, is unique.

Proof. Let x be any system. By (*Non-GMW-Entangled Systems*), there is some degree-1 projector P such that $E(x,\sigma^r_r(P))$, where $r = \dim(P)$. Suppose for *reductio* that there is some other degree-1 projector Q with $\dim(Q) = r$

such that $E(x,\sigma^r_r(Q))$. $Q \not\leq P$ and $P \not\leq Q$, since $P \neq Q$ and $\dim(P) = \dim(Q)$. So by (*Definition of Parthood*), $x \not\sqsubseteq x$, which contradicts Proposition 4.2 (*Reflexivity of* $\sqsubseteq$). □

This allows us to attribute to each system a pure state, as follows:

Definition 4.2 (Subsystem States) *For any system x, the state of x is the unique projector* $\sigma^r_r(P)$ *such that* $r = \dim(P) = \deg(x)$ *and* $E(x, \sigma^r_r(P))$.

With this definition we can extend application of (*Eigenstate-Eigenvalue Link*) to systems other than Ω.

Proposition 4.6 (Anti-Symmetry of $\sqsubseteq$) *For any two systems x, y, if* $x \sqsubseteq y$ *and* $y \sqsubseteq x$, $x = y$.

Proof. (*Definition of Parthood*) and Proposition 4.1 (*Unique Degree*) entail: if $x \sqsubseteq y$ and $y \sqsubseteq x$, then for all degree-1 projectors P and all $r \in \{1, 2, ..., N\}$, $E(x, \sigma^r_r(P))$ iff $E(y, \sigma^r_r(P))$. From (*Non-GMW-Entangled Systems*), there is some degree-1 projector Q such that $E(x,\sigma^r_r(Q))$ and $\dim(Q) = r$. So also $E(y,\sigma^r_r(Q))$. By Proposition 4.4 (*State-System Uniqueness 1*), $x = y$. □

Propositions 4.2, 4.3 and 4.6 entail that parthood for fermions is a partial ordering relation, thereby satisfying the first mereological axiom.

Corollary 4.7 (Criterion of Identity) *For any systems x, y,* $x = y$ *iff: for all projectors P and all* $r \in \{1, 2, ..., N\}$, $E(x, \sigma^r_r(P))$ *iff* $E(y, \sigma^r_r(P))$.

Proof. Left to Right: This follows from the indiscernibility of identicals. *Right to Left:* From (*Definition of Parthood*), If for all projectors P and all $r \in \{1, 2, ..., N\}$, $E(x, \sigma^r_r(P))$ iff $E(y, \sigma^r_r(P))$, then $x \sqsubseteq y$ and $y \sqsubseteq x$. From Proposition 4.6 (*Anti-Symmetry of* $\sqsubseteq$ it follows that $x = y$. □

In the particular case in which $\deg(x) = \deg(y) = 1$, i.e. for elementary fermions, this entails a satisfyingly straightforward statement of the Pauli Exclusion Principle:

$x = y$ iff for all degree-1 projectors P: $E(x, P)$ iff $E(y, P)$.

An important consequence of this is that in any non-GMW-entangled joint state, any two individual fermions are discernible by monadic predicates (a phenomenon which Muller and Saunders (2008) call *absolute discernibility*). This is contrary to the orthodoxy in the quantum literature, in which bosons and fermions are taken to be *either merely weakly discernible or utterly indiscernible* (French and Redhead 1988; Butterfield 1993; Huggett 2003; French and Krause 2006; Muller and Saunders 2008; Muller and Seevinck 2009; Caulton 2013).

This break with the orthodoxy comes down to the hard line I wish to urge regarding permutation invariance, according to which the invariance is to be construed as an indication of "gauge" (i.e. representationally redundant) quantities in the formalism. Here the "gauge" quantities in question are the single-particle factor Hilbert space labels (or, the order in which the single-particle Hilbert spaces appear in the tensor product). On the view that they represent nothing at all, permutation invariance follows as a compulsory requirement for any quantity's being genuinely physical.

According to the opposing, orthodox interpretative line in the quantum identity literature, these labels represent or denote the constituent elementary (i.e. rank-1) systems. Guided by this interpretation, one may recover the state of an elementary system x by performing a partial trace on the total system Ω's density operator, over all states of rank-1 systems disjoint to x. Given permutation invariance, one obtains the *same* reduced density operator for each elementary system; hence the celebrated indiscernibility results (see French and Redhead 1988, pp. 240–242).

However, this partial trace operation can have no direct physical meaning on the hard line interpretation, since it requires identifying each rank-1 system by its corresponding factor Hilbert space label. If the hard line interpretation is right, we cannot identify rank-1 systems this way.[8] Instead, given Corollary 4.7, the rank-1 systems may be identified according to their *qualitative* properties, represented in the formalism by degree-1 projectors; i.e. projectors which act on the single-particle Hilbert space. This method for identifying constituent systems is discussed by French and Krause (2006, §4.2.1), though rejected in the light of the indiscernibility results just mentioned. The rejection is sensible if, but only if, one takes the orthodox interpretative line on permutation invariance. The qualitative individuation strategy is pursued in detail by Ghirardi, Marinatto and Weber (2002), Ghirardi and Marinatto (2003, 2004, 2005), Dieks and Lubberdink (2011) and Caulton (2015).

Proposition 4.8 (Ω is Maximal) *Every system is a part of the total system Ω.*

[8] Since one obtains the same reduced density operator for each Hilbert space label, the result is permutation-invariant. Shouldn't there therefore be *some* physical meaning to the reduced density operator so obtained? Indeed there is: it can be interpreted as the *average* state of all the rank-1 systems. See Caulton (2015) for more details.

Proof. Take Ω. Its state is $|\psi_1\rangle \wedge |\psi_2\rangle \wedge \ldots \wedge |\psi_N\rangle$. By (*Eigenstate-Eigenvalue Link*), it follows that $E(\Omega, \sigma^N{}_N(P))$, where $\mathrm{ran}(P) = \mathrm{span}(\{|\psi_i\rangle\})$. Since $\dim(\sigma^N{}_N(P)) = 1$ and because of (*Eigenstate-Eigenvalue Link*), any other degree-1 projector Q such that $E(\Omega, \sigma^N{}_N(Q))$ must satisfy $\sigma^N{}_N(P)) \leq \sigma^N{}_N(Q)$, and so $P \leq Q$.

Now take any system x. From (*Non-GMW-Entangled Systems*), there is some $r \in \{1, 2, \ldots, N\}$ and some degree-1 projector R such that $E(x, \sigma^r{}_r(R))$ and $\dim(R) = r$. By Proposition 4.4 (*State-System Uniqueness 1*), x is unique. We may now use (*Uniqueness of Subsystems*) to infer $E(\Omega, \sigma^N{}_r(R))$. But from the previous paragraph, we must have $\sigma^N{}_N(P) \leq \sigma^N{}_r(R)$. So, by Proposition 3.1, $r \leq P$. From this and (*Eigenstate-Eigenvalue Link*), it follows that $E(x, \sigma^r{}_r(P))$. And by (*Eigenstate-Eigenvalue Link*) again, for any degree-1 projector Q such that $P \leq Q$, $E(x, \sigma^r{}_r(Q))$.

From the two preceding paragraphs it follows that, for any degree-1 projector Q such that $E(\Omega, \sigma^N{}_N(Q))$, we also have $E(x, \sigma^r{}_r(Q))$. So, by Proposition 4.1 (*Unique Rank*) and (*Definition of Parthood*), $x \sqsubseteq \Omega$.□

Definition 4.3 (System-Spaces) *For any system x, the system-space* $\mathfrak{s}(x)$ *associated with x is the range of the unique degree-1 projector P such that* $\dim(P) = \deg(x) =: r$ *and* $E(x, \sigma^r{}_r(P))$.

So, as anticipated in Section 2.2, any system x is associated with a subspace of the single-particle Hilbert space $\mathscr{H}$. Moreover, for any system x, $\dim(\mathfrak{s}(x)) = \deg(x)$. In general, for any system x, we may write x's state as $|\varphi_1\rangle \wedge |\varphi_2\rangle \wedge \ldots \wedge |\varphi_r\rangle$, where $r := \deg(x)$ and the $\{|\varphi_i\rangle\}$ are orthonormal; then $\mathfrak{s}(x) = \mathrm{ran}(\Sigma_{i=1}{}^r|\varphi_i\rangle\langle\varphi_i|)$.

Proposition 4.9 (Subspacehood Represents Parthood) For any systems x, y, $\mathfrak{s}(x) \subseteq \mathfrak{s}(y)$ iff $x \sqsubseteq y$.

Proof. Left to Right: Assume $\mathfrak{s}(x) \subseteq \mathfrak{s}(y)$. Given Proposition 4.5 (*State-System Uniqueness 2*), let P be the unique degree-1 projector such that $\dim(P) = \deg(x) =: r$ and $E(x, \sigma^r{}_r(P))$ and Q be the unique degree-1 projector such that $\dim(Q) = \deg(y) =: s$ and $E(y, \sigma^s{}_s(Q))$. Since $\mathfrak{s}(x) \subseteq \mathfrak{s}(y)$, $P \leq Q$ and $r \leq s$. $\dim(\sigma^s{}_s(Q)) = 1$, so any degree-1 projector R such that $E(y, \sigma^s{}_s(R))$ must be such that $Q \leq R$, whence $E(x,\sigma^r{}_r(R))$. From (*Definition of Parthood*), it follows that $x \sqsubseteq y$.

Right to Left: Assume $x \sqsubseteq y$. Given Proposition 4.5 (*State-System Uniqueness 2*), let P be the unique degree-1 projector such that $\dim(P) = \deg(x) =: r$ and $E(x, \sigma^r{}_r(P))$ and Q be the unique degree-1 projector such that $\dim(Q) = \deg(y) =: s$ and $E(y, \sigma^s{}_s(Q))$. From (*Definition of Parthood*) and Proposition 4.1 (*Unique Degree*), it follows that $E(x, \sigma^r{}_r(Q))$. But $\dim(\sigma^r{}_r(P)) = 1$, so $P \leq Q$; whence $\mathfrak{s}(x) \subseteq \mathfrak{s}(y)$. □

Corollary 4.10 (System-Subspace Link) *For any systems x, y,* $\mathfrak{s}(x) = \mathfrak{s}(y)$ *iff* $x = y$.

Proof. The Right to Left direction is trivial. *Left to Right:* Assume $\mathfrak{s}(x) = \mathfrak{s}(y)$. Then $\mathfrak{s}(x) \subseteq \mathfrak{s}(y)$ and $\mathfrak{s}(y) \subseteq \mathfrak{s}(x)$. It follows from Proposition 4.9 (*Subspacehood Represents Parthood*) and Proposition 4.6 (*Anti-Symmetry of* $\sqsubseteq$) that $x = y$. □

Propostion 4.11 (Each System-Space is a Subspace of $\mathfrak{s}(\Omega)$) *For any system x,* $\mathfrak{s}(x) \subseteq \mathfrak{s}(\Omega)$.

Proof. This follows straightforwardly from Proposition 4.8 (*Ω is Maximal*) and Proposition 4.9 (*Subspace Represents Parthood*). □

Proposition 4.12 (Each Subspace of $\mathfrak{s}(\Omega)$ is a System-Space) For any non-zero space $\mathfrak{x} \subseteq \mathfrak{s}(\Omega)$, there is a unique x such that $\mathfrak{s}(x) = \mathfrak{x}$.

Proof. Take any non-zero space $\mathfrak{x} \subseteq \mathfrak{s}(\Omega)$. This defines the degree-1 projector P for which $\mathrm{ran}(P) = \mathfrak{x}$. Let $r := \dim(P)$. $\mathfrak{s}(\Omega)$ similarly defines the degree-1 projector Q for which $\mathrm{ran}(Q) = \mathfrak{s}(\Omega)$ and $\dim(Q) = N$. We know from Definition 4.3 (*System-Spaces*) that $E(\Omega, \sigma^N{}_N(Q))$. And, since $\mathfrak{x} \subseteq \mathfrak{s}(\Omega)$, $P \leq Q$; and so, by Proposition 3.1, $\sigma^N{}_N(Q) \leq \sigma^N{}_r(P)$. Using (*Eigenstate-Eigenvalue Link*) we may infer that $E(\Omega, \sigma^N{}_r(P))$. From (*Uniqueness of Subsystems*) it follows that there is a unique system x such that $E(x,\sigma^r{}_r(P))$. From Definition 4.3 (*System-Spaces*) it follows that $\mathfrak{s}(x) = \mathfrak{x}$. □

The foregoing results show that, given our non-GMW-entangled N-fermion assembly Ω, the totality of all non-GMW-entangled systems in existence correspond one-to-one to the subspaces of $\mathfrak{s}(\Omega)$, i.e. the elements of the exterior algebra $\Lambda(\mathfrak{s}(\Omega))$ (except for the zero subspace). Using the fact that parthood is represented by subspacehood, we can infer the representations of the other mereological notions: overlap, product. Two objects overlap iff they possess a common part; so two systems overlap iff their systems spaces have a non-zero intersection. The mereological product $x \sqcap y$ of any two systems x and y (if it exists), is then a system whose system-space is the intersection of the two corresponding system-spaces. this greatly clarifies the compositional structure of non-GMW-entangled fermionic states.

Proposition 4.13 (Parthood Obeys Atomicity) *Every system has some part that has no proper parts.*

Proof. We use Propositions 4.9 (*Subspacehood Represents Parthood*), 4.11 (*Each System-Space is a Subspace of* $\mathfrak{s}(\Omega)$) and 4.12 (*Each Subspace of* $\mathfrak{s}(\Omega)$ is a System-Space). Take any system x. x has a system-space $\mathfrak{s}(x)$ which is a subspace of $\mathfrak{s}(\Omega)$. $\mathfrak{s}(x)$ is spanned by deg(x)-many 1-dimensional subspaces of $\mathfrak{s}(\Omega)$; each one corresponds to a degree-1 system. Since parthood is represented by subspacehood, degree-1 systems have no proper parts. □

Proposition 4.14 (Parthood Obeys Strong Supplementation) *For any systems x and y, if x is not a part of y, then some part of x is disjoint from y, i.e. there is some system z such that* $z \sqsubseteq x$ *and* $\mathfrak{s}(z) \cap \mathfrak{s}(y) = \varnothing$.

Proof. Assume that $x \not\sqsubseteq y$. So by Proposition 4.9 (*Subspacehood Represents Parthood*), $\mathfrak{s}(x) \not\subseteq \mathfrak{s}(y)$. Then there is some subspace $\mathfrak{z}$ of $\mathfrak{s}(x)$ such that $\mathfrak{z} \cap \mathfrak{s}(y) = \varnothing$. By Proposition 4.12 (*Each Subspace of* $\mathfrak{s}(\Omega)$ is a System-Space), $\mathfrak{z} = \mathfrak{s}(z)$ for some system z. By (*Subspacehood Represents Parthood*) again, z and y are disjoint. □

Thus we have proven all of our mereological axioms, except the axiom schema *Unrestricted Fusion*. The correspondence between systems and the elements of the exterior algebra $\Lambda(\mathfrak{s}(\Omega))$ leads to a surprising result:

Proposition 4.15 (Continuum-Many Atomic Parts) *For any system x, if* deg(x) ≥ 2*, then x has continuum-many atomic parts.*

Proof. If deg(x) ≥ 2, then dim($\mathfrak{s}(x)$) ≥ 2. So there are continuum-many 1-dimensional subspaces $\mathfrak{y} \subset \mathfrak{s}(x)$. Each one corresponds to a degree-1 (and therefore atomic) system. □

A vivid example is provided by the 2-fermion system Ω_2 in the spin-singlet state

(19) $|\uparrow\rangle \wedge |\downarrow\rangle = (1/\sqrt{2})(|\uparrow\rangle \otimes |\downarrow\rangle - |\downarrow\rangle \otimes |\uparrow\rangle)$

It is well-known that this state is spherically symmetric, in that $|\uparrow\rangle \wedge |\downarrow\rangle = |\leftarrow\rangle \wedge |\rightarrow\rangle = \ldots$, for any pair of oppositely-pointing spin-(1/2) states. In our framework, this spherical symmetry reflects the fact that Ω_2 possesses a continuum-multitude of atomic parts: one of which is spin-up, one is spin-down, one is spin-left, etc.

We are now ready to see the conflict with classical mereology. For simplicity, I use the example of degree-1 systems.

Proposition 4.16 (Non-Existence of Mereological Fusions 1) *For any two degree-1 systems x and y: if $\mathfrak{s}(x) \not\subseteq \mathfrak{s}(y)$, then there does not exist a system z which is the mereological fusion $x \sqcup y$ of x and y, i.e. which is such that $\forall w\, (w \circ z \leftrightarrow (w \circ x \vee w \circ y))$.*

Proof. Since x and y are degree-1 systems and therefore atomic, $\mathfrak{s}(x) \not\subseteq \mathfrak{s}(y)$ entails that x and y are disjoint. Since all systems' states are represented by subspaces, z's state is represented by a subspace. So we seek a subspace $\mathfrak{z} \subseteq \mathfrak{s}(\Omega)$ such that

(20) $\mathfrak{w} \subseteq \mathfrak{s}(\Omega)$, $\mathfrak{w} \cap \mathfrak{z} \neq \varnothing$ iff ($\mathfrak{w} \cap \mathfrak{s}(x) \neq \varnothing$ or $\mathfrak{w} \cap \mathfrak{s}(y) \neq \varnothing$)

$\mathfrak{z}$ must have dimension at least 2, since it must overlap both $\mathfrak{s}(x)$ and $\mathfrak{s}(y)$. But now consider some 1-dimensional subspace $\mathfrak{w}_0 \subseteq \mathfrak{z}$ which is skew to (i.e. neither coincident with nor orthogonal to) both $\mathfrak{s}(x)$ and $\mathfrak{s}(y)$; such a subspace will always exist (take e.g. the normalised sum proportional to $\mathfrak{s}(x) + \mathfrak{s}(y)$). $\mathfrak{w}_0$ overlaps $\mathfrak{z}$, yet overlaps neither $\mathfrak{s}(x)$ nor $\mathfrak{s}(y)$. So no $\mathfrak{z}$ exists such that (20) is satisfied. □

Proposition 4.17 (Non-Existence of Mereological Furions 2) *There are some satisfied 1-place formulas φ such that there is no system x for which $\mathscr{F}_\varphi(x)$.*

Proof. Let x and y be any two distinct degree-1 systems, and let P and Q be the degree-1 projectors such that $\dim(P) = \dim(Q) = 1$ and $E(x, P)$ and $E(y, Q)$. Recall that, for any φ, the fusion of the φs is defined by

(21) $\forall z\, (\mathscr{F}_\varphi(z) \leftrightarrow \forall w\, (w \circ z \leftrightarrow \exists t\, (\varphi(t)\ \&\ t \circ w)))$

We now set $\varphi(t) := (E(t, P) \vee E(t, Q))$. $\varphi(t)$ is satisfied by x and y only: for all systems are individuated by their states, given Corollary 4.7 (*Criterion of Identity*). So $\varphi(t)$ is equivalent to $(t = x \vee t = y)$. By (21), the fusion z of the φs satisfies $\forall w\, (w \circ z \leftrightarrow (w \circ x \vee w \circ y))$. By Proposition 4.16 (*Non-Existence of Mereological Fusions 1*), no such z exists. □

It is worth emphasising that the failure of (*Unrestricted Fusion*) entailed by the above proposition is particularly extreme. For *any* two distinct degree-1 systems, no system exists which is their fusion. This result may be generalised to higher-degree systems: in general, no fusion of x and y exists if the union of their subspaces $\mathfrak{s}(x) \cup \mathfrak{s}(y)$ is not itself a subspace; i.e., if $x \not\subseteq y$ and $y \not\subseteq x$.

This is inconsistent not only with (*Unrestricted Fusion*), but also with any plausible substitute which would allow fusions under restricted conditions. Let an axiom for the existence of fusions count as 'trivial' iff it permits the existence of the fusion of x and y only when $x \sqsubseteq y$ or $y \sqsubseteq x$ (in which case $x \sqcup y = x$ or $x \sqcup y = y$). Then the above result is inconsistent with any non-trivial existence axiom for mereological fusions.[9]

This extreme failure of (*Unrestricted Fusion*) might seem surprising, since we have a way of producing a fermionic joint state out of any collection of degree-1 fermion states: this is given by the wedge product, as discussed in Section 2.2. What is going on here is that the corresponding notion of composition is *not* mereological.

Take any systems x and y. There are two degree-1 projectors P and Q such that $\dim(P) = \deg(x) =: r$, $\dim(Q) = \deg(y) =: s$, $E(x,\sigma^r_r(P))$ and $E(y,\sigma^s_s(Q))$. Now let $\sigma(P, Q)$ be the degree-1 projector whose range is the span of the ranges of P and Q. Then we may define

Definition 4.4 (Fermionic Fusion) *For any systems x and y and associated degree-1 projectors P and Q, the fermionic fusion of x and y, denoted* $x +_f y$*, is the unique system z such that* $E(z, \sigma^t_t(\Sigma(P, Q)))$*, where* $t = \dim\Sigma(P, Q)$.

The existence and uniqueness of fermionic fusion is guaranteed by Propositions 4.12 (*Each Subspace of* $\mathfrak{s}(\Omega)$ *is a System-Space*) and 4.4 (*State-System Uniqueness 1*). This constitutes a fermionic analogue to *Unrestricted Fusion.*

To better understand fermionic fusion, we note that $x +_f x = x$, and, denoting the vector-state of any system x by $\mathfrak{v}(x)$,

Proposition 4.18 (Wedge product and fermionic fusion) If x and y are degree-1 systems such that $\mathfrak{v}(x) \perp \mathfrak{v}(y)$, then $\mathfrak{v}(x +_f y) = \mathfrak{v}(x) \wedge \mathfrak{v}(y)$.

Proof. Since $\mathfrak{v}(x) \perp \mathfrak{v}(y)$, $\mathfrak{v}(x) \wedge \mathfrak{v}(y)$ corresponds to a correctly normalised anti-symmetric vector, corresponding to the space spanned by $\mathfrak{v}(x)$ and $\mathfrak{v}(y)$. □

Proposition 4.19 (Failure of Distributivity) *For not all systems x, y, z:* $x \sqcap (y +_f z) = (x \sqcap y) +_f (x \sqcap z)$.

[9] I am grateful to Matteo Morganti for this observation.

Proof. It suffices to give an example of three degree-1 systems all of whose states are coplanar; see Figure 1. □

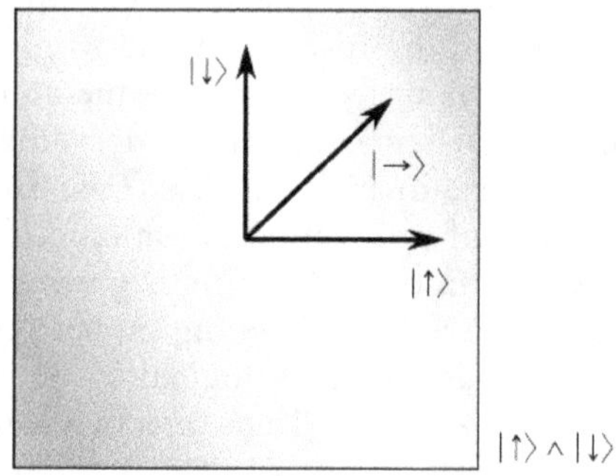

Figure 1: Three coplanar vectors, corresponding to degree-1 fermionic systems in the states $|\uparrow\rangle$, $|\downarrow\rangle$ and $|\rightarrow\rangle := (1/\sqrt{2})(|\uparrow\rangle + |\downarrow\rangle)$. The plane corresponds to a degree-2 fermionic system in the state $|\uparrow\rangle \wedge |\downarrow\rangle$, which is the fermionic fusion of any pair of the three degree-1 systems. Mereological product does not distribute over fermionic fusion; in this example represented by the fact that the intersection of $|\rightarrow\rangle$ with the span of $|\uparrow\rangle$ and $|\downarrow\rangle$ ($= |\rightarrow\rangle$) is not equal to the span of the intersections of $|\rightarrow\rangle$ with $|\uparrow\rangle$ and $|\downarrow\rangle$ ($= \mathbf{0}$).

At this point we see a strong analogy between the structure of fermionic composition and the quantum logic of Birkoff and von Neumann (1936). For a fuller discussion of quantum logic (particularly subtleties involving infinite-dimensional Hilbert space), see Dalla Chiara, Giuntini and Rédei (2007). Here it will suffice to draw the following analogies between relations and operations on objects and propositions, both classical and quantum:

Classical composition	*Classical logic*	*Fermionic composition*	*Quantum logic*
Classical system	Classical proposition	Fermionic system	Quantum proposition
Total system	Tautology	Total system Ω	Tautology
Parthood $\sqsubseteq$	Classical entailment	Parthood $\sqsubseteq$	Quantum entailment
Mereological product $\sqcup$	Classical conjunction $\wedge$	Mereological product $\sqcup$	Quantum conjunction
Mereological fusion $\sqcap$	Classical disjunction $\vee$	Fermionic fusion $+_f$	Quantum disjunction

Classical composition	*Classical logic*	*Fermionic composition*	*Quantum logic*
Classical complement	Classical negation ¬	Fermionic complement	Quantum negation (orthocomplement) ⊥

In each row, analogies between columns 1 and 2 and columns 3 and 4 respectively are exact insofar as they receive the same mathematical representation, in Boolean algebras and Hilbertian lattices, respectively. Analogies between columns 1 and 3 or between columns 2 and 4 are looser.

A final comment. For those of us who wish to think of the middle-sized dry goods of everyday life as "made up" of fermions, it might appear at first blush as something of a mystery how to reconcile the Hilbertian structure of fermionic composition with the Boolean structure of our heuristic understanding of the composition of middle-sized dry goods. In fact there need be no mystery here: macroscopic objects are individuated (at least approximately) by their spatial boundaries. This picks a preferred orthobasis in the Hilbertian lattice of fermionic states, and the subspaces spanned by the rays in this basis have the familiar structure of a Boolean algebra.

4.2. *Can Mereology Be Saved?*

The idea that quantum mechanics might prompt a revision in logic, a view argued by Putnam (1969, 1974), has received rather short shrift. For example, here is Jauch (1968), quoted in Dalla Chiara, Guintini and Rédei:

> The propositional calculus of a physics system has a certain similarity to the corresponding calculus of ordinary logic. In the case of quantum mechanics, one often refers to this analogy and speak of quantum logic in contradistinction to ordinary logic. ... The calculus introduced here has an entirely different meaning from the analogous calculus used in formal logic. Our calculus us the formalization of a set of *empirical* relations which are obtained by making measurements on a physical system. It expresses an objectively given property of the physical world. It is thus the formalization of empirical facts, inductively arrived at and subject to the uncertainty of any such fact. The calculus of formal logic, on the other hand, is obtained by making an analysis of the meaning of propositions. It is true under all circumstances and even tautologically so. Thus, ordinary logic is used even in quantum mechanics of systems with a propositional calculus vastly different from that of formal logic. The two need have nothing in common.

One might wish to say the same of mereological and fermionic composition. That is, although the mathematical theory associated with fermions suggests a particular calculus with similarities to – but crucial differences from – classical mereology, the option seems to be open simply to deny that anything other than mereological composition is composition worthy of the name. Such a strategy will be friendly, if not downright essential, to anyone who takes classical mereology to be 'perfectly understood, unproblematic, and certain' (Lewis 1991, p. 75).

In fact this strategy is possible, and could proceed by simply admitting the existence of the mereological fusions currently ruled out. That is, we expand the domain to include not only the fermionic systems, but also arbitrary fusions thereof. A typical such fusion will *not* be a system in the sense that its state, if one can be attributed to it at all, will not be representable as a vector in $\mathcal{A}(\otimes^r\mathcal{H})$, for any r.

To get a better idea of these non-system objects, recall that, following Kibble (1979) and Ashtekar and Schilling (1999), we may describe the possible states of a degree-1 quantum system not with the unit-vectors of the single-system Hilbert space $\mathcal{H}$, but rather by the points of the projective Hilbert space $\wp(\mathcal{H})$. Given the correspondence proven above between non-GMW-entangled N-fermion states and N-dimensional subspaces of $\mathcal{H}$, and the well-defined map between rays of $\mathcal{H}$ and points of $\wp(\mathcal{H})$, we may carry over the above results to represent arbitrary non-GMW-entangled fermion states as *regions* of $\wp(\mathcal{H})$ – indeed they will be regions that are also subspaces of $\wp(\mathcal{H})$.

The (singletons of) points and subspaces of $\wp(\mathcal{H})$ do not constitute a Boolean algebra: this is the mathematical expression of the failure of classical mereology.

But we can add more subsets of $\wp(\mathcal{H})$ *until* we achieve a Boolean algebra. For example, for any two singletons $\{\psi\}$, $\{\varphi\}$, where $\varphi, \psi \in \wp(\mathcal{H})$, we add their union $\{\psi, \varphi\}$. We associate this union with the mereological fusion of the degree-1 systems whose states are given by ψ and φ.

Given the fact that the mereological axioms we have been considering are first-order, we may recover the truth of *Unrestricted Fusion* without having to admit the full power set of $\wp(\mathcal{H})$; in fact arbitrary finite unions of (singletons of) points and subspaces will do.

It is hardly any objection that these non-system objects are somehow unnatural or that we have no practical use for them in any scientific theory: that is a familiar feature of arbitrary mereological fusions. However, the stubborn mereologist must still accept that a *rival* notion of something *like* composition, i.e. what I have called 'fermionic composition', still applies alongside the classical one; and whether or not we consider this a type of composition worthy of the name, it is unobjectionably the operation that produces familiar fermionic states from familiar fermionic states of lower rank. The bizarre non-system objects may be admitted or they may not; there is just no getting around that fact that the compositional structure of the fermionic *systems* is Hilbertian, rather than Boolean.

5. Conclusion

The foregoing arguments can be summarised as follows: *at least one* of the following three claims must be rejected:

(1) Permutation invariance reflects representational redundancy.
(2) Fusions of fermionic systems are always fermionic systems.
(3) Fermions compose (i.e. fuse) mereologically.

Premise 1 was crucial to our new way of thinking of entanglement for fermionic systems. By rejecting the hard line on permutation-invariance as representational redundancy, we may retrench to an identification of systems with factor Hilbert spaces. That way we avoid the permutation-invariant method for identifying subsystems, and the resulting failure of mereology. (We also must embrace the celebrated absolute indiscernibility results.) Rejecting premise 1 means giving up on an understanding of permutation invariance that, in Section 2.1, I claimed best explains it. That claim must now be weighed against the unpalatability of giving up on premise 2 or 3.

Premise 2 may be held by anyone who is sanguine about the possibility that our best theory of composition might be informed by empirical science. By rejecting it, we may take the saving strategy discussed in Section 4.2, and hang on to both our strong reading of permutation-invariance and our conviction that mereological composition is the only composition worth the name. But rejecting it is unpalatable, since it entails admitting new, strange objects into our ontology in whose existence we have no independent reason to believe.

We have seen that the natural mathematical structure of fermionic states poses a threat to premise 3. This threat is not compelling, insofar as premises 1 and 2 are not compelling. And rejecting premise 3 will be unpalatable to anyone who takes mereology to be 'perfectly understood, unproblematic, and certain'. The question of whether our traditional understanding of composition is immune to the deliverances of quantum mechanics hangs on which unpalatable claim one is prepared to accept.

University of Cambridge
Faculty of Philosophy
e-mail: adam.caulton@gmail.com

REFERENCES

Ashtekar, A., Schilling, T.A. (1999). Geometrical formulation of quantum mechanics. In: *On Einstein's Path*, pp. 23–65. New York.

Bacciagaluppi, G. (2009). Is Logic Empirical? In: K. Engesser, D. Gabbay and D. Lehmann (eds.), *Handbook of Quantum Logic and Quantum Structures: Quantum Logic*, pp. 49–78. Amsterdam: Elsevier.

Birkhoff, G., von Neumann, J. (1936). The logic of quantum mechanics. *Annals of Mathematics* **37**, 823–843.

Butterfield, J.N. (1993). Interpretation and identity in quantum theory. *Studies in the History and Philosophy of Science* **24**, 443–76.

Caulton, A. (2013). Discerning "indistinguishable" quantum systems. *Philosophy of Science* **80**, 49–72.

Caulton, A. (2015). *Qualitative Individuation in permutation-invariant quantum mechanics.* Preprint available online at: http://philsci-archive.pitt.edu/10990/.

da Costa, N.C.A. (1980). *Ensaio Sobre os Fundamentos da Logica.* Sao Paulo: Hucitec/EdUSP.

Dalla Chiara, M.L., Giuntini, R., Krause, D. (1998). Quasiset Theories for Micro-Objects: A Comparison. In: E. Castellani (ed.), *Interpreting Bodies: Classical and Quantum Objects in Modern Physics*, pp. 142–152. Princeton, NJ: Princeton University Press

Dalla Chiara, M.L., Giuntini, R., Rédei, M. (2007). The History of Quantum Logic. In: D. Gabbay and J. Woods (eds.), *Handbook of History of Logic*, Vol. 8: *The Many Valued and Nonmonotonic Turn in Logic*, pp. 205–283. North Holland: Elsevier.

Dummett, M. (1976). Is Logic Empirical? In: H.D. Lewis (ed.), *Contemporary British Philosophy*, 4th series, pp. 45–68. London: Allen and Unwin. Reprinted in M. Dummett, *Truth and other Enigmas*, pp. 269–289. London: Duckworth.

French, S., Krause, D. (2006). *Identity in Physics: A Historical, Philosophical and Formal Analysis.* Oxford: Oxford University Press.

French, S., Redhead, M. (1988). Quantum physics and the identity of indiscernibles. *British Journal for the Philosophy of Science* **39**, 233–46.

Ghirardi, G., Marinatto, L. (2003). Entanglement and Properties. *Fortschritte der Physik* **51**, 379–387.

Ghirardi, G., Marinatto, L. (2004). General Criterion for the Entanglement of Two Indistinguishable Particles. *Physical Review A* **70**, 012109–1-10.

Ghirardi, G., Marinatto, L. (2005). Identical Particles and Entanglement. *Optics and Spectroscopy* **99**, 386–390.

Ghirardi, G., Marinatto, L., Weber, T. (2002). Entanglement and Properties of Composite Quantum Systems: A Conceptual and Mathematical Analysis. *Journal of Statistical Physics* **108**, 49–122.

Hovda, P. (2009). What Is Classical Mereology? *Journal of Philosophical Logic* **38**, 55–82.

Huggett, N. (1999). On the significance of permutation symmetry. *British Journal for the Philosophy of Science* **50**, 325–347.

Huggett, N. (2003). Quarticles and the Identity of Indiscernibles. In: K. Brading and Castellani, E. (eds.), *Symmetries in Physics: New Reflections*, pp. 239–249. Cambridge: Cambridge University Press.

Jauch, J.M. (1968), *Foundations of Quantum Mechanics.* London: Addison-Wesley.

Kibble, T.W.B. (1979). Geometrization of Quantum Mechanics. *Communications in Mathematical Physics* **65**, 189–201.

Kochen, S., Specker, E.P. (1967). The Problem of Hidden Variables in Quantum Mechanics. *Journal of Mathematics and Mechanics* **17**, 59–87.

Krause, D. (1990). *Non-Reflexivity, Indistinguishability and Weyl's Aggregates.* PhD Thesis, Sao Paulo: University of Sao Paulo.

Krause, D. (1996). Axioms for collections of indistinguishable objects. *Logique et Analyse* **153–154**, 69–93.

Ladyman, J., Linnebo, Ø., Bigaj, T. (2013). Entanglement and non-factorizability. *Studies in History and Philosophy of Modern Physics* **44**, 215–221.

Leonard, H.S., Goodman, N. (1940). The Calculus of Individuals and Its Uses. *Journal of Symbolic Logic* **5**, 45–55.

Leśniewski, S. (1916/1992). *Podstawy ogólnej teoryi mnogości I.* Moskow: Prace Polskiego Koła Naukowego w Moskwie, Sekcya matematyczno-przyrodnicza. Trans. by D.I. Barnett (1992), *Foundations of the General Theory of Sets.* In: S.J. Surma, D. Srzednicki, D.I. Barnett and F.V. Rickey (eds.), *Collected Works*, Vol. 1, pp. 129–173. Dordrecht: Kluwer.

Lewis, D. (1968). Counterpart Theory and Quantified Modal Logic. *Journal of Philosophy* **65**, 113–126.

Lewis, D. (1991). *Parts of Classes.* Blackwell: Wiley-Blackwell.

MacLane, S., Birkoff, G. (1991). *Algebra.* Third Edition. Providence, RI: AMS Chelsea.

Margenau, H. (1944). The Exclusion Principle and Its Philosophical Importance. *Philosophy of Science* **11**, 187–208.

Maudlin, T. (1998). Part and Whole in Quantum Mechanics. In: E. Castellani (ed.), *Interpreting Bodies: Classical and Quantum Objects in Modern Physics.* Princeton: Princeton University Press.

Maudlin, T. (2005). The Tale of Quantum Logic. In: Y. Ben-Menahem (ed.), *Hilary Putnam: Contemporary Philosophy in Focus*, pp. 156–187. Cambridge: Cambridge University Press.

Messiah, A.M.L., Greenberg, O.W. (1964). Symmetrization Postulate and Its Experimental Foundation. *Physical Review* **136**, B248-B267.

Muller, F.A., Saunders, S. (2008). Discerning Fermions. *British Journal for the Philosophy of Science,* **59**, 499–548.

Muller, F.A., Seevinck, M. (2009). Discerning Elementary Particles. *Philosophy of Science,* **76**, 179–200.

Nielsen, M.A., Chuang, I.L. (2010). *Quantum Computation and Quantum Information.* 10th anniversary edition. Cambridge: Cambridge University Press.

Putnam, H. (1968). Is Logic Empirical?. In: R. Cohen and M. Wartofsky (eds.), *Boston Studies in the Philosophy of Science*, vol. 5, pp. 216–241. Dordrecht: Reidel. Reprinted as "The Logic of Quantum Mechanics," in: H. Putnam, *Mathematics, Matter, and Method. Philosophical Papers*, vol. 1, pp. 174–197. Cambridge: Cambridge University Press.

Putnam, H. (1974). How to Think Quantum-Logically. *Synthese* **29**, 55–61. Reprinted in: P. Suppes (ed.), *Logic and Probability in Quantum Mechanics*, pp. 47–53. Dordrecht: Reidel.

Russell, B. (1918/1985). *The Philosophy of Logical Atomism.* Chicago, IL: Open Court.

Tarski, A. (1929). *Les fondements de la géométrie des corps, Księga Pamiątkowa Pierwszego Polskiego Zjazdu Matematycznego, supplement to Annales de la Société Polonaise de Mathématique* 7, 29–33. Translated by J.H. Woodger, (1956). *Foundations of the Geometry of Solids.* In: A. Tarski, *Logics, Semantics, Metamathematics. Papers from 1923 to 1938*, pp. 24–29. Oxford: Clarendon Press.

Tung, W.-K. (1985). *Group Theory in Physics.* River Edge: World Scientific.

Varzi, A. (2014). *Mereology.* In: E.N. Zalta (ed.), *The Stanford Encyclopedia of Philosophy (Spring 2014 Edition)*, URL = <http://plato.stanford.edu/archives/spr2014/entries/mereology/>.

Andreas Hüttemann

PHYSICALISM AND THE PART-WHOLE RELATION

Abstract: In this paper I intend to analyse whether a certain kind of physicalism (Part-whole-physicalism) is supported by what classical mechanics and quantum mechanics have to say about the part whole relation. I will argue that not even the most likely candidates – namely cases of micro-explanation of the dynamics of compound systems – provide evidence for part whole-physicalism, i.e. the thesis that the behavior of the compound obtains *in virtue* of the behavior of the parts. Physics does not dictate part-whole-physicalism.

In this paper I intend to analyse whether a certain kind of physicalism (Part-whole-physicalism) is supported by what classical mechanics and quantum mechanics have to say about the part whole relation.

1. Physicalism

I will first characterize what I take to be the core physicalist intuition. Next I will disambiguate two physicalist claims and will then make one of the physicalist claims as precise as is necessary for the purposes of this paper.

Different authors use different vocabulary when they characterize what they take to be the core physicalist intuition. Jaegwon Kim, for instance, describes his own view (which he calls "physicalism" elsewhere) as follows:

> The broad metaphysical conviction that underlies these proposals is the belief that ultimately the world – at least, the physical world – is the way it is because the micro-world is the way it is [...]. (Kim 1984a, p. 100)

(The qualification in the parentheses has to be dropped for physicalism proper.) Kim uses 'because' to express that the macro-world *depends* on

In: Tomasz Bigaj and Christian Wüthrich (eds.), *Metaphysics in Contemporary Physics* (*Poznań Studies in the Philosophy of the Sciences and the Humanities*, vol. 104), pp. 323-344. Amsterdam/New York, NY: Rodopi | Brill, 2015.

the micro-world. A central tenet in the debate about physicalism is to say something informative about this dependence relation. Philip Pettit invokes political metaphors for this purpose:

> The fundamentalism that the physicalist defends gives total hegemony, as we might say, to the microphysical order: it introduces the dictatorship of the proletariat. (Pettit 1993, pp. 220–221)

And elsewhere:

> [M]icrophysicalism [...] is the doctrine that actually (but not necessarily) everything non-microphysical is composed out of microphysical entities and is governed by microphysical laws. (Pettit 1994, p. 253)

What is important in this context is that these metaphors characterise the dependence in question as *asymmetrical*. This will be essential for my later argument. Another expression that is sometimes used to characterize the asymmetric dependence relation is "in virtue of." Barry Loewer, for instance, writes:

> Physicalism claims that all facts obtain *in virtue* of the distribution of the fundamental entities and properties – whatever they turn out to be – of completed *fundamental physics*. (Loewer 2001, p. 37)

I will use Loewer's formulation as my starting point for an explication of physicalism.[1]

Before I approach the issue of clarifying the *in virtue*-claim I will disambiguate two different kinds of physicalism – levels-physicalism and part-whole-physicalism. The different issues at stake can be illustrated by an example.[2] Consider a case in which the state of a whole (the ferromagnetic state of a piece of iron) is explained in terms of the states of the parts (magnetic dipoles of the iron-atoms). Two questions/issues can be distinguished. First, we can ask whether the ferromagnetic state of the piece of iron corresponds to some microstate of the piece of iron for instance a state that can be described as a so-called spin-wave state of the piece of iron. This issue concerns the relation of two kinds of states of the same system – the ferromagnetic state and the spin-wave-state of the piece of iron. A second question is whether the spin-wave state of the piece of

[1] There are various problems I will bypass. One of these has been called Hempel's dilemma. Physicalism can either be defined via reference to contemporary physics, but then it is most probably false, or it can be defined via reference to a future or ideal physics, but then it is trivial in the sense of not falsifiable, because we are unable to predict what a future physics will contain (see Hempel 1969; Crane and Mellor 1990; Melnyk 2003, pp. 11–20; Stoljar 2009).

[2] For a more detailed analysis of the difference between levels-physicalism and part-whole-phyicalism see (Hüttemann and Papineau 2005).

iron can be explained in terms of the states of the individual atoms and certain relations and interactions among them. This question concerns the relation between the state of the whole piece of iron on the one hand and the states of its components on the other hand: how do the individual states of the atoms add up to the spin-wave-state of the whole? The latter issue concerns the relation between parts and wholes, not between two states of the same system.

More generally, one issue concerns levels. How do entities picked out by non-fundamental terminology, such as biological or psychological terminology (or "magnetization"), relate to fundamental physical entities? A physicalist with respect to levels claims:

Levels physicalism: Putatively non-physical properties obtain *in virtue* of (fundamental) physical properties.

A second issue concerns parts and wholes. A physicalist with respect to the part-whole-relation claims:

Part-whole-physicalism: The properties of compound systems are the way they are *in virtue* of the properties of their parts (and some further facts about how the parts interact and how they are related).[3]

In this paper I will be concerned with the question whether part-whole-physicalism is supported by what classical mechanics or quantum mechanics have to say about the part-whole relation.

2. Physicalism, Supervenience and Duplicates

Loewer's characterization of physicalism as well as my own characterizations of levels-physicalism and part-whole-physicalism contain the expression "in virtue." Very often the *in virtue*-claim is spelled out in terms of supervenience and related concepts such as duplicates. I will not go into the details of this discussion but only briefly indicate why this approach is not satisfactory.

Loewer discusses Frank Jackson's explication of physicalism. According to Jackson physicalists hold:

(P) Physicalism is true iff every world that is a minimal physical duplicate of the actual world is a duplicate simpliciter.[4]

---

[3] The term "physicalism" in this context is only appropriate if it is assumed that there are fundamental parts, which can be characterized as physical parts. This is clearly a contentious issue but nothing in what follows will depend on this choice of terminology.

[4] The formulation is due to Barry Loewer (Loewer 2001, p. 39). Frank Jackson defends his position in (Jackson 1998, Chapter 1).

Principle (P) is meant to capture the idea that once the physical facts of our world are fixed all the facts of our world are fixed. If (P) is true all non-physical facts globally supervene on the physical facts.

As Jackson acknowledges, definitions of physicalism have to capture asymmetry claims that are associated with it:

> Physicalism is associated with various asymmetry doctrines, most famously with the idea that the psychological depends in some sense on the physical, and not the other way round. (Jackson 1998, p. 14)

However, as Loewer points out Jackson's principle (P) fails to capture the asymmetry or *in virtue*-claim (Loewer and Jackson discuss what I have called "levels-physicalism"):

> The worry is that (P) may not exclude the possibility that mental and physical properties are distinct but necessarily connected in a way that neither is more basic than the other. In this case it doesn't seem correct to say that one kind of property obtains in virtue of the other's obtaining. (Loewer 2001, p. 39)

Claims about supervenience and duplicates do not *entail* that properties of one kind obtain *in virtue* of properties of another kind. Loewer acknowledges this problem without providing a solution:

> if considerations about the nature of necessity do not rule this possibility out then we must admit that (P) is not quite sufficient for physicalism. However, it seems to me that if we had good reasons to believe (P), then, unless we also had some reason to believe that despite (P) mental facts (or some other kind of facts) do not hold in virtue of physical facts, we have good reason to accept physicalism. (Loewer 2001, p. 39)

In the remainder of this paper I will argue that classical and quantum mechanics fail to provide good reasons for the claim that in the case of part-whole-physicalism the *in virtue*-claim does hold. Physics does not dictate part-whole-physicalism. This argument, however, presupposes that something more is said about the *in virtue* relation.

3. The In Virtue-Relation

Recently various authors have attempted to explicate such expressions as "fact F obtains in virtue of fact G" or "fact F is grounded in fact G" (Rosen 2010; Audi 2012). The terminology developed in this context allows me to define part-whole-physicalism as precise as is necessary for arguing against it.

What needs to be analysed are sentences like "The fact that *p* obtains *in virtue* of (is grounded in) the fact that *q*" where '*p*' and '*q*' stand for propositions. Following Rosen, I will introduce some notation:

[p]: the fact that p
[p] $\leftarrow$ [q]: "[p] is grounded in [q]"
[p] $\leftarrow \Gamma$: "The fact that p is grounded in the collection of facts Γ."
[p] $\twoheadleftarrow$ [q] $=_{def}$ for some Γ: [p] $\leftarrow \Gamma$, [q]: "[p] obtains partially in virtue of (is partially grounded in) [q]"

We can now reformulate the doctrine of part-whole-physicalism in terms of this terminology. The claim

> The fact that a compound has certain properties obtains *in virtue* of (is grounded in) the facts that the parts have certain properties and some further facts about how the parts interact and how they are related.

can be reformulated in terms of the following abbreviations:

[w]: the fact that the compound/whole has certain properties,
[p_1]: the fact that part p_1 has a certain property, etc,
Δ: further facts about how the parts interact and how they are related.

Part-whole-physicalism can now be written as the claim that for all wholes w there are parts $p_1 \ldots p_n$ and further facts Δ such that

$$[w] \leftarrow [p_1], [p_2], \ldots [p_n], \Delta$$

Furthermore, we can reformulate claims like the following: "The fact that a whole has certain properties *partially* obtains in virtue of (is *partially* grounded in) the fact that part [p_1] has certain properties." and similar claims for [p_2] etc.:

[w] $\twoheadleftarrow$ [p_1]
[w] $\twoheadleftarrow$ [p_2]
etc.

Rosen's approach in developing a theory of the *in virtue*- or *grounding*-relation is to distil certain principles, which we hold to be true in all those cases where we seem to understand *in virtue*-talk. The first such principle is asymmetry (and that is all I will need):

asymmetry: if [p] $\twoheadleftarrow$ [q] then: not [q] $\twoheadleftarrow$ [p]

To give an example: When we claim that semantic facts obtain in virtue of non-semantic facts we (implicitly) deny that non-semantic facts obtain in virtue of semantic facts. (As a matter of fact the asymmetry principle is controversial among grounding-theorists (see for instances Wilson forthcoming). However, since I intend to explicate the *in virtue*-expression as it is used in the limited debate about part-whole-physicalism, where – as we have seen – it is used as expressing some kind of asymmetry, there is no problem accepting this principle for the purpose of this paper.)

We have seen in sections 1 and 2 that part-whole-physicalism is associated with asymmetry-claims. Rosen's terminology provides us with the means to make this claim sufficiently precise so as to work with it.

4. Micro-Explanation

Part-whole-physicalism claims that the properties of compound systems are the way they are *in virtue* of the properties of their parts (and some further facts about how the parts interact and how they are related). There is an *asymmetrical dependence* of the behavior of the compound on that of the parts. Physics seems to provide ample evidence for this claim. Robert Klee, for instance, argues:

> Micro-explanation is powerful in virtue of the fact that when a level of organization within a system can be explained in terms of lower-levels of organization this must be because the lower-levels (i.e. the micro-properties) determine the higher-levels (i.e. the macro-properties). This is why micro-explanation makes sense – the direction of explanation recapitulates the direction of determination. (Klee 1984, pp. 59–60)

So, the argument runs like this: The fact that we can explain the behavior of compound systems (wholes) in terms of the behavior of its parts supports the claim that there is a direction of determination from the micro-level to the macro-level. The fact that determination is directed warrants the claim that what happens at the macro-level happens *in virtue* of what happens at the micro-level.

In what follows I will take a closer look at this kind of argument from physics to physicalism.

Explaining the behavior of compound systems in terms of their parts may mean more than one thing. So what does 'behavior' mean in this context? With respect to the behavior of a physical system, we can distinguish the state of the system, its constants, and its temporal evolution. Some quantities of a physical system are constant; others vary with time. In the case of classical particles, we can, for instance, distinguish their positions and momenta as changing quantities, while other quantities (that might be relevant for the system under consideration) such as the gravitational constant remain constant. The values of the variable quantities at a particular time are called the state of the physical system at this time. However, the constants and the state of a system at a particular time do not exhaust what is commonly understood as the system's behavior. Furthermore, we have laws that describe the connections between the various quantities involved, and in particular, they describe how the state of the system develops in time. What these laws describe is the temporal evolution or dynamics of the system. Explaining the behavior of compound systems in terms of their parts may either refer to the state or to the dynamics.

Micro-explanation of the *state* of a compound system explains the state at a certain time in terms of the states of the parts at the same time. Thus, we might explain why a compound system, such as an ideal gas, has the determinate energy value E^* (the macro-state) by pointing out that the constituents have the determinate energy values E^1 to E^n (the states of the parts).

Quantum entanglement is a prominent counterexample to this kind of micro-explanation. It is not, in general, possible to explain the state of compound quantum mechanical systems in terms of the states of the parts because quantum mechanics does not, in general, specify such states for the parts (see e.g. Maudlin 1998).

This is bad news for the part-whole-physicalist (assuming that the evidence for part-whole-physicalism consists in successful micro-explanations), but not as bad as it might seem. There is another dimension to micro-explanation – micro-explanation of the dynamic of the compound system – that is not confronted with counterexamples from quantum mechanics (see Hüttemann 2005 for this distinction).

Micro-explanation of the dynamics of a compound specifies the temporal evolution or dynamics of the system in terms the dynamics of the parts (Plus interactions among the parts). This is why it is appropriately considered as a form of micro-explanation: the behavior of the compound (the dynamics of the system) is explained in terms of the behavior (dynamics) of the parts.

In what follows I will focus exclusively on the micro-explanation of the dynamics of a system, because it is the only option for the part-whole-physicalist.

So, how does this kind of micro-explanation work? By way of illustration, a simple example is a non-interacting two-particle system. The first step in the explanation or analysis of the dynamics of this system is the identification of its parts, i.e. the two (isolated) one-particle systems.

The second step consists in the determination of the dynamics of the isolated one-particle system. According to classical mechanics the complete behavior of a one-particle system is specified by its path in six-dimensional phase-space. A point in phase-space represents a state of a classical system. The Hamilton equations specify the system's time-evolution or dynamics and thus its path in phase-space. These equations in turn require a classical Hamilton-function. The dynamics of an isolated particle, for instance, can be described by a classical Hamilton-function of the form

$H = p^2/2m$, where p is the momentum and m the mass of the isolated particle.

For a non-interacting two-particle system we first need to specify two six-dimensional phase-spaces, one for each of the particles as well as a classical Hamilton-function of the above form for each of them. That, however, is not yet a description of a two-particle system. It is a description of two separate one-particle systems.

What we furthermore need is something that tells us how the descriptions of the behavior of subsystems have to be combined so as to obtain the description of the behavior of the compound system. We basically need the following information: 1) The phase-space for a compound system is the direct sum of the phase-spaces of the subsystems. Thus, for the two-particle system we obtain a twelve-dimensional phase-space. 2) The Hamilton-function for the compound system is the sum of those for the isolated constituents. Thus the dynamics of the system of two non-interacting particles in classical mechanics is described by a Hamilton-function of the form: $H = p_1^2/2m_1 + p_2^2/2m_2$.

This is the third and final step of the explanation or analysis of the dynamics of the non-interacting two-particle system: adding up the contributions of the parts according to laws of composition.

In the presence of interactions we have to introduce a further term into the Hamiltonian, e.g., a term for gravitational interaction such as $-Gm_1m_2/r$, where G is the gravitational constant and r the distance between the two particles.

Let me add an example from quantum mechanics: carbon monoxide molecules consist of two atoms of mass m_1 and m_2 at a distance x. Besides vibrations along the x-axis, they can perform rotations in three-dimensional space around its centre of mass. This provides the motivation for describing the molecule as a rotating oscillator, rather than as a simple harmonic oscillator. The compound's (the molecule's) behavior is explained in terms of the behavior of two subsystems, the oscillator and the rotator. These parts are not spatial parts, they are sets of degrees of freedom. The physicist Arno Bohm, who discusses this example in his textbook on quantum mechanics, describes this procedure as follows:

> We shall therefore first study the rigid-rotator model by itself. This will provide us with a description of the CO states that are characterised by the quantum number $n = 0$, and will also approximately describe each set of states with a given vibrational quantum number n. Then we shall see how these two models [The harmonic oscillator has already been discussed in a previous chapter. Author] are combined to form the vibrating rotator or the rotating vibrator. (Bohm 1986, p. 128)

This is a perfect illustration of a quantum-mechanical micro-explanation. It is in carrying out this programme that Bohm considers the following subsystems: (1) a rotator, which can be described by the Schrödinger equation with the Hamiltonian: $H_{rot} = L^2/2I$, where L is the angular momentum operator and I the moment of inertia. (2) an oscillator, which can be described by the Schrödinger equation with the following Hamiltonian: $H_{osc} = P^2/2\mu + \mu\omega^2Q^2/2$, where P is the momentum operator, Q the position operator, ω the frequency of the oscillating entity and μ the reduced mass.

He adds up the contributions of the subsystem by invoking a law of composition:

> IVa. Let one physical system be described by an algebra of operators, A_1, in the space R_1, and the other physical system by an algebra A_2 in R_2. The direct-product space $R_1 \otimes R_2$ is then the space of physical states of the physical combinations of these two systems, and its observables are operators in the direct-product space. The particular observables of the first system alone are given by $A_1 \otimes I$, and the observables of the second system alone are given by $I \otimes A_2$ (I = identity operator). (Bohm 1986, p. 147)

The explanatory strategy both in the quantum and the classical case can be summarized as follows:

The dynamic (temporal evolution) of a compound system is micro-explainable if it is – at least in principle – possible to deduce (to explain) it on the basis of

(i) general laws concerning the dynamics (temporal evolution) of the components considered in isolation,
(ii) general laws of composition, and
(iii) general laws of interaction.

The following point is essential: laws concerning constituents considered in isolation are never sufficient to explain even the simplest kinds of compound systems. We always need a law of composition.[5]

On the basis of this analysis of micro-explanation I will now examine whether micro-explanation provides evidence for part-whole-physicalism – more precisely: whether successful micro-explanation of the temporal evolution of compound systems provides evidence for the claim that the behavior of compound systems are the way they are *in virtue* of the behavior of their parts (and some further facts about how the parts interact and how they are related).

[5] In this sense the behavior of wholes always transcends that of the isolated parts.

5. Determination and the *In Virtue*-Relation

Let us return to Klee's argument quoted at the outset of section 4. He claimed that explanation presupposes determination.

The intuition behind this is that when we have something explained to us we understand it, and a large part of understanding something is knowing how it is determined (Klee 1984, p. 60).

This is a claim I will concede. But much depends on how we understand "determined" in this context. I will concede, first that if we have an explanation we have to assume that, e.g. the event that the *explanans* refers to determines the event that the *explanandum* refers to and, second, we know why this determination relation holds. I understand determination as bare determination, i.e. as a modal notion, such that, for instance, the values of x determine those of y iff for any value i of x there is some value j of y such that, necessarily, if x has i, y has j. The exact sense of "necessarily" depends on whether the determination relation holds in virtue of laws of nature, causation or something else. To give an example: For a (deterministic) causal explanation to work we have to assume that the cause determines the event to be explained (assuming certain factors can be held fixed) and we furthermore have to assume that there is some kind of relation in nature (causation) that underlies a given explanation and makes the determination relation feasible.

If we make the above concession, the case of micro-explanation has the following implication: because we are able to explain the behavior (dynamics) of the compound system in terms of that of the parts, we can conclude that the parts *determine* the behavior of the compound.

Isn't that exactly the conclusion the part-whole-physicalist was looking for? Doesn't the concession imply that the behavior of compound systems is the way it is *in virtue* of the behavior of the parts?

As we will see bare determination will not be sufficient to establish an *in virtue*-relation and thus part-whole-physicalism (this relates back to our discussion in section 2). For the argument from micro-explanation to part-whole-physicalism to be successful the relation between parts and wholes that has to be presupposed in micro-explanation has to qualify as something stronger than bare determination, it has to qualify as an *in virtue*-relation, i.e. minimally as bare determination plus the principle of asymmetry. So the question we have to answer is whether the relation that obtains between parts and wholes is indeed such that not only bare determination but also the asymmetry principle obtains.

In what follows I will argue that this is not the case. The relation between parts and wholes is *mutual* and thus fails to comply with the principle of asymmetry. The relation between parts and wholes is thus no *in virtue*-relation. The success of micro-explanation therefore fails to establish part-whole-physicalism.

I will first argue for this claim by considering non-interacting parts and will then take into consideration the more general case of interacting parts of a compound.

5.1. *The Non-Interaction Case*

In the last section I characterized micro-explanation as the explanation of the behavior of compound systems in terms of (a) general laws about how the constituents would behave in isolation and (b) general laws of composition and (c) general laws of interaction. On the basis of this analysis we are now in the position to pin down the exact nature of the relation between parts and wholes that is involved in micro-explanation. The behavior of the compound is determined by the behavior of the parts and the general laws of composition. (For the sake of simplicity I will disregard interaction terms in this sub-section). Given the behavior of the parts it is the laws of composition that make the behavior of the compound nomologically necessary.

Clearly, there is a direction of explanation from the parts to the whole. Whenever we explain the behavior of compound systems in quantum mechanics on the basis of the Schrödinger equation, our starting point is the set of Hamiltonians for the subsystems. This is an asymmetry with respect to explanation: We do not (at least not generally) explain the behavior of the parts in terms of the behavior of the compound. While it is an interesting question why there is this explanatory asymmetry, it on its own does not give us an ontological *in virtue*-relation that we need for part-whole-physicalism.[6] But what about the underlying part-whole relation? Does it, as Klee suggested, mirror the explanatory asymmetry? Does it obey the asymmetry principle?

Let us take a look at the law of composition. The law of composition for quantum mechanics gives us a prescription for the Hamiltonian that describes the temporal evolution of a compound system. In the absence of interactions we have, strictly speaking, the following.

$$H_{comp} = H_1 \otimes I_2 \otimes I_3 \otimes \ldots \otimes I_n + I_1 \otimes H_2 \otimes I_3 \otimes \ldots \otimes I_n + \ldots I_1 \otimes I_2 \otimes I_3 \otimes \ldots H_n$$

[6] The explanatory asymmetry might, for instance, be due to pragmatic reasons.

The index *i* ranges over all subsystems and I_n is the identity operator for the *n*-th subsystem's Hilbert-space. That looks somewhat cumbersome. Instead we typically encounter the considerably simpler

$$H_{comp} = H_1 + H_2 + \ldots + H_n$$

Let us consider the case of a compound consisting of three subsystems. Thus we have

$$H_{comp} = H_1 + H_2 + H_3$$

The law of composition gives rise to this formula for the Hamiltonians. It ensures that the behavior (dynamics) of the subsystems (represented by H_1, H_2 and H_3 respectively) determines the behavior (dynamics) of the compound (represented by H_{comp}).

A bare determination relation between the behavior of the parts and the behavior of the compound holds because we are dealing with an equation, and once the three Hamiltonians on the right hand side are specified, so is the fourth for the compound on the left hand side. But obviously the same is true for any of the other Hamiltonians as well. If H_{comp}, H_1 and H_2 are given, H_3 is determined according to the equation $H_3 = H_{comp} - H_1 - H_2$, and so forth.

Each of the four is determined as soon as the other three are fixed. The relation between the subsystems and the compound is *mutual*. Let me be very clear on one point: I am not claiming that the behavior of the compound on its own determines the behavior of any of the parts. The claim is rather, that if H_{comp} is given and two of the other Hamiltonians for the parts, the Hamiltonian for the third part is determined. The parts' behavior determine the behavior of the compound and any part's behavior is determined by the compound's behavior plus the behavior of the other parts. This is what I mean by "mutual determination" and it suffices to reject the *in virtue*-claim.

The result of these considerations is: The relation that has to be presupposed in order to understand the success of the micro-explanation cannot be an *in virtue* relation as it is presupposed in the discussion about part-whole-physicalism. The reason is that *both* of the following claims come out as true:

> $[w] \twoheadleftarrow [p_3]$, because the compound's behavior is partially determined by that of the third component or part. (The other determining factors are the fact that the law of composition obtains as well as $[p_1]$ and $[p_2]$.)

$[p_3] \twoheadleftarrow [w]$, because the behavior of the third component is partially determined by that of the compound. (The other determining factors are the fact that the law of composition obtains as well as $[p_1]$ and $[p_2]$.)

By appealing to laws of composition we are appealing to relations of mutual determination not to *in virtue*-relations.

To sum up: micro-explanations in physics essentially invoke laws of composition. Laws of composition describe the relations that obtain between parts and wholes (they underlie the micro-explanations). These relations are relations of mutual determination. Because laws of composition describe relation of mutual determination they fail to establish the principle of asymmetry and thus an *in virtue*-relation. Therefore, appeal to micro-explanations provides no evidence for part-whole-physicalism.

5.2. *The Interaction-Case*

One may object that the non-interaction case is rather trivial and not very interesting. Taking into account interactions does indeed complicate the picture. But the complications have to do with the question what to consider as the parts in a part-whole-explanation with interactions – rather than with the *nature* of the relation between parts and wholes. When the physicalist argues that micro-explanations provide evidence for the claim that the behavior of the compound obtains in virtue of the behavior of the parts, the physicalist has to specify what she means by "the behavior of the parts." I will consider two specifications and argue that in both cases the same conclusions as in the non-interaction case hold.

Let us take a classical case with interaction. In the presence of interactions we have to introduce a further term into the Hamiltonian, e.g., a term for gravitational interaction such as $-Gm_1m_2/r$, where G is the gravitational constant and r the distance between the two particles. In such a case the physicalist probably has two options of describing what an explanation in terms of the behavior of the subsystems might mean. According to the first (very natural) option the relevant subsystems are the isolated particles in the absence of any forces acting on them. In order to explain the compound's behavior we do not only rely on the general law of composition. Furthermore the term for the gravitational field potential has to be added. This reading of 'the behavior of the parts' accords with the claim that the compound's behavior is explained in terms of the behavior of the parts *and their interactions*. This yields the following Hamilton-function for the compound system:

$$H_{1+2} = p_1^2/2m_1 + p_2^2/2m_2 - Gm_1m_2/r$$

or

$$H_{1+2} = H_1 + H_2 - Gm_1m_2/r$$

The bare determination relation holds because we are dealing with an equation, and once the three terms on the right hand side are specified, so is the fourth for the compound on the left hand side. But, as before, the same is true for any of the other terms as well. If H_{1+2}, $-Gm_1m_2/r$ and H_2 are given, H_1 is determined according to the equation $H_1 = H_{1+2} - H_2 + Gm_1m_2/r$.

Each of the four terms is determined as soon as the other three are fixed. The relation between the subsystems, the interaction and the compound with respect to determination is *mutual.*[7]

We get the same conclusion as in the non-interaction case: Both of the following claims come out as true:

> $[w] \twoheadleftarrow [p_1]$, because the compound's behavior is partially determined by that of the first component or part. (The other determining factors are the fact that the law of composition obtains, the fact that the law of gravitation obtains as well as $[p_2]$.)
>
> $[p_1] \twoheadleftarrow [w]$, because the behavior of the first component is partially determined by that of the compound. (The other determining factors are the fact that the law of composition obtains, the fact that the law of gravitation obtains as well as $[p_2]$.)

As in the non-interaction case this result is incompatible with the principle of asymmetry, which is constitutive for the *in virtue* relation as presupposed in the discussion about part-whole-physicalism.

The physicalist might hold that there is a different reading of "the behavior of the parts." It is not the behavior of the particles considered on their own, but rather the particles' actual behavior in the field that is generated by the other particle. (The other particle itself is not part of the subsystem.) Thus, the behavior of the first subsystem consists of the

---

[7] The claim that the determination relations that underly physical laws are mutual has already been invoked by Bertrand Russell. He famously argued that the fundamental physical laws provide no room for an asymmetrical casual relation. Russell observed that "the future 'determines' the past in exactly the same sense in which the past 'determines' the future." (Russell 1912/13, p. 15). The determination relation that is described or presupposed by the fundamental laws of physics implies (given that the universe is closed and we are dealing with the physics of 1912/13) that past and future determine each other mutually and does not give rise to any kind of asymmetry. While Russell's claim about the determination relation pertains to the temporal development of systems my analogous claim concerns the synchronic part-whole relation.

first particle's behavior in an external gravitational field generated by the second particle. The second subsystem is described analogously. The two subsystems behave according to the Hamilton equations with the following Hamilton functions:

$$H_{1*} = p_1^{\,2}/2m_1 - (Gm_1m_2/r)\,|_1$$
$$H_{2*} = p_2^{\,2}/2m_2 - (Gm_1m_2/r)\,|_2$$

'$|_i$' indicates that the function (Gm_1m_2/r) is restricted to the phase-space of particle i. Let me stress that I am not committed to the claim that this can in general be consistently done. The physicalist who takes this option is confronted with a dilemma here: Either the particle's actual behavior (i.e. the particle's behavior in the external field) cannot be individuated as indicated above. Then it is not clear in what sense part-whole explanations provide evidence for the in virtue-claim because it remains unclear what the parts' behavior is. Or there is some way of individuating the parts' behavior in this sense, but then it wouldn't help the physicalist's argument. What we would end up with is a Hamiltonian that has the same form as in the non-interaction case:

$$H_{1+2} = H_{1*} + H_{2*}$$

So, by the same kind of argument as in the non-interaction case the determination relation would turn out to be mutual.

To conclude: Whether we consider non-interaction cases of part-whole explanations or interaction cases: The relations between parts and wholes invoked in micro-explanations turn out to be mutual. Therefore, an *in virtue*-relation between parts on the one hand and the compounds are not presupposed. Micro-explanations provide no evidence for part-whole-physicalism. Physics does not dictate part-whole-physicalism.

6. Objections and Replies

For the part-whole-physicalist there are various possible ways to react to the argument just presented. First, one might object to the argument by pointing out that there might be genuinely metaphysical relations that obtain between parts and wholes, but are not dealt with in physics. Answer: While there might be such relations they are not my concern in this paper. My aim is merely to figure out whether part-whole-physicalism is supported by what classical mechanics and quantum mechanics have to say about the part whole relation.

Second one might argue that the equations of physics that I relied on do not capture all that classical and quantum mechanics have to say about the part whole relation. An analogous position is sometimes attributed to Nancy Cartwright with respect to causation (Field 2003, p. 443). However, while there is no a priori argument against this possibility, there is no account that I know of that tells us what additional physical facts concerning the part whole relation there might be (that is over and above those captured in the equations of classical and quantum mechanics). In the absence of such a positive account it is difficult to evaluate this objection and I will refrain from doing so.

Finally, and maybe most importantly, a physicalist might doubt that what I have presented is what anyone ever meant when they were thinking that the properties of the whole are determined by the properties of the parts in an asymmetrical way. After all, we are dealing with microscopic physics, and not just with two or three particles. So the objection is to point to further physical relations between parts and wholes that I have not taken account of. The objections dealt with in the following sections, in particular those in sections 6.2 and 6.3 will consider the possibility of further candidates for the in virtue-relation.

6.1. *Flagpole*

In the literature on explanation there is the well-known case of the height of a flagpole and the length of its shadow. According to the laws of geometrical optics the length of the shadow is determined by the height of the flagpole holding fixed certain circumstances like the position of the sun. At the same time, these circumstances plus the length of the shadow determines the height of the flagpole. So we have a case of mutual determination. With respect to this determination relation the principle of asymmetry does not hold. However, we do nevertheless believe that the fact that the shadow has a certain length obtains partially in virtue of the fact that the flagpole has a certain length but not *vice versa*. By analogy, even though the determination relation between parts and wholes might fail to obey the principle of asymmetry, it might still be true that the behavior of the compound obtains *in virtue* of the behavior of the parts.

The reply is that the two cases are in a relevant way disanaloguous. In the case of the flagpole we can give an account of how the asymmetry arises, whereas we cannot do the same in the case of the relation of parts and wholes.

Here is one way of explaining the origin of the asymmetry in the case of the flagpole. Geometrical optics is a simplified model of the situation at hand. A more detailed description would mention the propagation of the light waves. In the more complete picture it is possible to explain in what sense the length of the shadow is the dependent variable. Gerhard Schurz suggested that what's essential in this context is the fact that a change in the dependent variable is brought about *later*:

> The crucial idea [...] is that the distinction between those variables which are directly influenced by an allowed intervention, in contrast to those which are only indirectly influenced by it, is possible by considering the delays of time in the process of disturbing the system's equilibrium state. (Schurz 2001, p. 61)

And with respect to our example:

> Hence in every intervention allowed by C [circumstances like the position of the sun, Author] which disturbs the equilibrium state of the systems variables, the length variation of the shadow will take place slightly after the variation of the pole's length – because of the finite velocity of light. (Schurz 2001, p. 61)

I will not discuss whether this suggestion does indeed give a complete account of the asymmetry in this example. The essential point is that this strategy to break the symmetry cannot be applied in the case of parts and wholes. What is essential for Schurz's strategy is that we supplement the original description of the relation of the length of the shadow and the height of the flagpole by *additional physical facts* such as the propagation of the light wave. The simultaneous and mutual determination of the height of the flagpole and the length of its shadow is only apparent. It is a feature of a simplified and incomplete description of the situation only. Breaking the symmetry relies on a better and more detailed description.

However, the case of parts and wholes is different in this respect. There are no additional physical facts. For all we know the description of the part-whole relation given in section 4 is the most complete we have.

6.2. *One-To-Many-Relation*

However, even though our account of the part whole-relation as described in classical and quantum mechanics may be complete, the account may give room for the obtaining of asymmetries that have been overlooked so far. Frank Jackson, for instance, argues – in the context of levels-physicalism – that the asymmetry characteristic for the physicalist claim is due to an asymmetry of determination:

> For the physicalist, the asymmetry between physical and psychological (or semantic, or economic, or biological, ...) lies in the fact that the physical fully determines the psychological (or semantic, ...), whereas the psychological (or semantic, ...) grossly underdetermines the physical. (Jackson 1998, p. 15)

An analogous argument in the case of part-whole-physicalism runs as follows: While the behavior of the parts fully determines that of the compound, the behavior of the compound grossly underdetermines that of the parts. In other words: The relation between the whole and the parts surely seems asymmetrical insofar as to a certain behavior of the whole (dynamic or state) there correspond many different arrangements of the parts.

However, as I will argue, even though there is this one-to-many-relation, it does not suffice to establish an asymmetry claim. Let me illustrate this through a simple example. Suppose we are dealing with a massive compound system consisting of three subsystems. We are only interested in mass. Leaving out relativistic effects we know that the mass of the compound (m_4) adds up as follows:

(M) $m_1 + m_2 + m_3 = m_4$

Thus, (M) is our law of composition for our three masses. m_4 characterizes the compound or macro-system whereas m_1 to m_3 characterize the constituents or micro-systems. Let us assume that the compound system has a mass of 17 kg. This value is compatible with a plethora of values for m_1 to m_3. 1 kg/5 kg/11 kg, 6 kg/6 kg/5 kg, 7 kg/6 kg/4 kg – all of these micro-states are compatible with a macro-state of 17 kg. We have a one-to-many-relation between the compound and its constituents, which seems to support an asymmetry claim and therefore (maybe) the obtaining of an *in virtue*-relation (asymmetry being a necessary condition for the obtaining of an *in virtue*-relation). However, the same kind of one-to-many-relation occurs if we fix a value for one of the constituents, say m_1. If m_1 is fixed at 5 kg, that is compatible with an infinite number of values for m_2 to m_4: 5 kg/5 kg/15 kg, 6 kg/6 kg/17 kg, 3 kg/7 kg/15 kg – all of them will do. The fact that the compound has a certain mass value is compatible with lots of value distributions for the subsystems. But that does not single it out as something special.

The laws of composition give rise to equations that allow calculating the behavior of the compound on the basis of the behavior of the constituents. (Calculation presupposes determination of the relevant magnitudes.) However, they equally allow calculating the behavior of a constituent given the relevant information about the compound and the other constituents. Whenever we have three values in (M) we can calculate the fourth value. In this respect there is nothing special about m_4, the value for the macro-state. With respect to determination all of the values are on a par. In this sense the laws of composition (in quantum mechanics as well as in classical mechanics) are impartial with respect to the micro and the macro. It is true that the behavior of the parts fully determines that of the compound and the behavior of the compound grossly underdetermines that of the parts. It is however also true that the behavior of the first and second part together with that of the compound fully determine the behavior of the third part, while the third part on its own grossly underdetermines that of the rest. If the issue of full determination by the behavior of the parts vs.

gross underdetermination by the behavior of the compound were sufficient for the obtaining of an in virtue-relation between parts and wholes *both* of the following claims would come out as true:

$$[w] \leftarrow [p_1], [p_2], [p_3], \Delta$$

because the compound's behavior is fully determined by that of the third parts (*P*lus some compositional facts).

$$[p_3] \leftarrow [w], [p_1], [p_2], \Delta$$

because the behavior of the third part is fully determined by that of the compound, the first two parts (*P*lus some compositional facts).

As a consequence the following two claims about partial grounding/ partial obtaining *in virtue of* would hold:

$$[w] \twoheadleftarrow [p_3],$$
$$[p_3] \twoheadleftarrow [w].$$

Again, this result, is incompatible with the principle of asymmetry which is constitutive for the *in virtue* relation as presupposed in the discussion about part-whole-physicalism.

6.3. *Coarse Concepts*

When it comes to the thermodynamics of, say, ideal gases, we not only encounter the one-to-many-relation as discussed in the previous section. There seems to be a further candidate for an asymmetrical relation.

The macro-description in terms of pressure (p), volume (V) and temperature (T) plus the exact specification of $N - 1$ particles doesn't determine the state of the 'last' particle (the N-th particle). There are various possible states that are compatible with the given constraints. On the other hand, the specification of all particles does determine the values for p, V and T. Is that an asymmetrical relation of the relevant kind?

For example, if temperature is mean kinetic energy, the velocities and positions of $N - 1$ particles and the temperature of the gas don't determine the velocities and position of the Nth particle. There is a whole set of velocities of the Nth particle compatible with a certain temperature of the gas plus the velocities and positions of the $N - 1$ particles.

Rejoinder:

For a start I will leave out the thermodynamic description of the ideal gas and focus on the mechanical description. Let's assume we have a complete description of the compound system (the gas). The state of the compound can be represented as a point in $6N$-dim phase-space. Given the state of the compound as well as the states of $N - 2$ parts, the state of the second but last particle is not yet completely determined, because it can get into either the $N-1$-slot or the N-slot. However, given the state of the compound and the states of $N-1$ particles, the state of the N-th particle is determined. Of course the particles' states also determine the state of the compound. In this sense we have mutual determination of parts and wholes on the level of a purely mechanical characterization.

When we describe the ideal gas in terms of thermodynamic properties such as temperature and pressure, we use a coarser description of the compound system. It is coarse in the sense that a lot of micro-states are compatible with given values for p, V and T. Because we use this coarse terminology, i.e. p, V, T for the compound system, the states of $N-1$ particles plus the state of the compound fail to determine the state of the Nth particle. Strictly speaking, this is a case where the variables representing the behavior of the compound system are determined by the variables representing the behavior of the parts, whereas it does not hold that the variables representing the behavior of $N-1$ parts plus the variable(s) representing the behavior of the compound determine the variable for the N-th particle's behavior.

However, I think we have good reasons not to take this asymmetry at face value, i.e. not to read it realistically as telling us something about the underlying the ontology. The reason is that the asymmetry is generated by our choice of coarse-grained variables for the compound system. The asymmetry disappears if we choose the more precise mechanical description. Furthermore, asymmetries that are due to coarse-grained variables can be generated at will. This can be illustrated by the following example: Let's define an object as *heavy* if it weighs, say 150, 151, ... or 200 kg. If the object has N parts then the masses of the N parts determine whether or not the object is heavy. But the object being heavy plus the masses of $N-1$ parts do not determine the mass of the N-th part. The parts determine the whole, but the whole plus $N-1$ parts do not determine the remaining part.

However, the same kind of coarse concept can be defined for one of the parts. Take part no. 7. Part no. 7 is *quite heavy* if it weighs 50 or 51 or 52 or 53 kg. If the compound that no. 7 is a part of has N parts, then the mass of the compound plus all the masses of the other parts determine whether or not no. 7 is quite heavy. However, the mass of the compound is not determined by no. 7 being quite heavy plus the masses of the other parts (because of the coarseness of 'quite heavy').

What this shows is that we can generate asymmetries at will wherever we introduce coarse-grained variables. Therefore we should not read these asymmetries realistically. They are entirely due to the choice of coarse rather than precise variables and do not seem to have any implication with respect to the question what kind of ontological relations obtains between parts and wholes.

7. Conclusion

To sum up: Part-whole-physicalism is not supported by what classical mechanics and quantum mechanics have to say about the part whole relation. Not even those cases in classical and quantum mechanics, which are most favorable to the part whole physicalist (in the sense of *prima facie* support) – namely cases of micro-explanation of the dynamics of compound systems – provide evidence for the thesis that the behavior of the compound obtains *in virtue* of the behavior of the parts (and some further facts about how the parts interact and how they are related). Physics does not dictate part-whole-physicalism.

Universität zu Köln
Philosophisches Seminar
e-mail: ahuettem@uni-koeln.de

REFERENCES

Audi, P. (2012). A Clarification and Defense of the Notion of Grounding. In: F. Correia and B. Schnieder (eds.), *Metaphysical Grounding: Understanding the Structure of Reality*, pp. 101–121. Cambridge: Cambridge University Press.

Bohm, A. (1986). *Quantum Mechanics: Foundations and Applications*. New York: Springer.

Carnap, R. (1932). Die physikalische Sprache als Universalsprache der Wissenschaft. *Erkenntnis* **2**, 432–465.

Crane, T., Mellor, D.H. (1990). There is no Question of Physicalism. *Mind* **99**, 185–206.

Field, H. (2003). Causation in a Physical World. In: M. Loux and D. Zimmerman (eds.), *The Oxford Handbook of Metaphysics*, pp. 435–460. Oxford: Oxford University Press.

Hempel, C. (1969). Reduction: Ontological and Linguistic Facets. In: S. Morgenbesser, P. Suppes and M. White (eds.), *Philosophy, Science and Method*, pp. 179–199. New York: St. Martin's Press.

Hüttemann, A. (2005). Explanation, Emergence and Quantum-entanglement. *Philosophy of Science* **72**, 114–127.

Hüttemann, A., Papineau, D. (2005). Physicalism Decomposed. *Analysis* **65**, 33–39.

Kim, J. (1984). Epiphenomenal and Supervenient Causation. *Midwest Studies in Philosophy* **9**, 257–270.

Klee, R. (1984). Micro-Determinism and Concepts of Emergence. *Philosophy of Science* **51**, 44–63.

Loewer, B. (2001). From Physics to Physicalism. In: C. Gillet and B. Loewer (eds.), *Physicalism and Its Discontents*, pp. 37–56. Cambridge: Cambridge University Press.

Maudlin, T. (1998). Part and Whole in Quantum Mechanics. In: E. Castellani (ed.), *Interpreting Bodies*, pp. 46–60. Princeton: Princeton University Press.

Melnyk, A. (2003). *A Physicalist Manifesto.* Cambridge: Cambridge University Press.

Papineau, D. (2001). The Rise of Physicalism. In: C. Gillet and B. Loewer (eds.), *Physicalism and Its Discontents*, pp. 3–36. Cambridge: Cambridge University Press.

Pettit, P. (1994). Microphysicalism without Contingent Micro-Macro Laws. *Analysis* **54**, 253–257.

Rosen, G. (2010). Metaphysical Dependence: Grounding and Reduction. In: B. Hale and A. Hoffmann (eds.), *Modality: Metaphysics, Logic, and Epistemology*, pp. 109–136. Oxford: Oxford University Press.

Russell, B. (1912/13). On the Notion of Cause. *Proceedings of the Aristotelian Society* **13**, 1–26.

Schurz, G. (2001). Causal Asymmetry, Independent versus Dependent Variables and the Direction of Time. In: W. Spohn, M. Ledwig and M. Esfeld (eds.), *Current Issues in Causation*, pp. 47–67. Paderborn: Mentis.

Stoljar, D. (2009). Physicalism. In: E.N. Zalta (ed.), *The Stanford Encyclopaedia of Philosophy*, http://plato.stanford.edu/entries/physicalism/#10, retrieved 08/04/2010.

Wilson, J. (forthcoming). No work for a Theory of Grounding.

Jessica Wilson

METAPHYSICAL EMERGENCE: WEAK AND STRONG

ABSTRACT. Motivated by the seeming structure of the sciences, metaphysical emergence combines broadly synchronic dependence coupled with some degree of ontological and causal autonomy. Reflecting the diverse, frequently incompatible interpretations of the notions of dependence and autonomy, however, accounts of emergence diverge into a bewildering vari-

Here I argue that much of this apparent diversity is superficial. I first argue, by attention to the problem of higher-level causation, that two and only two strategies for addressing this problem accommodate the genuine emergence of special science entities. These strategies in turn suggest two distinct schema for metaphysical emergence – 'Weak' and 'Strong' emergence, respectively. Each schema imposes a condition on the powers of (features of) entities taken to be emergent: Strong emergence (associated with British emergentism) requires that higher-level features have more token powers than their dependence base features, whereas (following Wilson 1999) Weak emergence (associated with non-reductive physicalism) requires that higher-level features have a proper subset of the token powers of their dependence base features. Importantly, the notion of 'power' at issue here is metaphysically neutral, primarily reflecting commitment just to the plausible thesis that what causes an entity may (perhaps only contingently) bring about are associated with how the entity is – that is, with its features.

1. Introduction

Why care about what emergence is, and whether there is any? To start, many complex entities of our acquaintance – tornados, plants, people and the like – appear to be composed of less complex entities, and to have features which depend, one way or another, on features of their composing entities. Yet such complex entities also appear to be to some extent autonomous, both ontologically and causally, from the entities upon which they depend. Moreover, and more specifically, many 'higher-level' entities (particulars, systems, processes) treated by the special sciences appear to be broadly synchronically dependent on 'lower-level' (and ultimately

In: Tomasz Bigaj and Christian Wüthrich (eds.), *Metaphysics in Contemporary Physics Poznań Studies in the Philosophy of the Sciences and the Humanities*, vol. 104), pp. 345-402. Amsterdam/New York, NY: Rodopi | Brill, 2015.

fundamental physical) entities.[1] Yet, as is suggested by the associated special science laws, many higher-level entities appear also to be ontologically and causally autonomous, in having features in virtue of which they are *distinct from* and *distinctively efficacious relative to* the lower-level entities upon which they depend, even taking into account that the latter stand in configurational or aggregative relations. An account of emergence making sense of these appearances would vindicate and illuminate both our experience and the existence and tree-like structure of the special sciences, as treating distinctively real and efficacious higher-level entities and their features.

Reflecting these motivations, nearly all accounts of emergence take this to involve both broadly synchronic dependence[2] and (some measure of) ontological and causal autonomy.[3] Beyond this agreement, however, ac-

---

[1] Talk of 'higher-level' and 'lower-level' entities is relative, and reflects the pre-theoretic and theoretic appearances. Here I treat as at the same 'level' both individual entities treated by a given science, and certain combinations of such entities, where the allowable modes of combination include aggregations of relations which may hold between individual entities, as well as mereological and certain boolean combinations of such individuals or relational entities. So, for example, both atoms and relational entities consisting of atoms standing in atomic relations are taken to be at the same level, as are mereological or disjunctive combinations of atoms or relational atomic entities.

[2] Some accounts of emergence present this as diachronic, but most such accounts can be translated into synchronic terms, and those that cannot are aimed at characterizing single-level, not higher-level, emergence, and so can be put aside here. Mill (1843/1973) suggests that certain ('heteropathic') effects emerge from temporally prior causes, but also suggests that entities having powers to produce such effects synchronically emerge from lower-level entities (see §3.3). O'Connor (1994) and O'Connor and Wong (2005) take emergence to be diachronic, on grounds that emergent features are caused by lower-level features (sometimes in combination with other emergent features), and causation is diachronic; but here again diachronic emergence can be understood in terms of the synchronic emergence of features having the powers to produce the effects in question, and in any case the essentials of a causal account of dependence are preserved whether or not (the relevant) causation is synchronic (see §3.3). And Rueger (2001) takes emergence to be diachronic since involving temporally extended processes; but the emergence of such processes is compatible with these 'synchronically' depending on a temporally extended base (compare spatiotemporally global supervenience). Humphreys (1997) characterizes an irreducibly diachronic emergence, involving the exhaustive (non-mereological) 'fusion' of lower-level entities into another lower-level entity; but such same-level emergence is besides the point of accommodating the existence of higher-level entities.

[3] These core components are occasionally explicitly flagged (see Bedau 1997), but more typically are encoded in specific accounts of dependence and autonomy, as when Kim (2006, p. 548) says "two [...] necessary components of any concept of emergence that is true to its historical origins [...] are supervenience and irreducibility". Here and throughout I distinguish ontological autonomy (distinctness) from causal autonomy (distinctive efficacy), and I assume that both are required of an account of metaphysical emergence aiming to vindicate

counts of emergence diverge into a bewildering variety, reflecting that the core notions of dependence and autonomy have multiple, often incompatible interpretations.

In particular: candidate conceptions of (broadly) synchronic dependence include composition (Mill 1843, Stephan 2002); supervenience/necessitation (Broad 1925, Van Cleve 1990, Kim 2006, Noordhof 2010); causation or causal dependence (Searle 1992, O'Connor and Wong 2005); and functional or other realization (Putnam 1967, Boyd 1980, Antony and Levine 1997, Yablo 1992, Poland 1994, Wilson 1999, Shoemaker 2000/2001, Gillett 2002a, Melnyk 2003). Candidate conceptions of ontological and causal autonomy are even more various. Metaphysical accounts of autonomy include ontological irreducibility (Silberstein and McGeever 1999, Pereboom 2002, Kim 2006); novelty – e.g., of entities, properties, powers, forces, laws (Anderson 1972, Humphreys 1996, Crane 2002, Pereboom 2002); fundamentality – e.g., of entities, properties, powers, forces, laws (the British Emergentists, Cunningham 2001, O'Connor 2002, Wilson 2002, Barnes 2012); non-additivity (the British Emergentists, Newman 1996, Bedau 1997, Silberstein and McGeever 1999); 'downward' causal efficacy (Sperry 1986, Searle 1992, Klee 1984, Schroder 1998); multiple realizability (Putnam 1967, Fodor 1974, Klee 1984, Wimsatt 1996, Aizawa and Gillett 2009); elimination in degrees of freedom (Wilson 2010a); and the holding of a proper subset relation between token powers (Wilson 1999, Shoemaker 2000/2001, Clapp 2001). And epistemological accounts of autonomy include in-principle failure of deducibility, predictability, or explicability (Broad 1925 and other British Emergentists, Klee 1984, Lepore and Loewer 1989); predictability, but only by simulation (Newman 1996, Bedau 1997); lack of conceptual or representational entailment (Chalmers 1996, Van Gulick 2001), and theoretical/mathematical singularities (Batterman 2002). No surprise, then, that many recent articles on emergence are devoted mainly to taxonomizing its many varieties (Klee 1984, Van Gulick 2001, Stephan 2002, O'Connor and Wong 2015).

---

special science entities as entering into distinctive (typically causal) laws; this assumption also reflects that causal as well as ontological autonomy is constitutive of the distinctively emergentist responses to the problem of higher-level causation that we will later consider. Of course, causal autonomy entails ontological autonomy, by Leibniz's law. Ontological autonomy is compatible with an absence of causal autonomy, however, as with epiphenomenalist accounts of higher-level entities; correspondingly, though epiphenomenalist accounts are occasionally presented as accounts of 'emergence' (see Chalmers 2006), they are not so in the sense at issue here.

Though in general a thousand flowers may fruitfully bloom, this much diversity is unuseful for purposes of illuminating the structure of natural reality. Different accounts often disagree on whether an entity is emergent; and when they agree, there is often no clear basis for this agreement. Hence it is said that references to emergence "seem to have no settled meaning" (Byrne 1994, p. 206), that accounts of emergence are "not obviously reconcilable with one another" (O'Connor 1994, p. 91), and that "those discussing emergence, even face to face, more often than not talk past each other" (Kim 2006, p. 548). Moreover, and importantly for the relevance of emergence to contemporary debate, different accounts often disagree over whether emergence is compatible with *Physicalism*, according to which all broadly scientific entities are 'nothing over and above' physical entities. So, to take just one example, Kim (1999) takes physical realization to be incompatible with emergence, while Gillett (2002) takes such realization to be required.

I'll argue here that much of this apparent diversity is superficial. I'll start by showing, by attention to the available responses to the problem of higher-level causation, that there are two and only two schematic conceptions of higher-level metaphysical emergence of broadly scientific entities: *Strong* and *Weak* emergence, respectively (§2). The two schemas are similar in each imposing a condition on the powers of entities taken to be emergent, relative to the powers of their base entities. *For purposes of appreciating the generality of the schemas, it is of the first importance to register that the notion of 'power' here is metaphysically almost entirely neutral, reflecting commitment just to the plausible thesis that what causes an entity may potentially bring about (perhaps only contingently) are associated with how the entity is – that is, with its features.*[4] As I'll later discuss, even a categoricalist contingentist Humean can accept powers in the weak sense at issue in the schemas. Though similar in each involving a condition on powers, the schemas are also crucially different – a difference reflected in the fact that (given the physical acceptability of the lower-level entities) one schema is compatible with *Physicalism* and the other is not. (The results here generalize to distinguish two basic forms of higher-level emergence from lower-level entities, whether or not the latter are physically acceptable.) I will then consider the main accounts of emergent dependence (§3) and emergent autonomy (§4 and §5), and argue that all such accounts intended as characterizing metaphysical emergence are appropriately interpreted as targeting one or

[4] Here and elsewhere, nominalists are invited to interpret talk of features (properties, states) in their preferred terms.

the other schema. The two schemas thus unify and clarify the many apparently diverse accounts of higher-level metaphysical emergence, while explaining controversy over whether emergence is compatible with *Physicalism*.

Others have observed that accounts of emergence may be broadly sorted into 'weak' and 'strong' varieties, that are and are not compatible with *Physicalism*, respectively (see, for example, Smart 1981, Bedau 1997, Chalmers 2006, and Clayton 2006). My powers-based treatment (the key features of which were first proposed in Wilson 1999[5]) goes beyond these (typically gestural) treatments in explicitly cashing the distinction between *Weak* and *Strong* emergence in metaphysical rather than epistemological terms, in more specifically identifying the differing schematic metaphysical bases for these two types of emergence, and in explicitly locating the schemas in a representative spectrum of existing accounts of emergent dependence and emergent autonomy. My treatment also goes beyond previous taxonomic descriptions of the varieties of emergence, in that the schemas for *Weak* and *Strong* emergence arguably exhaust the available ways in which higher-level, broadly scientific entities might synchronically metaphysically emerge from lower-level such entities, and in that identification of what is key and crucial to such emergence indicates that certain accounts have more work to do if they are to ensure satisfaction of the conditions in the intended schema.

2. Two Schemas for Emergence

2.1. *The Target Cases*

Accounts of emergence tend to focus on emergence of features (e.g., either tokens or types of properties or states) from lower-level features, it being supposed (as per the background contrast with substance dualism; see §3.1) that emergence of entities (systems, processes, particulars) may be understood in terms of emergent features.[6] The lower-level features are typically taken to be physically acceptable relational features – that is, physically acceptable features of relational lower-level (and ultimately

[5] See note 13 for discussion of the genesis of this treatment.

[6] Hence Bedau (2002, p. 6) says: "[A]n entity with an emergent property is an emergent entity and an emergent phenomenon involves an emergent entity possessing an emergent property – and they all can be traced back to the notion of an emergent property".

physical) entities.[7] So, for example, a discussion of emergence might target the seeming autonomous dependence of:

- the higher-level property/state (of a complex system) of being in the basin of a strange attractor, on the (lower-level, relational) property/state (of a system of molecules) of having parts with certain positions and momenta;
- the property/state (of a plant) of being phototropic on the (lower-level, relational) property/state (of the plant's cellular walls) of being such as to undergo certain cellular wall weakenings and cellular expansions;
- a mental property/state (of a person) on a (lower-level, relational) neurophysiological property/state (of certain neurons standing in certain neuronal relations).

Emergence, as applying to such cases, is treated (multiple dependence or realizability to one side) as a one-one relation between higher-level and lower-level features. This treatment presupposes that certain relational lower-level entities exist and have features serving as a dependence base for the associated emergent features. The presupposition is useful, in encoding (as had by the posited lower-level relational entity) the sorts of features of complex entities that are assumed by all parties *not* to be emergent, in any interesting sense. Alternatively, one might (following Gillett 2002) dispense with the relational lower-level middleman and take the dependence base to consist in collections of comparatively non-relational lower-level features (say, features of individual molecules and pairwise relations between individual molecules), understood as combinable via certain ontologically 'lightweight' compositional principles, including additive causal combination (see §4.3), and certain boolean or mereological operations (see note 1). In any case, it's clear that the 'one-one' and 'many-one' approaches target the same phenomena: the latter considers the nature of the dependence of higher level entities/features on comparatively

---

[7] 'Physically acceptable' here refers to entities/features that are (taken to be, in some or other sense) 'nothing over and above' physical entities/features, where physical entities/features are, roughly and commonly, the relatively non-complex, not-fundamentally-mental entities/features that are the proper subject matter of fundamental physics (see Wilson 2006). Interestingly, the question of which features of lower-level physically acceptable entities should also count as lower-level physically acceptable features is usually left at a heuristic or intuitive level, with the assumption seeming to be that something akin to the combinatorial strategy generating the class of lower-level physically acceptable entities (discussed in note 1) also serves to properly identify the lower-level physically acceptable features of such entities. The heuristic assumption serves well enough for most purposes; see Wilson (2010a) for a more precise account.

non-relational lower-level entities/features given certain allowable combinatorial principles, whereas the former considers the nature of the dependence of higher-level entities/features on relational lower-level entities/features allowed by the combinatorial principles. By default I'll take the one-one perspective, but as we'll see some accounts of emergent dependence and autonomy take the many-one perspective.

2.2. *The Problem of Higher-Level Causation*

The primary challenge to the claim that higher-level entities and features may be metaphysically emergent – dependent on, yet also distinct from and distinctively efficacious with respect to, configurations of lower-level entities and features – is posed by the problem of higher-level causation, articulated most prominently by Jaegwon Kim.[8] This problem starts with a question: how can special science entities cause effects, given their strong synchronic dependence on lower-level (ultimately physical) entities? And the initial pressing concern is that, on the face of it, no answer satisfies all of certain intuitive or otherwise well-motivated premises.

First, some setup. Following common practice, I assume that the efficacy of entities lies in their having efficacious features; talk of entities themselves is thus suppressed. Moreover, given that causation is in the first instance a relation between spatiotemporally located goings-on, reference to 'features' in what follows is to be understood as reference to spatiotemporally located tokens (e.g., property instances, states, events) potentially of a type (property, state type, event type).[9]

Six premises lead to the problem.[10] Four of these concern special science features:

(1) *Dependence*. Special science features depend on lower-level physically acceptable features (henceforth, 'base features') in that, at a minimum, special science features (at least nomologically) require and are (at least nomologically) necessitated by base features.

(2) *Reality*. Both special science features and their base features are real.

(3) *Efficacy*. Special science features are efficacious.

---

[8] See, e.g., Kim (1989, 1993, 1998, 2005).

[9] That said, I will sometimes gloss the type/token distinction – e.g., when discussing necessitation of features, below.

[10] What follows reflects my preferred way of presenting the problem and slate of candidate resolutions, as set out in Wilson (2009, 2011), and elsewhere. Kim's own presentations more specifically target motivating reductive over non-reductive versions of physicalism, via denial of the fourth premise (*Distinctness*).

(4) *Distinctness*. Special science features are distinct from their base features.

And two concern causation:

(5) *Physical Causal Closure*. Every lower-level physically acceptable effect has a purely lower-level physically acceptable cause.
(6) *Non-overdetermination*. Apart from 'firing squad' cases, effects are not causally overdetermined.

On to the problem. There are two cases to consider, in each of which a special science feature *S* depends, on a given occasion, on base feature *P* (*Dependence*). There are two cases to consider: one in which special science feature *S* is assumed to cause a special science feature *S**, and one in which *S* is assumed to cause a lower-level physical feature *P**. In Kim's classic presentation, *S* is taken to be a mental feature (e.g., a token state of being thirsty); *P* is taken to be a lower-level (neurological, and ultimately fundamental physical) physically acceptable feature upon which mental state *S* depends, on a given occasion; and mental state *S* is taken to cause either another mental state *S** (e.g., a desire to quench one's thirst) or a lower-level physically acceptable state *P** (e.g., a physical reaching for a glass of water). But the considerations about to be raised more generally apply to raise a concern about how any real and distinct higher-level feature might be unproblematically efficacious.

First, suppose that *S* causes special science feature *S** on a given occasion (compatible with *Efficacy*). *S** is dependent on some base feature *P** (*Dependence*), such that *P** necessitates *S**, with at least nomological necessity. Moreover, *P** has a purely lower-level physically acceptable cause (*Physical Causal Closure*) – without loss of generality, *P*. If *P* causes *P**, and *P** (at least nomologically) necessitates *S**, then it is plausible that *P* causes *S**, by causing *P**. So, it appears, both *P* and *S* cause *S**, and given that *P* and *S* are both real and distinct (*Reality*, *Distinctness*), *S** is causally overdetermined; moreover (given *Dependence*) this overdetermination is not of the firing-squad variety (contra *Non-overdetermination*).

Second, suppose that *S* causes some base feature *P** on a given occasion (compatible with *Efficacy*). *P** has a purely lower-level physically acceptable cause (*Physical Causal Closure*) – without loss of generality, *P*. So, it appears, both *P* and *S* cause *P**, and given that *P* and *S* are both real and distinct (by *Reality* and *Distinctness*), *P** is causally overdetermined; moreover (given *Dependence*) this overdetermination is not of the firing-squad variety (contra *non-Overdetermination*).

So goes the argument that real, distinct and efficacious higher-level features induce problematic overdetermination. Kim sees the argument as motivating rejection of the premise that special science features are distinct from their base features – that is, he goes for reductionism. For present purposes, however, it is useful to more generally note that rejection of each of the premises of the (valid) argument is associated with one or other fairly comprehensive position in the metaphysics of science. The first four are as follows:

- *Substance dualism* or *Pan/proto-psychism*. Deny *Dependence*: avoid overdetermination by denying that *S* depends on physically acceptable *P*.
- *Eliminativism*. Deny *Realism*: avoid overdetermination by denying that *S* and/or *S** is real.
- *Epiphenomenalism*. Deny *Efficacy*: avoid overdetermination by denying that *S* is efficacious.
- *Reductive physicalism*. Deny *Distinctness*: avoid overdetermination by denying that *S* is distinct from *P*.

None of these strategies makes sense of the seeming emergence of higher-level features: *Substance dualism* and *Pan/proto-psychism* fail to accommodate dependence; *Eliminativism* and *Reductive physicalism* fail to accommodate ontological autonomy; *Epiphenomenalism Reductive physicalism* fail to accommodate causal autonomy.

2.3. *The Two 'Emergentist' Strategies*

The remaining strategies do better by way of accommodating emergence. These are:

- *Strong emergentism*. Deny *Physical Causal Closure*: avoid overdetermination by denying that every lower-level physically acceptable effect has a purely lower-level physically acceptable cause.
- *Non-reductive physicalism*. Deny *Non-overdetermination*: allow that there is overdetermination, but deny that it is of the firing squad variety that would be intuitively problematic as generally characterizing higher-level causation.

I'll now argue, for each of these strategies, that the strategy may be perspicuously understood as imposing one or another condition on the causal powers (henceforth, just 'powers') of a given special science feature, and that satisfaction of the associated condition provides a plausible principled basis for taking the feature to be emergent, in ways that proponents of each strategy would endorse.

2.3.1. *A Metaphysically Neutral Understanding of Powers*

Before getting started, let us ask: What are powers? Here, talk of powers is simply shorthand for talk of what causal contributions possession of a given feature makes (or can make, relative to the same laws of nature) to an entity's bringing about an effect, when in certain circumstances. That features are associated with actual or potential causal contributions ('powers') reflects the uncontroversial fact that what entities do (can do, relative to the same laws of nature) depends on how they are (what features they have). So, for example, a magnet attracts nearby pins in virtue of being magnetic, not massy; a magnet falls to the ground when dropped in virtue of being massy, not magnetic. Moreover, a feature may contribute to diverse effects, given diverse circumstances of its occurrence (which circumstances may be internal or external to the entity possessing the feature). Anyone accepting that what effects a particular causes (can cause, relative to the same laws of nature) is in part a function of what features it has – effectively, all participants to the present debate – is in position to accept powers, in this shorthand, metaphysically neutral and nomologically motivated sense.[11]

Besides commitment to the platitude that what entities can do (cause), relative to the same laws of nature, depends on how they are (what features they have), only one metaphysical condition is required in order to make sense of the powers-based conditions to follow; namely, that one's account of (actual or potential) causal contributions (powers) has resources sufficient to ground the identity (or non-identity) of a token causal contribution associated with a token of a higher-level feature, with a token causal contribution associated with a token of a lower-level feature. Here again, effectively all participants to the debate can make sense of such identity (non-identity) claims as applied to token (actual or potential) causal contributions (token powers).[12]

---

[11] For example, even a contingentist categoricalist Humean can accept powers in the neutral sense here: for such a Humean, to say that a (ultimately categorical) feature has a certain power would be to say that, were a token of the feature to occur in certain circumstances, a certain (contingent) regularity would be instanced. Of course, contemporary Humeans will implement more sophisticated variations on this theme.

[12] For example, suppose a contingentist categoricalist Humean wants to take a physicalist approach to the problem of higher-level causation, and so aims (as I will expand on below) to identify every token power of a token higher-level feature with a token power of its lower-level base feature. As previously, such a Humean understands powers in terms of actual or potential instances of a (contingent) regularity. Where the aim is to avoid overdetermination, the Humean may suppose, to start, that the (relevant instances of the) regularities overlap, both with respect to the (single) effect, and with respect to the (single) circumstances in

Of course, beyond the neutral characterization of powers, understood as tracking the nomologically determined causal contributions associated with a given feature, philosophers disagree. It is of the first importance, in order to appreciate the generality of the upcoming schemas for emergence, to see that no commitment to any controversial theses about powers (or associated notions such as property or law) will be required payment in what follows. Three key points of non-commitment, to be further defended in §2.3.3, are worth highlighting.

First, nothing in what follows requires accepting that it is essential to features that they have the powers they actually have. Maybe powers are essential to features; maybe they aren't. As we will shortly see, it suffices to characterize the strong emergentist and non-reductive physicalist strategies, and associated schemas for emergence, that powers are contingently had by the features at issue.

Second, nothing in what follows requires accepting that features are exhaustively individuated by powers. Maybe they are, maybe they aren't; perhaps features are also or ultimately individuated by quiddities or other non-causal aspects of features. In any case, the presence or absence of quiddities, which primarily serve to locate actually instanced features in worlds with different laws of nature, plays no role in actually individuating broadly scientific features in either scientific theorizing or practice. As such, the presence or absence of non-causal aspects of the features at issue can play no interesting role in a metaphysical account aiming to vindicate the scientific appearances supporting higher-level emergence; and nor does it, in the schemas to come.

Third, nothing in what follows requires accepting that powers are or are not reducible to categorical features, or that attributions of powers are or are not reducible to certain conditionals or counterfactuals, etc. Maybe powers, or talk of them, are reducible to other entities or terms; maybe they aren't. Again, scientific theorizing and practice is transparent to such

---

which the two token features occur. If the Humean aims to be a reductive physicalist, they may suppose that such overlap motivates identifying the token features at issue, and hence the associated powers. If the Humean aims to be a non-reductive physicalist, they can reject this identification of features, on difference-making grounds (e.g., of the sort associated with Mill's methods). Such a Humean will suppose that attention to broader patterns of regularities can provide a basis for identifying token powers of token features, even when the token features are not themselves identical. Whether reductive or non-reductive, the contingentist categoricalist Humean can make sense of the claim that some, all, or none of the token powers of token features are identical. This case is like the case of New York: if we can make it (out) here, we can make it (out) anywhere.

further metaphysical details, and so too should be our associated conceptions of emergence.

2.3.2. *Strong Emergentism*

As above, strong emergentists maintain that some special science features are real, distinct, and distinctively efficacious as compared to their physically acceptable base features. The threat of overdetermination posed by the problem of higher-level causation is avoided by denying, contra all varieties of physicalism, that every lower-level physically acceptable effect has a purely lower-level physically acceptable cause – that is, by denying *Physical Causal Closure*. Rather, it is maintained that at least some higher-level features have fundamentally novel powers to produce effects – powers not had by their physically acceptable base features, or (more weakly) had only derivatively by these base features, in virtue of base features' being preconditions for the emergent features which are more directly implicated in causing the physical effects in question. Whether the effect in question is a special science feature (S^*) or is rather a lower-level physically acceptable feature (P^*), either way the rejection of *Closure* blocks the route to P's causing E by way of causing P^*.

The strong emergentist strategy for avoiding the problem of higher-level causation can thus be put in terms of fundamentally novel powers, with *Physical Causal Closure* being denied on grounds that either (a) lower-level feature P does not have the power to cause the effect in question, or more weakly, (b) that while P does have the power to cause the effect, P has this power only derivatively, in virtue of being a dependence base for higher-level feature S, which more directly causes S^* (if the effect is a special science feature) or P^* (if the effect is a lower-level physically acceptable feature). Implementing either (a) or (b) requires that the powers of the higher-level feature satisfy the following condition:

New Power Condition: Token higher-level feature S has, on a given occasion, at least one token power not identical with any token power of the token lower-level feature P on which S synchronically depends, on that occasion.

Suppose, for example, that the special science feature at issue is a state of being thirsty, which in appropriate circumstances causes a physical effect – say, a physical reaching for a nearby glass of water. On the assumption that the state of being thirsty is strongly emergent, then the movement would not, contrary to the assumption of *Physical Causal Closure*, have a purely lower-level physical cause; hence even if the physically acceptable base feature in some sense causes the movement, this would only be in virtue of its being a precondition for the emergent state which more directly causes the movement. The effect is thus not overdetermined: even granting that the lower-level dependence base feature does cause the effect, the causal relation here goes through the higher-level emergent feature and its powers, so that there is only one causing, not two.

It is clear that satisfaction of this condition guarantees that *S* is both ontologically and causally autonomous from *P*: since *S* has a token power that *P* doesn't have, *S* is distinct from *P* (by Leibniz's law) and can do at least one thing that *P* can't do, or in any case cannot do in the same way as *S*.

2.3.3. *Non-Reductive Physicalism*

Like the strong emergentist, the non-reductive physicalist maintains that (some) special science features are real, distinct, and distinctively efficacious with respect to their base features. Problematic overdetermination is avoided, consistent with *Physical Causal Closure*, by denying *Non-overdetermination*, with the suggestion being that higher-level and base features stand in one or other 'realization' relation that, while not identity, is intimate enough to avoid overdetermination of the firing squad variety. A number of such relations have been proposed, including functional realization, the part/whole relation, and the determinable/determinate relation. This seeming diversity hides a deeper unity of strategy, however, which again can be put in terms of a certain condition on powers.

To start, the non-reductive physicalist maintains, as does the reductive physicalist, that every token power of *S*, on a given occasion, is identical to a token power of the base feature *P* upon which it depends, on that occasion. They moreover maintain that in such a case the token powers of *S* are a non-empty proper subset of the token powers of *P*, as per:

Subset of Powers Condition: Token higher-level feature *S* has, on a given occasion, a non-empty proper subset of the token powers of the token lower-level feature *P* on which *S* synchronically depends, on that occasion.[13]

[13] This approach to characterizing emergence of a physically acceptable variety (a.k.a. 'non-reductive realization'), is sometimes inaccurately called 'Shoemaker's strategy' or 'Shoemaker's account' of realization, following Shoemaker 2000/2001 – inaccurately, since my (1999) paper (first written for a Spring 1998 seminar with Richard Boyd on naturalism, during my third year of graduate school at Cornell, and submitted to a Philosophical Quarterly competition later that year) was the first published paper presenting and defending the subset-of-powers strategy for making sense of non-reductive physicalism. There I motivated the strategy as required to block the strong emergence of higher-level features from lower-level

As will be discussed in more detail below, the holding of the Subset of Powers Condition is typically (though not exclusively) motivated on grounds of multiple realizability, with the idea being that the powers of a multiply realized type are those in the intersection of the sets of powers of its realizing features, and that the associated proper subset relation between powers is preserved at the level of token features and powers. Satisfaction of the condition clearly blocks problematic overdetermination: when a power of *S* manifests in a given effect on a given occasion, there is only one causing (as between *S* and *P*), not two. Satisfaction of this condition also guarantees conformity to *Physicalism*, compatible with both ontological and causal autonomy.

Let's start with conformity to *Physicalism*. To start, note that the recipe for avoiding overdetermination accommodates the core physicalist claim (*Physical Causal Closure*, above) that every lower-level physically acceptable effect has a purely lower-level physically acceptable cause.

Moreover, imposition of the condition blocks all the usual routes to physical unacceptability. The main concern about physical acceptability turns on the possibility that higher-level feature *S* might be strongly emergent, such that, as above, either (a) *P* does not have the power to cause the effect *E* in question, or (b) that while *P* does have the power to cause

---

physical features; I moreover argued that apparently diverse accounts of non-reductive physicalism are more similar than they appear, in having in common that the proffered realization relations each arguably satisfy the subset condition on powers. More generally, my paper directed attention to powers as suitably metaphysical means, going beyond appeals either to supervenience or to explanation, of distinguishing reductive from non-reductive versions of physicalism, and non-physicalist accounts from any form of physicalism. The powers-based approach I endorse has certain advantages over Shoemaker's – importantly, as I'll rehearse down the line, it is not required to implement the strategy that one accept Shoemaker's (1980) view of properties as essentially and exhaustively characterized by their powers.

The pedigree of the proper subset strategy ultimately traces back to John Heil's 1996 NEH summer seminar in the metaphysics of mind, which took place at Cornell following my first year of graduate school, and which Heil graciously allowed me to attend. During the course of the seminar, Michael Watkins struck upon the idea of treating the problem of mental causation by taking the powers of the mental feature to be a proper subset of those of its physical realizer(s). The original idea for the subset-of-powers approach to non-reductive realization is thus Watkins's; however, he did not go on to much develop his view, whereas both I and Shoemaker (chair of my dissertation) did so, in parallel. Unfortunately, though I cited Shoemaker's then work-in-progress, he did not and has never cited any of my work on this topic, which has, perhaps predictably, led to its being commonly assumed that he was the sole originator of the view. It didn't help that the title of my 1999 paper ('How Superduper does a Physicalist Supervenience Need to Be?') was less than informative about the key results therein. Be all this as it may, I hope that those informed about this citation and priority issue will do what they can to ensure that my contribution to the original and subsequent development of the proper subset approach to realization is appropriately tracked.

this effect, this power is not identical with that had by *S* (it is manifested differently, or in different conditions). Satisfaction of the subset condition blocks both (a) and (b). Satisfaction of the condition also blocks the other live routes to physical unacceptability, associated with *S*'s being non-natural (see Moore 1903) or supernatural: such designations plausibly require the having of non-natural or supernatural powers, which are ruled out by satisfaction of the subset condition (assuming, as we are, that the base feature *P* has no such powers).

Now, as it stands (and remaining broadly neutral on the metaphysics of features) satisfaction of the proper subset condition is compatible with *S*'s having a non-causal aspect not had by *P* – say, a non-causal quiddity or an epiphenomenal quale. But, as discussed above, and as is reflected in the dispute between strong emergentists and physicalists, any non-causal aspects of *S* are irrelevant to broadly scientific goings-on: scientific truths do not in any way depend on or otherwise track whether scientific features have non-causal aspects (much less track how any such aspects are related). Hence that *S* has such aspects (whether or not shared by *P*) cannot undermine *S*'s physical acceptability, given *P*'s physical acceptability.

This point bears emphasizing, since many have supposed – following the assumptions of certain advocates of a powers-based approach to non-reductive realization (e.g., Shoemaker 2000/2001 and Clapp 2001) – that such an approach requires commitment to an account of features on which these are essentially or exhaustively individuated by their powers. Hence Melnyk (2006, pp. 141–143) suggests that unless features are identified with clusters of token powers, satisfaction of the proper subset condition will not guarantee conformity to *Physicalism*, since such satisfaction will not guarantee that physically realized entities are constituted by physical entities, or that truths about physically realized entities are made true by physical goings-on. More specifically, Melnyk claims that if realized features have non-causal aspects, then even given that an entity's having *P* entails that it has (bestowed upon it) the token powers associated with having *S*, it won't follow from satisfaction of the proper subset condition that the entity's having *P* constitutes its having *S*, or that the entity's having *P* (along with physical laws, etc.) makes *S* truly attributed to it. But Melnyk's claims are incorrect: truths about physical constitution or truthmaking, being broadly scientific truths, are neutral as regards whatever non-causal aspects of features there might be; hence the grounds of such truths must also be neutral on whether properties have non-causal aspects. It follows that satisfaction of the proper subset condition suffices for conformity with *Physicalism* independent of whether states or features are exhaustively individuated by their associated powers.[14]

The general pattern, blocking any route to *S*'s physical unacceptability, is as follows: if *P* is physically acceptable, and every token power of *S*, on an occasion, is identical with a token power of *P*, on that occasion, then any causal aspects of *S* are guaranteed to be physically acceptable; non-causal aspects of *S* are irrelevant to *S*'s physical acceptability; hence a realization relation satisfying the proper subset condition on powers guarantees *S*'s physical acceptability, in conformity to *Physicalism*, independent of what account of properties one endorses.[15]

Let's turn now to the question of autonomy. Satisfaction of the subset condition clearly accommodates ontological autonomy: if *S* has only a proper subset of *P*'s powers, then *S* is distinct from *P*, by Leibniz's law (see, however, Morris 2011 for reasons to think that this much ontological autonomy is compatible with reducing *S* to some other lower-level property). The strategy arguably also makes room for *S*'s being causally autonomous (pace, e.g., Ney 2010, Walter 2010), with the key idea being that causal autonomy does not require that *S* have a distinctive power. Rather, it is enough that *S* have a distinctive set (collection, plurality) of powers – that is, a *distinctive power profile.*

How might the having of a distinctive power profile suffice for causal autonomy? One case for this appeals to difference-making or other 'proportionality' considerations, of the sort reflecting that *S* (or *S*'s type) is multiply realizable. Suppose *S* is a state of feeling thirsty, which causes an effect *E* – say, a reaching for a glass of water. Now suppose that *S* (or another instance of *S*'s type, etc.) had been realized by *P'* rather than *P*. Would the (or a) reaching still have occurred? Intuitively, yes, because the additional powers possessed by *P*, in virtue of which it differs from

[14] Similarly for Melnyk's claim (2006, pp. 138–140) that unless realized entities are identified with clusters of powers, the condition's satisfaction will not guarantee satisfaction of the 'necessitation' condition, according to which a physically acceptable realized entity must (Perhaps together with physical laws, etc.) metaphysically necessitate the realized entity: "Why should it? Why assume that along with possession of power-tokens of certain types there automatically comes possession of a property [...] that would have conferred them?" (140). Given that truths about broadly scientific entities are transparent to facts about non-causal aspects of entities, from an entity's possession of power-tokens of a type it follows that the entity has the feature, whether or not features have non-causal aspects.

[15] It is also worth noting that in assuming that only powers are relevant to investigations into the physical acceptability of features, there is no danger of 'leaving out' what is relevant to, e.g., qualitative mental experience; for qualitative and other aspects of mentality do have causal implications (e.g., to produce awareness of qualitative aspects in experiencing subjects), as per the rejection of epiphenomenalism.

P' – say, to produce a specific reading on a neuronal state detector – don't matter for the production of the (or a) reaching. Rather, all that matters for this are the powers associated with *S*. That *S*'s distinctive power profile contains just the powers crucial for *E* provides a principled reason for taking *S* to be efficacious *vis-à-vis E* in a way that is distinctive from *P*'s efficacy *vis-à-vis E*. Note that nothing in this line of thought requires that one accept a difference-making account of causation or relatedly, that one reject *P* as being a cause of *E* – indeed, physicalists, who accept *Physical Causal Closure*, will plausibly maintain that *P* does cause *E*, either directly (if *E* is a lower-level physically acceptable cause) or indirectly (via production of a lower-level physically acceptable realizer or *E*). The suggestion is simply that attention to difference-making considerations provides a principled ground for *S*'s being distinctively efficacious *vis-à-vis E*, in that *S*'s power profile tracks those powers that are counterfactually relevant to the production of *E*.

Another case for taking distinctive power profiles to (at least sometimes) suffice for causal autonomy appeals to the connection between sets of powers and distinctive systems of laws (e.g., the special science laws governing entities of *S*'s type). Plausibly, systems of laws track causal joints in nature. Correspondingly, *S*'s distinctive power profile may be indicative of a distinctive causal joint in nature (and this may be the case even if S is only singly realized). Causal joints may overlap – in particular, in respect of *S*'s and *P*'s token power to cause effect *E*. Still, if the joints as a whole are different, this provides a principled basis for taking *S* to be distinctively efficacious *vis-à-vis E*, in that *S* produces *E* as part of a different system of laws than *P*.[16]

2.4. *Strong and Weak Emergence*

The strong emergentist and non-reductive physicalist responses to the problem of higher-level causation are the only responses aiming to accommodate the metaphysical emergence – dependence with ontological and causal autonomy – of higher-level entities; and as just argued, there are cases to be made that satisfaction of either of the associated conditions on powers would fulfill this aim. Moreover, and independent of the specifics of the problem (in particular, independent of the shared assumption that the base entities are physically acceptable), attention to these conditions makes clear the relatively limited ways in which, most crucially, the causal (hence also ontological) autonomy of a higher-level feature *vis-à-vis* its

[16] See Wilson (2010a) for further defense of this claim.

base feature may be gained. To wit: the feature may (as per strong emergentism) have *more* powers than its base feature; or the feature may (as per non-reductive physicalism) have *fewer* powers that its base feature. Since complete coincidence of powers doesn't make room for causal autonomy, these routes to emergence exhaust the options.

We may thus take the responses as exhaustive representative bases for two schematic conceptions of metaphysical emergence. The first schema is that associated with strong emergentism:

Strong emergence: Token higher-level feature *S* is strongly metaphysically emergent from token lower-level feature *P*, on a given occasion, just in case (i) *S* synchronically depends on *P* on that occasion; and (ii) *S* has at least one token power not identical with any token power of *P* on that occasion.[17]

The first condition minimally specifies synchronic dependence; the second (reflecting the *New Power Condition*) captures the comparatively strong sense in which an emergent feature may be causally, hence ontologically, autonomous *vis-à-vis* the lower-level base feature upon which it synchronically depends.

The second schema is that associated with non-reductive physicalism:

Weak emergence: Token higher-level feature *S* is weakly metaphysically emergent from token lower-level feature *P* on a given occasion just in case (i) *S* synchronically depends on *P* on that occasion; and (ii) *S* has a non-empty proper subset of the token powers had by *P*, on that occasion.[18]

---

[17] The schema is relativized to occasions, but it suffices for the strong emergence of *S*, *simpliciter*, that the condition is ever satisfied; and it suffices for the strong emergence of *S*'s feature type from lower-level physically acceptable feature types that any token feature *S* on any occasion (either actual, or counterfactually compatible with the actual laws of nature) satisfies the condition. These complications won't play a role in what follows.

[18] Again, the condition is relativized to occasions. If one wants to maintain that token feature *S* is weakly metaphysically emergent, one needs to generalize the condition to apply, as follows:

Weak emergence simpliciter: Token higher-level feature *S* is weakly metaphysically emergent, from lower-level physically acceptable features just in case for at least one occasion for at least one occasion on which *S* (actually, or counterfactually, compatible with the actual laws of nature) exists, *S* satisfies Weak emergence; and for every such occasion, (i) *S* synchronically depends on some token lower-level physically acceptable *P* on that occasion; and (ii) *S* has a proper or improper subset of the token powers had by *P*, on that occasion.

Further quantification over all actual or counterfactual (nomologically possible) tokens of *S*'s type would be required to establish that *S*'s type was weakly emergent from (any) lower-level

Again, the first condition minimally specifies synchronic dependence; the second (reflecting the *Subset of Powers Condition*) captures the comparatively weak sense in which an emergent entity is causally, hence ontologically, autonomous *vis-à-vis* its base entity.

Each schema encodes a different way in which a higher-level feature might be dependent on, yet ontologically and causally autonomous from, a base feature; and each is thus promising, so far as accommodating the motivations for emergence is concerned. And again, attention to the available responses to the problem of higher-level causation, and the associated relations between powers that might serve as a basis for dependent causal autonomy, indicate that these schemas encode the *only* options for characterizing the metaphysical emergence of higher-level, broadly scientific entities (henceforth, typically, just 'emergence').

Let's now turn to seeing how specific accounts of emergent dependence and emergent autonomy, properly disambiguated and interpreted, aim to conform to one or other schema. In what follows, I'll usually leave off the qualifier 'aim to', since my primary goal is not to assess the success of these accounts for purposes of characterizing emergence, but to make explicit their underlying theoretical intentions for doing so. That said, as prefigured, my discussion will track certain concerns about whether a given account presently satisfies its aim. One final remark before getting started: reflecting the role of emergence in the physicalism debates, accounts of emergent dependence and autonomy frequently presuppose that the base entities at issue are physically acceptable; the morals to be drawn, however, are broadly independent of this presupposition.

3. Emergent Dependence

Four main accounts of emergent dependence are on offer: material composition, modal covariation, causation or causal dependence, and functional or other realization. The first two are not exclusive of the last two: effectively all accounts of higher-level emergence take both material composition and modal covariation to be some part of emergent dependence. Where accounts primarily differ, as we will see, is in the assumed strength of modal covariation, and (relatedly) in whether broadly causal or rather realization-based dependence is (tacitly or explicitly) assumed. As I'll

---

physically acceptable feature types. These complications won't make a difference what follows.

argue, accounts of emergent dependence differing in these respects conform to either *Strong* or *Weak* emergence, respectively.

3.1. *Material Composition*

Accounts of emergence typically suppose that special science entities (again: systems, processes, particulars) depend on lower-level, ultimately physical entities at least in that the former are exhaustively composed of the latter:

> All organised [living] bodies are composed of parts similar to those composing inorganic nature, and which have even themselves existed in an inorganic state; but the phenomena of life which result from the juxtaposition of those parts in a certain manner bear no analogy to any of the effects which would be produced by the action of the component substances considered as mere physical agents. (Mill 1843, p. 243)

> The first feature of contemporary theories of emergence, the thesis of physical monism, is a thesis about the nature of systems that have emergent properties (or structures). The thesis says that the bearers of emergent properties are made up of material parts only. It denies that there are any supernatural components responsible for a system having emergent properties. Thus, all substance-dualistic positions are rejected [...]. (Stephan 2002, p. 79)

The assumption of compositional dependence reflects the intended contrast with dualist accounts on which higher-level features depend on the existence of physically unacceptable entities (e.g., souls, entelechies, conscious or proto-conscious fundamental particles). Compositional dependence is, however, compatible with either *Weak* or *Strong* emergence, and indeed, with the absence of emergence, since it is a further question, concerning any exhaustively physically composed particular, what features it has and whether any of these are emergent in either schematic sense.

3.2. *Modal Covariation*

A further common baseline assumption is that emergent features depend on base features in standing in certain relations of (at least nomologically) necessary covariation, reflecting that emergent features both require (for their occurrence) and are upwardly necessitated by base features. For example, Broad (1925) maintains that emergent features of a compound are functionally dependent on features of the compound's parts (pp. 54–55), and that emergent features are "completely determined" by such lower-level features, in that "whenever you have a whole composed of these [...] elements in certain proportions and relations you have something with the [compound's] characteristic properties" (p. 64); Van Cleve (1990) concurs that "an emergent property of *w* is one that *depends on* and is *determined by* the properties of the parts of *w*" (p. 222). The holding of both directions of necessary correlation may be expressed by (a version of) supervenience (see Kim 1990) that I'll call 'minimally nomological supervenience', according to which an emergent feature (at least nomologically) requires some base feature, and a given base feature (at least nomologically) necessitates any associated emergent feature.[19]

Understood as an asymmetric relation (see Kim 1998, p. 11), minimally nomological supervenience distinguishes reductive from emergent dependence. Without further specification, however, such a conception is compatible with either *Strong* or *Weak* emergence. Broad and other strong emergentists typically maintain that emergent features minimally nomologically supervene on base features. And the schema for *Strong emergence* makes sense of such claims: laws of nature, after all, express what broadly scientific entities can do – that is, what powers they have; hence if an emergent feature has a power not had by its base feature per (as *Strong emergence*), it is plausible to suppose that the features stand in some sort of nomological connection (see §3.3). Minimally nomological supervenience is also compatible with *Weak emergence*, for some relations satisfying *Weak emergence* (e.g., the determinable/determinate relation) entail that the higher-level entities supervene with metaphysical, hence with nomological, necessity (see §3.4).

It remains to consider whether strengthening of the modal covariation relations – pertaining specifically to the strength of upward necessitation – distinguishes *Strong* from *Weak* emergence. Indeed, many accept as characteristic of physically unacceptable emergence that emergent features would supervene with *only* nomological necessity on base entities, in contrast with relations (like identity or the determinable-determinate relation) which plausibly preserve physical acceptability. So, for example, Chalmers (2006) says,

> [C]onsciousness still supervenes on the physical domain. But importantly, this supervenience holds only with the strength of laws of nature (in the philosophical jargon, it is natural or nomological supervenience). (p. 247)

Van Cleve (1990) similarly characterizes emergence of the sort intended to contrast with *Physicalism*:

> If *P* is a property of *w*, then *P* is emergent iff *P* supervenes with nomological necessity, but *not* with logical necessity, on the properties of the parts of *w*. (p. 222)

(See also Noordhof 2010.) Though common, the supposition that *Strong* and *Weak* emergence contrast with respect to modal strength of dependence

[19] The notion of upward necessitation may be stochastic (see Kim 2006, p. 550); emergent dependence need not be deterministic.

relation is problematic, for two reasons. First, a physically acceptable feature might supervene with only nomological necessity on a physically acceptable base feature. For example, the subset condition in *Weak* emergence could be satisfied even if features are essentially individuated by non-causal quiddities and only contingently associated with their actual powers. Second, a physically unacceptable feature might supervene with metaphysical necessity on a physically acceptable base feature (see Wilson 2005). This would be the case if, for example, a consistent Malbranchean God brings about certain higher-level features upon the occasion of certain lower-level features in every possible world; or if features are essentially constituted by (all) the laws of nature into which they directly or indirectly enter; or if some strongly emergent features are grounded in non-physical interactions, and all the fundamental interactions are unified.

These considerations lead to a dilemma for anyone aiming to distinguish physically acceptable from unacceptable emergence by appeal to modal correlations alone. Those characterizing strong emergence in terms of mere nomological supervenience sometimes reject counter-cases whereby *Strong* emergent features supervene with metaphysical necessity on base properties, as violating Hume's Dictum, according to which there are no metaphysically necessary connections between (wholly) distinct entities (but see Howell 2009, Noordhof 2010, and Kim 2011 for some alternative strategies). As it happens, *Strong* emergent features need not be wholly distinct from base features (see Stoljar 2007), and in any case post-Humean reasons for believing Hume's Dictum are in short supply (see Wilson 2010b and elsewhere). But suppose that Hume's Dictum is accepted, and grant that it ensures that *Strong* emergent dependence holds with only nomological necessity. It remains, as per the first counter-case, that physically acceptable features might supervene on base features with only nomological necessity – if, as above, features are essentially individuated by non-causal quiddities, not powers. To block this case, non-causal quiddities must be rejected as individuating powers. But – here's the dilemma – proponents of Hume's Dictum arguably must (and typically do) accept non-causal quiddities as essentially individuating features, since after all (as per their denial that there are no metaphysically necessary causal connections) they cannot take features to be essentially individuated by powers. The means of blocking the two counter-cases are thus incompatible with each other, and so a modal characterization of the distinction between physically acceptable and unacceptable emergence cannot be maintained, even if one is willing to commit to certain controversial metaphysical theses.

Moving forward, it's worth noting that, though covariation accounts officially aim to characterize emergent dependence in purely correlational terms, they rely for their plausibility on the underlying contrast between certain nomological relations (e.g., causation) and certain metaphysical relations (e.g., the determinable/determinate relation). The next two proposals each cash out emergent dependence by explicit appeal to such relations, so as to both plausibly and determinately target either *Strong* or *Weak* emergence.

3.3. *Causation or Causal Dependence*

Yablo (1992) notes "a subtle interpretive question about supervenience", according to which

> On the emergence interpretation, a thing's physical properties are metaphysically prior to its mental properties and bring them into being. To caricature emergentism just slightly, supervenience is a kind of "supercausation" which improves on the original in that supercauses act immediately and metaphysically guarantee their supereffects [...]. (pp. 256–257)

The suggestion that emergent dependence is in some sense causal ranges back to Mill (1843/1973), the father of British Emergentism. Here it is important to be clear concerning how emergent features are considered causally dependent on base features. Mill's discussion initially focuses on a distinction between 'homopathic' and 'heteropathic' effects of a composite entity, where the former but not the latter effects are broadly additive combinations of effects of the sort that would have been produced were the component entities acting separately. Such a conception of emergence aims ultimately to characterize emergent autonomy in terms of a failure of additivity of causal influences, where such failure, in turn, is criterial of the composite entity's having a new power (to produce the heteropathic effect); see McLaughlin 1992. Hence it is ultimately not (heteropathic) effects, but rather features of complex entities having powers to produce such effects, which are emergent by Mill's lights. (See §4.3.1.)

That said, the question remains whether emergence of such features might itself be a causal phenomenon. Indeed, there are two ways in which emergent features might be causally dependent on base features. First, base features might act as synchronic nomologically necessary preconditions for the operation, or coming into play, of certain nomological features – i.e., fundamental forces or interactions – associated in turn with new powers; even if the relation here is not causation as traditionally understood, it nonetheless involves broadly causal lawful dependence (see Wilson 2002). This is the sort of causal dependence that is generally operative in British Emergentist accounts. Second, base features might more straightforwardly cause emergent features, as some contemporary emergentists (e.g. O'Connor and Wong 2005) suppose. The two approaches (causal dependence vs.

causation) are close variants, with the primary difference being that, if one supposes that causation is diachronic, one might further suppose that emergence is diachronic (as do O'Connor and Wong).

Whether emergent dependence is synchronic or diachronic, a conception in terms of causation or causal dependence will make good sense of *Strong emergence*. Either way an emergent feature has powers different from its base features: if caused, because effects typically have powers different from those of their causes; if causally dependent, because the operation of new fundamental forces or interactions serves as a (perhaps partial) ground for the having of new powers. The precise nature of the ground for the new powers varies depending on the preferred account of causal autonomy (see §4.2).

Seeing how causation and causal dependence make sense of *Strong emergence* sheds light on Kim's (2006, p. 558) claim that "the emergence relation from [*P*] to *S* cannot properly be viewed as causal". Kim asks, rhetorically, "How can there be a causal chain from [e.g.] pain to the hand motion that is separate and independent from the physical causal chain from the neural state to the motion of the hand?" (fn. 7). This would indeed be strange against the assumption of *Physicalism*, and the associated closure claim that every lower-level physically acceptable effect has a purely lower-level physically acceptable cause; however, the strong emergentist's strategy as encoded in *Strong emergence* just is to deny the closure claim, rather maintaining that the production of some physically acceptable effect requires (the manifestation of) powers not had by any lower-level physically acceptable feature. That said, Kim is clearly right that causation and causal dependence cannot characterize physically acceptable emergence, since such a nomologically generative connection does not ensure that the powers of emergent and lower-level features stand in the proper subset relation requisite for *Weak emergence*.

3.4. *Non-Reductive Realization*

The second metaphysically robust notion of emergent dependence is in terms of realization. There are many accounts of this notion (for surveys, see Polger 2007, Morris 2010, and Baysan forthcoming); all have in common the aim of characterizing a realized entity as 'nothing over and above' its realizing entity (or entities), compatible (given the physical acceptability of base entities) with *Physicalism*. Some physicalists moreover think that such nothing over-and-aboveness is compatible with a realized feature's being emergent. Hence Gillett (2002) sees the project of establishing the possibility of emergence as "deeply interwoven with the project of vindicating non-reductive physicalism as a viable position" (p. 102).

A realization-based conception of emergent dependence is indeed well-suited for physicalist purposes, in that the standard accounts of realization each have understandings on which their holding guarantees satisfaction of the conditions of *Weak emergence*. Here I consider a representative sample.

First, consider a 'functionalizability' account, according to which realized features are second-order features, having causal roles played by the lower-level features that realize them on a given occasion (see Putnam 1967, Fodor 1974, Papineau 1993, Antony and Levine 1997, Melnyk 2003, and others). Now, to be associated with a distinctive causal role is just to be associated with a distinctive set of powers; hence if the distinctive causal role of a realized feature is, on a given occasion, played by a lower-level realizing feature, every token power of the higher-level feature, on that occasion, will be numerically identical with a token power of the feature upon which it synchronically depends, on that occasion. This much suffices, as previously argued, for the physical acceptability of a functionally realized feature, as per *Physicalism*. Still, one might think that functional realization is incompatible with *Weak emergence*, on grounds that a functionally realized feature inherits *all* of the token powers of its realizing feature:

> A functional reduction of pain has the following causal and ontological implications: Each occurrence of pain has the causal powers of its neural realizer; thus if pain occurs by being realized by *N*, this occurrence of pain has the causal powers of *N*. [...] In general, if *M* occurs by being realized by *N* on a given occasion, the *M*-instance has the causal powers of the *N*-instance. (Kim 2006, p. 554)

Where a functional role may be played by multiple realizers, however, there is a case to be made that a functionally realized feature has, on a given occasion, only a proper subset of the token powers of the feature realizing it on that occasion. To see this, recall the analogy initially motivating functionalism (see, e.g., Putnam 1967), to cases where multiple hardware systems may implement the instructions associated with a given piece of software. Here the realizing systems are similar in each having whatever powers are needed to implement the software, but are different in having other powers associated with their distinctive hardware bases. More generally, in cases where a type of functionally characterized higher-level feature may be multiply realized, it is plausible that each of its realizing types will have all of the powers associated with its functional role, and more besides.[20] Correspondingly, a proper subset relation will hold between the

[20] See, e.g., the discussion in Antony and Levine (1997, p. 93) of how "realization indifferent" regularities may lead to a functionally specified property's being associated "with a

powers of the realized type and those of any of its realizing types. This relation between powers will hold on any occasion of realization involving tokens of the types; hence an account of emergent dependence in terms of functional realization will conform to *Weak emergence*.

Second, consider powers-based accounts of realization (see Wilson 1999, Shoemaker 2000/2001, Clapp 2001).[21] Wilson first argues, by attention to the intended contrast with British emergentism, that a physicalist account of higher-level features must satisfy the following condition:

> Condition on Causal Powers: each individual causal power associated with a supervenient property is numerically identical with a causal power associated with its base property. (p. 42)

She goes on to observe:

> Conceiving of a physicalist supervenience in terms of causal powers would [...] provide a method for non-reductivists to establish that a given supervenient property is distinct from its base property, by showing that the base property has (one or more) causal powers different from those of the supervenient property. In this case the set of causal powers associated with the supervenient property would be a proper subset of the set of causal powers associated with the base property, thus providing a clear-cut account (which has too often been lacking) of how a higher-level (say, mental) property could be distinct from, and yet 'nothing over an above', a lower-level physicalistically acceptable property. (p. 45)

As Shoemaker puts it:

> Property *X* realizes property *Y* just in case the conditional powers bestowed by *Y* are a subset of the conditional powers bestowed by *X* (and *X* is not a conjunctive property having *Y* as a conjunct). (p. 26)

Shoemaker moreover claims:

> Where the realized property is multiply realizable, the conditional powers bestowed by it will be a proper subset of the sets bestowed by each of the realizer properties. (pp. 8–9)

His motivations here parallel those used to motivate the same claim for functionally realized properties. In brief, higher-level features are associated with distinctive sets of powers; if such a feature is multiply realized, then its realizing types will share the powers of the realized type, but will differ in respect of further powers. This relation will plausibly hold on any occasion of realization of tokens of the types; hence an account of

distinctive set of causal powers".

[21] Strictly speaking, in Wilson (1999) and following I do not endorse a powers-based account of realization, but rather identify the proper subset of powers condition as key to any adequate account of (non-reductive) realization, including ones based in the determinable/determinate relation or Pettit's (1993) micro-physicalist account.

emergent dependence in terms of powers-based realization will conform to *Weak emergence*.[22]

Finally, consider accounts of non-reductive realization in terms of the determinable/determinate relation (see Yablo 1992, Wilson 1999, Wilson 2009), the relation of increased specificity paradigmatically holding between colors and their shades. Yablo (1992) expected the suggestion that, e.g., mental features stand to their physical realizations in the relation that colors bear to their shades to be met with some incredulity. One way to make his conjecture more plausible is to put the point in terms of the causal powers of the properties involved (see Wilson 1999 and Wilson 2009). Consider a patch that is red, and more specifically scarlet. Sophie the pigeon, trained to peck at any red patch, is presented with the patch, and she pecks. The patch's being red caused Sophie to peck after all, she was trained to peck at red patches. But the patch's being scarlet also caused Sophie to peck – after all, to be scarlet just is to be red, in a specific way. Nonetheless, Sophie's pecking was not problematically overdetermined. Plausibly, this is because each token power of the determinable red instance is numerically identical to a token power of its determining scarlet instance. Similarly, the proponent of determinable/determinate-based account of realization maintains, for the case of special science features *vis-a-vis* their dependence base features.

Again, one might be concerned that such an account of realization is incompatible with *Weak emergence*, on grounds that instances of determinables and associated determinates are token-identical (see MacDonald and MacDonald 1986 and Ehring 1996); for in that case a higher-level feature will inherit *all* of the token powers of the feature that realizes it on that occasion. But here too, there is a case to be made that instances of determinables have only a proper subset of the token powers of the features that determine them on a given occasion (see Wilson 1999). Plausibly, a given determinable will be associated with a distinctive set of powers; moreover, this determinable will typically be 'multiply determined' by associated determinates; distinct determinates of the determinable will share the powers of the determinable, but will differ in respect of other of their powers. Moreover, insofar as determinables are distinctively unspecific,

[22] Note that nothing in the preceding line of thought requires acceptance of any particular account of the metaphysics of properties. As previously, the physical acceptability of a higher-level feature hinges solely on the relations between its token powers and those of its base feature on a given occasion; as such, issues of physical realization are independent of whether features have non-causal quiddities; and one may correspondingly also maintain that issues of physical realization are independent of whether the actual powers of a given feature are essentially or exhaustively individuative of it.

this characteristic should be preserved in their instances; but if a determinable token is identical with a determinate token on a given occasion, the former will have all the token powers of the latter, and this distinctive lack of specificity will be lost. This provides another reason to suppose that a determinable token will have only a proper subset of the powers of their associated determinate token on any given occasion, in conformity with *Weak emergence*.

3.5. *Results*

We have arrived at the following results concerning accounts of emergent dependence:

- Conceptions of emergent dependence in terms of material composition are compatible with either *Weak* or *Strong* emergence, as well as with ontological reduction.
- Conceptions in terms of asymmetrical minimally nomological supervenience rule out ontological reduction, and are compatible with either *Weak* or *Strong* emergence.
- Conceptions in terms of mere nomological supervenience aim to conform only to *Strong emergence*, and conceptions in terms of metaphysical supervenience aim to conform only to *Weak emergence*; however, there are cases to be made that either strength of modal correlation is compatible with either schema; blocking all the cases requires endorsing controversial theses (the rejection of quiddities, Hume's dictum) which appear to be incompatible.
- Conceptions in terms of causation and causal dependence aim to conform to *Strong emergence*.
- Conceptions in terms of realization aim to conform to *Weak emergence*.

4. Emergent Autonomy: Metaphysical Conceptions

I turn now to considering metaphysical accounts of emergent autonomy in light of the two schemas for emergence.

4.1. *Ontological and Causal Autonomy*

Causal autonomy (distinctive efficacy) guarantees ontological autonomy (distinctness), by Leibniz's law. But for reasons previously noted, ontological autonomy does not guarantee causal autonomy. Causal autonomy is necessary, however, for vindicating the ontological *and* causal autonomy of special science entities, and relatedly, for solving the problem of higher-level causation in a way preserving both the dependence and the distinctive efficacy of higher-level entities. Hence an account of metaphysical emergence aiming to accomplish these goals must do so in virtue of causal differences between higher-level and base features, rather than in virtue of any bare ontological differences there may be between these features.

This observation is crucial in appropriately interpreting accounts of emergent autonomy. Consider, for example, the conception of emergent entities as being new or genuinely novel with respect to their base entities:

> [Emergence involves] a new kind of relatedness. (Morgan 1923, p. 19)

> [Emergence involves] a new quality [...] distinctive of the higher-complex. (Alexander 1920, p. 45)

> [A]t each new level of complexity entirely new properties appear. (Anderson 1972, p. 393)

> What seems to be central to our conception of emergent phenomena is the idea that something genuinely novel is present in the emergent entity that is not present in entities that are prior to it. (Humphreys 1996, p. 53)

All such conceptions need to make explicit that the novelty/difference at issue has causal as well as ontological implications. Note that mere adherence to Alexander's Dictum (a.k.a. the Eleatic Principle) – that real (broadly scientific) properties have powers – will not in itself establish that a novel/different feature has the desired causal autonomy. A stronger conception of emergent autonomy is needed, establishing that novel/different emergent features have either powers or power profiles different from those had by their base features.

Here I will consider five common ways in which emergentists fill in the notions of novelty/difference so as to gain causal as well as ontological autonomy, by appeal to: (1) fundamental powers, forces, laws; (2) non-additivity of effects; (3) downward efficacy; (4) imposition of lower-level constraints; and (5) multiple realizability and its variants. As we will see, individual variants on these strategies aim to characterize emergent autonomy as involving either *fundamental* or *non-fundamental* novelty/difference, along lines encoded in *Strong* and *Weak* emergence, respectively.

4.2. *Fundamental Powers, Forces, Laws*

The notions of ontological novelty or difference are sometimes supplemented by appeal to fundamentality (ontological basicness):

> A fundamental property is an ontologically basic property of a basic entity [...] An ontologically-emergent property is an ontologically basic property of a complex entity. (Cunningham 2001, p. S67)

> [Emergence involves] a fundamentally new kind of feature. (O'Connor and Wong 2005, p. 665)

> My central thesis is this: that there is ontological emergence is the claim that some things which are fundamental are not ontologically independent. (Barnes 2012, p. 882)

An appeal to fundamentality is in the right direction, but still does not make the requisite causal implications explicit, since a feature might be fundamentally new in having a fundamentally new non-causal quiddity.

It is appropriate, then, that accounts of emergent autonomy as involving fundamental novelty/difference typically take this more specifically to involve fundamentally new powers, forces, or laws.[23] Such conceptions are characteristic of British Emergentism as "the doctrine that there are fundamental powers to influence motion associated with types of structures of particles that compose certain chemical, biological, and psychological kinds" (McLaughlin 1992, p. 52). As McLaughlin goes on to note, these powers were typically taken to be powers to "generate fundamental forces not generated by any pairs of elementary particles" (p. 71). Relatedly, British Emergentists commonly took emergent features to be governed by fundamental laws (tracking or otherwise associated with the having of new powers to produce fundamental forces, etc.). Hence Broad (1925) says:

> [T]he law connecting the properties of silver-chloride with those of silver and of chlorine and with the structure of the compound is, so far as we know, an unique and ultimate law. (pp. 64–65)

Appeal to fundamentally new powers, forces, or laws is similarly a theme in contemporary accounts of emergent autonomy. So, for example, Silberstein and McGeever (1999) understand emergent features as having irreducible causal capacities (that is, fundamentally new powers):

> Ontologically emergent features are features of systems or wholes that possess causal capacities not reducible to any of the intrinsic causal capacities of the parts nor to any of the (reducible) relations between the parts. (p. 186)

O'Connor and Wong (2005) similarly make explicit that emergent features are fundamentally new specifically in having new causal capacities:

[23] Barnes's (2012) conception just in terms of fundamentality and dependence is an exception, and is consequently overly general.

> [A]s a fundamentally new kind of feature, [an emergent feature] will confer causal capacities on the object that go beyond the summation of capacities directly conferred by the object's microstructure. (p. 665)

And reflecting that powers are plausibly grounded in fundamental forces/interactions, Wilson (2002) offers a fundamental interaction-relative account of emergence, according to which (in present terms) a dependent higher-level feature *S* is strongly emergent from its base feature *P*, relative to a set of fundamental interactions *F*, just in case *S* has (on an occasion, etc.) a token power different from any token powers of *P* grounded only in forces/interactions in *F*.

Accounts on which emergent autonomy involves fundamentally new powers, forces or laws all conform to *Strong*, and not *Weak* emergence. Accounts on which emergent features have fundamentally new powers explicitly do so, and the other accounts implicitly do so, since these accounts entail that emergent features will have new powers to generate fundamental forces/interactions, and in virtue of which they will enter into fundamental laws.

4.2.1. *The Flip Side: Failure of Realizability*

Under the rubric of emergent autonomy as involving fundamental powers, forces, or laws we may also place negative conceptions of emergent autonomy as involving a failure of realizability. So, for example, Kim (2006) identifies irreducibility of emergents as a necessary condition of emergence, where this is understood in terms of failure of functional realizability:

> Property *M* is emergent from a set of properties $N_1, \ldots, N_n$ only if *M* is not functionally reducible with the set of the *N*s as its realizer. (p. 555)[24]

As above (§3.4), standard accounts of (non-reductive) realization all guarantee satisfaction of the condition, in *Weak emergence*, that the token powers of emergent and base features stand in the proper subset relation. Putting aside epiphenomenalism, then, an account of emergent autonomy as involving failure of (any such account of) realization will entail that an emergent entity has a new power, as per *Strong emergence*.

[24] Note that Kim here, somewhat uncharacteristically, takes the 'one-many' perspective on emergence.

4.3. *Non-Additivity*

Mill characterized emergent autonomy in terms of a failure of causal additivity. As we'll shortly see (§4.3.1), in the British Emergentist tradition such appeals are aimed at providing a (negative) metaphysical criterion for fundamental powers (and associated forces or laws); such conceptions of emergent autonomy thus conform to *Strong emergence*. As we'll also see, however, certain contemporary understandings of non-additivity, as grounded in non-linearity associated with, e.g., chaotic dynamical systems (§4.3.2), or in powers that latently exist at the microphysical level (§4.3.3), have been associated with *Weak emergence*. I'll address each of these approaches, in turn.[25]

4.3.1. *Non-Additivity as a Criterion For Fundamentality*

As previously discussed, Mill ("On the Composition of Causes", 1843/1973) distinguishes two types of effects of joint or composite causes. 'Homopathic' effects conform to the principle of 'composition of causes' in being (in some sense) mere sums of the effects of the component causes when acting in relative isolation, as when the weight of two massy objects on a scale is the scalar sum of their individual weights, or when the joint operation of two forces conforms to vector addition in bringing an object to the same place it would have ended up, had the forces operated sequentially. 'Heteropathic' effects violate the principle in not being mere sums in the previous sense, and are therefore indicative of the operation of new laws. Mill says:

> This difference between the case in which the joint effect of causes is the sum of their separate effects, and the case in which it is heterogeneous to them; between laws which work together without alteration, and laws which, when called upon to work together, cease and give place to others; is one of the fundamental distinctions in nature. (pp. 408–409)

And he offers chemical compounds and living bodies as entities that are capable of producing heteropathic effects.

Mill did not use the term 'emergence' (evidently Lewes 1875 first did so), but his notion of heteropathic effects serves as a basis for characterizing *Strong emergence*. To start: given the reciprocal connection between powers and effects, to say that an effect of a feature of a composite entity is non-additive, relative to effects of features of the parts acting separately, is just to say that the higher-level feature has a power not had by its

[25] See Wilson (2013) for a fuller discussion of the bearing of non-linearity or non-additivity on metaphysical emergence.

lower-level base features when in additive combination (taking the many-one perspective) or, equivalently, that the higher-level feature has a power not had by its relational lower-level base feature (taking the one-one perspective). Mill himself moves seamlessly from talk of heteropathic effects to talk of new properties of and laws governing entities capable of causing such effects:

> [W]here the principle of Composition of Causes [...] fails [...] the concurrence of causes is such as to determine a change in the properties of the body generally, and render it subject to new laws, more or less dissimilar to those to which it conformed in its previous state (p. 435).

Both Mill's reference to "new laws" and his taking such cases to contrast with "the extensive and important class of phenomena commonly called mechanical" indicate that Mill's appeal to non-additivity of effects is aimed at identifying a criterion for a higher-level feature's having a new fundamental power, enabling it (or its possessing "body") to override the usual composition laws in the production of certain effects. As McLaughlin (1992) notes, "Mill holds that collocations of agents can possess fundamental force-giving properties" (p. 65). All this is in conformity with *Strong*, and not *Weak*, emergence.

Most other British Emergentists followed Mill in characterizing emergent autonomy as involving violations of broadly additive composition laws, including Alexander (1920), who characterized emergent properties as having powers to produce heteropathic effects; Morgan (1923), who contrasted resultant with emergent features as being "additive and subtractive only"; and Broad (1925), who offered scalar and vector addition as paradigms of the compositional principles whose violation was characteristic of emergence. An interesting exception to this rule is found in Lewes' (1875) characterization of emergent autonomy as involving any failure of "general mathematizability", with emergence being correspondingly harder to come by. As in Mill's case, and following the standard British Emergentist conception of emergent autonomy as involving fundamental powers, forces, or laws, these appeals to non-additivity are best seen as attempts to provide a substantive metaphysical criterion of fundamental novelty, in conformity with *Strong*, and not *Weak*, emergence.

4.3.2. *Non-Additivity and Non-Linearity*

Though British Emergentists saw non-additivity as characteristic of *Strong emergence*, some contemporary accounts of emergent autonomy (see Newman 1996 and Bedau 1997) take non-additivity of the sort associated with non-linear features of complex systems (e.g., being in the basin of a strange attractor) as motivating a conception of emergence compatible with *Physicalism*. What accounts for this discrepancy in the status as physically acceptable, or not, of non-additive higher-level features?

We should start by noting that certain motivations for taking non-linear phenomena to be physically acceptable do not establish this claim. Newman (1996), for example, cites the supposition that complex systems are strictly deterministic in support; but strict determinism of non-linear systems does not rule out such systems as being *Strongly* emergent, for in the first instance such determination is a matter of nomological necessity, and as previously, all emergentists agree that emergent features (and associated powers to produce systemic behaviors) are (at least) nomologically necessitated by base features. Relatedly, that macro-states of non-linear systems are derivable from non-linear equations and initial (more generally, external) conditions does not establish physical acceptability, since it remains to consider the metaphysical basis for non-linearity (and associated equations). Bedau (1997) claims that features of non-linear systems are physically acceptable because they are 'structural' (effectively: because they are features of relational lower-level entities); but given that non-linear phenomena do not consist solely in additive combinations of micro-level goings-on, the claim that such features are merely structural needs to be established, not assumed. What is needed to warrant taking non-linear phenomena to be physically acceptable is specific attention to the metaphysical basis for the non-linearity, and some argument to the effect that this basis does not involve new fundamental powers (or associated forces/interactions or laws).

Along these lines, it is worth noting that some accounts of the metaphysical basis for non-linearity are compatible with *Strong emergence*, contra *Physicalism*. Consider, for example, cases where the non-linear phenomena involves feedback between the micro-entities constituting the base, associated with strange attractors and other dynamic phenomena. As Silberstein and McGeever (1999) note, one metaphysical account of non-linearity (again compatible with strict determinism) appeals to a kind of system-level holism:

> What is the causal story behind the dynamics of strange attractors, or behind dynamical autonomy? The answer, it seems to us, must be the non-linearity found in chaotic systems. [...] But why is non-linearity so central? [...] Non-linear relations may be an example of what Teller calls 'relational holism' [...]. (p. 197)

As above (§4.2), Silberstein and McGeever take the associated holism as indicative of emergent features' possessing fundamentally new powers ("irreducible causal capacities"). Such an account of the metaphysical basis of non-linear emergence is again in line only with *Strong emergence*.

Proponents of non-linearity as characteristic of *Weak* emergence have a different interpretation in mind, typically illustrated by attention to one or more specific examples (often involving cellular automata). The general moral to be drawn from these examples is that (*pace* traditional appeals to failures of additivity) a metaphysical account of non-linearity need not involve fundamental higher-level powers or laws, but rather only micro-level goings-on (notwithstanding that the aggregative result of such micro-interactions can be very surprising), compatible with *Physicalism*.

Granting this moral, a remaining, underappreciated, and more serious problem for taking non-linearity to be characteristic of *Weak emergence* concerns whether the higher-level features at issue are plausibly understood as being ontologically and causally autonomous from their base entities, in having only a proper subset of the powers of their base entities. Indeed, both Newman and Bedau maintain that non-linear features are in-principle reducible to micro-level phenomena, though Bedau attempts to ground a measure of higher-level autonomy in certain broadly metaphysical constraints on the predictability of non-linear and other phenomena supposed to instance weak emergence. We'll consider Bedau's account of such autonomy down the line (§5.2). Here I want to focus on another aspect of non-linear phenomena, also noted by Bedau:

> [T]here is a clear sense in which the behaviors of weak emergent phenomena are autonomous with respect to the underlying processes. The sciences of complexity are discovering simple, general macro-level patterns and laws involving weak emergent phenomena. [...] In general, we can formulate and investigate the basic principles of weak emergent phenomena only by empirically observing them at the macro-level. In this sense, then, weakly emergent phenomena have an autonomous life at the macro-level. (Bedau 1997, p. 395)

That non-linear phenomena associated with complex dynamical systems give rise to "simple, general macro-level patterns" may indeed provide a basis for the ontological and causal autonomy of the associated higher-level features, compatible with *Physicalism*, quite apart from how such patterns may be discovered.

Here we are motivated to attend to a second way in which higher-level phenomena may be ontologically novel or different – namely, as being *non-fundamentally* novel or different. And given that this form of difference must have causal implications, if it is to be characteristic of emergence, the strategy for establishing that features entering into higher-level patterns have the desired form of autonomy is clear: one must establish, first, that the macro-level patterns are different from (in being, plausibly, more general or less specific than) those at the micro-level, and second, that the correct account of this difference entails that the target (token) higher-level features have, on a given occasion, only proper subsets of the powers of their (token) base features, as per *Weak emergence*. One strategy for establishing that the requisite proper subset relation is in place might appeal to the higher-level features' being functionally or otherwise multiply realizable, and so having causal roles that are indeed more general than those of their realizers, in being associated with fewer of the latter's powers. Another strategy, which I will discuss in §4.5, may be implemented even if a given non-linear feature is only singly realizable.

In any case, proponents of non-additivity as a basis for physically acceptable metaphysical emergence need to establish that the requisite autonomy is in place, and, it seems clear, should dispense with claims of in-principle ontological and causal reducibility. Such claims of reducibility may be motivated by thinking that in-principle ontological reducibility is required for *Physicalism*; but this motivation is suspect, given the seeming viability of the non-reductive physicalist's strategy for resolving the problem of higher-level causation, encoded in the schema for *Weak emergence*.

4.3.3. *Non-Additivity and Micro-Latency*

Yet another understanding of the source of non-additivity is as involving the manifestation of powers that are existent, but latent, at the micro-physical level. For example, Shoemaker (2002) distinguishes between 'micro-manifest' and 'micro-latent' powers of lower-level entities, and suggests that emergent features have ('Type-2') powers that are latent at the micro-physical level:

> When micro-entities are combined in an emergence engendering way, the resulting object will apparently have two sorts of micro-structural properties. One sort, call these provisionally Type-1 micro-structural properties, will consist of properties that can be specified entirely in terms of the micro-manifest powers of the constituent micro-entities together with how these micro-entities are related – i.e., in terms of what could be known about them prior to their entering into emergence engendering combinations. [...] The other sort, which I will provisionally call Type-2 micro-structural properties, will be properties that are specified in terms of *all* of the powers, micro-latent and micro-manifest, of the constituent micro-entities. [...] Type-2 micro-structural properties, although they are micro-structural, will be emergent properties. [...] If emergentism is false, manifest causal powers are the only ones the micro-entities have, and physical micro-structural properties are the only ones macro-objects have – and the other properties of macro-objects are realized in their physical micro-structural properties. (p. 55)

The underlying suggestion here is that, while emergent features may be non-additive (have powers to produce non-additive effects) relative to micro-manifest powers, this need not impugn their physical acceptability; Gillett (2002) offers a similar account as "vindicating non-reductive physicalism as a viable position" (p. 102). Interestingly, Shoemaker traces the suggestion to Broad (1925), who seems to have taken the view that the powers of emergent features are micro-latent as a variant of the view that emergence involves violation of composition laws and associated coming into play of 'trans-physical' laws, as per *Strong*, and not *Weak*, emergence. So here again the question arises whether emergent autonomy as involving non-additivity is or is not compatible with *Physicalism*.

In answering this question, we should first note that the mere existence of micro-latent powers does not suffice to render emergent features physically acceptable, for proponents of *Strong emergence* will generally agree that in some broad sense physical entities have latent powers to bring about emergent features:

> [I]t is true in an emergentist scenario that everything that occurs rests on the complete dispositional profile of the physical properties prior to the onset of emergent features. For the later occurrence of any emergent properties are contained (to some probabilistic measure) within that profile, and so the effects of the emergent features are indirectly a consequence of the physical properties, too. [...] The difference that emergence makes is that what happens transcends the immediate [...] interactions of the microphysics. (O'Connor and Wong 2005, p. 669)

Such a weak dispositional understanding of micro-latent powers is compatible with micro-goings-on' being preconditions for the occurrence of new fundamental powers, forces/interactions, or laws at the higher-level, contra *Physicalism*. Indeed, Broad's assumption that emergence has anti-materialist implications indicates that he has such a weak dispositional sense in mind, in allowing that micro-physical entities have latent powers that become manifest when in emergence-engendering combinations. Physicalist proponents of micro-latent powers as a metaphysical basis for failures in additivity thus need to identify a more substantive understanding of micro-latency, capable of blocking a *Strong* emergent reading of apparent failures of additivity.

The prospects for doing this are unclear, however. To start, it isn't enough to specify, as Shoemaker does, that the effects of micro-level dispositions also be micro-level, since this is compatible with the conditions of manifestation of the micro-level disposition involving physically unacceptable goings-on. Gillett (2002) more explicitly recognizes the concern, and attempts to block it, as follows:

> In our broached scenario [...] the fundamental micro-physical properties have such conditional powers which they contribute conditionally upon instantiating certain realized properties. In such a case, a realized property instance thus determines that one of its realizer properties contributes a certain power that it would not otherwise contribute. It is important to mark the non-causal nature of the determination exerted by the realized property in such a scenario, for this suggests that there will likely be no new ontologically fundamental forces (or other properties). The relevant realized property instance, '*H*', is not causing a microphysical property instance, '*P*', to contribute certain powers. Causal relations typically are mediated by forces and/or the

> transfer of energy – thus if *H* causally determined *P*'s contribution of powers then there might well be a new force. But in the scenario, *H* is exerting a non-causal determinative influence. (p. 113)

One problem here is that, even on the *Strong* emergentist interpretation of non-additivity, the weak dispositional micro-latent powers will not be caused by the higher-level feature: new fundamental interactions do not cause, but rather enter into constituting, the new powers for which they serve as a (perhaps partial) ground. The deeper problem, however, is that Gillett is stipulating that, rather than explaining how or why, the powers occurent in emergence-engendering combinations might not involve (either causally or constitutively) any fundamental higher-level interactions or the like. The question remains: how are we to make sense of the claim that such powers are compatible with *Physicalism*, given that these powers do not make an appearance in the laws of fundamental physics and given that they cannot be understood as additive combinations of powers which do make such an appearance?

What the proponent of micro-latency needs to do in order to establish that non-additivity is compatible with *Physicalism* is to make a case that fundamental physical laws *might themselves entail violations in broadly additive composition laws* when micro-entities enter into emergence-engendering combinations. It is unclear how this can be established however, since composition laws (incorporating, e.g., scalar and vector addition, along with other 'ontologically lightweight' – boolean, mereological – modes of combination plausibly preserving physical acceptability) appear to exhaustively encode the broadly additive ways in which micro-manifest entities might combine while preserving physical acceptability. At the very least, at present it remains unclear how a 'micro-latent' understanding of non-additivity is supposed to conform to the usual understanding of *Physicalism* as the thesis that all broadly scientific goings-on are nothing over and above the goings-on explicitly (and not just latently) at issue in fundamental physics.

Relatedly, there is reason to avoid characterizing physically acceptable emergence in terms of micro-latent features. Traditionally (and as per O'Connor and Wong, above), the dispute between physicalists (of all stripes) and strong emergentists has turned on whether or not all broadly scientific goings-on are nothing over and above goings-on that are *manifest* at the micro-level, when micro-entities are not in emergence-engendering combinations. From this perspective, a characterization of *Weak emergence* as involving micro-latent powers is not in the spirit of *Physicalism*. As Clarke (1999) notes, if higher-level features have token powers not identical with those of their base features...

> ... emergent causal powers would be due to (bestowed by) some macro-level, structural properties possessed by the complex object [...] It matters little whether the macro-level properties that are acknowledged to carry emergent powers are said to be physical properties or whether the emergent laws are said to be physical laws; if there are emergent powers, then the kind of micro-explanation that is the ambition of most physicalists, an explanation of the behavior of all objects in terms of micro-level properties and relations and micro-level laws, will be impossible. (p. 309)

As such, it is no surprise that Broad did not feel the need to rule out the micro-latent interpretation in taking apparent violations of composition laws to have anti-materialist implications.

4.4. *Downward Efficacy*

Many accounts of emergent causal autonomy require that such autonomy be specifically with respect to lower-level goings-on. Hence Morgan (1923) says:

> But when some new kind of relatedness is supervenient (say, at the level of life), the way in which the physical events which are involved run their course is different in virtue of its presence – different from what it would have been if life had been absent. (p. 15)

In a series of papers, Sperry (1969, 1986, 1976) suggests that conscious mental phenomena are emergent in causally affecting underlying neurophysical states, as does Searle (1992). More generally, as Kim (2006) observes, "downward causation is of paramount importance to the emergentists. For they want to claim that the emergence of consciousness and rational thought has made a fundamental difference to the world at the physical level" (p. 558).[26]

Unclarity over whether downward causation is compatible with *Physicalism* is a main source of unclarity over whether emergence is compatible with *Physicalism*. There is, perhaps, a *prima facie* appearance of incompatibility:

> Of all the marks of emergence [downward causation] is the one which presents the clearest and most direct challenge to micro-determinism. (Klee 1984, p. 58)

On the other hand, commentators disagree – sometimes as regards a single account – over whether downward causation is so incompatible.

Sperry's account is a nice case in point. On the one hand, Sperry (1976) speaks of downward influence as involving higher-level powers:

[26] That said, Kim thinks that the supposition of downward causation is problematic, for reasons we will consider in §4.4.1.

> The conscious subjective properties in our present view are interpreted to have causal potency in regulating the course of brain events; that is, the mental forces or properties exert a regulative control influence in brain physiology. (p. 44)

McLaughlin (1992) interprets such talk as committing Sperry to an account involving fundamental configurational forces, hence as incompatible with *Physicalism*. On the other hand, Sperry (1969) describes downward influence as analogous to that involved when the atoms in a wheel must go where the wheel goes:

> The subjective mental phenomena are conceived to influence and to govern the flow of nerve traffic by virtue of their encompassing emergent properties. Individual nerve impulses and other excitatory components of a cerebral activity pattern are simply carried along or shunted this way and that by the prevailing overall dynamics of the whole active process (in principle – just as drops of water are carried along by a local eddy in a stream or the way the molecules and atoms of a wheel are carried along when it rolls downhill [...]. (p. 532)

Schroder (1998) (following Klee's 1984 suggestion) interprets this analogy as suggesting that downward causation involves not new powers, but lower-level constraints: "we can see what is wrong with a critique of emergentism that castigates it for assuming 'configurational forces' [...]. Emergentists who adopt downwards causation as a criterion for emergent properties need assume no such force. [...] In order to produce live and mindful beings, what is needed is not special laws but special structures that constrain the sequence of possible events in special ways" (p. 449). Searle's (1992) account of "radical" emergence, which is supposed both to involve new powers and to be no more physically problematic than, say, liquidity, has produced a similar degree of interpretive confusion.

A plausible diagnosis of this confusion reflects that there are two ways for a higher-level feature to be downwardly efficacious: one conforming to *Weak emergence* and one conforming to *Strong emergence*. Confusion concerning Sperry's and Searle's accounts is then plausibly located in these authors' failing to be sufficiently clear about which form of downward causation (hence of emergence) they have in mind, as with Sperry's remarks, above.

That downward causation may be interpreted in line with *Strong* emergence is clear: one simply additionally requires that the new power associated with a *Strongly* emergent feature be associated with the production of lower-level effects. Similarly for a version of emergence discussed by Chalmers (2006) involving "a sort of incompleteness of physical laws even in characterizing the systematic evolution of low-level processes" (p. 248) and which he thinks is best understood "as involving a sort of downward causation" (p. 249). Here the appeal to downward causation may be seen as providing an account of the specific way in which *Strongly* emergent features are fundamentally novel or distinct.

Alternatively, downward causation may be interpreted along *Weak* emergent lines, as involving the holding of certain physically acceptable constraints on lower-level entities; here the appeal to downward causation may be seen as providing an account of the specific way in which *Weakly* emergent features are non-fundamentally novel or distinct. That said, as with appeals to non-linearity it is not obvious that such downwardly efficacious features have the requisite ontological or causal autonomy, even granting that they inherit the physical acceptability of their base features. To the *prima facie* contrary: might not a given (token) feature of the wheel in virtue of which it rolls, on a given occasion, be identified with a (token) feature of the relational lower-level entity (consisting of atoms standing in atomic relation) constituting the wheel on that occasion? We will revisit this issue when considering conceptions of emergent autonomy that more directly appeal to the imposition of constraints (§4.5); there I will sketch a strategy that may work to gain autonomy in at least some cases of constraint-based downward causation. The broader moral at present is that additional work needs to be done to establish that features associated with lower-level constraints are non-fundamentally autonomous in the way required for physically acceptable emergence.

4.4.1 *Kim's Concerns About Downward Causation*

The latter issue is key to Kim's concerns about downward causation and his associated exclusion argument (previously discussed in §2.2), which he sees as establishing that non-reductive physicalism must collapse either into reductive physicalism or expand into strong emergentism (see, e.g., his 1989, 1993, and 1998). As Kim correctly notes, blocking the strong emergentist's understanding of higher-level features requires accepting what he calls the "Causal Inheritance Principle", according to which every token power of a realized property instanced on a given occasion is numerically identical with a token power of the property instance realizing it on that occasion; and the challenge he has offered to the non-reductive physicalist is to show how, if token higher-level features have no powers not already had by token realizers, the former may be ontologically and causally (in particular, downwardly) efficacious. Certainly it is hard to see how such autonomy might be gained if higher-level instances inherit *all* the powers of their realizing instances. However, the explicit identification of *Weak emergence* as encoding the non-reductive physicalist's distinctive approach to higher-level causal autonomy at least makes clear what the non-reductive physicalist needs to do in order to address Kim's concerns about downward causation. First, the non-reductive physicalist must establish that it suffices for causal autonomy that a higher-level feature has a distinctive power profile, as per the proper subset condition in *Weak emergence*; second, they must establish that at least some higher-level features in fact have distinctive power profiles. As previously argued, non-reductive physicalists do have resources along these lines, but whether these strategies succeed is the subject of ongoing debate.

4.5. *The Imposition of Constraints*

Closely related to physicalist conceptions of emergent autonomy in terms of downward causation are conceptions on which such autonomy is taken to reflect the imposition of lower-level constraints (see Klee 1984, Schroder 1998, and Wilson 2010a).

To repeat, granting that features associated with the imposition of lower-level constraints conform to *Physicalism*, it is not obvious that such features have the requisite ontological or causal autonomy; hence additional argument is needed to show that this conception conforms to *Weak emergence*. I provide a detailed such argument, for a special class of features associated with lower-level constraints, in Wilson (2010a). Here I sketch, very briefly, the strategy of that argument.

To start, I consider the notion of a degree of freedom (DOF) – roughly, one of a minimal set of independent parameters needed to characterize the states upon which the law-governed features of a (token of a given type of) entity (including systems) functionally depends.[27] Attention to DOF is useful in the present context, because the imposition of constraints at a lower-level generally affects, one way or another, the DOF needed to characterize the higher-level entities whose existence is to some extent determined by the holding of the constraints. Some sorts of changes in DOF resulting from the imposition of lower-level constraints may not be indicative even of *Weak emergence* – for example, cases where the DOF needed to characterize a higher-level entity (e.g., a rigid body, or a molecule) are the same as those needed to characterize lower-level relational entities realizing such higher-level entities, but where the latter DOF can take on only constant or a restricted range of

[27] So, for example, specifying the configuration state for a free point particle requires 3 independent parameters (e.g., x, y, and z; or r, ρ, and θ); hence a free point particle has 3 configuration DOF, and a system of N free point particles has $3N$ configuration DOF. And specifying the kinematic state for a free point particle requires 6 independent parameters: one for each configuration coordinate, and one for the velocity along that coordinate; hence a free point particle has 6 kinematic DOF, and a system of N free point particles has $6N$ kinematic DOF.

values. However, sometimes the imposition of lower-level constraints does not just reduce or restrict (values of) lower-level DOF, but moreover *eliminates* certain lower-level DOF from those needed to characterize the associated higher-level entity. This is the case, for example, with certain features of quantum, statistical-mechanical, and complex dynamical entities or systems. (Note that the present strategy, supposing it works, would vindicate accounts of *Weak* emergence appealing to non-linearity.)

In such cases of elimination of DOF, I argue, there are reasons to think that the associated higher-level feature satisfies *Weak emergence*, in having only a proper subset of the token powers of the relational lower-level feature upon which it depends. As above, the usual strategy for showing this appeals to *S*'s (functional or other) multiple realizability. Attention to DOF suggests a means of establishing satisfaction of *Weak emergence* even if *S* is only singly realized. Suppose *S* is singly realized by a base feature *P*. Now, again, what powers an entity has are plausibly a matter of what it can do; and the sciences are plausibly in the business of expressing what the entities they treat can do. It follows that, plausibly, what powers an entity has are expressed by the laws in the science treating it. The powers of *S* are thus those expressed by the laws in the theory treating (constrained) entity *S*, while the powers of *P* are those expressed by the laws in the more fundamental theory treating the (relatively unconstrained) lower-level constituents of *P* – that is, the constituents of *P* as existing both inside and outside the constraints associated with *S*. Consequently, the laws of the theory treating *S* express what happens when certain lower-level entities stand in relations associated with certain lower-level constraints, and the laws treating *P* express what happens when certain lower-level entities stand both in these relations and in other relations not associated with the constraints. Hence the relational base feature *P* has more powers than *S*, and the proper subset relation between powers in *Weak emergence* is thus in place.[28]

Of course, this is only a sketch of how higher-level autonomy may be gained via the imposition of lower-level constraints, when these are

[28] For example, suppose that *P* is a quantum relational entity, and *S* is a classical entity singly realized by *P*. Then the causal powers of *S* include all those powers to produce, either directly or indirectly, effects that can occur in the macroscopic limit. The realizing entity *P* has all these causal powers, and in addition has all those powers to produce, either directly or indirectly, effects that can occur in circumstances that are not so constrained, and in which quantum physics is operative – for example, effects occurring in circumstances where no macro-entities can exist.

associated with eliminations in DOF (see my paper for details). The larger point for present purposes is that this or some other work needs to be done if such constraints are to serve as the basis for *Weak* emergence.

4.6. *Multiple Realizability and Its Variants*

As previously discussed (§3.4), non-reductive physicalists commonly appeal to multiple realizability in service of establishing the ontological and causal autonomy of higher-level entities. Related conceptions are in terms of 'dynamical autonomy', where micro-level changes do not make a causal difference at the level of a system's dynamics (Wimsatt 1996), and 'compositional variance', where the base entities of a given higher-level system exhibit "a much greater degree of variance and fluctuation from moment to moment than does the level of organization where [the higher-level entity] occurs" (Klee 1984, p. 48).

Why should multiple realizability, dynamical autonomy, or compositional variance support ontological and causal autonomy? Making the case for autonomy is crucial, since a now-standard reductionist strategy for accommodating multiple realizability and its variants proceeds by identifying multiply realized types with the disjunctions of their realizing types (see Kim 1992; though see Clapp 2001 and Antony 2003 for arguments that disjunctive features satisfy the proper subset condition). Plausibly, instances of a disjunctive type, on a given occasion, are identical with instances of whatever disjunct is instanced on that occasion; hence disjunctive identification blocks conformity to either *Strong* or *Weak* emergence.

That said, as above this reductive strategy for accommodating multiple realizability and its variants may be resisted, in service of establishing that some higher-level realized features are non-fundamentally novel or distinct, in a way having appropriate implications for their causal autonomy. In particular, non-reductionists may understand multiple realizability, and its dynamical and compositional variants, as tracking the higher-level feature's association with a distinctive causal role – that is, with a distinctive set of powers. Each lower-level realizer will have these powers (else it would not be a realizer), and some others besides, reflecting lower-level causal potentialities which differ between it and other lower-level realizers. Hence one may reasonably maintain that any instance of a multiply realizable feature has only a proper subset of the token powers of the base feature realizing it on that occasion, as *Weak emergence* requires. Correspondingly, conceptions of emergent autonomy appealing to multiple realizability, dynamical autonomy and compositional variability are best understood as providing a plausible basis for establishing that the proper subset condition in *Weak emergence* is met.

4.7. *Results*

We have arrived at the following results concerning metaphysical accounts of emergent autonomy:

- Conceptions of autonomy in terms of mere ontological novelty/difference or fundamental novelty/difference guarantee ontological autonomy (distinctness) but not causal autonomy.
- Conceptions in terms of fundamentality of powers, forces, laws (and relatedly, conceptions in terms of failure of realization) aim to conform to *Strong* emergence.
- Conceptions in terms of non-additivity of effects aim to conform to either *Strong* or *Weak* emergence, depending on whether the source of the non-additivity (non-linearity) involves new powers. A pressing need here is for those taking non-linearity as a basis for physically acceptable emergence to establish that higher-level non-linear features are ontologically and causally autonomous from their base features, in satisfying the proper subset condition on powers in *Weak emergence*.
- Conceptions in terms of downward efficacy aim to conform to either *Strong* or *Weak* emergence, depending on whether the source of the downward efficacy involves new powers, or rather merely involves the imposition of lower-level constraints. Here too, it remains for those characterizing physically acceptable emergence in terms of downward efficacy to establish that the requisite ontological and causal autonomy is in place.
- Conceptions in terms of the imposition of lower-level constraints aim to conform to *Weak* emergence. Here too, it remains for those characterizing physically acceptable emergence in terms of lower-level constraints to establish that the requisite ontological and causal autonomy is in place (though see Wilson 2010a).
- Conceptions in terms of multiple realizability, dynamical autonomy and compositional variance aim to conform to *Weak* emergence.

5. Emergent Autonomy: Cognitive Conceptions

Many historical and contemporary accounts of emergent autonomy involve appeals to the failure to hold of certain epistemological, representational, or conceptual connections, including in-principle failure of predictability or deducibility (Broad 1925), predictability, but only by simulation (Bedau 1997), and failure of representational or conceptual entailment (Smart 1981, Chalmers 1996, Van Gulick 2001). Such accounts are broadly cognitive in that they appeal to one or other failure on the part of creatures like us (or suitably idealized versions of us) to recognize certain connections as holding between certain higher-level and base features. For convenience, then, I will speak broadly of such conceptions as 'cognitive' conceptions.

With few exceptions, cognitive conceptions of emergent autonomy aim to characterize metaphysical emergence. Typically, the relevant failures of cognitive connections are supposed to be concomitants of novelty or ontological irreducibility (or both). This is characteristic of, for example, Alexander's (1920) understanding of emergent phenomena as admitting no explanation because involving "brute empirical fact"; Kekes' (1966) understanding of emergence as involving a priori unpredictability of (claims about) higher-level features from (claims about) lower-level structure, due to novelty of higher-level property; and Kim's (1999) characterization of emergence as involving the joint failure of explanatory, predictive, and ontological reduction. Such conceptions may fall under the rubrics of *Weak* or *Strong* emergence, respectively, depending on which ontological aspect is at issue (as per §4).

Here I want to focus attention on accounts of emergence that are primarily or in any case officially cashed in cognitive terms. Along the way, we will confirm both that those endorsing cognitive conceptions typically aim to characterize metaphysical autonomy, and that they take themselves to have reason to think this can be done in epistemological or other cognitive terms. This is not true across the board, however; and I'll close (§5.4) with discussion of certain accounts of "non-reductive" physicalism which are explicitly cashed in terms of failure of conceptual connection, and which are better seen as ontologically reductive physicalist accounts aiming to make sense of our seeming inabilities to bridge certain explanatory gaps.

5.1. *Failure of In-Principle Deducibility*

Broad's official formulation of emergence (1925) is as follows:

> The emergent theory asserts that there are certain wholes, composed (say) of constituents *A*, *B*, and *C* in a relation *R* to each other [...] and that the characteristic properties of the whole *R(A, B, C)* cannot, even in theory, be deduced from the most complete knowledge of the properties of *A*, *B*, and *C* in isolation or in other wholes which are not of the form *R(A, B, C)*.[29] (p. 64)

[29] Note that Broad's formulation appeals both to a 'one-one' and a 'many-one' perspective on the relata of emergence, with an uncharacteristically flexible understanding of what features may enter into the deduction, as going beyond the holding of pairwise (or other relatively

Though this formulation is in epistemological terms, the discussion preceding the formulation makes clear that Broad's appeal to failure of deducibility aims to characterize a metaphysical notion of emergent autonomy.

Broad begins his discussion of emergence by observing a distinction between two kinds of inter-level ('trans-ordinal') laws, which distinction is also presented in seemingly epistemological terms. First are trans-ordinal laws holding "between physical properties and properties at higher levels of the hierarchy which, while deducible in principle from a theory of the physical properties alone, are not deducible in fact". Second are trans-ordinal laws that are moreover 'trans-physical', holding "between physical properties and properties at higher levels which are not deducible, even in principle, from a theory of the physical properties alone". Broad's official formulation of emergence thus obliquely characterizes the holding of trans-physical laws, by reference to the associated in-principle failure of deducibility that he assumes attaches to such laws.

In turn, for Broad, the existence of trans-physical laws has clear metaphysical consequences. That Broad supposes that trans-physical laws are at odds with a "mechanistic" (materialistic, physicalistic) view is some indication of this. Yet more telling are Broad's previously cited remarks to the effect that such laws are "unique and ultimate" (pp. 64–65) – that is, fundamental. That Broad understands trans-physical laws as indicative of metaphysical emergence is confirmed in passages such as the following:

> On the emergent theory we have to reconcile ourselves to much less unity in the external world and a much less intimate connexion between the various sciences. At best the external world and the various sciences that deal with it will form a kind of hierarchy. (p. 78)

Emergence has implications for the unity of "the external world" and for the unity of the sciences "that deal with" the external world. These are clearly claims about metaphysical emergence; no failures of cognitive connection are ultimately at issue.

Similar remarks apply to other British Emergentists (e.g., Alexander), who, like Broad, sometimes characterize emergence as involving a failure of predictability. More generally, as McLaughlin (1992, p. 73) notes, "Emergentists often speak of emergent properties and laws as unpredictable from what they emerge from. But [...] the Emergentists do not maintain that something is an emergent because it is unpredictable. Rather, they maintain that something can be unpredictable because it is an emergent" (p. 73).

---

non-complex) relations between the composing entities, to include relations between lower-level relata in any other (possibly complex) situations besides that at issue.

5.1.1. *Why (Failure of) Deducibility?*

Since Broad's concern is clearly metaphysical emergence and more specifically *Strong* emergence (as involving fundamental laws and associated powers and forces), why does he characterize emergence in epistemological terms?

I speculate that this reflects a felt need to clarify the notion of fundamentality at issue, since certain ways of understanding this notion will not make sense of the characteristic dependence of emergent phenomena. In particular, we cannot here understand 'fundamental' as 'basic', 'independent', or 'axiomatic'. Relatedly, Broad may have wanted to provide a substantive criterion of fundamentality, for purposes of applying his account. Insofar as it will plausibly be the case that goings-on governed (in part) by fundamental trans-physical laws will not be deducible from goings-on governed by physical laws, it would be natural to look to deducibility as a means of clarifying the distinctively dependent sort of fundamentality in *Strong* emergence. And Broad might reasonably have thought that the immediate concern with characterizing metaphysical emergence in epistemological terms – namely, that creatures as limited as we are might not be cognitively situated to recognize metaphysical connections that in fact exist – could be overcome by additionally qualifying the failure of deducibility as being 'in-principle'.

That said, the concern remains that even an ideal reasoner might fail to recognize metaphysical connections that in fact exist, in which case the criterion will produce false negatives. The procedure might also produce false positives, if certain uncontroversially physically acceptable phenomena (say, complex dynamical phenomena, of which Broad wasn't aware) are in-principle as well as in-practice unpredictable (perhaps because the sensitivity of such systems to initial conditions would require in-principle unavailable resources for predictability into the indefinite future). Supposing so, then Broad's criterion will inappropriately deem some physically acceptable features of complex phenomena *Strongly* emergent, hence physically unacceptable. In-principle failure of deducibility is thus best seen as a good though not infallible epistemological guide to the metaphysical features (involving fundamental powers and laws) characterizing *Strong* emergence.[30]

---

[30] That said, we will shortly consider whether in-principle failure of the broader notion of *a priori* entailment might do better along these lines.

5.2. *Failure of In-Practice Deducibility*

Failure of deducibility or predictability also enters into some accounts of emergent autonomy aiming to characterize physically acceptable emergence (see Newman 1996, Bedau 1997, Rueger 2001; here I also subsume Batterman 2002, though see Wilson 2013 for fuller discussion of Batterman's view); I'll focus on Bedau's work as representative in what follows. Bedau's (1997) account applies under conditions where a system *S* is composed of micro-level entities having associated micro-states, and where a microdynamic *D* governs the time evolution of *S*'s microstates:

> Macrostate *P* of *S* with microdynamic *D* is *weakly emergent* iff *P* can be derived from *D* and *S*'s external conditions but only by simulation. (p. 378)

Derivation of a system's macrostate "by simulation" involves iterating the system's microdynamic, taking initial and any relevant external conditions as input. A broadly equivalent conception takes emergent autonomy to involve "explanatory incompressibility", where there is no "short-cut" explanation of macro-features of a system with emergent features (see Bedau 2008). In being derivable by simulation from a micro-physical dynamic, associated macrostates are understood to be physically acceptable; as Bedau (1997) says, such systems indicate "that emergence is consistent with reasonable forms of materialism" (p. 376).[31]

Though Bedau sometimes speaks of such systems as "epistemologically weakly emergent", he is explicit that the emergence involved is also metaphysical. He signals that "the modal terms in this definition are metaphysical, not epistemological" (1997, p. 379); he states his aim of capturing a form of "metaphysical autonomy" (2002, p. 11); he emphasizes that "weak emergence is not just in the mind; it is real and objective in nature" (2008, p. 444). Such claims would seem to be in tension with Bedau's taking it to be characteristic of physically acceptable emergence that "the macro is ontologically and causally reducible to the micro in principle" (2008, p. 445); but Bedau thinks this implication can be resisted:

[31] As a referee pointed out, the failure of deducibility at issue in Bedau's account(s) differs from that at issue in Broad's account in being diachronic (involving the evolution of the system over time), and in that there might be such failures even in the absence of higher-level patterns. Still, to the extent that such failures can be associated with macro-patterns (as per Bedau's motivating examples from the Game of Life, and as is reflected in his saying, as above, that such patterns emerge from "underlying processes" (Bedau 1997, p. 395), and to the extent that diachronic emergence can be recast in synchronic terms as involving (in a suitably broad sense) powers of configurations to give rise to such patterns, it is worth considering whether and how Bedau's account can serve as a basis for making sense of synchronic metaphysical emergence of higher-level entities and features.

> [W]eak emergence exhibits a kind of macro autonomy because of the incompressibility of the micro-causal generative explanation of the macro structure. Because the explanation is incompressible, it is useless in practice (except in so far as it serves as the basis for a good simulation of the system). (2008, p. 449)

But it is unclear how usefulness in practice of explanations appealing to complex micro-phenomena might be relevant to establishing the ontological and/or causal autonomy of higher-level features, even granting that there is a metaphysical fact of the matter about when a feature has or does not have a compressible explanation. Effectively, such facts, though perfectly objective, are not of the right sort to ground the requisite ontological and causal autonomy. There is a parallel here to the failure of mere ontological distinctness to successfully capture emergent autonomy: what is needed for such autonomy is not just some or other metaphysical distinction between the higher-level and base features, but moreover one which plausibly serves as a basis for the causal as well as ontological autonomy of the former.

There are, however, resources at least potentially available for making sense of genuine autonomy in the cases Bedau aims to characterize, to which Bedau himself sometimes gestures. We saw previously how Bedau's (1997) observation that non-linear phenomena may enter into "simple, general, macro-level patterns" might serve as a basis for establishing genuine emergent autonomy of a physically acceptable variety: if, more generally, explanatorily incompressible phenomena enter into different, higher-level systems of laws, this might serve to support an understanding of the associated features as having only a proper subset of the token powers of their lower-level base features, and hence as genuinely (Weakly) metaphysically emergent. Relatedly, Bedau (1997) observes: "Interesting macrostates [of the sort at issue in weak emergence] typically average over microstates and so compress microstate information" (p. 377). If such compression of information involves an elimination in degrees of freedom (see §4.5), this would provide another route to ontological and causal autonomy. Alternatively, one might argue that compression of information is indicative of multiple realizability and/or difference-making considerations, of the sort that, as we have seen, plausibly motivate taking the requisite proper subset relation to be in place (here Bedau's 2002, p. 25 remarks concerning glider guns and their variable constituents are evocative). Hence it may be that, while Bedau's broadly epistemological account of emergent autonomy does not itself serve to characterize metaphysical emergence, an account based on the relevant metaphysical features of "interesting" cases of explanatory incompressibility may do so.

5.2. *Failure of Conceptual Entailment*

Next, consider Chalmer's notion of emergence in terms of a failure of *a priori* or conceptual entailment. Chalmers (1996) characterized (physically unacceptable) emergence in terms of a failure of broadly logical (conceptual) supervenience; in recent work (see Chalmers 1999 and Chalmers and Jackson 2001) he has developed the suggestion that one aspect of meaning is appropriately seen as tracking *a priori* connections. The notion of *a priori* entailment here goes beyond deducibility or any other syntactic notion, rather being linked to ideal conceivability and associated judgments about what is true in situations that are fully described along a certain (i.e., fundamental physical) dimension. So, for example, Chalmers argues that, upon contemplation of a scenario in which exists a creature functionally and physically identical to an actually conscious creature, an ideal reasoner would positively conceive that such a creature might not be conscious; Chalmers moreover argues that such ideal conceivability suffices for establishing the metaphysical possibility in question. The precise nature of the possibility that is established by so-called 'zombie' arguments is subject to different broadly dualist interpretations (including substance dualism, strong emergence, and pan- or proto-psychism); but perhaps in combination with possibilities established by other ideal conceivings (namely, that there could be no conscious entities that were not dependently embodied, some way or other, at least in worlds relevantly like ours) one might so aim to establish the truth of *Strong emergence*.

It remains controversial whether conceivability, even of the highly idealized and nuanced variety, suffices for establishing the truth of various possibilities (see, e.g., Block and Stalkaker 1999); and the additional concern remains that such an idealized account is unuseful for or irrelevant to our gaining insight into the structure of natural reality (see, e.g. Melnyk 2008). Here I want just to call attention to two points. First, Chalmers, like the other proponents of cognitive conceptions of emergent autonomy we have discussed, endorses this conception in service of establishing the holding (or failure to hold) of a metaphysical dependence relation. Second, supposing the strategy works and the appropriate conceivings are in place, and putting aside the concern that even an idealized conceiver might fail to discern certain Weakly emergent connections, the conception conforms to *Strong*, and not *Weak*, emergence. *Strong* emergence involves fundamental powers (forces/interactions, laws), and such fundamentality makes room for and sense of the failures of conceptual entailment present even to idealized conceivers.

5.4. *Mere Failures in Cognitive Connection*

Though most epistemological accounts of emergent autononomy are aimed at characterizing metaphysical emergence (either *Weak* or *Strong*), this is not uniformly the case. In particular, a not-uncommon way of formulating (a version of what is sometimes called) non-reductive physicalism is as combining both in-principle ontological reduction with failure of one or other variety of cognitive connection (see, e.g., Smart 1981 and Van Gulick 2001). On such accounts, the 'non-reduction' at issue is understood in purely epistemological terms, having no metaphysical implications; on the contrary, ontological reduction is assumed. Metaphysically speaking, such accounts are best understood as versions of ontologically reductive, not ontologically non-reductive, *Physicalism*, which aim to makes sense of the presence and seeming intractability of various explanatory gaps, and show that these do not pose a threat to such reductionism (see, e.g., Perry 2000). Such accounts may still be seen as addressing the initial motivations for attending to emergence, of understanding and accounting for the appearances of dependence and ontological and causal autonomy of higher-level entities, and associated hierarchical relations between (entities and features treated by) special and more fundamental sciences. But the account they offer will be importantly deflationary, from a metaphysical point of view, in denying that the appearance of autonomy is genuine (which is not to say that the appearances themselves are not grounded in objective facts). Given the desirability of providing a metaphysical ground for the ontological and causal autonomy of higher-level entities, however, proponents of reductive accounts would do well to consider whether the epistemological failures in question might, as with Bedau's understanding of emergence as involving in-practice failure of deducibility, be at least sometimes understood in terms compatible with *Weak*, if not *Strong*, metaphysical emergence.

5.5. *Results*

We have arrived at the following results concerning epistemological accounts of emergent autonomy:

- Conceptions of emergent autonomy in terms of failure of cognitive connection typically aim to conform to metaphysical emergence.
- Conceptions in terms of in-principle failure of deducibility aim to conform to *Strong emergence*, and may do so, assuming that there are no barriers to the in-principle deducibility of *Weakly* emergent phenomena.

- Conceptions in terms of in-practice failure of deducibility (due to, e.g., explanatory incompressibility) aim to conform to *Weak emergence*, and may do so if the assumption of in-principle ontological and causal reducibility is dropped and the requisite ontological and causal autonomy established.
- Conceptions in terms of failure of ideal conceivability aim to conform to *Strong* emergence, and again may do so, assuming that there are no barriers to idealized conceivability of *Weakly* emergent phenomena.
- Accounts of *Physicalism* characterized in terms of one or other failure of cognitive connection coupled with ontological and causal reducibility fail to characterize any variety of emergence; these are best seen as versions of ontologically reductive *Physicalism*.

6. Concluding Remarks

The problem of higher-level causation acts as a crucial constraint on feasible accounts of synchronically dependent higher-level features; and though Kim presented the problem in service of motivating a reductive physicalist stance, consideration of the spectrum of available responses to the problem provides, more generally, a convenient way of seeing what our options are, so far as making sense of the metaphysical emergence of such higher-level features is concerned.

There are only two responses to the problem which make sense of higher-level features as both appropriately dependent on, and ontologically and causally autonomous from, lower-level features. Correspondingly, there are only two schemas for metaphysical emergence, which like the associated responses to the problem, turn on the two available ways in which dependent higher-level features may be causally autonomous *vis-à-vis* their base features: either by having more powers, as per *Strong emergence*, or by having fewer powers, as per *Weak emergence*. Again, the notion of power here is almost entirely metaphysically neutral, requiring nothing much more than acceptance of the view that what entities can do is a matter of what features they have. There are no other options for gaining the causal autonomy of synchronically dependent higher-level features; hence these two schemas exhaust the available options for the metaphysical emergence of such features and the entities that have them.

Flexibility remains in filling in the schemas, however, via suitable accounts of emergent dependence and emergent autonomy. As I have argued, the many seemingly diverse accounts of these notions, when properly understood, individually aim to conform to one or the other schema. And though my task here was not to assess the success of these aims, I have pointed out where more work needs to be done if certain accounts of emergent dependence or autonomy are to satisfy the conditions of the intended schema. Perhaps most crucially, it largely remains to establish that accounts of *Weakly* emergent autonomy in terms of non-linearity, lower-level constraints, and/or explanatory compressibility characterize higher-level features as having the ontological and causal autonomy requisite for genuine metaphysical emergence. That proponents have not realized that this work needs to be done likely reflects, I submit, that the powers-based conditions on (broadly synchronic, higher-level) metaphysical emergence have not previously been made fully explicit.

Hence it is, I hope, that the two schemas do more than systematize and unify the seeming diversity of accounts while explaining their different stances on *Physicalism*. Additionally, and perhaps more importantly, with the schemas on the table we are in better position to consider and assess the available ways of filling them in, in ultimate service of better understanding the potentially diverse – but after all, not all *that* diverse – structure of natural reality.

ACKNOWLEDGMENTS

Thanks to Benj Hellie, Umut Baysan, Kevin Morris, an anonymous referee for this collection, students in my seminars on emergence at the University of Toronto, and members of audiences at the many talks on different aspects of emergence that I have given over the past decade, for helping to shape and improve my views on this topic.

University of Toronto
Department of Philosophy
e-mail: jessica.m.wilson@utoronto.ca

REFERENCES

Aizawa, K., Gillett, C. (2009). The (Multiple) Realization of Psychological and Other Properties in the Sciences. *Mind and Language* **24**, 181–208.

Alexander, S. (1920). *Space, Time, and Deity.* London: Macmillan.

Anderson, P.W. (1972). More is Different. *Science* **177**, 393–396.

Antony, L.M. (2003). Who's Afraid of Disjunctive Properties? *Philosophical Issues* **13**, 1–21.

Antony, L.M., Levine, J.M. (1997). Reduction with Autonomy. *Philosophical Perspectives* **11**, 83–105.

Barnes, E. (2012). Emergence and Fundamentality. *Mind* **121**, 873–901.

Bedau, M. (1997). Weak Emergence. *Philosophical Perspectives 11: Mind, Causation and World* **11**, 375–399.

Batterman, R. (2002). *The Devil in the Details: Asymptotic Reasoning in Explanation, Reduction, and Emergence.* Oxford: Oxford University Press.

Baysan, U. (2015). Realization Relations in Metaphysics. *Minds and Machines* **25**, 1–14.

Bedau, M.A. (2002). Downward Causation and the Autonomy of Weak Emergence. *Principia* **6**, 5–50.

Bedau, M.A. (2008). Is Weak Emergence Just in the Mind? *Minds and Machines* **18**, 443–459.

Block, N., Stalnaker, R. (1999). Conceptual Analysis, Dualism, and the Explanatory Gap. *Philosophical Review* **108**, 1–46.

Boyd, R. (1980). Materialism without Reductionism: What Physicalism Does Not Entail. In: N. Block (ed.), *Readings in the Philosophy of Psychology*, Vol 1, pp. 1–67. Cambridge: Harvard University Press.

Broad, C.D. (1925). *Mind and Its Place in Nature.* Cambridge: Kegan Paul.

Byrne, A. (1994). *The Emergent Mind.* Princeton: Princeton University.

Chalmers, D. (1996). *The Conscious Mind.* Oxford: Oxford University Press.

Chalmers, D. (1999). Materialism and the Metaphysics of Modality. *Philosophy and Phenomenological Research* **59**, 473–496.

Chalmers, D., Jackson, F. (2001). Conceptual Analysis and Reductive Explanation. *The Philosophical Review* **110**, 315–60.

Chalmers, D. (2006). Strong and Weak Emergence. In: The Re-Emergence of Emergence. Oxford: Oxford University Press.

Clapp, L. (2001). Disjunctive Properties: Multiple Realizations. *Journal of Philosophy* **98**, 111–136.

Clarke, R. (1999). Nonreductive Physicalism and the Causal Powers of the Mental. *Synthese* **51**, 295–322.

Clayton, P. (2006). Conceptual Foundations of Emergence Theory. In: *The Re-Emergence of Emergence*. Oxford: Oxford University Press.

Crane, T. (2001). The Significance of Emergence. In: C. Gillett and B. Loewer (eds.), *Physicalism and Its Discontents*, pp. 207–224. Cambridge: Cambridge University Press.

Cunningham, B. (2001). The Reemergence of Emergence. *Philosophy of Science* **68**, S62–S75.

Ehring, D. (1996). Mental Causation, Determinables, and Property Instances. *Noûs* **30**, 461–480.

Fodor, J. (1974). Special Sciences (Or, The Disunity of Science as a Working Hypothesis). *Synthese* **28**, 77–115.

Gillett, C. (2002). The Dimensions of Realization: A Critique of the Standard View. *Analysis* **62**, 316–323.

Gillett, C. (2002). The Varieties of Emergence: Their Purposes, Obligations and Importance. *Grazer Philosophische Studien* **65**, 95–121.

Howell, R. (2009). Emergentism and Supervenience Physicalism. *Australasian Journal of Philosophy* **87**, 83–98.
Humphreys, P. (1996). Aspects of Emergence. *Philosophical Topics* **24**, 53–70.
Humphreys, P. (1997). How Properties Emerge. *Philosophy of Science* **64**, 1–17.
Kekes, J. (1966). Physicalism, the Identity Theory, and the Concept of Emergence. *Philosophy of Science* **33**, 360–375.
Kim, J. (1989). The Myth of Nonreductive Materialism. *Proceedings and Addresses of the American Philosophical Association* **63**, 31–47.
Kim, J. (1990). Supervenience as a Philosophical Concept. *Metaphilosophy* **21**, 1–27.
Kim, J. (1992). Multiple Realization and the Metaphysics of Reduction. *Philosophy and Phenomenological Research* **52**, 1–26.
Kim, J. (1993a). The Non-Reductivist's Troubles with Mental Causation. In: J. Heil and A. Mele (eds.), *Mental Causation*, pp. 189–210. Oxford: Oxford University Press.
Kim, J. (1993b). *Supervenience and Mind: Selected Philosophical Essays*. Cambridge: Cambridge University Press.
Kim, J. (1998). *Mind in a Physical World.* Cambridge: MIT Press.
Kim, J. (1999). Making Sense of Emergence. *Philosophical Studies* **95**, 3–36.
Kim, J. (2005). *Physicalism, or Something Near Enough.* Princeton: Princeton University Press.
Kim, J. (2006). Emergence: Core Ideas and Issues. *Synthese* **151**, 547–559.
Kim, J. (2011). From Naturalism to Physicalism: Supervenience Redux. *Proceedings of the American Philosophical Association* **85**, 109–134.
LePore, E., Loewer, B. (1989). More on Making Mind Matter. *Philosophical Topics* **17**, 175–191.
Klee, R. (1984). Micro-Determinism and Concepts of Emergence. *Philosophy of Science* **51**, 44–63.
Lewes, G.H. (1875). *Problems of Life and Mind.* London: Kegan Paul, Trench, Turbner & Co.
MacDonald, C., MacDonald, G. (1986). Mental Causes and Explanation of Action. *Philosophical Quarterly* **36**, 145–158.
McLaughlin, B. (1992). The Rise and Fall of British Emergentism. In: A. Beckerman, H. Flohr and J. Kim (eds.), *Emergence or Reduction? Essays on the Prospects of Non-Reductive Physicalism*, pp. 49–93. Berlin: De Gruyter.
Melnyk, A. (2003). *A Physicalist Manifesto: Thoroughly Modern Materialism.* New York: Cambridge University Press.
Melnyk, A. (2006). Realization-Based Formulations of Physicalism. *Philosophical Studies* **131**, 127–155.
Melnyk, A. (2008). Conceptual and Linguistic Analysis: A Two-Step Program. *Noûs* **42**, 267–291.
Noordhof, P. (2010). Emergent Causation and Property Causation. In: C. Macdonald and G. Macdonald (eds.), *Emergence in Mind*, pp. 69–99. Oxford: Oxford University Press.
Mill, J.S. (1843). *A System of Logic.* Toronto: University of Toronto Press.
Moore, G.E. (1903). *Principia Ethica.* Cambridge: Cambridge University Press.
Morgan, C.L. (1923). *Emergent Evolution.* London: Williams & Norgate.
Morris, K. (2010). Guidelines for Theorizing About Realization. *Southern Journal of Philosophy* **48**, 393–416.
Morris, K. (2013). On Two Arguments for Subset Inheritance. *Philosophical Studies* **163**, 197–211.
Newman, D. (1996). Emergence and Strange Attractors. *Philosophy of Science* **63**, 245–261.
Ney, A. (2010). Convergence on the Problem of Mental Causation: Shoemaker's Strategy for (Nonreductive?) Physicalists. *Philosophical Issues* **20**, 438–445.

O'Connor, T. (1994). Emergent Properties. *American Philosophical Quarterly* **31**, 91–104.

O'Connor, T. (2002). *Free Will.* In: E.N. Zalta (ed.), *The Stanford Encyclopedia of Philosophy* (Fall 2014 Edition), URL = <http://plato.stanford.edu/archives/fall2014/entries/freewill/>.

O'Connor, T., Wong, H.Y. (2005). The Metaphysics of Emergence. *Noûs* **39**, 658–678.

Papineau, D. (1993). *Philosophical Naturalism.* Oxford: Basil Blackwell.

Pereboom, D. (2002). Robust Nonreductive Materialism. *Journal of Philosophy* **99**, 499–531.

Perry, J. (2001). *Knowledge, Possibility, and Consciousness.* Cambridge: MIT Press.

Poland, J. (1994). *Physicalism: The Philosophical Foundations.* New York: Oxford University Press.

Polger, T.W. (2007). Realization and the Metaphysics of Mind. *Australasian Journal of Philosophy* **85**, 233–259.

Putnam, H. (1967). Psychological Predicates. In: *Art, Mind, and Religion*, pp. 37–48. Pittsburgh: University of Pittsburgh Press.

Rueger, A. (2001). Physical Emergence, Diachronic and Synchronic. *Synthese* **124**, 297–322.

Schroder, J. (1998). Emergence: Non-Deducibility or Downward Causation? *The Philosophical Quarterly* **48**, 433–452.

Searle, J.R. (1992). *The Rediscovery of the Mind.* Cambridge: MIT Press.

Shoemaker, S. (1980). Causality and Properties. In: P. van Inwagen (ed.), *Time and Cause*, pp. 109–135. Dordrecht: Reidel.

Shoemaker, S. (2000/2001). Realization and Mental Causation. *Proceedings of the 20th World Congress in Philosophy*. Cambridge: Philosophy Documentation Center.

Shoemaker, S. (2002). Kim on Emergence. *Philosophical Studies* **58**, 53–63.

Silberstein, M., McGeever, J (1999). The Search for Ontological Emergence. *Philosophical Quarterly* **50**, 182–200.

Smart, J.J.C. (1981). Physicalism and Emergence. *Neuroscience* **6**, 109–113.

Sperry, R. (1986). Discussion: Macro- Versus Micro-Determination. *Philosophy of Science* **53**, 265–270.

Sperry, R.W. (1969). A Modified Concept of Consciousness. *Psychological Review* **76**, 532–536.

Sperry, R.W. (1976). Mental Phenomena as Causal Determinants in Brain Function. In: G. Globus, G. Maxwell, I. Savodnik (eds.), *Consciousness and the Brain*, pp. 163–177. New York: Plenum Press.

Stephan, A. (2002). Emergentism, Irreducibility, and Downward Causation. *Grazer Philosophische Studien* **65**, 77–93.

Stoljar, D. (2007). Distinctions in Distinction. In: J. Kallestrup and J. Hohwy (eds.), *Being Reduced: New Essays on Causation and Explanation in the Special Sciences*, pp. 263–279. Oxford: Oxford University Press.

van Cleve, J. (1990). Mind-dust or Magic? Panpsychism versus Emergence. *Philosophical Perspectives* **4**, 215–226.

Van Gulick, R. (2001). Reduction, Emergence and Other Recent Options on the Mind/Body Problem: A Philosophic Overview. *Synthese* **8**, 1–34.

Walter, S. (2010). Taking Realization Seriously: No Cure for Epiphobia. *Philosophical Studies* **151**, 207–226.

Wilson, J. (1999). How Superduper does a Physicalist Supervenience Need to Be? *The Philosophical Quarterly* **49**, 33–52.

Wilson, J. (2002). Causal Powers, Forces, and Superdupervenience. *Grazer Philosophische Studien* **63**, 53–78.

Wilson, J. (2006). On Characterizing the Physical. *Philosophical Studies* **131**, 61–99.

Wilson, J. (2009). Determination, Realization, and Mental Causation. *Philosophical Studies* **145**, 149–169.

Wilson, J. (2010a). Non-reductive Physicalism and Degrees of Freedom. *British Journal for the Philosophy of Science* **61**, 279–311.

Wilson, J. (2010b). What is Hume's Dictum, and Why Believe It? *Philosophy and Phenomenological Research* **80**, 595–637.

Wilson, J. (2011). Non-reductive Realization and the Powers-based Subset Strategy. *British Journal for the Philosophy of Science* **94**, 121–154.

Wilson, J. (2013). Nonlinearity and Metaphysical Emergence. In: S. Mumford and M. Tugby (eds.), *Metaphysics and Science*, pp. 201–235. Oxford: Oxford University Press.

Wilson, J. (2014). Hume's Dictum and the Asymmetry of Counterfactual Dependence. In: Wilson, A. (ed), *Chance and Temporal Asymmetry*, pp. 258–279. Oxford: Oxford University Press.

Wimsatt, W. (1996). Aggregativity: Reductive Heuristics for Finding Emergence. *Philosophy of Science* **64**, 372–384.

Yablo, S. (1992). Mental Causation. *The Philosophical Review* **101**, 245–280.

Mauro Dorato
Michael Esfeld

THE METAPHYSICS OF LAWS: DISPOSITIONALISM VS. PRIMITIVISM

ABSTRACT. The paper compares dispositionalism about laws of nature with primitivism. It argues that while the distinction between these two positions can be drawn in a clear-cut manner in classical mechanics, it is less clear in quantum mechanics, due to quantum non-locality. Nonetheless, the paper points out advantages for dispositionalism in comparison to primitivism also in the area of quantum mechanics, and of contemporary physics in general.

1. Introduction

There are three main stances with respect to laws of nature in current philosophy of science: Humeanism, primitivism and dispositionalism.[1] Roughly speaking, according to Humeanism, the world is a mosaic of local matters of particular fact – such as the distribution of point-particles in a background spacetime – and laws are the axioms of the description of this mosaic that achieve the best balance between simplicity and informativeness or empirical content (see e.g. Lewis 1994 as well as Cohen and Callender 2009 and Hall unpublished).

According to *primitivism*, over and above there being local matters of particular fact – such as an initial configuration of point-particles in a background spacetime – there are in each physically possible world irreducible nomic facts instantiated by the world in question, according to which the corresponding laws hold in that world. The laws, *qua* instantiated in a

[1] A fourth notable view about laws, defended in Cei and French (2014) and French (2014), will be introduced in the next section, when discussing primitivism.

In: Tomasz Bigaj and Christian Wüthrich (eds.), *Metaphysics in Contemporary Physics* Poznań *Studies in the Philosophy of the Sciences and the Humanities*, vol. 104), pp. 403-424. Amsterdam/New York, NY: Rodopi | Brill, 2015.

world or by the world, fix the temporal development of an initial configuration of matter (in a deterministic manner if the law is deterministic, in a probabilistic manner if the law is probabilistic) (for primitivism see, notably, Carroll 1994 and Maudlin 2007). Laws are therefore not made true by locally instantiated properties or local matters of particular facts; on the contrary, such properties are discovered and determined by the laws that hold in a world.

According to *dispositionalism*, the local matters of particular fact – such as an initial configuration of point-particles in a background spacetime – instantiate a property (or a plurality of properties) that fixes the behaviour of these local matters of particular fact, for example the temporal development of an initial configuration of particles (either in a deterministic or in a probabilistic manner, the property being a propensity in the latter case). This property thus is a disposition or a power, and the behaviour of the local matters of particular fact is its manifestation. This property grounds a law in the sense that the latter is made true by the former, so that a law describes how objects that instantiate the property in question behave or would behave under various circumstances (if the property is a propensity, it grounds a probabilistic law that describes how objects that instantiate the property in question behave or would behave under various circumstances; note that propensities are not probabilities, but are that what grounds probabilities) (see notably Bird 2007 and Suárez 2014 for propensities). As we will see, dispositionalism can be further divided into a realistic *vs.* an antirealistic position about laws; the latter has been defended in particular by Mumford (2004, 2005a, 2005b).

In this paper we will assume that the main dividing line runs between Humeanism on the one hand and primitivism as well as dispositionalism on the other. Humeanism has to accept the whole distribution of the local matters of particular fact as a primitive, since the laws, being the axioms of the description of that distribution that achieve the best balance between simplicity and empirical content, supervene only on that entire distribution. In a nutshell, thus, what the laws of nature are, is fixed only "at the end of the world." It is not the laws that determine the development of the world, but it is the development of the world, in the sense of its spatiotemporal arrangement, that determines what the laws are (see Beebee and Mele 2002, pp. 201–205). By contrast, primitivism and dispositionalism have only to accept the initial conditions of the world – such as an initial configuration of point-particles in a background spacetime – as a primitive. The initial conditions, plus the fact that (i) certain laws are instantiated in the world in question (Primitivism about laws) or that (ii) the instantiation of certain properties (dispositions) is part and parcel of the initial conditions (dispositionalism), fix the further development of the world.

The reason for this divergence is that Humeanism eschews a commitment to objective *modality*, whereas both primitivism and dispositionalism subscribe to it. According to Humeanism, there is nothing about any proper part of the distribution of the local matters of particular fact in a world that fixes what is physically possible and what is not possible as regards the rest of the distribution of the local matters of particular fact in the world under consideration. The physical modality in question is not "in re," but belongs to the model or is a purely linguistic feature of nomic *statements*. According to both primitivism and dispositionalism, by contrast, there is something about a proper part of the distribution of the local matters of particular fact in a world that fixes what is physically possible and what is not possible in the world at issue, because either laws or a set of dispositional properties respectively are instantiated everywhere in the world.

Consequently, not only on primitivism, but also on dispositionalism, modality is not grounded in anything that is not itself modal. Thus, the dispositions that ground the laws according to dispositionalism are not themselves grounded in non-dispositional properties, but are basic properties. Their modal nature is therefore fundamental. In other words, both primitivism and dispositionalism are committed to a primitive modality. The difference between primitivism and dispositionalism is that the former position spells out primitive modality in terms of laws being primitive, whereas the latter position traces laws back to properties that display an ungrounded – and hence primitive – modality.

In this paper we will try to adjudicate the dispute between dispositionalism and primivitism by taking for granted that both parties believe in the existence of laws as well as in that of properties. The difference between these two positions lies in the fact that the dispositionalist regards laws as secondary to the properties, while the nomic primitivist considers properties to be ontologically secondary in a sense to be further specified below. For instance, by looking at the debate from a dynamical perspective, dispositionalism locates modal aspects in matters of particular fact by taking them to instantiate properties that are dispositions or powers and hence modal properties, while primitivism holds that it is the universal validity of laws in space and time that determines the temporal development of parts of the world or of the world itself by determining which properties exist.

After presenting primitivism and its possible formulations in the next section, we move on to discussing two case studies, by contrasting in each of them primitivism and dispositionalism (we take for granted that dispositional essentialism does not need a further presentation on our part – see Bird 2007). The first case study (section 3) is about laws in classical mechanics and is meant to illustrate the central feature of dispositionalism, namely grounding something that looks like a governing character of laws in properties that are localised in entities that there are in the world. The second case study then casts doubt on this straightforward picture by showing that properties that are supposed to ground the laws of our world as it is described by contemporary physics cannot be *local* properties, but have to be global and holistic. Such a non-locality seems to imply that the distinction between primitivism and dispositionalism becomes blurred, at least to the extent that primitivism is committed to the idea that what happens locally in region R in virtue of what properties are instantiated in R depends on what holds globally in the world, in virtue of the spatio-temporal universality of laws. We will therefore investigate whether this distinction can be upheld and if so, whether it provides a reason to prefer dispositionalism to primitivism (or the other way round) (section 4).

2. What does Primitivism about Laws Mean?

Let us first of all distinguish *conceptual* primitivism about laws from *ontological* primitivism. The former amounts to the claim that the notion of law cannot be analyzed or reduced in terms of counterfactuals, causation, regularity, explanatory or predictive power, and the like, since all of these notions presuppose it. The latter claims that laws exist in a primitive way, so that the existence of properties is grounded, supervenient or dependent on the existence of laws.

Here we are interested in spelling out what it means to claim that laws are ontological prior to properties (*ontic* nomic primitivism), since this problem seems to have been left in the background even by nomic ontic primitivists: "My analysis of law is no analysis at all. Rather I suggest we accept laws as fundamental entities in our ontology. Or, speaking at the conceptual level, the notion of law cannot be reduced to other more primitive notions" (Maudlin 2007, p. 18). Since Maudlin's clarification here moves from ontic priority to conceptual priority, in order to understand the sense in which the existence of properties may be grounded in that of laws, it will be opportune to start our discussion by clarifying the main features of *conceptual* nomic privimitism.

The latter view, defended among others also by Carroll (2004) insists on the "conceptual centrality of the nomic":

> If there were no laws, then there would be no causation, there would be no dispositions, there would be no true (nontrivial) counterfactual conditionals. By the same token, if there were no laws of nature, there would be no perception, no actions, no persistence. There wouldn't be any tables, no red things, no things of value, not even any physical object." (Carroll 2004, p. 10)

In this sense, conceptual primitivism about laws implies that the notion of law is necessary (and sufficient) to explain (note, an epistemic notion) the notion of physical possibility and therefore to specify the set of models that are consistent with the law (Maudlin 2007, p. 18). In a clear sense then, a law is more fundamental than the notion of model because different models can share the same law (think of different cosmological models sharing the same laws, namely Einstein field equations). In addition, once we have the notion of law, that of counterfactual can also be explained: if "all *A*s are *B*s" is fundamental, then the fact that "if *x* were an *A*, it would be a *B*" follows.[2] The latter notion of counterfactual, in its turn, would provide an analysis of causation (a counterfactual theory of causation), and of dispositions: if a certain stimulus were to be applied to a glass or to a flammable match, the glass and the match would manifest their dispositions to break and to catch fire respectively. Also the notion of property (say, "being charged"), according to nomic conceptual primitivism, cannot but be analyzed by using nomic concepts as primitives. For instance, what charge *is* (its causal role) and what it *does* (its behaviour) depends or is derivative (for short, is grounded) on the particular laws in which the property of charge figures: the Coulomb law defines the behaviour of electrostatic charges, the Lorentz law fixes the behaviour of a charged particle entering an electromagnetic field, while the motion of charges creating an electromagnetic field is governed by the relevant Maxwell equation (see Roberts 2008, p. 65). In a word, not only are natural properties discovered by finding out what the laws are (*epistemic* priority of laws), but their causal role is also fixed by the laws (*ontic* priority of laws).

We can now move on to clarify what ontic nomic primitivism amounts to. There are at least *two* senses in which laws can be ontically prior to dispositional properties, which we will discuss in turn: the first is spelled out in terms of supervenience, the second in terms of a structuralist viewpoint on laws (Cei and French 2014, French 2014).

1) According to a first way to spell out the failure of supervenience of laws on properties, "two worlds could differ in laws but not in any observable respect" (Maudlin 2007, p. 17). Suppose that "observable respect" is read as "observable properties." Two worlds could have different laws but could share all observable, non-quiddistic properties. Of course, one could block this failure of supervenience if one *defined* the notion of property as something that essentially plays a certain nomic role, so that a difference in laws would automatically imply a difference in properties. But since this move would beg the question against nomic ontic primitivism, the real

[2] For a contrary view, see Lange (2009).

issue at stake is which of the two positions, dispositionalism or nomic conceptual primitivism, is more suitable to perform an explanatory role with respect to the other notions that are typically associated with that of law (necessity, possibility, model, causation, counterfactual, regularity, etc). The next step would then be to ask whether such an explanatory role due to conceptual priority suffices for an inference to the best explanation *vis à vis* ontological priority. But also an inference to the best explanation in this case would be suspicious for both primitivism and dispositionalism. In fact, one could argue that even though the concepts of law or of dispositional property are non-reducible to non-nomic concepts and are furthermore explanatory primary in being indispensable to analyse causation, dispositions and counterfactuals, etc., there is nothing in the world that corresponds to laws of nature (nomic antirealism about laws) or dispositional properties. In sum, it seems that antirealists about any type of modal notions may coherently recognize that laws or dispositional properties are primitive *only* on the conceptual level.

We are convinced that stalemates of this kind in the metaphysics of science can best be settled by appeal to specific examples. Given the case study that we will discuss in the next section, for now it is appropriate to prepare the ground for the discussion by introducing a specific example, involving Einstein's and Newton's laws of gravity vis à vis the dispositional property "being massive." Suppose along with the ontic nomic primitivist that a difference of laws in a spacetime region *R* did *not* require a difference in the properties *P* instantiated in *R*. Since Einstein's and Newton's laws of gravity are *different*, it follows that the ontic primitivist must argue that this difference is compatible with the fact that the property of being massive in the two cases (or the two possible worlds in which these laws hold) stays *the same*. For instance, granting that "causing acceleration" is essential to "being massive," the nomic ontic primitivist might insist that the latter property has the same causal role in the two different laws. On the other hand, the reply of the dispositionalist might consist in pointing out that the mediating role in the manifestation of a dispositional property is essential for establishing the causal role played by a property. In Newton's case, the acceleration is mediated by a force, but in general relativity, masses accelerate via the mediation of the curvature of spacetime, and also in this case gravitational forces are non-existent. So the property "being massive" in the two worlds would be different. Given that we will discuss these two different readings of the property "being massive" in the next section, we can move on to the second way of cashing nomic ontic primitivism, which is in terms of the identity of properties.

2) The second way of cashing out the priority of laws over properties instantiated by local matters of facts, is suggested by the structuralist

understanding of laws proposed in Cei and French (2014) and French (2014). Despite the fact that these authors do not interpret their position as a kind of *nomic* ontic primitivism, quotations as the following seem to authorize this interpretation: "objects, whatever their status might be, do not enter certain lawlike relations in virtue of certain ontological aspects of their properties; rather *their properties present certain ontological aspects because of the relations they enter*" (Cei and French 2014, p. 36). The relations in questions are the nomic structures, which can be regarded as epistemic or ontic, as in the case of the distinction between epistemic and ontic structural realism. Also in the nomic case in fact, we can have an *epistemic* structural primitivism about laws and an *ontic* structural primitivism about them.

The former insists on the fact that all we know about the world are nomic relations, and locally instantiated properties are discovered by discovering the laws.[3] This might be regarded as meaning that the nomic structure in question is fixed by certain spatiotemporal symmetries that the world instantiates (consider the conservation laws as they are explicated by Noether's theorem), and objects and their properties are discovered in terms of what is left invariant by these symmetries. The ontic priority of laws can be construed either in an eliminationist fashion (there is just nomic structure and no local matters of facts instantiating natural properties) or as moderate form of the structural primitivity of laws: both laws and properties are real, but the latter are grounded in the former, whatever grounded may mean in this case.

The connection between this way of construing the priority of laws with respect to dispositions and the first just sketched in 1) is given by the fact that if the properties get their *identities* from the laws in which they occur (as in Ramsey-style versions of structuralism), then the properties' identities supervene on laws. Worlds with the same laws must have the same properties. Of course, the problem in this second version of ontic primitivism is to characterize the nomic *ontic* structure in a clear way, a problem that notoriously besets the ontic version of structural realism and that here cannot be discussed.

In a word, primitivism about laws of nature is the view that there are nomic facts holding in each possible worlds that determine or at least put a constraint on the distribution of the local matters of particular facts in each of the worlds.

[3] In this perspective, for example, Chakravartty's detecting properties (2007) would depend on the nomic structure, and not conversely.

3. Dispositions and Laws in Classical Mechanics

According to dispositionalism, it is in virtue of having a mass m that particles exert a force of attraction F upon each other as described by the law of gravitation:

(1) $F = G\,(mm')\,/\,r^2$

On dispositionalism therefore, this law tells us that if two masses change their velocity due to the action of forces on them, the generated forces can be traced back to, or explained in terms of, the properties of the particles.

In other words, mass is a disposition that manifests itself in the mutual attraction of massy objects. The presence of another mass m' acts as a stimulus on m (and conversely) for the manifestation of the disposition in terms of a mutual acceleration. As soon as there are at least two massive objects in a world, that disposition is triggered. It is essential for the property of gravitational mass to manifest itself in the mutual attraction of the objects that instantiate this property. That's what gravitational mass *is* – the property that makes objects accelerate in a certain manner. It is in this sense that the dispositional property "having a mass" grounds the law of gravitation. More precisely, mass as a property type grounds the law (1), with the concrete values of mass – the mass tokens – determining, together with the square of the distance between the massy objects, how these objects attract each other in virtue of possessing each a certain value of mass. Hence, that law reveals and describes what objects do in virtue of possessing a mass, and, crucially, in Newton's mechanics this property depends on its manifestation (namely, the acceleration) on the existence of a force. Since in Einstein's theory the notion of force is jettisoned, we cannot consider that the notion of mass in the two theories is the same, since in the latter case the manifestation depends on the curvature of spacetime: therefore, different laws imply different properties, and laws supervene on the dispositional property of mass.

We can now take up the questions mentioned in the first section: assume that the local matters of particular fact consist in the distribution of point-particles in a background spacetime. Given an initial configuration of particles, that configuration develops in time in such a way that the particles trace out certain trajectories in space according to the laws of classical mechanics. Dispositionalism about Newton's laws (the first one in particular) maintains that the particles have the disposition to continue to move with constant velocities on straight lines in space (or to continue to be at rest), *unless* external forces act on them. This dispositional property grounds Newton's first law, and also grounds, or is identical with, the tendency to resist acceleration (inertial mass). The possible non-existence of inertially moving systems (nothing can be screened off from gravitational forces) makes the posit of a disposition to continue with the same speed rather plausible or perhaps even indispensable. Such a disposition is in fact the *truth-maker* of Newton's first law, regarded as the statement found in textbooks and used for the construction of the mathematical model given by Newtonian spacetime. Positing instead a primitive nomic fact about inertially moving bodies (along with the primitivist) seems inappropriate, since the fact in question might be, and most probably is, *uninstantiated.* How can a primitivist justify her position with non-instantiated laws? Note that this is a major problem also for Humean regularists, since they rely on the existence of concrete regularities in order to justify the existence of patterns of local facts, even if one claims that Newton's laws are axioms that maximize simplicity and informativeness. In any case, the burden of proof is on the side of the Humean to show how an *uninstantiated* regularity can be part of the regularities in a given mosaic of local matters of particular fact that allows to simplify that mosaic while being informative about it. In this respect, dispositionalism seems to fare much better than its two rivals.

The primitivist may raise at least two objections against this argument.[4]

(1) Requiring a truth-maker for laws begs the question against primitivism. After all, positing a truth-maker automatically implies that laws are not primitive, as they must be grounded in something else!

(2) According to primitivism, laws determine the physically possible models. Models in their turn are often idealized representations of the properties of physical systems: in this way, primitivism can account for the fact that the first law might be uninstantiated (if indeed it is) as models are a limiting case of the behaviour of real systems.

The response to the first objection lies in two counterobjections. First, the truth-maker truth-bearer distinction is widely shared, especially among philosophers inclined toward scientific realism, and is therefore completely *neutral vis à vis* the debate we are interested in. Second, we need to disambiguate «law» in terms of a distinction between laws of science (*statements* expressing scientific laws) and what they denote, namely laws of nature (Weinert 1995), which according to *both* camps exist independently of any such statements. We take it that these two distinctions are quite reasonable and should therefore by endorsed by both camps. But then, if one rejects the instrumentalist view according to which statements expressing

[4] We owe these objections to the anonymous referee.

scientific laws are neither true nor false, and if one endorses the first distinction, the fact that no *real* physical system obeys the first law has the consequence that any statement expressing it must be considered to be strictly speaking *false*. If regarding the first law of mechanics as false is the price to pay for primitivism, it must be admitted that dispositionalism in this account fares better. The second distinction reveals the *confusion* on which the first objection is based: it is not laws of nature that need grounding, but laws of science, and primitivism, unlike dispositionalism, cannot offer any grounding for statements expressing the fist law.

The response to the second objection, related to the first, involves the notion of models as idealizations of real physical situations. As is well known, physical models are often regarded as *mediators* between the theory and the world (among others, see Morgan and Morrison 1999). Here we will assume that this is indeed the case also in our context. Then the question once again is: how can something that does not exist (an uninstantiated law of nature) ground abstract, idealized models by determining them? Such models would be grounded on *nothing* real. Furthermore, assuming that models are mediators between the theory and the world seems to imply that the idealizations of reality that feature in the model are fixed by our theories of the physical world and not by laws of nature. At least if theories are not deducible by the facts (in our case, primitive *nomic* facts) but are, as Einstein put it, «free inventions of the human mind». But while in dispositionalism the constraint on theories is given by real dispositions, in primitivism such a constraint would be rather weak to say the least, at least in the case of uninstantiated laws.

Of course, the fact that the first law might be non-instantiated (only a system that were completely removed from any gravitational mass would obey Newton's first law) does not imply that in classical mechanics there is no empirical distinction between an inertial and an accelerated system. On the contrary, the distinction in question cannot be explained by the primitivist, because the law in question, unlike the related disposition, does not or might not exist in the actual, concrete world.[5]

From this viewpoint, one could *prima facie* take *two* positions: anti-realism and realism about laws. According to the first position, laws do not exist in nature, since dispositions do all the work that the latter are supposed to do (Mumford 2004, 2005a, 2005b) and laws of science are

[5] When it comes to classical general relativity, inertial motion is explained by the vanishing of the covariant divergence of the stress-energy tensor, but the geodesic principle can still be interpreted in a dispositionalist fashion and not only within the dynamical approach to relativity favoured by Brown (1995, pp. 160 ff.). We have no room to argue in favour of this claim here.

true descriptions of what dispositions do. As Ellis put it: "Laws are not superimposed on the world, but grounded in the natures of the various kinds of things that exist" (2006, p. 435). As such, they cannot govern at all, because they do not exist. Or, secondly, one can endorse *realism* about laws, but *analyze* it and ground it by using the existence of dispositions: that is, the fact that laws exist is tantamount to the fact that dispositions or relations among them manifest themselves in a certain way. As long as the dependence of laws on dispositions is clear, we think that it is not important to choose between these two positions: they both agree that laws are grounded in dispositions and then take different stances with respect to the ontological status of non-fundamental entities

However, already in this paradigmatic example of a disposition grounding a law, other complications arise. Considering the formula (1), it seems that one can hold the masses m and m' fixed, but conceive a possible world in which the gravitational constant G has another value, or a world in which the force of gravitation does not decrease with the square of the distance r^2 among the particles, but only with the distance r, or with the cube of the distance r^3, etc. It seems that in all these possible worlds, there is mass as in the actual world, but the law of gravitation is different, although it still is a law of gravitation.

Both Humeanism and primitivism admit such a scenario. On Humeanism, it is a contingent matter of fact that in the actual world the property which we refer to by using the term "mass" plays a role that is described by the law of gravitation that holds in the actual world. The role that this property plays can vary from one possible world to another. On primitivism, it is a primitive matter of fact that the law (1) is instantiated in the actual world. In other logically possible worlds, different, but similar laws are instantiated, which can also be considered as laws of gravitation.

In order for the dispositionalist to maintain that the law (1) is grounded in the property of mass (so that, whenever in a possible world there are objects that instantiate the property of mass, the law (1) applies), the dispositionalist has to hold that the property of mass includes not only what is represented by the variable m in the formula (1), but also the gravitational constant having a certain value and the fact that the force of acceleration that objects exert upon each other in virtue of possessing a mass decreases with the square of the distance. In other words, the dispositionalist has to pack everything that the law of gravitation says about the interaction of massive objects into the property of mass, in such a way that this property can ground the law. Making this move has the following consequence: since according to dispositionalism the role that a property exercises is the essence of the property, the dispositionalist is committed to maintaining that in a possible world in which the gravitational constant has another value, or in which the force of gravitation does not decrease with the square of the distance r^2, the property of mass is not instantiated. In such other possible worlds, another property is instantiated which is similar to the property of mass that is instantiated in the actual world. However, it is not mass, but only its counterpart. There is mass if and only if the law of gravitation as expressed in formula (1) holds. Note that this is an ontological issue; we may of course be in error about the dispositional essence of mass and therefore misconceive the law of gravitation, or amend our conception of that law through theory change. These epistemological matters have no bearing on the fact that according to dispositionalism, there is mass if and only if a particular law holds, whether or not we are right about what that law is.

Making this move has a clear advantage: it grounds everything that there is in a world for the law of gravitation to hold in an intrinsic property of the objects that there are in the world, namely the massive particles. Hence, thanks to this move, dispositionalism is committed in this case, like Humeanism, only to local matters of particular fact, namely particles instantiating certain properties and not also to locally instantiated laws fixing those properties. What distinguishes dispositionalism from Humeanism is that dispositionalism conceives these properties as modal, so that a world in which the gravitational force decreases with the inverse cubic power of r is only logically but not physically possible.

Let us briefly turn to another paradigmatic example from classical mechanics, namely charge. According to dispositionalism, in virtue of possessing a charge (positive or negative), particles exert a force of attraction or repulsion upon each other as described by the laws of electromagnetism (Lorentz's equation included). In the case of Coulomb's law for example, each particle acts as a stimulus for the manifestation of the disposition of the other one to be attracted or repelled, and therefore to accelerate: the electrostatic force in this respect is fully analogous to the gravitational force. The classical theory of electromagnetism, however, distinguishes itself from Newton's theory of gravitation in that the force that particles exert upon each other in virtue of possessing a certain property is mediated by a *field*, so that the effects of this property are typically retarded (but advanced solutions exist, given the time-symmetric character of Maxwell's equations). In the case of Lorentz's force law, for example, the magnetic field triggers the disposition of the charge to be accelerated by the field, and the deviation of its trajectory is the manifestation of the disposition, while the motion of charges in a circuit manifests itself as a change in the magnetic field, which is disposed to be changed by a current, etc. In a word, as in the case of the gravitational law, the typical fingerprint of dispositions is present: a trigger mechanism and the manifestation of the disposition, either of the charge or of the field.

In sum, despite the mentioned difficulties and despite many details that have to be filled in, the dispositionalist can make a case for the fundamental laws of *classical* physics being grounded in dispositions that are intrinsic and thus local properties of particles or regions of fields. As we have seen, a first advantage of dispositionalism over primitivism is rather evident in the case of non-instantiated laws, of which we discussed only the law of inertia. Furthermore, since we have seen that the property of mass is given by its causal power, the mediation of force in one case and of the gravitational curvature in the other makes it the case that the property of mass is different in Newton's and in Einstein's theory of gravity, so that laws supervene on properties. Finally, the dispositionalist might claim to take up the advantages of both Humeanism and primitivism, while avoiding the drawbacks of each of these positions: like the Humean, the dispositionalist is committed only to local matters of particular fact; however, since these local matters of particular fact instantiate modal properties in the guise of dispositions or powers (such as mass, charge and local regions of the electromagnetic field), these properties ground laws in the sense of primitivism about modality, namely laws that implement an objective and irreducible modality. Since simplicity, however, need not be a guide to truth, the final balance between the two camps *vis à vis* the last requirement must be drawn in the next section.

4. Dispositions and Laws in Quantum Mechanics

Let us now turn to quantum physics and focus on what is known as "primitive ontology" approaches.[6] These are approaches that admit an ontology of matter distributed in three-dimensional space or four-dimensional spacetime as the referent of the formalism of quantum mechanics and propose a law for the temporal development of this distribution of matter. The motivation for doing so is to obtain an ontology that can account for the existence of measurement outcomes – and, in general, the existence of the macroscopic objects with which we are familiar before doing science. Here we will focus on three different primitive ontology approaches that have

[6] Here the term "primitive" is confusing, but clearly there are two senses of "primitive" at play, one referring to what exists concretely in spacetime, the second to laws as being conceived as conceptually and ontically prior to properties or dispositions.

been developed in the philosophical literature on non-relativistic quantum mechanics (see notably Allori et al. 2008).

There is in the first place Bohmian mechanics, which is committed to an ontology of particles. This theory conceives a law, known as the guiding equation, that employs the quantum mechanical wave-function in such a way that, in brief, the temporal development of the wave-function according to the Schrödinger equation supplies the temporal development of the configuration of particles in three-dimensional space, by yielding a velocity field along which the particles move (see the papers in Dürr, Goldstein and Zanghì 2013 for the dominant contemporary version of the theory going back to de Broglie 1928 and Bohm 1952).

There is furthermore the amendment of the Schrödinger equation proposed by Ghirardi, Rimini and Weber (1986) (GRW). The GRW equation has the purpose to modify non-linearly the Schrödinger equation in such a way that it can describe the temporal development of matter that is localized in three-dimensional space. As regards matter, there are two different proposals for a primitive ontology in physical space put forward in the literature that use the GRW equation: according to the proposal set out by Ghirardi himself (Ghirardi, Grassi and Benatti 1995), matter is "gunky," there being a continuous distribution of matter in space, namely a matter density field. That field can contract spontaneously in order to form well-localized macroscopic objects (where the stuff is more dense), as described by the spontaneous localization (the "collapse") of the wave-function in configuration space. Bell (1987) took up the GRW modification of the Schrödinger equation in another manner, proposing an ontology of events in spacetime, which in today's literature are known as flashes (that term goes back to Tumulka 2006, p. 826). According to this ontology, there is an event (a flash) in four-dimensional spacetime whenever the wave-function in configuration space spontaneously localizes ("collapses") as described by the GRW equation. Consequently, these events are sparsely distributed in spacetime, there being no continuous sequences of events. Nonetheless, the distribution of these events can be quite dense in certain regions of spacetime, so that well-localized macroscopic objects are accounted for also in the flash ontology.

The structure of all these proposals is such that (i) an ontology of matter in space or spacetime is admitted as the referent of the quantum formalism and (ii) a law is proposed that describes the temporal development of the configuration of matter in physical space. The universal wave-function – that is, the wave-function of the whole configuration of matter in physical space – is nomological in the sense that it is part of the law of the development of the primitive ontology, by contrast to being a *concrete* physical entity on a par with the primitive ontology (see notably Dürr,

Goldstein and Zanghì 2013, ch. 12, in the context of Bohmian mechanics). The reason is that the wave-function could not be an entity that exists in three-dimensional or four-dimensional spacetime: it could not be a field in physical space, since it does not assign values to spacetime points. If it were a field, it could only be a field on the very high-dimensional configuration space of the universe (if there are *N* particles, the dimension of the corresponding configuration space is 3*N*).

While it is an option to regard the universal wave-function as a field and thus as a physical entity existing in configuration space, this option is not plausible within the primitive ontology approach: it is unclear to say the least how the universal wave-function could perform the task that it has according to the primitive ontology approach to quantum physics – namely to determine the temporal development of the primitive ontology – if it were a physical entity on a par with the primitive ontology, but existing in another space. How could a field on a very high-dimensional space make matter move in three-dimensional space or four-dimensional spacetime? It seems that anything doing so has to be situated in the same space as the matter whose motion it determines. Furthermore, according to configuration space realism, the high-dimensional configuration space of the universe is fundamental, being the space in which the physical reality, namely the wave-function, plays itself out and evolves (see notably Albert 1996 and 2013). On the primitive ontology approach, by contrast, three-dimensional space or four-dimensional spacetime is the domain in which the physical reality is situated. Everything else that is admitted in this approach then is introduced through the role that it plays in the law that describes the physical reality in three-dimensional space or four-dimensional spacetime. It is therefore well motivated to regard the universal wave-function as nomological in the primitive ontology approach to quantum physics, by contrast to being a physical entity on a par with the primitive ontology, but existing in another space.

When it comes to spelling out what it means the claim that the universal wave-function is nomological, the three general stances on laws mentioned above are available and defended in the literature: on primitivism, a law is instantiated in the world over and above the primitive ontology, incorporating the universal wave-function or the quantum state (see Maudlin 2007, in particular ch. 2). On dispositionalism, as we shall elaborate on below, the configuration of matter in physical space instantiates at any time a holistic property that grounds the law of motion and that is *represented* by the universal wave-function, as the mass or the charge variable in the laws of classical mechanics represent dispositional properties of the particles. On Humeanism, the universal wave-function is nothing in addition to the distribution of the primitive ontology (the particles, the matter density field, the flashes) throughout the whole of space-time; it supervenes on that distribution, figuring in the Humean best system, that is, the system that achieves the best balance between being simple and being informative in describing the distribution of the primitive ontology throught the whole space-time (see Miller 2014, Esfeld 2014, Callender forthcoming).[7]

Although there is a good reason to regard the wave-function as nomological in contrast to being a physical entity on a par with the primitive ontology, one has to bear in mind the following two facts: at least as a *law of science* as it is formulated in the model, the universal wave-function develops itself in time according to a law, namely the Schrödinger equation (or the GRW equation), which for realists about laws does refer to a *law of nature* – unless one assumes that the universal wave-function will eventually turn out to be stationary, for instance in a quantum theory of gravitation that replaces the Schrödinger equation with the Wheeler-de-Witt equation. But even if the universal wave-function were stationary, there would still remain the fact that there are many different universal wave-functions possible, all of which fit into the same law (the Schrödinger equation, the GRW equation, or the Wheeler-deWitt equation). We will show below how dispositionalism is in a better position to accommodate these facts than primitivism.

Let us first point out two important differences between classical and quantum dispositions. The first one is that both in the Bohmian picture and (even more) in the dynamical reduction models, the dispositions of the quantum objects manifest themselves in a *spontaneous* manner, that is, they are not triggered by anything external. One may object that the lack of a clearly identified stimulus for the manifestation of a disposition makes the property in question non-dispositional.[8] However, this charge is unjustified, especially for those dispositions that are also propensities and are indeterministic in nature. The dispositional property to decay possessed by radioactive material manifests itself spontaneously, since the time at which the manifestation of the disposition occurs is utterly indeterminate, and uncaused by anything external. For the Bohmian, deterministic case, one should simply note that, given non-locality, there cannot be any external triggering mechanism for the manifestation of the particles' disposition to fix their velocity field since – as we are about to see in the next paragraph – the disposition in question is a holistic property instantiated by

---

[7] As regards the ontology of the wave-function, see the essays in Albert and Ney (2013). Unfortunately, this book ignores the Humean supervenience view of the wave-function, even though it contains papers that, like the Humean, are antirealist about the wave-function (notably French 2013, Monton 2013).

[8] We owe this objection to Steven French and Juha Saatsi.

the whole particle configuration. Furthermore, why should spontaneous dispositions not qualify as such? If the dispute here is not purely semantic, disqualifying spontaneous dispositions seems question-begging. Consider David Miller's example of the disposition or propensity of today's world "to develop in a year's time into a world in which I am still alive" (Miller 1994, p. 189). This disposition obviously does not require an external stimulus to be manifested, because it is a global one too.

The second difference has to do with the main feature of the quantum mechanical wave-function – the feature that marks the distinction between quantum and classical mechanics – which is its entanglement. That is to say, whenever one considers a configuration of matter that comprises more than one particle, it is in general not possible to attribute to each particle a wave-function that, when put into the dynamical law, correctly describes its temporal development (for an interesting attempt to defend the contrary view, see Norsen 2010). Only the universal wave-function, that is, the wave-function of the *whole configuration* does so. The entanglement of the wave-function accounts for quantum non-locality: the temporal development of any part of the configuration of matter in physical space depends on all the other parts (although, as shown by the decoherence of the wave-function in configuration space, that dependence is in many cases negligible for all practical purposes).

Despite these differences, dispositionalism can be applied to quantum mechanics in the same way as in classical mechanics. The quantum law that describes the temporal development of matter in physical space (the Bohmian guiding equation, the GRW equation) is grounded in a property of matter that is a disposition, manifesting itself in the way in which the distribution of matter in space develops in time. The main difference between classical and quantum mechanics is that in the latter, the law can only be grounded in a property of the configuration of matter as a whole, that is, a global and holistic by contrast to a local property.

Thus, on dispositionalism applied to Bohmian mechanics, the configuration of all the particles in the universe at any given time t (recall that we are presupposing Newtonian spacetime) instantiates a dispositional property that manifests itself in the velocity of each particle at t; *the universal wave-function at t represents that property* and, within the guiding equation, the wave-function expresses how that property manifests itself in the temporal development of the position of the particles (that is, their velocity) (see Esfeld et al. 2014, sections 4–5).

On dispositionalism applied to the GRW quantum theory, on the matter density version, the matter density as a whole instantiates a dispositional property (more precisely, a propensity) that manifests itself in the temporal development of the matter density – notably in its spontaneous concentration around certain points in space – and that is represented by the universal wave-function and the probabilities for the temporal development of the matter density that the universal wave-function yields if it is plugged into the GRW equation. On the flash version of GRW, the configuration of flashes as a whole in its turn instantiates a dispositional property (more precisely, a propensity) that manifests itself in the occurrence of further, later flashes and that is represented by the universal wave-function and the probabilities for the occurrence of further flashes that the universal wave-function yields once it is plugged into the GRW equation (see Dorato and Esfeld 2010 for dispositions in GRW and grounding the GRW probabilities in propensities).

Hence, when it comes to quantum mechanics, dispositionalism loses a characteristic feature by which it distinguishes itself from primitivism about laws as far as classical mechanics is concerned: it is no longer possible to maintain that the laws are grounded in local or intrinsic properties of particles. If the dynamical law of quantum mechanics is grounded in a property of matter, that property can only be a global or holistic property of the configuration of matter as a whole. Dispositionalism thereby comes close to primitivism in the following respect: on primitivism, each possible world instantiates a global fact – a world-fact so to speak – that a certain dynamical law holds in the world in question. In a nutshell, quantum mechanics compels dispositionalism to join primitivism in going global, at least to the effect that primitivism, by relying on the nomic structural realism presented in section 2, is best formulated as a view that stresses the existence of universal spatiotemporal symmetries as the bedrock for the primitive existence of laws, and therefore for the supervenient existence of properties (Cei and French 2014).

However, by contrast to primitivism, dispositionalism has no problem in accommodating the fact that the quantum mechanical wave-function develops itself in time as Schrödinger's equation prescribes (whereas fundamental dispositional properties in classical physics – such as mass and charge – do not develop in time, since a particle always possesses the same values of mass and charge). The temporal development of the wave-function tracks or describes in the mathematical model the temporal development of the dispositional property that the configuration of matter as a whole instantiates at a time. Thus, this disposition manifests itself not only in a certain temporal development of the configuration of physical entities that instantiate this property (a development described by the guiding equation in Bohmian mechanics), but in inducing or causing such a temporal development, this property also causes its own temporal development (described by the Schrödinger equation in Bohmian mechanics) (in the

GRW theory, the GRW equation incorporates both these developments).[9] In general, a dispositional property can change in time without the law in which the property in question figures being subject to a temporal development. By contrast, a law-fact instantiated in the world is not supposed to change in time, and it is difficult to see how primitivism about laws could accommodate the difference between a universal wave-function changing in time and the law in which it figures not changing in time.

Furthermore, dispositionalism has no problem in accommodating the fact that there are many different universal wave-functions possible, all of which fit into the same law. Consider, for example, two possible worlds described by Bohmian mechanics and assume that in these worlds there is the same initial particle configuration, but different initial wave-functions applying to the initial configurations, leading hence to different trajectories of the particles in these two possible worlds. Dispositionalism accounts for this case by maintaining that there are different values of the holistic, dispositional property of the particle configuration instantiated in these two worlds, so that these two worlds differ in the initial quantum state represented by the universal wave-function. But there is no nomological difference between two such worlds. By the same token, there can be the same initial distribution of particle positions in two possible worlds of classical mechanics and different distributions of mass or charge, leading to different trajectories of the particles. Hence, by grounding the law in a dispositional property of the particles – be it a local property, be it a global, holistic one – dispositionalism can admit different values that this property can take without these differences amounting to any nomological difference. This fact points to the failure of supervenience of properties on laws; since different properties do not entail different laws, the dependence of laws on properties invoked by primitivists in this case fails. By contrast, on primitivism, different universal wave-functions amount to a difference in the law facts instantiated in the worlds in question. In a nutshell, primitivism, in these quantum examples in particular, faces a dilemma: either it has to bite the bullet of conceiving the law as developing itself in time and as including differences that correspond to different initial wave-functions, or it has to conceive the universal wave-function as a physical entity.

In sum, comparing quantum mechanics to classical mechanics, the distinction between dispositionalism and primitivism about laws of nature is much less sharp in the former than in the latter: due to the entanglement of the wave-function, it is no longer possible in quantum mechanics to

[9] Recall that the spacetime presupposed by these non-relativistic theories admits a privileged foliation, i.e., simultaneity is absolute.

ground laws in local or intrinsic properties of particles. Nonetheless, laws can still be grounded in properties, albeit global ones (so the term "intrinsic" does not really apply), and doing so can still be regarded as an argument in favour of dispositionalism: this position can make intelligible how laws can "govern" the behaviour of objects – they are our epistemic access to what can causally influence the behaviour of objects in a clear and straightforward way, namely certain properties of objects. In short, it is the essence of the properties that objects instantiate to influence their behaviour in a certain manner: this claim of dispositionalist essentialism holds independently of whether the properties are local, being instantiated by the physical objects taken individually, or global, being instantiated by a configuration of objects as a whole. By contrast, it is unclear how the fact of certain laws being instantiated in a world could influence the behaviour of the objects in the world in question. The Humean objection against the governing conception of laws of nature hits primitivism, but it does not apply to dispositionalism, at least if it is legitimate to assume a primitive modality. We take it to be the decisive advantage of dispositionalism over primitivism to make the governing conception of laws of nature intelligible by anchoring the laws in the properties of physical objects, which also allows dispositionalism to maintain that there can be different initial values of these properties and that they can develop in time, without these variations touching the laws that the properties in question ground.

University of Rome III
Department of Philosophy, Communication and Media Studies
e-mail: dorato@uniroma3.it

University of Lausanne
Department of Philosophy
e-mail: michael-andreas.esfeld@unil.ch

ACKNOWLEDGMENTS

We thank an anonymous referee for penetrating comments on two previous versions of this paper.

REFERENCES

Albert, D.Z. (2000). *Time and chance.* Cambridge: Harvard University Press.

Albert, D.Z. (2013). Wave function realism. In: D.Z. Albert and A. Ney (eds.), *The wave function: Essays in the metaphysics of quantum mechanics*, pp. 52–57. Oxford: Oxford University Press.

Albert, D.Z., Ney, A., eds. (2013). *The wave function: Essays in the metaphysics of quantum mechanics.* Oxford: Oxford University Press.

Allori, V., Goldstein, S., Tumulka, R., Zanghì, N. (2008). On the common structure of Bohmian mechanics and the Ghirardi-Rimini-Weber theory. *British Journal for the Philosophy of Science* **59**, 353–389.

Beebee, H., Mele, A. (2002). Humean compatibilism. *Mind* **111**, 201–223.

Bell, J.S. (1987). *Are there quantum jumps*? In: C.W. Kilmister (ed.), *Schrödinger: Centenary celebration of a polymath*, pp. 41–52. Cambridge: Cambridge University Press.

Bird, A. (2007). *Nature's metaphysics: Laws and properties.* Oxford: Oxford University Press.

Bohm, D. (1952). A suggested interpretation of the quantum theory in terms of 'hidden' variables. *Physical Review* **85**, 166–193.

Brown, H. (2005). *Physical Relativity.* Oxford: Oxford University Press.

Callender, C. (forthcoming). One world, one beable. *Synthese*, DOI 10.1007/s11229–014-0582–3.

Carroll, J.W. (1994). *Laws of Nature.* Cambridge: Cambridge University Press.

Cei, A., French, S. (2014). Getting away from governance: a structuralist approach to laws and symmetries. In: G.B. Masera and G. Lini (eds.), *A World of Structures, Structures of a World. Methode: Analytic Perspectives* **3** (2), 25–48.

Chakravartty, A. (2007). *A metaphysics for scientific realism.* Cambridge: Cambridge University Press.

Cohen, J., Callender, C. (2009). A better best system account of lawhood. *Philosophical Studies* **145**, 1–34.

de Broglie, L. (1928). *La nouvelle dynamique des quanta. Electrons et photons. Rapports et discussions du cinquième Conseil de physique tenu à Bruxelles du 24 au 29 octobre 1927 sous les auspices de l'Institut international de physique Solvay.* Paris: Gauthier-Villars. English translation in G. Bacciagaluppi and A. Valentini, (2009). *Quantum theory at the crossroads. Reconsidering the 1927 Solvay conference.* Cambridge: Cambridge University Press.

Dieks, D. (2006). Becoming, relativity and locality. In: D. Dieks (ed.), *The ontology of spacetime*, pp. 157–176. Amsterdam: Elsevier.

Dürr, D., Goldstein, S., Zanghì, N. (2013). *Quantum physics without quantum philosophy.* Berlin: Springer.

Earman, J. (2011). Sharpening the electromagnetic arrow of time. In Callender, C. (ed.), *The Oxford Handbook in the Philosophy of Time*, p. 185–527. Oxford: Oxford University Press.

Ellis, B. (2006). Looking for Laws. *Metascience* **15**, 437–438.

Esfeld, M. (2014). Quantum Humeanism, or physicalism without properties. *The Philosophical Quarterly* **64**, 453–470.

Esfeld, M., Lazarovici, D., Hubert, M., Dürr, D. (2014). The ontology of Bohmian mechanics. *British Journal for the Philosophy of Science* **65**, 773–796.

French, S. (2013). *Whither wave-function realism?*. In: D.Z. Albert and A. Ney (eds.), *The wave function: Essays in the metaphysics of quantum mechanics*, pp. 76–90. Oxford: Oxford University Press.

French, S. (2014). *The Structure of the World.* Oxford: Oxford University Press.

Ghirardi, G.C., Grassi, R., Benatti, F. (1995). Describing the macroscopic world: Closing the circle within the dynamical reduction program. *Foundations of Physics* **25**, 5–38.

Ghirardi, G.C, Rimini, A., Weber, T. (1986). Unified dynamics for microscopic and macro scopic systems. *Physical Review D* **34**, 470–491.

Hall, N. (unpublished). Humean reductionism about laws of nature. http://philpapers.org/rec/HALHRA

Lange, M. (2009). *Laws and lawmakers.* Oxford: Oxford University Press.

Lewis, D. (1994). Humean supervenience debugged. *Mind* **103**, 473–490.

Mach, E. (1919). *The science of mechanics. A critical and historical account of its develop ment.* Chicago: Open Court.

Maudlin, T. (2007). *The metaphysics within physics.* Oxford: Oxford University Press.

Miller, D. (1994). *Critical rationalism.* La Salle: Open Court.

Miller, E. (2014). Quantum entanglement, Bohmian mechanics, and Humean supervenience *Australasian Journal of Philosophy* **92**, 567–583.

Monton, B. (2013). Against 3N-dimensional space. In: D.Z. Albert and A. Ney (eds.), *The wave function: Essays in the metaphysics of quantum mechanics*, chapter 7. Oxford: Ox ford University Press.

Morgan M., Morrison M., eds. (1999). *Models as Mediators.* Cambridge: Cambridge Uni versity Press.

Mumford, S. (2004). *Laws in Nature.* London: Routledge.

Mumford, S. (2005a). Laws and Lawlessness. *Synthese* **144**, 397–413.

Mumford, S. (2005b). Author's reply. Symposium on Laws in nature with Alexander Bird, Brian Ellis and Stathis Psillos. *Metascience* **15**, 462–469.

Norsen, T. (2010). The theory of (exclusively) local beables. *Foundations of Physics* **40** 1858–1884.

Psillos, S. (2005). Critical notice: Laws in nature. *Metascience* **15**, 437–469.

Roberts, J. T. (2008). *The law-governed universe.* Oxford: Oxford University Press.

Savitt, S. (2001). A limited defense of passage. *American Philosophical Quarterly* **38**, 261–270.

Suárez, M. (2014). A critique of empiricist theories of propensities. *European Journal for the Philosophy of Science* **4**, 215–231.

Tumulka, R. (2006). A relativistic version of the Ghirardi-Rimini-Weber model. *Journal of Statistical Physics* **125**, 825–844.

Weinert, F. (1995). *Laws of nature. Essays on the scientific, philosophical and historical dimensions.* Berlin: de Gruyter.

Marek Kuś

CLASSICAL AND QUANTUM SOURCES OF RANDOMNESS

ABSTRACT. Is there a place for genuine randomness in nature, independent of our possibilities and limitations in acquiring knowledge about the external world? Or in other words, what is the ontological status of randomness and probability? The problem has been present in philosophy nearly since its advent. Not astonishingly, no satisfactory answer has been proposed, the problem recurs in different forms and gains new insights in the course of the development of physical sciences. Undoubtedly, quantum mechanics gave a revolutionary turn to the issue. What is even more interesting, it constantly provokes new problems. It is thus worth returning to the basic questions of the nature of randomness in the light of recent developments of quantum theory and increasing experimental possibilities to check its predictions. Here I want to compare the possibilities given by classical and quantum physics to accommodate and prove the existence of 'genuine randomness' and discuss recent results concerning 'amplification of randomness', showing how quantum physics 'outperforms' classical mechanics in approaching an ultimate goal of exhibiting 'truly random process'.

1. Metaphysical Character of Randomness and Its Practical Significance

An ever-recurrent problem of determinism can be disguised as a question about the metaphysical status of randomness in physical theories. Truly nondeterministic theories should be based on the assumption that randomness has an ontic character, or in other words, it is their intrinsic property, whereas the only way in which random behavior demanding probabilistic description appears in a deterministic theory is due to lack of knowledge. Usually such an ignorance concerns initial conditions since fundamental theories of physics aim at explaining dynamics i.e. evolution of a state of a system in time.

The two conflicting views concerning the origin of randomness in nature are commonly traced back to ancient atomists. For Leucippus and

In: Tomasz Bigaj and Christian Wüthrich (eds.), *Metaphysics in Contemporary Physics* (*Poznań Studies in the Philosophy of the Sciences and the Humanities*, vol. 104), pp. 425-436. Amsterdam/New York, NY: Rodopi | Brill, 2015.

Democritus randomness had an apparently epistemic character, an event is random for us due to our lack of knowledge (Diels 1906; Freeman 1948; Laertius 1925). Hundred years later Epicurus stressed ontic sources of odds and chances. The deterministic motion of atoms is interrupted, without a cause, by "swerves" which give rise to indeterminacy on higher levels (Cicero 1933).

In general, showing that the world is (non-)deterministic seems to be a hopeless task. It is even quite hard to imagine how to attack such a problem. On the other hand, the answer would have an important practical meaning. Randomness is a basic resource for variety of applications. For example provably unbreakable cryptographic systems are based on the assumption that we are able to produce strings of perfectly random, uncorrelated digits. In practice such strings are generated by specialized computer programs able to produce satisfactory 'pseudo-random' outputs. The results are not really random, they are obtained by running a completely deterministic computer program, their 'randomness' is certified by passing several 'randomness tests' checking the degree to which they resemble truly random processes (see Knuth 1969 for a thorough discussion and theory of random numbers generators and randomness tests). Reliability of random number generators depends mainly on the power of employed algorithms, nevertheless it can be degraded by failures of devices, attacks of adversaries having more computational power at their disposal etc. It would be thus desirable to design a random number generator that uses an unpredictable physical process and does not require additional assumptions about the internal structure of the used device. In other words, the task is to find in nature an 'intrinsically' random proces. Random, by its own, and not because we do not know how it came into existence or do not fully control it. Again, gaps in our (or our adversaries') knowledge can be, in principle, closed, as long as there is no genuine randomness in nature. Thus, security of our bank deposits and credit cards secured by cryptographic algorithms, depends on the ontological status of randomness.

In the following I am going to discuss a few concepts, stemming both from classical and quantum mechanics, where to look for sources of genuine randomness. Despite murky prospects of finding a satisfactory answer to the problem, some new developments in quantum information theory and theoretical quantum cryptography, are worth discussing from this point of view.

Let us start with some obvious, albeit necessary, definitions and remarks. The problem of defining determinism is discussed thoroughly in the literature (a reference of choice is here Earman 1986, see also Wüthrich 2011). Shortly, the world ('a world', when possible worlds are admissible) or a physical system is deterministic if and only if to its state at time t there corresponds a unique state at time t' compatible with the laws of evolution. Often one assumes $t' > t$, but extending the conditions to arbitrary t' is, usually, harmless and useful.

At first sight it seems reasonable to distinguish (in)determinism on the level of a real world or a real physical system on one side and as a feature of a physical theory pertaining to it on the other, i.e., we may, in principle, distinguish (in)deterministic worlds from (in)deterministic theories. Unfortunately, definitions of a determinism on the level of a theory rarely avoid epistemic flavor,[1] again blurring the ontic-epistemic distinction. For the purposes of this essay I suggest abandoning the 'world-theory' distinction.[2] A proper theory should adequately represent features of the world. It does not mean that it must be 'exact' in any reasonable sense. As always, it can give an approximate description of reality; what we demand is only that it properly reflects the ontology.

2. Is Classical Mechanics Really Deterministic?

It is commonly believed (or at least, I would say that this is an orthodox view among physicists) that classical mechanics is a deterministic theory in which randomness has an epistemic character, i.e. one adopts (sometimes tacitly) Democritean attitude. A purportedly random process is in fact completely determined and can be predicted with desired accuracy once we improve our measuring devices and the computational power of our computers. A perfect example of a physical theory based on such principles is statistical physics, where measurable quantities like pressure or temperature are determined by mean values of microscopical 'hidden variables' – positions and momenta of gas particles. These hidden variables, however, are completely determined at each instant of time by the laws of classical mechanics, and with enough money and effort can be, in principle, measured with arbitrary accuracy.

The same principle is recovered in the theory of 'deterministic chaos'. where predictions about future states are practically impossible due to sensitive dependence of initial conditions – inaccuracies in determination of the initial state grow exponentially in time. Such phenomena can be used as arguments for indeterminism in classical mechanics (Popper 1982), but

---

[1] 'A theory T is deterministic just in case, given the state description $s(t_1)$ at any time t_1, the state description $s(t_2)$ at any other time t_2 is deducible [in principle] from T' (Earman 1986, p. 20).

[2] The usefulness of such a distinction is also put in doubts by Earman (1986), ch. II-11.

again only of a Democritean character. In comparison with statistical physics, where limits on predictions are due to a large number of constituents, here unpredictability occurs also in systems with only few degrees of freedom. For the problems discussed here this difference is immaterial.[3]

The definition invoking only uniqueness of future states carefully and deliberately avoids any association with predictability. It allows to stay away from the pitfalls of having too many indeterministic theories. Indeed, adopting the unpredictability as the definition of indeterminism we have to admit that every physical system in which the phenomenon of sensitive dependence of dynamics on initial conditions occurs should be treated as indeterministic. Although there is nothing fundamentally wrong in such a concept, it is hard to argue that this gives an ontological status to the chance. Hence, any definition connecting chance or indeterminism (certainty and indeterminism) with predictability is unsatisfactory since it does not allow to tell apart ontologically indeterministic worlds and worlds in which probabilities appear as a consequence of our epistemic weakness.

The determinism of classical mechanics can be decreed by *fiat*. This is an attitude commonly adopted (and defensible) when classical mechanics is presented as a closed deductive system (in fact, something like a specific part of mathematics). Thus Arnol'd in his treatise on ordinary differential equations, after endorsing the very definition of determinism advocated above,[4] continues to write: 'Thus for example, classical mechanics considers the motion of systems whose past and future are uniquely determined by the initial positions and velocities of all points of the system' (ibid). The same can be found in his fundamental exposition of mathematical methods of classical mechanics.[5] Here, however, a deeper motivation is given, 'It is hard to doubt this fact, since we learn it very early' (ibid). But the role of the argument is, it seems, considerably modest. It is invoked to convey the observation that the states of a mechanical system are uniquely determined solely by the positions and momenta of its constituents, as the rest of the paragraph makes clear: 'One can imagine a world in which to determine

[3] In fact the sensitiveness to changes of initial conditions was treated as a source of chance by those who tried, back at the beginning of the 20th century, to give a firm basis to statistical physics on the ground of classical mechanics (Poincaré 1912, Smoluchowski 1918). However, it was clear for Poincaré that such a method of grounding statistical reasoning has purely epistemic character. In the same Introduction to his *Calcul des probabilités* he remarks, '*Il faut donc bien que le hasard soit autre chose que le nom que nous donnons à notre ignorance*' (Poincaré 1912, p. 3).

[4] 'A process is said to be *deterministic* if its entire future course and its entire past are uniquely determined by its state at the present instant of time' (Arnol'd 1973).

[5] 'The initial state of a mechanical system (the totality of positions and velocities of its points at some moment of time) uniquely determines all of its motion' (Arnol'd 1989).

the future of a system one must also know the acceleration at the initial moment, but experience shows us that our world is not like this' (ibid.). Very similar statements can be found in Landau and Lifschitz's *Mechanics*:

> If all the co-ordinates and velocities are simultaneously specified, it is known from experience that the state of the system is completely determined and that its subsequent motion can, in principle, be calculated. Mathematically, this means that, if all the co-ordinates q and velocities $\dot{q}$ are given at some instant, the accelerations $\ddot{q}$ at that instant are uniquely defined. (Landau 1960).

Here, again, 'experience' concerns merely a specific feature of classical mechanics, which needs only positions and velocities and not higher time-derivatives of these to determine states of a system.

All this is in a perfect accordance with Newton's laws of mechanics, in particular with the mathematical form of the Second Law, under an additional assumption that the differential equations connecting accelerations and forces have unique solutions giving trajectories of the motion. Technically such a uniqueness can be guaranteed by imposing certain conditions (e.g. the Lipschitz condition) limiting the variability of the forces with respect to the positions. Breaking them can lead to circumstances where for some initial positions and velocities the future trajectory is not uniquely determined.

A seemingly technical detail that some differential equations do not always have unique solutions, attracted recently considerable attention. A simple mechanical system admitting non-unique solutions for some initial conditions was presented by Norton in a remarkable paper (Norton 2007, see also Norton 2008). Norton's example is that of a massive point particle sliding on a dome of a particular shape. The non-uniqueness of solutions is reflected by the possibility of initiating, at any arbitrary moment of time and without any external cause, a movement of a resting particle. The example triggered a sequence of papers (Korolev 2007, Korolev 2008, Kosyakov 2008, Malament 2008, Roberts 2009, Wilson 2009, Zinkernagel 2010, Fletcher 2012, Laraudogoitia 2013), discussing several aspects of the model, especially its practical relevance. However, the real problem is whether classical mechanics admits situations where uniqueness is not guaranteed.[6] The virtue of the Norton example stems from the construction of a model which is 'very close' to what can be encountered in reality and fits very well to all schemes of approximating real physical systems by collections of point particles, etc. To defend determinism, each similar

[6] For the particular case of Norton's dome the question would be 'does classical mechanics allows the force $F(x) = \sqrt{|x|}$ as legitimate?', since mathematically this is the essence of the Norton Dome model.

example should be treated separably by attacking its weak points. Without any guiding principle such a baring of 'individual monsters' lacks methodological refinement.

Clearly a world in which there are systems governed by equations admitting non-unique solutions is not deterministic according to our definition – we are brutally confronted with a situation when the laws of evolution allow for various futures.[7] Consequently, classical mechanics is an incomplete theory leaving undetermined the fate of such systems. This does not mean that there is necessarily a room for introducing a probabilistic ingredient and the awaited 'intrinsic randomness' to the reasoning. To this end we need to postulate an additional law supplementing classical mechanics by attributing probabilities to different solutions of non-Lipschitzian equations.[8] It is hard to see how to discover (or even look for) such a law, and how to check its validity.

Summarizing, classical mechanics in its commonly accepted form, i.e. without imposing *ad hoc* additional (usually technical) constraints, is an intrinsically non-deterministic theory according to the definition adopted in this essay. This is not because mechanics admits systems which are chaotic due to instabilities, but because of indeterminacies occurring in some well defined systems. As such, it leaves some room for 'genuine randomness' but without any clear prospects of proving that such a randomness is really present in the world and can be used to extract 'truly random' sequences from experiments.[9]

---

[7] Arguments that such situations are in some sense 'rare' are misguided. Estimation how rare are unacceptable cases is not possible without imposing a concrete measure in the space of functions representing forces. There is no 'natural' measure in the space of functions and, consequently, no genuine and unconditioned notion of rarity of undesirable forces.

[8] Thus in the case of Norton's dome, the new law of nature should, in particular, ascribe a probability $p(T)$ to the event that the point staying at rest at the apex starts to move at time T.

[9] The standard theorems concerning the existence and uniqueness of solutions ensure these properties only locally, i.e. it can happen that a solution exists only for finite times. Trajectories of N point particles moving under the laws of Newtonian physics can become unbounded and reach infinity in a finite time even if locally the solutions of equations of motion are unique (McGehee 1980). Again, using time reversal symmetry of classical mechanics we can put in doubt its deterministic character – if particles are able to reach infinity at any instant of time, we should expect that at some instants 'space invaders' may unexpectedly appear in our vicinity. Here, however, classical mechanics is at variance with a superseding more universal theory, the special theory of relativity, which does not allow reaching boundaries of an infinite universe in a finite time due to limitations it poses on the maximal possible velocity of motion. The situation may change if the universe is finite, but a more detailed analysis taking into account relativistic effects is needed.

3. Is Quantum Mechanics Really Non-Deterministic?

With the advent of the quantum theory it became clear that we can count only on a probabilistic description of reality, but initially there were no reasons to switch to the Epicurean view. It seemed we were merely confronted with an incomplete theory admitting deterministic hidden variables, values of which were beyond our control. Bell's theorem showed incompatibility of hidden-variable theories with quantum mechanics (Bell 1964, Bell 1966). Impossibility of instantaneous, or breaking the speed of light limit communication between spatially separated systems ('no-signaling') and full determinism imply that all correlations between results of measurements must be local, i.e. have to obey Bell's inequalities. Let us remind that Bell inequalities impose limitations on the values of particular combinations of joint probabilities $P(x, y, \dots , z \mid X, Y, \dots , Z)$ of obtaining $x, y, \dots , z$ when we chose to measure observables $X, Y, \dots , Z$. Non-signaling is guaranteed by the fact that the measurements of different observables are performed on different spatially separated subsystems, hence the choice of an observable to be measured in a subsystem is not influenced by what and with which results is measured in other places.

Reversing the argument we conclude that, again under the assumption of non-signalling, the existence of non-local (i.e. breaking Bell's inequalities) correlations implies that there is no deterministic hidden variables theory explaining observed quantum phenomena. We are thus forced to admit that quantum mechanics is intrinsically random and only superfluously resembles statistical physics.

At first sight physics provided thus a model situation of two theories making completely different assumptions about the metaphysical status of chance and probability. Democritean classical mechanics and enforced by our ignorance statistical physics are contrasted with Epicurean, 'intrinsically random' quantum mechanics. The chances of proving the existence of 'intrinsic randomness' in the world seem thus to be much higher when we switch to quantum mechanics. What we need is to show experimentally that Bell's inequalities can be violated. This would constitute a proof of a non-deterministic nature of quantum mechanical reality and allow certifying the existence of truly random processes. Such experiments were indeed performed (Aspect et al. 1982). The problem is that to provide conclusive results, such experiments require random adjustments of measuring devices (Bell 1964). We need a truly random process controlling the choice of what we measure at any given run of the experiment. Otherwise, as shown by (Brans 1988), if measurement settings and measured states of a system are determined by some hidden variables sufficiently early,[10] one can find a distribution of the hidden variables giving the same statistical predictions as quantum mechanics, but on a completely deterministic basis.

It is important to stress that resorting to some stopgaps, for example relaxing slightly the requirement that measurements settings can be chosen completely freely, may have damaging consequences for certifying randomness *via* such experiments. In (Hall 2010 and Koh et al. 2012) the authors show that reducing the 'amount of free will' i.e. the possibility of choosing freely what is measured below some calculable threshold reduces the possibility to produce a truly random sequence of bits from the results of the performed measurements. Summarizing, if the choice of measurements is not free enough, quantum correlations can be explained by some hidden variable deterministic theory like in the case of thermodynamics and classical statistical physics. This, ironically, closes an unavoidable *circulus vitiosus*. We can check the indeterministic character of the physical reality only assuming that it is, in fact, indeterministic.

What can be done instead? It seems that the optimal result one can hope to establish is the following (Gallego et al. 2013). Assume we have a source producing an arbitrarily small randomness (what it means will be specified below). Is it possible to prove that there exist in nature a completely random process? In other words, is it possible to 'amplify randomness', i.e., to produce a perfectly random process using as input a source producing a process arbitrary close to a deterministic one? Does quantum mechanics outperforms here, in providing such a certification of randomness, the classical physics? If the answer to the first question is positive, we may claim to establish the following alternative: either the world is completely deterministic (there is no, even arbitrarily tiny amount of randomness in dynamical processes), or there are events which are completely random, thus, for practical reasons not only unpredictable, but unpredictable in principle, independently of an available computational power and other means or resources.

To make the reasoning more concrete one considers the following scenario. Let us assume we have a source producing a sequence b_1, b_2, ... , of random bits (random numbers taking two possible values, say 0 and 1). The consecutive outputs may depend on the previous ones and also on some other external factors, in any case at the moment of producing the

[10] The events of production of correlated subsystem like, e.g., decay of a quantum system into two subsystems, as well as all relevant events of measurements should be within the future light cone of the event at which the whole laboratory equipment was prepared and its future determined by some distribution of the hidden variables.

output b_k we know all the preceding results and the state of the rest of the universe possibly influencing the value of b_k. Nevertheless we assume that there is some amount of randomness in each event, i.e. the probability that b_k is equal 0 fulfills

(1) $1/2 - \varepsilon \leq P(b_k = 0 \mid \Lambda) \leq 1/2 + \varepsilon$

Here $P(b_k = 0 \mid \Lambda)$ is the conditional probability that $b_k = 0$ under the assumption of Λ, symbolizing all results of previous events and values of all other relevant parameters. The value $\varepsilon = 0$ corresponds to full randomness, when the probabilities of having $b_k = 0$ and $b_k = 1$ are both equal 1/2. On the other extreme, for $\varepsilon = 1/2$, the condition (1) is trivially fulfilled by arbitrary probabilities $P(b_k = 0 \mid \Lambda)$. We do not know anything about the source, in principle it can be even fully deterministic. The process of randomness amplification described in the preceding paragraph consists of starting from some $\varepsilon > 0$ (the closer the value of ε is to 1/2, the smaller the amount of randomness in the process is) and using the obtained outputs $b_1, b_2, \ldots$, to produce a similar random sequence, but with $\varepsilon' < \varepsilon$, i.e. with a larger amount of randomness. If the output probability distribution of the source were explicitly known and not merely restricted by inequalities (1), it would be possible to extract a sequence of fully random bits of the length limited by the entropy of the source. This is a consequence of Shannon source coding theorem (Shannon 1948). A model situation is that of producing consecutive bits by flipping a biased coin with a constant bias i.e. a fixed difference between probabilities of outputting ‘heads’ (corresponding to, say, 0) and ‘tails’ (corresponding to 1). In a general case encompassed by (1) we are confronted with a more complicated situation. Here we may imagine that the bias is set before a consecutive flip by an ‘adversary’ who has the full access to the history of the whole experiment and tries to destroy our efforts to produce fully random bits, limiting himself only by conditions (1). It was proved by Santha and Vazirani (1986) that an amplification of randomness in the previously described sense is not possible in this situation using only classical means. More precisely, they showed that there is always a strategy that can be adopted by our ‘adversary’ to thwart our efforts to produce fully random bits *via* an algorithm applied to the sequence of results obtained up to know. Such an amplifying algorithm can be thought as some function,

(2) $f: \{0, 1\}^n \rightarrow \{0, 1\}$,

from the set $\{0, 1\}^n$ of the possible n outputs obtained from the source to the set $\{0, 1\}$ of the values of the fully random bit. Thus we would like f to produce 0 or 1 from the values of all previously outputted bits in such a way that the obtained bit is ‘more random’, i.e., the probability of obtaining 0 is restricted by $\varepsilon' < \varepsilon$. A strategy of the adversary consists of applying a map

(3) $S : \{0, 1\}^k \rightarrow [1/2 - \varepsilon, 1/2 + \varepsilon]$

from the set of possible previous k outputs (to which the adversary has the full access) to the interval $[1/2 - \varepsilon, 1/2 + \varepsilon]$ such that

(4) $S(x)$ = Probability(next bit is 0 | until now the outputted sequence was x)

For an arbitrary function f there exists a strategy such that the probabilities for new sequence of bits produced by f will obey (1) with the same value of ε. The new sequence will not be better than the original one. In general thus one can not amplify randomness by classical means, i.e., by manipulating bits using a boolean function (2).

It is tempting to interpret this result as a consequence of the fact that since the classical world is deterministic, there is no source of 'additional' randomness that can be used to amplify an imperfect one. Going one step further we can even claim that the possibility of amplifying randomness is the defining property of an 'intrinsically indeterministic' (or 'intrinsically random') system. Such a definition is, admittedly, a bit cavalier but at least it draws a clear border between classical and quantum indeterminism.

Indeed, Colbeck and Renner (2012) showed that quantum mechanics offers a possibility of randomness apmplification. Originally, the authors were able to prove that if we start above some threshold of initial amount of randomness ($\varepsilon_0 < 0.086$), we are able to perform the amplification procedure up to an arbitrary small ε thus obtaining a perfectly random sequence. Recently (Gallego et al. 2013), the threshold was pushed to an arbitrary $\varepsilon < 1/2$ justifying the above mentioned alternative of a fully deterministic world or one admitting fully random events. The main idea is to use the sequence of bits obtained from an imperfect source (1) to steer a collection of Bell-type experiments. The outputs from the source are used to choose measurements settings among two possibilities (e.g. of measuring one of two observables) encoded as 0 and 1. After a sequence of measurements it is checked whether the appropriate Bell's inequalities are violated. If the verification is positive one combines the measurement results to obtain a new bit, hopefully 'more random' than the ones we have at our disposal up to now. The authors of (Gallego et al. 2013) prove that if we have at our disposal at least five 'laboratories' in which such experiments could be performed, i.e., measuring devices can be put in two settings chosen in accordance with outputs of the source, and the non-signalling principle is obeyed, one can construct a protocol producing fully random bits from the measurement outputs.[11] In the spirit of the above proposed definition, quantum mechanics is thus 'intrinsically random'.

Summarizing, classical physics can accommodate genuine randomness but not without a heavy price of imposing new laws of nature, moreover without giving hints how and where we should look for such laws. Quantum mechanics does it more happily, but without ultimate conclusiveness; we are not in a comfortable situation where security is warranted by genuine randomness,[12] but it is probably as close as possible to it. In any case quantum realm seems to give a warmer environment for random processes by permitting unconstrained amplification of randomness forbidden in the classical world.

Polish Academy of Sciences
Center for Theoretical Physics
e-mail: marek.kus@cft.edu.pl

ACKNOWLEDGEMENTS

This publication was made possible through the support of a grant from the John Templeton Foundation.

REFERENCES

Arnol'd, V.I. (1973). *Ordinary differential equations.* Cambridge: MIT Press.

Arnol'd, V.I. (1989). *Mathematical methods of classical mechanics.* New York: Springer.

Aspect, A., Dalibard, J., Roger, G. (1982). Experimental Test of Bell's Inequalities Using Time-Varying Analyzers. *Physical Review Letters* **49**, 1804–1808.

Bell, J.S. (1964). On the Einstein-Podolsky-Rosen paradox. *Physics* **1**, 195–200.

Bell. J.S. (1966). On the problem of hidden variables in quantum mechanics. *Reviews of Modern Physics* **38**, 447–452.

Brans, C.H. (1988). Bell's theorem does not eliminate fully causal hidden variables. *International Journal of Theoretical Physics* **27**, 219–226.

Colbeck, R., Renner, R. (2012). Free randomness can be amplified. *Nature Physics* **8**, 450–453.

Diels, H. (1906). *Die fragmente der Vorsokratiker griechisch und deutsch.* Berlin: Weidmannsche Buchhandlung.

[11] It is possible to give a general scheme for amplifying randomness in systems consisting of *N* subsystems each on which one can perform *K* different measurements (Kuś, forthcoming).

[12] If we feel that the proposed definition connecting 'intrinsic randomness' with the possibility of amplifying imperfect randomness is unsatisfactory.

Diogenes Laertius. (1925). *Lives of Eminent Philosophers, translated by R.D. Hicks*. Vol. 2. Cambridge: Harvard University Press.

Earman, J. (1986). *A Primer on Determinism*. Dordrecht: Reidel.

Fletcher, S.C. (2012). What counts as a Newtonian system? The view from Norton's dome. *European Journal for Philosophy of Science* **2**, 275–297.

Freeman, K. (1948). *Ancilla to the pre-Socratic philosophers*. Cambridge: Forgotten Books.

Gallego, R.L., et al. (2012). Full randomness from arbitrarily deterministic events. *Nature Communication* **4**, 2645.

Hall, M.J.W. (2010). Local Deterministic Model of Singlet State Correlations Based on Relaxing Measurement Independence. *Physical Review Letters* **105**, 250404.

Knuth, D.E. (1969). *The Art of Computer Programming II: Seminumerical Algorithms*. Reading: Addison-Wesley.

Koh, D.E. (2012). Effects of reduced measurement independence on Bell-based randomness expansion. *Physical Review Letters* **109**, 160404.

Korolev, A. (2007). Indeterminism, asymptotic reasoning, and time irreversibility in classical physics. *Philosophy of Science* **74**, 943–956.

Korolev, A. (2008). The Norton-type Lipschitz-indeterministic systems and elastic phenomena: Indeterminism as an artefact of infinite idealizations. *Phil. of Sci. Assoc. 21st Biennial Mtg. Pittsburgh, PA: PSA 2008 Contributed Papers*.

Kosyakov, B.P. (2008). Is classical reality completely deterministic? *Foundations of Physics* **38**, 76–88.

Kuś, M. (forthcoming). A General Scheme for Randomness Amplification.

Landau, L.D., Lifshitz, E.M. (1960). *Classical mechanics*. Oxford: Pergamon Press.

Laraudogoitia, J.P. (2013). On Norton's Dome. *Synthese* **190**, 2925–2941.

Malament, D.B. (2008). Norton's slippery slope. *Philosophy of Science* **75**, 799–816.

Marcus Tullius Cicero. (1933). *De Natura Deorum, English translation by H. Rackham*. Cambridge: Harvard University Press.

McGehee, R.R. (1980). Singularities in classical celestial mechanics. *Proceedings of the International Congress of Mathematicians Helsinki 1978*.

Norton, J.D. (2003). Causation as folk science. *Philosophers' Imprint* **3** (4). Reprinted in: H. Price and R. Corry (eds.), *Causation, physics, and the constitution of reality: Russell's republic revisited*, pp. 11–44. New York: Oxford University Press.

Norton, J.D. (2008). The dome: An unexpectedly simple failure of determinism. *Philosophy of Science* **75**, 86–798.

Poincaré, H. (1912) *Calcul des probabilités*. Paris: Gauthier-Villars.

Popper, K.R. (1982). *The Open Universe: An Argument for Indeterminism*. London and New York: Routledge.

Roberts, B.W. (unpublished). Wilson's case against the dome: Not necessary, not sufficient.

Santha, M., Vazirani, U.V. (1986). Generating quasi-random sequences from semi-random sources. *Journal of Computer and System Sciences* **33** (1), 75–87.

Shannon, C.E. (1948). A Mathematical Theory of Communication. *Bell System Technical Journal* **27**, 379–423.

Smoluchowski, M. (1918). Über den Begriff des Zufalls und den Ursprung der Wahrscheinlichkeitsgesetze in der Physik. *Naturwissenschaften* **6**, 253–263.

Wilson, M. (2009). Determinism and the mystery of the missing physics. *The British Journal for the Philosophy of Science* **60**, 173–193.

Wüthrich, C. (2011). *Can the world be shown to be indeterministic after all? Probabilities in Physics*. Oxford: Oxford University Press, Oxford.

Zinkernagel, H. (2010). Causal fundamentalism in physics. In: M. Suárez, M. Dorato, M. Rédei (eds.), *EPSA Philosophical Issues in the Sciences*, vol. 2, pp. 311–322. Dordrecht: Springer.

Jeremy Butterfield
Nazim Bouatta

RENORMALIZATION FOR PHILOSOPHERS

ABSTRACT. We have two aims. The main one is to expound the idea of renormalization in quantum field theory, with no technical prerequisites (Sections 2 and 3). Our motivation is that renormalization is undoubtedly one of the great ideas – and great successes – of twentieth-century physics. Also it has strongly influenced, in diverse ways, how physicists conceive of physical theories. So it is of considerable philosophical interest. Second, we will briefly relate renormalization to Ernest Nagel's account of inter-theoretic relations, especially reduction (Section 4).

One theme will be a contrast between two approaches to renormalization. The old approach, which prevailed from ca. 1945 to 1970, treated renormalizability as a necessary condition for being an acceptable quantum field theory. On this approach, it is a piece of great good fortune that high energy physicists can formulate renormalizable quantum field theories that are so empirically successful. But the new approach to renormalization (from 1970 onwards) explains why the phenomena we see, at the energies we can access in our particle accelerators, are described by a renormalizable quantum field theory. For whatever non-renormalizable interactions may occur at yet higher energies, they are insignificant at accessible energies. Thus the new approach explains why our best fundamental theories have a feature, viz. renormalizability, which the old approach treated as a selection principle for theories.

That is worth saying since philosophers tend to think of scientific explanation as only explaining an individual event, or perhaps a single law, or at most deducing one theory as a special case of another. Here we see a framework in which there is a space of theories. And this framework is powerful enough to deduce that what seemed "manna from heaven" (that some renormalizable theories are empirically successful) is to be expected: the good fortune is generic.

We also maintain that universality, a concept stressed in renormalization theory, is essentially the familiar philosophical idea of multiple realizability; and that it causes no problems for reductions of a Nagelian kind.

In: Tomasz Bigaj and Christian Wüthrich (eds.), *Metaphysics in Contemporary Physics* (*Poznań Studies in the Philosophy of the Sciences and the Humanities*, vol. 104), pp. 437-485. Amsterdam/New York, NY: Rodopi | Brill, 2015.

1. Introduction

We have two aims. The main one is to expound the idea of renormalization in quantum field theory, with no technical prerequisites (Sections 2 and 3). Our motivation is that renormalization is undoubtedly one of the great ideas – and great successes – of twentieth-century physics. Also it has strongly influenced, in diverse ways, how physicists conceive of physical theories. So it is of considerable philosophical interest. Second, we will briefly relate renormalization to Ernest Nagel's account of inter-theoretic relations, especially reduction (Section 4).[1]

One main point will turn on a contrast between two approaches to renormalization. The traditional approach was pioneered by Dyson, Feynman, Schwinger and Tomonaga in 1947–50: they showed how it tamed the infinities occurring in quantum electrodynamics, and also agreed with experiments measuring effects due to vacuum fluctuations in the electromagnetic field – even to several significant figures. After these triumphs of quantum electrodynamics, this approach continued to prevail for two decades. For this paper, the main point is that it treats renormalizability as a necessary condition for being an acceptable quantum field theory. So according to this approach, it is a piece of great good fortune that high energy physicists can formulate renormalizable quantum field theories that are so empirically successful; as they in fact did, after about 1965, for forces other than electromagnetism – the weak and strong forces.

But between 1965 and 1975, another approach to renormalization was established by the work of Wilson, Kadanoff, Fisher etc. (taking inspiration from ideas in statistical mechanics as much as in quantum field theory). This approach explains why the phenomena we see, at the energies we can access in our particle accelerators, are described by a renormalizable quantum field theory. In short, the explanation is: whatever non-renormalizable interactions may occur at yet higher energies, they are insignificant at accessible energies. Thus the modern approach explains why our best fundamental theories have a feature, viz. renormalizability, which the traditional approach treated as a selection principle for theories. (So to continue the metaphor above: one might say that these theories' infinities are not just tamed, but domesticated.)

[1] That Section is brief because one of us (J.B.) discusses Nagelian themes more fully in a companion paper (2014). From now on, it will be clearest to use 'we' for a contextually indicated community e.g. of physicists, as in 'our best physical theories'; and 'I' for the authors, e.g, in announcing a plan for the paper like 'In Section 2, I will'.

That point is worth making since philosophers tend to think of scientific explanation as only explaining an individual event, or perhaps a single law, or at most deducing one theory as a special case of, or a good approximation of, another. This last is of course the core idea of Nagel's account of inter-theoretic reduction. The modern approach to renormalization is more ambitious: it explains, indeed deduces, a striking feature (viz. renormalizability) of a whole class of theories. It does this by making precise mathematical sense of the ideas of a *space* of theories, and a flow on the space. It is by analyzing this flow that one deduces that what seemed "manna from heaven" (that some renormalizable theories are empirically successful) is to be expected: the good fortune we have had is generic.

But I will urge that this point is not a problem for Nagel's account of inter-theoretic relations. On the contrary: it is a striking example of the power of Nagelian reduction. And I will end with an ancillary point, which also strikes a Nagelian note. I will argue that universality, a concept stressed in renormalization theory, is essentially the familiar philosophical idea of multiple realizability; and I will claim (following Sober, Shapiro and others) that multiple realizability does not cause problems for reductions of a Nagelian kind.

The plan is as follows. I sketch the old and new approaches to renormalization in Sections 2 and 3.[2] Then in Section 4, I shall maintain that these developments accord with Nagel's doctrines.

2. Renormalization: The Traditional Approach

2.1. *Prospectus: Corrections Needed*

Consider a classical point-particle acting as the source of a gravitational or electrostatic potential. There is no problem about using the measured force F felt by a test-particle at a given distance r from the source, to calculate the mass or charge (respectively) of the source particle.

[2] I will give very few references to the technical literature; as perhaps befits a primer. But I recommend: (i) Baez's colloquial introductions (2006, 2009), of which Sections 2 and 3 are an expansion into more academic prose; (ii) Wilson's *Scientific American* article (1979) and Aitchison (1985)'s introduction to quantum field theory, especially its vacuum, which discusses renormalization in Sections 3.1, 3.4, 3.6, 5.3, 6.1; (iii) Teller (1989) and Hartmann (2001) as philosophical introductions; (iv) Kadanoff's masterly surveys (2009, 2013), which emphasize both history and aspects concerning condensed matter – here treated in Section 3.3.

Thus in the electrostatic case, for a test-particle of unit charge, the force is given by minus the derivative of the potential energy V with respect to the distance r between the source and the test-particle. In symbols, this is, for a source of charge e (neglecting constants): $F = -\nabla V(r) \sim -\nabla - e/r \equiv -e/r^2$. We then invert this equation to calculate that the source's charge is: $e = -F.r^2$. (Adding in the constants: $F = -e/(4\pi\varepsilon_0 r^2)$, where ε_0 is the permittivity of free space (electric constant), implies that $e = -F(4\pi\varepsilon_0 r^2)$.)

This straightforward calculation of the source's mass or charge does not work in quantum field theory! There are complicated corrections we must deal with: perhaps unsurprisingly, since it amounts to trying to characterize one aspect of an interacting many-field system in a way that is comparatively simple and independent of the rest of the system. The corrections will depend on the energy and-or momentum with which the test particle approaches the source. A bit more exactly, since of course the test particle and source are equally minuscule: the corrections depend on the energy (or momentum) with which we theoretically describe, or experimentally probe, the system I called the 'source'. We will write μ for this energy; and the corrections depending on μ will be centre-stage in both the traditional and the modern approaches to renormalization (this Section and the next).

This Section lays out the traditional approach in four subsections. In Section 2.2, I introduce the idea that the medium or field around the source can affect the observed value of its mass or charge. In quantum field theory, this is often expressed in the jargon of "virtual states" or "virtual particles." Again: it is a matter of the energy-scale μ at which we describe or probe the source.

Then in Section 2.3, I report that to get finite predictions, quantum field theory needs a regularization scheme. The archetypal scheme is to neglect energies above a certain value Λ; equivalently, one neglects variations in fields that occur on a spatial scale smaller than some small length d. I also adopt length as the fundamental dimension, so that I express regularization as a cut-off length d, rather than an energy Λ.

In Section 2.4, I present the core task of the traditional approach to renormalization. Since the theory assumes spacetime is a continuum, while d is our arbitrary choice, we need to show consistency while letting d tend to 0. That is: we must find an assignment of intrinsic charges (electric charge, mass etc.: called *bare coupling constants*), to the sources, as a function of the diminishing d, which delivers back the observed value of the charges: i.e. the values we in fact measure at the energy-scale μ at which we probe the system. These measured values are called the *physical coupling constants*. If we can do this, we say our theory is *renormalizable*.

This requirement is weak, or liberal, in two ways. First: we even allow that the assigned intrinsic charge is infinite at the limit $d \to 0$. (It is this allowance that a bare coupling constants be infinite that makes many – including great physicists like Dirac – uneasy.)

Second (Section 2.5): we allow that we might have to add other terms to our theory (to be precise: to the Lagrangian or Hamiltonian), in order to make a consistent assignment. But we only allow a finite number of such terms: this reflects the fact that our framework of calculation is perturbative.

Then in Section 2.6, I report (the simplest rendition of) Dyson's criterion for when a theory is renormalizable: the dimension (as a power of length) of the bare coupling constant(s) needs to be less than or equal to zero. Finally, I report the happy news that our theories of the electromagnetic, weak and strong forces are in this sense renormalizable. *Why* we should be so fortunate is a good question: which, as announced in Section 1, I will take up in Section 3.

2.2. *Renormalizing a Coupling Constant*

Underlying the details to come, there is a simple idea, which is valid classically and indeed an everyday experience. Imagine a ping pong ball under the surface of the water in a bath. It is buoyant: in terms of the gravitational field, it has a negative mass. So the idea is: the medium in which a system is immersed can alter the parameters associated with the system, even the parameters like an electric or gravitational charge, i.e. its coupling constants.

Agreed, in this example we of course retain the notion that the ball has a positive intrinsic mass, not least because it can be taken out of the water and then will fall under gravity. But three factors make this notion more problematic in fundamental physics, especially in quantum theory.

(i) We cannot take the system out of the field of force, which is all-pervasive, though varying in strength from place to place.

(ii) Even in classical physics, there are important differences between the electromagnetic and gravitational fields: differences that make it wrong, or at least more questionable, to conceptually detach an electric charge from the electromagnetic field, than it is to detach a mass from the gravitational field. In short, the difference is that the gravitational field, but not the electromagnetic field, can be thought of as a mere mathematical device giving the disposition of a test-particle to accelerate: *cf.* the field V above. On the other hand, the electromagnetic field has energy and momentum, and effects propagate through it at a finite speed: this leads to subtle issues about the self-energy of a classical charged particle.[3]

(iii) In quantum field theory, matter is represented by a quantum field, just as radiation (electromagnetism) is. The matter is represented by a fermionic field; e.g. in quantum electrodynamics, the electron field. And interactions (forces) between matter are represented by a bosonic field; e.g. in quantum electrodynamics, the quantized electromagnetic field, whose excitations are photons. In short: the physical system is an interacting many-field system, so that it makes little sense to conceptually detach one of the fields from the others. All the more so, if we think of our fields as effective, not fundamental: I return to this in Section 3.

In short: we need to take seriously, in our theoretical description as much as our experimental practice, that the system of interest, e.g. an electron (or excitation of the electron field), is immersed in a wider system through which we "access" it. This has two aspects which we need to spell out. The second is more briefly stated, and is fundamental: it will dominate the sequel. But the first sets the stage.

2.2.1. *Virtual Particles and Perturbation Theory*

First, we need the ideas of: *virtual states*, also often called *virtual particles*, which arise in the context of *perturbation theory*.

In quantum theory, we typically solve a problem by finding the states of definite energy and their corresponding values of energy. These are the energy eigenstates, i.e. eigenstates of the Hamiltonian (energy-function), and their eigenvalues. For once these are known, most of what we might want to know can be calculated. Usually, we cannot exactly calculate the eigenstates: the Hamiltonian is intractable. (The Hamiltonian H is essentially a matrix, and calculating the eigenstates and eigenvalues is a matter of changing bases in the vector space of states so as to render the Hamiltonian matrix diagonal i.e. to make all non-diagonal entries zero.) But often enough, H is "close" to another Hamiltonian, H_0 say, which *is* tractable, in that we can calculate H_0's eigenstates. Here, closeness means, roughly, that there is an additional Hamiltonian matrix H_i such that $H = H_0 + \varepsilon H_i$ where ε is a small number. Since ε is small, H_0 and H are approximately equal, $H_0 \approx H$. We can then write the desired eigenstates of H as superpositions (weighted sums of) of the eigenstates of H_0 (which, recall, we

[3] For these issues, cf. Zuchowski (2013) and references therein. Broader philosophical aspects of classical fields are discussed by Hesse (1965), McMullin (2002) and Lange (2002).

can calculate). Thus $|\psi_a\rangle = \Sigma_j c_j |\psi^0{}_j\rangle$: where a labels the real Hamiltonian's eigenvalue (meaning just that $H|\psi_a\rangle = a|\psi_a\rangle$); j labels the various eigenvalues of H_0, whose eigenstates are the $|\psi^0{}_j\rangle$; and our task is to calculate the complex numbers c_j. It is these eigenstates of H_0 that are called *virtual states* or *virtual particles*.

This jargon is especially used in quantum field theory, where the real Hamiltonian is usually complicated enough to force us to appeal to *perturbation theory*. This is a general framework for solving intractable problems by treating them as small adjustments to tractable problems: e.g. by adding a term, εV say, where ε is a small number and V a potential function, to the governing equations of the tractable problem. One then tries to calculate the quantity of interest (in our example, one of the c_j) by expressing it as a power series $\Sigma^\infty{}_n \alpha^n A_n$, where α is small, i.e. less than one, so that $\alpha^n \to 0$ as $n \to \infty$. Here, α may be the original ε, or some related number. The hope is that α^n tending to 0 will make the terms $\alpha^n A_n$ for higher values of n go to 0. If we are lucky, the first few terms of the series will give us an answer that is accurate enough for our purposes; and if we are very lucky, the series may even converge to the exact answer (i.e. the limit of the successive partial sums $\Sigma^N{}_n \alpha^n A_n$ is finite and is the exact answer). Whether these hopes are realized will of course depend on the A_n not growing too quickly.

I should stress immediately that in quantum field theory, the success of this sort of perturbative analysis is mixed. On the one hand, there is astounding success: in some cases, in our best theories, the first few terms of such a series give an answer that is astonishingly accurate. It matches the results of delicate experiments to as much as ten significant figures, i.e. one part in 10^{10}. That is like correctly predicting the result of measuring the diameter of the USA, to within the width of a human hair! For example, this accuracy is achieved by the prediction in quantum electrodynamics of the magnetic moment of the electron; (Feynman 1985, pp. 6–7, 115–119; Schweber 1994, pp. 206f.; Lautrup and Zinkernagel 1999).

On the other hand, there are serious mathematical problems. It is not just that, in general, the power series expressions we use, even in our best theories, are not known to converge, and are sometimes known not to converge. There are two deeper problems, which I have not yet hinted at.

The first concerns the mathematical definition of the interacting quantum field theory, which our perturbative approach with its various power series is aiming to approximate. Unfortunately, we do not at present have a rigorously defined interacting quantum field theory, for a four-dimensional spacetime. There are such theories for lower spacetime dimensions; and there has been much effort, and much progress, towards the goal. But so far, it remains unattained. One brief way to put the chief difficulty is to say that the central theoretical notion of a quantum field theory, the *path integral* (also known as: *functional integral*) – which is what our power series aim to approximate – has at present no rigorous mathematical definition, except in special or simple cases such as there being no interactions.[4]

The second problem is related to the first; it indeed, it is part of it. But the second problem specifically concerns the perturbative approach with its power series, and will be centre-stage in this paper. So it is best stated separately. In short: not only do the power series usually fail to converge; also, the factors A_n (in the successive terms $\alpha^n A_n$) are often *infinite*. Thus the worry that the A_n might 'grow too quickly' for the power series to converge, as I put it above, was a dire under-statement. Nevermind A_n being so large for large n that the series might diverge: the problem is really that each *term* $\alpha^n A_n$ is infinite! This is quantum field theory's notorious *problem of infinities*: which, as we will see, is addressed by renormalization. Why the A_n are infinite, and how renormalization addresses this by introducing a cut-off and then analysing what happens when the cut-off tends to a limit, will be taken up in Section 2.3 et seq. For the moment, I just confess at the outset that the overall problem of infinities will *not* be fully solved by renormalization, even by the modern approach (Section 3). The infinities will be tamed, even domesticated: but not completely eliminated.[5]

As an example of treating interactions in quantum theory using perturbation theory, let us consider an electron immersed in, and so interacting with, an electromagnetic field. Here, the electron need not be described by a quantum field; it can be described by elementary quantum mechanics; but we consider the electromagnetic field to be quantized. We take it that we can solve the electron considered alone: that is, we can diagonalize its Hamiltonian H_e say – this is essentially what Schrödinger did in 1925.

And we take it that we can solve the field considered alone: that is, we can diagonalize its Hamiltonian H_f – this is essentially what Dirac did in 1927. But the interaction means that the Hamiltonian of the total system, electron plus field, is not just the (tractable!) sum of H_e and H_f, call it H_0: $H_0 := H_e + H_f$. In terms of eigenstates: the energy eigenstates of the total system are not just products of those of the electron and the field; and so the total system's energy eigenvalues are not just sums of the individual eigenvalues.

[4] For a glimpse of these issues, cf. e.g. Jaffe (1999, 2008), Wightman (1999).

[5] But that is perhaps unsurprising since, as I said, this second problem is part of the first. So if it were fully solved, so might be the first problem also.

But happily, the interaction is rather weak. We can write the total Hamiltonian as $H = H_0 + \varepsilon H_i$, where H_i represents the interaction and ε being a small number represents its being weak; and then embark on a perturbative analysis. In particular, we may expand an energy eigenstate in terms of the eigenstates of H_0, which are each a product of an electron eigenstate and field eigenstate: which latter are states with a definite number of photons (i.e. excitations of the field). So according to the jargon above: these photons will be called 'virtual photons'.[6] And as I stressed, the theory that treats both the electron and the electromagnetic field as quantum fields which interact with each other, i.e. the theory of quantum electrodynamics, is amazingly empirically accurate. Such an accurate theory is surely getting something right about nature: despite the issues about renormalization, to which we now turn (and some of which, as we shall see later in the paper, are not yet resolved).

2.2.2. *Energy Scales*

I said, just before Section 2.2.1, that this second aspect is more briefly stated than the first, but is fundamental and will dominate the sequel. It amplifies the basic point I announced at the start of Section 2: that while in classical physics, there seems no problem about using the measured force felt by a test particle so as to calculate the charge or mass (coupling constant) of the source, this straightforward approach fails in quantum theory – we need to add complicated corrections.

Thus the general, or ideal, classical situation is that our theory says a measured quantity, e.g. a force F on a test particle, is a function of the charge (coupling constant) g of the source: $F = F(g)$; (the function being given by our theory, as in the electrostatic formula, $F = -\nabla - e/r$). So the task is to measure F and invert this equation to calculate g as a function of F: $g = g(F)$. But in quantum field theory, this approach breaks down: perhaps unsurprisingly, since it amounts to trying to characterize one aspect of an interacting many-field system in a way that is comparatively simple and independent of the rest of the system: recall (iii) at the start of this Section.

Broadly speaking, the corrections depend on the energy and-or momentum with which the test particle approaches the source. A bit more exactly, recognizing that the test particle and source are not conceptually distinguished, since e.g. they might both be electrons: writing μ for the energy

---

[6] So NB: 'virtual' does *not* connote 'illusory', nor 'merely possible', as it does in the jargon of 'virtual work' etc. in classical mechanics – for which cf. Butterfield (2004).

(or momentum) with which we theoretically describe, or experimentally probe, the system, the corrections depend on μ.

So let us write $g(\mu)$ for the *physical coupling constant*, i.e. the coupling constant that we measure: more exactly, the coupling constant that we calculate from what we *actually* measure, in the manner of $g = g(F)$ above, in the simple electrostatic example. Then the notation registers that $g(\mu)$ is a function of μ. But also: it is a function of the *bare coupling constant*, g_0 say, that appears in the theory's fundamental equations (like e) in the electrostatic example). So we can write $g(\mu) \equiv g(\mu, g_0)$.

Using the details in Section 2.2.1 about virtual states and perturbation theory, we can fill this out a bit. The Hamiltonians of our successful interacting quantum field theories, such as quantum electrodynamics, are indeed intractable, because they include terms (cf. H_i in Section 2.2.1) for interactions between the various fields, e.g. the electron field and the electromagnetic field. So we often analyse problems using perturbation theory, and in particular the eigenstates of the free Hamiltonian. Similarly if we formulate our theories using the alternative *Lagrangian*, rather than Hamiltonian, framework. The Lagrangian function (which is essentially a difference of energy functions) is intractable, because it contains interaction terms; and so again, we turn to perturbation theory.

Usually, for both frameworks and for most problems, perturbation theory yields, as its approximate answer to the problem, a power series in the coupling constant, i.e. $\Sigma_n g^n A_n$; or a power series in some closely related number. Note that getting a power series in the coupling constant is unsurprising, given Section 2.2.1's remark that one often gets a power series in the parameter ε, which in the interaction term $H_i = \varepsilon V$, looks like a coupling constant.

Besides, as one would guess for a power series in the coupling constant: increasing the exponent n from term to term corresponds to considering corrections which take account of successively more interactions between the fields concerned. So the other factor in each term, viz. A_n, encodes many ways that there can be n such interactions. In particular as we will see in Section 2.3: it allows for these interactions occurring in various places in space and time (by A_n being an integral over the various places, typically over all of space between two times), and at various energies and momenta (by integrating over energies and momenta).[7]

[7] For some details, cf. e.g. Aitchison (1985: Sections 3.4–3.6, 5.3) and Feynman (1985, especially pp. 115–122). They both discuss the phenomenon of *vacuum polarization*, and so *screening*: the intuitive idea is that $g(\mu)$ will be greater at high energies because the test particle penetrates further past the cloud of virtual particles that surround the source, and

2.3. *The Cut-off Introduced*

In the closing remark of Section 2.2.2, that A_n integrates over various possible energies and momenta, lurks the notorious problem of quantum field theory's infinities: the second problem of Section 2.2.1 which, as announced there, is addressed by renormalization.

Typically, A_n is a (multiple) integral over energy or momentum k, extending from some value, say k_0 (maybe zero), upto infinity of a function that increases at least as fast as a positive power of k, say k^a. So A_n looks like $\int^{\infty}_{k0} dk\ k^a$. If $A > 0$, this integral is infinite; as $k \rightarrow \infty$, so does k^a, for positive a, making the integral infinite.

So to get a finite answer from our formulas, we impose a *cut-off*: we replace the upper limit in the integral, ∞, by a suitably high energy or momentum, written Λ. (There are other 'less crude' ways to secure a finite answer – called *regularizing* the integrals – but I will only consider cut-offs.)

I have introduced the cut-off as an energy Λ. But in quantum theory, energy is like the reciprocal of distance; in the jargon, 'an inverse distance': energy $\sim$ 1/distance. (And so distance is like an inverse energy.) This is due to the fundamental equations, due to de Broglie, relating 'particle' and 'wave' properties: viz. that momentum p is inversely proportional to wavelength λ with proportionality constant H. That is: p = h/λ. (NB: Λ and λ are very different: an unfortunate notational coincidence, but a widespread one...) Wavelength is the number of units of length per complete cycle of the wave. So writing k for the reciprocal, called *wave-number*, i.e. the number of wave-cycles per unit-length, we have: p = hk. So high momenta (and high energies) correspond to high wave-number, which means short wavelengths and high frequencies.

So the cut-off energy Λ corresponds, in terms of distance, to a cut-off at a small distance d. That is: imposing the cut-off, i.e. requiring by *fiat* that $\int^{\infty}_{\Lambda} dk \ldots \equiv 0$ means ignoring contributions to the integrand that vary on a distance shorter than d. In other words: to get finite answers, we are declaring that the theory claims there are no fields varying on scales less than d. At least: any such fields do not contribute to the specific process we are calculating. Or at least: the theory claims this, unless and until we consider the limit $d \rightarrow 0$ – which we will do in Section 2.4.

So returning to the notation of Section 2.2.2: the physical coupling constant, $g(\mu)$, is a function, not only of the bare coupling constant g_0 and of μ itself of course, but also of the cut off d. Thus:

so "feels" a higher coupling constant. In Section 3.2, we will see the opposite phenomenon, *anti-screening* or *asymptotic freedom*, where $g(\mu)$ is a decreasing function of energy.

(2.1) $g(\mu) \equiv g(\mu, g_0, d)$.

So we can now state our task, introduced at the start of Section 2, more precisely. We are to measure $g(\mu)$ (better: to calculate it from what we really measure, like the force F in the simple electrostatics example) and then invert eq. 2.1, i.e. write $g_0 = g_0(g(\mu), d)$, so as to calculate which value of the bare constant would give the observed $g(\mu)$, at the given d. *This task is the core idea of the traditional approach to renormalization.*

It is sometimes convenient, for the sake of uniformity, to express all dimensions in terms of length. Section 2.6.1 will give more details. But for the moment, just note that we can trade in the energy-scale μ for an inverse length, say $\mu \sim 1 / L$ where L is a length. NB: *L is not the cut-off d*! We can think intuitively that L is experimental, and d is theoretical. That is: L is our choice about how microscopically to describe or probe – to peer into – the system. On the other hand, d is a (generally much smaller) length below which we are taking our theory to say there are no contributions. So we re-express the physical coupling constant $g(\mu)$ as a function of L: we will use the same letter g for this function, so that we write $g(L) \equiv g(\mu)$. Thus eq. 2.1 becomes:

(2.2) $g(L) \equiv g(L, g_0, d)$.

In the next two sections, I turn to two further aspects of the task just described: of inverting eq. 2.2 and assigning values to g_0 that give the observed $g(\mu)$. These aspects concern:

(i) letting $d \rightarrow 0$;
(ii) needing to allow for some extra terms in the equations: which will return us to the analogy of the ping pong ball, at the start of Section 2.2.

2.4. *Letting the Cut-off d Go To Zero*

Broadly speaking, the exact value of the cut-off d is up to us. Agreed: for some of the troublesome infinities – some of the infinite terms A_n in perturbative analyses of some problems – the physics of the problem will suggests a range of values of d that are sensible to take. That is: the physics suggest that no phenomena on scales much smaller than d will contribute to the process we are analysing. One such example, very famous for its role in the establishment of quantum electrodynamics, is the Lamb shift, where the electron's Compton wavelength seems a natural lower limit to d; cf. Aitchison (1985, Section 3.5–3.6), Schweber (1994, pp. 223–247).

But of course, we would like the theory and its predictions to be independent of any human choice. So generally speaking, it is natural to take d

smaller and smaller, at fixed μ; and to consider how g varies as a function of d, while preserving the observed $g(\mu)$.

More precisely: if we believe that:

(a) spacetime is continuum, and
(b) our theory holds good in principle at arbitrarily short lengths,

then we surely also believe that at fixed μ (or at least at some, maybe theoretically judicious or appropriate, μ: such as the observed μ), g_0 goes to a limit: that is:

(2.3) there exists a limit of g_0, as $d \rightarrow 0$; ($g(\mu)$ fixed at observed value)

We will later treat the issues that (i) since we can vary μ, there are various observed values $g(\mu)$, and therefore (ii) whether we should require eq. 2.3 for *all* the observed values $g(\mu)$. We will also treat letting μ go beyond the observed range, even letting it go to infinity, although we cannot measure $g(\mu)$ above some technological (financial!) maximum value of μ.

If the limit in eq. 2.3 exists and is finite, i.e. $\in \mathbb{R}$, we say: *the theory is finite*. As the label suggests: in the face of the troublesome infinities, such a conclusion would be a relief. But I should stress some limitations of this conclusion. There are three obvious general ones (cf. the references in footnote 4):

(i) this limit being finite does not imply that *any* of the power series which our perturbative analysis provides for specific physical problems converges;
(ii) even if such a series, for a problem at a given value (or range) of μ, does converge, this does not settle the behaviour of the corresponding series at other μ; and that behaviour might be bad – in particular, arbitrarily high μ might produce a troublesome infinity;
(iii) even if all goes well in a perturbative analysis, i.e. the various series converge for the ranges of parameters for which we might hope, there remains a gap between a perturbative analysis and the original physics problem or problems.

But even with these prospective limitations, some successful quantum field theories are *not* finite. The paradigm case is QED. For QED, the limit in eq. 2.3 is infinite. That is: for arbitrarily high cut-offs, the bare charge g_0 is arbitrarily high (and remains so for yet higher cut-offs). Mathematically, this is like elementary calculus where we say that some function $f(x)$ tends to infinity as x tends to infinity, e.g. $\lim_{x \rightarrow \infty} \sqrt{x} = \infty$. But of course this last is 'just' the infinity of pure mathematics. But here we face (assuming (a) and (b) above) a *physically real infinity* viz. as the value of the bare coupling constant.

The consensus, on the traditional approach to renormalization, is that this physically real infinity is *acceptable*. After all: since by construction we do not measure (nor calculate from measurements) the bare constant, we do not need to allow an 'Infinity' reading on our apparatus' dials. To reflect this consensus, the adjective 'renormalizable', with its honorific connotations, is used. That is: If the limit in eq. 2.3 exists, albeit perhaps being $\pm\infty$, we say the theory is *renormalizable*. So in particular: QED is renormalizable in this sense, though not finite. (This definition of 'renormalizable' will be filled out in the next two subsections.)

But I should add that despite this consensus, most physicists would admit to some discomfort that the bare constant should be infinite in the continuum theory that, according to (a) and (b), we are thinking of as fundamental. Thus great physicists like Dirac have been very uncomfortable (cf. the citations in Cao 1997, pp. 203–207); and Feynman himself calls renormalization 'a dippy process' and 'hocus-pocus' (1985, p. 128); (Teller 1989 is a philosophical discussion).

Besides, this discomfort does not depend on believing exactly (a) and (b) above. Suppose that instead we merely believe the corresponding claims of possibility:

(a′) spacetime might be a continuum, and
(b′) our theory should be a consistent description (of the interactions in question) at arbitrarily short lengths.

In short: we merely believe that our theory, formulated with a continuum spacetime as background, is a "way the world could be." Then we surely are committed to believing that at fixed μ (or at least some, maybe theoretically judicious or appropriate, μ), g_0 goes to a limit. And again, it is uncomfortable that this limit is infinity. Although this yields, not an *actual* physical infinity, but only a *possible* physical infinity: surely a philosopher should be uncomfortable at such a possibility. (Section 3.1.3 will return to this, suggesting a way to ease the discomfort.)

But despite this discomfort, the fact remains that after facing the troublesome infinities, it is obviously a great intellectual relief to find one's theory to be renormalizable, even if not finite. It means we can succeed in our task: namely, to consistently assign bare constants (albeit perhaps infinite ones) so as to recover the observed physical coupling – and do so independently of the cut-off d we adopt so as to make integrals finite.

This relief prompts the idea that even if one does not explicitly endorse (a) and (b) (or perhaps, even (a′) and (b′)), one should adopt renormalizability as a reasonable *criterion for selecting theories*. Thus the idea is: a good theory of whatever interactions, should make sense, albeit perhaps with an infinite bare coupling constant, when formulated with a continuum spacetime as background. This is indeed how renormalizability was regarded in the traditional approach to renormalization, which reigned ca. 1950 to 1970: acceptable theories should be remormalizable.

2.5. *The Need For Extra Terms*

The main issue in trying to write down a renormalizable theory is that we may need to add (to the Lagrangian or Hamiltonian function) one or more terms to represent extra interaction(s) between the fields, even though we believe the bare coupling constant for the extra interaction(s) are zero. I first describe (i) the physical rationale for this; and then (ii) the refinement it prompts in the definition of renormalizability, and thereby in what the task of formulating a renormalizable theory involves.

2.5.1. *The Physical Rationale*

Thus suppose we think the bare coupling constant of some interaction is zero. That is, we think that in our fundamental theory, a certain interaction – which would typically be represented by some product of field operators – does not happen. Then we will be tempted to have no term representing this interaction in our theory (as a summand in our Lagrangian or Hamiltonian). For whatever the form, say $\mathcal{F}$, of the interaction (i.e. of the product of operators), we will think we can leave the term out of all equations. For if $g = 0$ then the term $g.\mathcal{F}$ equals zero, and surely contributes nothing to any equations.

But this might be a mistake! Despite the zero *bare* coupling, the interaction might have a non-zero *physical* coupling constant at some scale μ; i.e. $g(\mu) \neq 0$. Indeed, this situation can arise not only for:

(a) the strength of a certain interaction between given fields; but also for

(b) the mass or charge of a given field, or as people say: the mass or charge of a given particle (treated as an excitation of a field).

In case (b), we would be tempted to omit as pointless terms for all possible interactions of the given field (Particle) that depend on that mass or charge, since the terms are apparently zero, and so surely contribute nothing to any equations. But this might be a mistake: the physical coupling constant may be non-zero at some scale μ. In such a case, we say: 'the field (or particle) acquires a mass/charge at the scale μ'.[8]

[8] Intuitively, case (b) seems more problematic than case (a): for the mass or charge of a given field seems more "intrinsic" to it than is participation in a certain interaction with other

The analogy of the ping pong ball, mentioned at the start of Section 2.2, may help. There, the fact that it falls in a vacuum (or air) but is buoyant in water – i.e. exhibits a positive gravitational mass in vacuum and air, but a negative one in water – illustrated the general idea that the coupling constants associated to a system can be altered by the medium in which the system is immersed. But now imagine the ping pong ball is so light as to be massless in air, so that in air, it does not fall under gravity but floats, weightless; yet when immersed in water, it 'acquires a mass' (in fact a negative one, since it moves upwards, opposite to the force of gravity). Thus a system with $g_0 = 0$ might have at some scale μ a non-zero physical coupling constant, $g(\mu) \neq 0$, which you could measure; (better: calculate from actual measurements).

So faced with this situation, whether case (a) or case (b): we should of course include the corresponding term or terms in our fundamental equations. For recall that our basic task is to find values of the bare constants that (at the given μ and d) imply the measured values of $g(\mu)$. Here it will help to generalize the notation slightly to reflect the fact that there are of course several, even many, coupling constants to consider; as well as several, even many, possible interactions (terms in the Lagrangian or Hamiltonian that are typically products of operators). So suppose that there are in all N physical coupling constants, $g_1(\mu)$, $g_2(\mu)$, ... , $g_n(\mu)$, occurring in the various terms/interactions in our theory. Then we cannot expect to have them implied by less than N bare constants, even if we think some of the bare constants are zero. After all, to fit N numbers, we expect to need N numbers.

2.5.2. *A Refined Definition of Renormalizability*

So we now envisage a number N of different coupling constants; and we recognize that we might have to allow extra terms for interactions, in particular those whose bare couplings are zero (at least in the limit of greatest interest, viz. $d \to 0$). This suggests a more sophisticated, indeed more flexible, task than I stated before (*cf.* after eq. 2.1 in Section 2.3). The task is still to assign bare constants so as to recover the measured physical constants, and in particular so as to secure the limit in eq 2.3. But now we are allowed to add (if need be) extra terms: terms which can be judiciously selected by us the theorist.

It seems reasonable to say that such extra terms are legitimate hypotheses to add to our initial theory (our initial collection of terms), *provided*

---

fields, and our habituation to mass and charge in classical physics makes us think such properties are "given" prior to any interactions, rather than acquired from them.

that all the terms taken together, together with the limiting values of the bare constants given by eq 2.3, imply the measured values of the various $g(\mu)$. After all: we have at least 'saved the phenomena' with our theory formulated on a spacetime continuum, albeit perhaps with the cost of judiciously selected extra terms. And this seems legitimate, even if (as conceded in Section 2.4) some of the limiting values of the bare constants are ∞. Indeed: this seems legitimate, even if some of the limiting values of the bare constants in the new additional terms selected by us theorists are ∞ – even though we originally motivated such terms by the case where the limiting value of the additional bare constant is zero.[9]

In any case, whether or not you would call it 'reasonable': this is the consensus, on the traditional approach to renormalization, under one proviso. Namely, that there should be only a *finite* number of extra terms. The idea is: our theory should not qualify as 'saving the phenomena' if we have to make infinitely many such augmentations to it. That is: a theory which secures the limit in eq 2.3, using either no extra terms, or only a finite number of them, is given the honorific adjective: *renormalizable*.

2.6. *Which Theories Are Renormalizable?*

I end this Section's review of the traditional approach to renormalization by very briefly reporting: (i) the criterion for when a theory is renormalizable, and (ii) that our empirically successful quantum field theories satisfy this criterion. The good fortune in (ii) will prompt the question: why should we be so lucky? Section 3 will take up this question (using the criterion in (i)).

2.6.1. *Dyson's Criterion*

Suppose we focus, not on a whole theory as given by a Lagrangian or Hamiltonian, i.e. by a sum of terms for the various sorts of energy of the various fields and their various interactions; but on a single such term. If you like, we imagine a theory so simple as to contain only one term. It turns out that the criterion for this theory, i.e. term, to be renormalizable, can be simply stated.

To do so, we should first express all dimensions in terms of length. We saw in Section 2.3 that, thanks to de Broglie's relation $p = h / \lambda$, we

[9] By focussing on renormalization, I have of course set aside the other requirements that the theory must satisfy, if we are to talk of 'legitimate hypotheses' and 'saving the phenomena'. To include those requirements, I should of course add something like: 'and provided that the theory is otherwise empirically successful'.

can trade in a cut-off in energy Λ for a distance d, and similarly the energy-scale μ for a distance L; with higher energies corresponding to shorter distances, so that e.g. $\mu \sim 1/L$. (Recall that L is not the cut-off d.) But it turns out that we can go much further: not only energy but all quantities can be expressed as powers of length, by invoking the idea (due to Planck) that two fundamental constants, such as the speed of light c and Planck's constant H, can be declared to be *dimensionless*, instead of (as usual) thinking of them as having respectively dimensions 'length divided by time' and 'length times momentum'. The idea is that after making this declaration, we 'back-calculate' what dimension some familiar quantity such as an electric charge must have, so that our equations come out as dimensionally consistent. In this sort of way, a quantity turns out to have as its dimension some power of length: it has dimension length^D. Here, the *power* (also called: *exponent*) D can be positive or negative. For example, $L^{-1} \equiv 1/L$, so that with H declared dimensionless, de Broglie's relation $p = h/\lambda$ implies that momentum has dimension length^{-1}. For brevity, this is often shortened by dropping mention of length, so that we say: 'momentum has dimension -1'.

We can now state the criterion for a term (in the Lagrangian) to be renormalizable. It turns out that this is so iff: the bare coupling constant which the term contains has dimensions of length^D, with $D \leq 0$. This is called *Dyson's criterion*, or the *power-counting criterion*.

More precisely: suppose that the bare coupling constant g_0 has dimensions of length^D. Then the corresponding physical coupling constant $g(\mu) \equiv g(L)$ will scale roughly like L^{-D}. That is:

(2.4) $g(L)/g_0 \sim (L/d)^{-D}$

Thus if $D > 0$, the exponent on the right-hand side will be negative; so when L is very small, i.e. much smaller than d, the right hand side is very large. That is: the physical coupling constant will be large compared with the bare one. That is a sign of bad behaviour at small distances L, i.e. high energies. At least, it is bad in the sense that the large coupling constant will prevent our treating the interaction represented by the term as a small perturbation. Thus it is perhaps unsurprising that such a term is non-renormalizable in the sense sketched informally at the end of Section 2.5.[10]

Eq. 2.4 will also be important in the modern approach to renormalization. To anticipate a little: Section 3.1 will examine the case $D > 0$, i.e.

[10] This bad behaviour is not to say that a non-renormalizable theory is mathematically inconsistent: e.g. the Gross-Neveu model is non-renormalizable.

non-renormalizability, for large distances; L and Section 3.2 will examine the "happy" case of $D \leq 0$, even of small L.

2.6.2. *Our Good Fortune*

So much for the general ideas. How do the quantum field theories we "believe in" or "take seriously" fare? That is: are the theories which are our best descriptions of the electromagnetic, weak and strong forces, renormalizable in the sense just discussed?

In short, they fare very well. For first: quantum electrodynamics (QED) is renormalizable in this Dyson sense. As to the other two forces: we have since the 1970s had:

(i) a unified theory of the electromagnetic and weak forces (the *electro-weak theory* of Weinberg and Salam; also called '(Quantum) flavour-dynamics' (QFD); and
(ii) a theory of the strong force (Quantum chromodynamics, QCD).

(Like QED, these theories are so far defined only perturbatively; but unlike QED, they each use a non-abelian gauge group: QFD uses $SU(2) \times U(1)$ and QCD uses $SU(3)$.) And indeed: both of these are renormalizable.

So all three – QED, QFD and QCD – are renormalizable. But we should recall that all three theories are defined only perturbatively: recall that we do not have a rigorously defined interacting quantum field theory in four spacetime dimensions (Section 2.2.1), and that even a finite theory is defined only perturbatively and may harbour divergent series (Section 2.4). Because of these limitations, a more modest jargon is appropriate. So a qualifying adverb is often added to the honorific 'renormalizable'. Namely, we say these three theories are *perturbatively/superficially renormalizable.*

It seems a piece of great good fortune that our best theory of some force of nature be renormalizable (even perturbatively): let alone our theories of three such forces. At least, it is a relief after (a) having to admit that we can so far only define the theory perturbatively, and (b) having to face, from Section 2.1 onwards, complicated corrections: a relief that the theory can in the above sense 'save the phenomena', even if it is not finite in Section 2.4's sense.

But we will now see that according to the modern approach to renormalization, this great good fortune is not so surprising. In a certain sense, *renormalizability is generic* at the low-ish energy scales we can access – cf. the next Section.

3. The Modern Approach to Renormalization

The key initial idea of this approach, initiated in the mid-1960s by the work of Wilson, Fisher, Kadanoff and others (with important earlier work by e.g. Stueckelberg, Gell-Mann and Low) is that instead of being concerned with good limiting behaviour as the cut-off $d \rightarrow 0$, we instead focus on how $g(\mu)$ varies with μ. In terms of the ping pong ball analogy at the start of Section 2.2, and Section 2.2.2's discussion of energy scales: we now focus, not on regularizing integrals with a cut-off d, but on how the parameters of a system, e.g. the mass of a ping pong ball, depend on the energy or momentum scale at which we describe it.

Indeed, if we envisage a number of coupling constants, say N for N possible interactions, then the "vector" of coupling constants $(g_1(\mu), ..., g_n(\mu))$ represents a point in an N-dimensional space; and as μ varies, this point flows through the space. And accordingly: if we envisage a theory as given by a Lagrangian (or Hamiltonian) which is a sum of terms representing different possible interactions, then this space is a space of theories. Jargon: we say the coupling constants *run*, and the flow is called the *renormalization group flow*.

As we shall see, this simple idea leads to a powerful framework. I shall first (Section 3.1) report how it explains why a theory (like QED) that concerns phenomena at the low (or low-ish!) energy scales that we can access, is renormalizable. That is: it explains why the good fortune noted in Section 2.6.2 is generic. Then in Section 3.2, I discuss high-energy, i.e. short-distance, behaviour. Finally, I discuss insights about the renormalization group that come from thinking about statistical mechanics (Section 3.3). All three Subsections will introduce some jargon, indeed "buzz-words," such as (respectively): fixed points, asymptotic freedom and universality.

3.1. *Good Fortune Explained: Non-Renormalizable Terms Dwindle at Longer Distances*

To explain "our good fortune" in the sense introduced in Section 2.6.2 is to explain why a theory about phenomena at the low, or moderate, energy scales that we can access should be renormalizable. There are of course various philosophical controversies about explanation. But I take it to be uncontroversial that one very satisfying way to explain this would be to show: not merely that some given theory is renormalizable; but that *any* theory, or more modestly, any of a large and-or generic class of theories, is renormalizable. To the extent that the class of theories is indeed large and-or generic, such an argument would demonstrate that our good fortune was "to be expected." (Admittedly, such an explanation, whether for a single theory, or for a class of them, will have to make some other assumptions about the theory or theories: a point I will stress in Section 3.1.1. So it is only relative to those assumptions that the good fortune is explained, and to be expected.)

This is indeed what the modern approach to renormalization gives us, with its idea of a space of theories, on which there is a flow given by varying the energy-scale μ. More precisely and modestly, but also more practically: I admit that this approach does not show that any of a large and-or generic class of theories has, at the comparatively low energies and large length-scales we can access, literally *no* non-renormalizable terms. Rather, the approach shows that for any such theory – "with whatever high-energy behaviour, e.g. non-renormalizable terms, you like" – the non-renormalizable terms dwindle into insignificance as energies become lower and length-scales larger. That is, in Section 2's notation: the physical coupling constant for non-renormalizable terms shrinks. For such terms: as $\mu \to 0$ (i.e. $L \to \infty$), $g(\mu) \equiv g(L) \to 0$.

Indeed, this explanation is already clear from Section 2.6.1's discussion of eq. 2.4: which, to repeat it, was:

$$(3.1) \quad g(L)/g_0 \sim (L/d)^{-D}$$

In Section 2.6.1, we focussed on the case where L is very small, so that a non-renormalizable term's positive exponent (in the dimension of length) makes for a large physical coupling constant. *But just look at the other side of the same coin.* When L is large (much larger than the cut-off d), and $D > 0$ (i.e. the term in question is non-renormalizable), then the right hand side of eq. 3.1 is very small. That is: the physical coupling constant is very small. So at large distances, the non-renormalizable interaction is weak: "you will not see it."

There are four main points I should make about this explanation, before discussing short-distance behaviour (Section 3.2). The first point is about how non-trivial the *explanans*, i.e. eq. 3.1, is. The second point is about the explanation not depending on spacetime being a continuum. This will prompt the third point, about *effective theories*. (The second and third points, in Sections 3.1.2 and 3.1.3, are developed in more detail in Sections 5.2.1 and 5.2.2 of the companion paper 2014.) The fourth point will somewhat generalize the discussion, from a physical not philosophical viewpoint; and will introduce some more jargon.

3.1.1. *Decoupling High-Energy Behaviour*

That at large distances, a non-renormalizable interaction is weak follows immediately from eq. 3.1. But that does *not* make it obvious! A good deal of theory needs to be assumed in order to deduce eq. 3.1. After all, there is of course no *a priori* guarantee that interactions that are strong at short distances should be weak at long distances. To show this "decoupling" of high-energy behaviour from the low-energy behaviour was a major achievement of Wilson, the other authors mentioned, and indeed many other physicists, e.g. Symanzik (1973), Applequist and Carazzone (1975). I will not go into details, but just make three general remarks.

(i) It can be shown under very general conditions, even within the confines of a perturbative analysis.
(ii) Looking ahead: Section 3.1.3 will mention Weinberg's perspective, based on a result roughly to the effect that, even without assuming the framework of quantum field theory *ab initio*, any relativistic quantum theory's description of physics at low enough energies must look like the description given by a quantum field theory.
(iii) Again, looking ahead: Section 3.2 will say a bit more about how the limiting high-energy behaviour of $g(\mu)$ is encoded in a function, the beta-function, which can be calculated perturbatively.

3.1.2. *Spacetime Need Not Be A Continuum*

Notice that this explanation does not depend on our theory (with all its terms, including non-renormalizable ones) being true, or even approximately true, at arbitrarily short distances. It only needs to be approximately true at suitable intermediate distances. We can put the same point in more physical terms, and in terms of energies. Maybe at very high energies, spacetime does not behave like a continuum. But provided the theory is "true enough" at some high, maybe even inaccessible, energies in the sense that it validates eq. 3.1, then we can deduce that at much lower, in particular accessible, energies, "we see only renormalizable interactions." That is: our theory's predictions have significant contributions only from renormalizable interactions.

Here we meet a widespread jargon. A theory that is taken to be approximately true in a given regime (of energy and-or length, and-or some other parameters) is called *effective.* So we can sum up the above explanation of what I called 'our good fortune' by saying: from studying the renormalization group flow, we deduce (subject to the theoretical assumptions gestured at in Section 3.1.1) that effective low-energy theories are renormalizable. The idea of effective theories leads in to the next point.

3.1.3. *Effective Theories Only?*

I ended Section 2.4 by reporting that on the traditional approach, renormalizability functioned as a criterion for selecting theories. But the explanation at the start of Section 3.1 undermines this stance. For it says that, although non-renormalizable terms induce bad behaviour, i.e. a large coupling, at short distances, this bad behavour is invisible at the larger distances we can access. So why worry? That is: why not countenance non-renormalizable terms, at least for inaccessibly high energies?

Of course, the words 'worry' and 'countenance' are vague. What you are inclined to worry about, and correspondingly what you are willing to countenance, will depend on your background attitudes to quantum field theory: for example, on how confident you are about using it at high energies, and about accepting results obtained from a heuristic formalism, rather than by rigorous mathematical proofs. So there are bound to be several possible positions. Here I will briefly develop one position, often called the *effective field theory programme* (or: approach). It is based, not so much on confidence about the two topics above, as on an opportunistic or instrumentalist attitude to being *unconfident* about them. (In Section 3.2, I will describe a less opportunistic or instrumentalist attitude, based on results showing some quantum field theories' good behaviour at arbitrarily short distances.)

The idea is that, even below the energy scale at which the entire framework of quantum field theory presumably breaks down (owing to, say, quantum gravity, as perhaps described by some version of string theory), there may, for all we know, *not* be any single quantum field theory which is more fundamental than the others, in the sense that each of them is derived from it by assuming extra conditions that specify the derived theory's regime (of energies and types of interaction considered etc.). That is: as the energy scale gets higher and higher (while remaining below the scale at which the entire framework of quantum field theory breaks down), physics might be described by a succession of quantum field theories, each of which accurately describes the phenomena at a certain range of energies, but becomes inaccurate above that range. And when it becomes inaccurate, it may also become even more badly behaved, mathematically.

This scenario is often called the *tower of effective field theories*. It gets some support from this Section's explanation of "our good fortune," viz. that any non-renormalizable interactions (terms), though they would be important at higher energies, will make a dwindling contribution to all processes, as the energy scale is reduced. For this explanation implies that we cannot get evidence about which non-renormalizable interactions, if any, operate at inaccessibly high energies. Whatever they are – and whatever the bad short-distance behaviour they suffer (cf. the end of Section 2.6.1) – we will not see them. So why worry about non-renormalizable interactions (terms)? And for all we know, or could ever know, the scenario of the tower holds good: there is no fundamental quantum field theory, and various such interactions operate at various inaccessibly high energies.

There is a further, and substantial, point to make. So far, my exposition of the effective field theory scenario has had the spirit of epistemic modesty: "for all we know." A true and worthy sentiment, if a bit dull. But Weinberg has developed a stronger and more positive perspective on the matter. It provides an answer to the question why physics at accessible energies should be described by a quantum field theory at all, *even if* the framework breaks down higher up, e.g. because of gravity. And this answer yields the narrative strategy for his magisterial exposition of quantum field theory (1995; *cf.* pp. xx-xxi, 1–2, 31–38; 1999, pp. 242–247). In short, there is the following result; (admittedly, with 'result' understood by the standards of heuristic quantum field theory, not pure mathematics). Any quantum theory that at low enough energies is Lorentz-invariant and satisfies one other main assumption, called 'cluster decomposition' (which is plausible, since it has the flavour of a locality assumption), must at low enough energies be a quantum field theory (Weinberg 1999, p. 246).

So much by way of sketching the effective field theory programme. We can sum it up as urging that, regardless of how and why quantum field theory might break down at very high energies (as it presumably does, thanks to gravity, if nothing else): we have no reason in theory, nor experimental data, to deny the scenario of the tower – a succession of theories, each accurately describing physics in its energy range, and inaccurate beyond it.

As I said at the start of this Subsection, this suggests a rather opportunistic or instrumentalist attitude to quantum field theories. I will return to this briefly at the start of Section 4. Meanwhile, in Section 3.2, I will describe how results showing some quantum field theories' *good* behaviour at arbitrarily high energies foster a less opportunistic or instrumentalist attitude. More precisely: the results suggest that there are good prospects that these theories can be rigorously defined.

3.1.4. *The Renormalization Group Flow*

So far, my talk of the renormalization group flow has been restricted in three ways; which we need to overcome. The most important is that we need to consider, not just the flow as energy μ decreases (length L increases), but also the flow in the other direction: as μ increases (L decreases). This needs a separate subsection: Section 3.2. Here I overcome two smaller restrictions: this will also introduce more jargon.

(a) A flow can have a *fixed point*, i.e. a point that is not moved by the flow: think of sources and sinks in elementary discussions of fluid flow. In our context (the renormalization group flow), this would mean a set of physical coupling constants $(g_1(\mu), \ldots, g_n(\mu))$ that is unchanged as μ decreases further (as the length-scale increases further). Jargon: the behaviour of the system is *scale-invariant*: "you see the same behaviour/theory/physical coupling constants, at many different length-scales." This can indeed happen; and we will see a vivid physical reason for this, related to statistical mechanics, in Section 3.3. Such a point is called an *infra-red fixed point*. Here, 'infra-red' is used on analogy with light: infra-red light has a longer wavelength, lower frequency and lower energy, than visible light.

(b) So far, we have had in mind one trajectory, maybe leading to a fixed point. But many trajectories might lead to the same fixed point; or at least enter and remain in the same small region of the space. If so, then the 'vectors' $(g_1(\mu), \ldots, g_n(\mu))$ at diverse early points on a pair of such trajectories representing dissimilar theories lead, as μ decreases, to the same fixed point, or at least to the same small region, and so to similar theories. That is: when you probe at low energies/long distances, "you see the same physical coupling constants/behaviour/theory." Jargon: This is called *universality*. And the set of 'vectors' that eventually lead, as μ decreases, to the given fixed point is called, on analogy with elementary discussions of fluid flow, the point's *basin of attraction*. But note that universality should really be called 'commonality' or 'similarity': for there can be different fixed points, each with their own basin of attraction. But jargon aside: Section 3.3 will give more physical details about universality, and Section 4.3 will assess whether it is different from the familiar philosophical idea of multiple realizability.

Finally, we can summarize this Subsection's main point, that non-renormalizable interactions dwindle at large length-scales, by combining the jargon we have just introduced with the previous jargon that a *free* theory is a theory with no interactions. Namely: the infra-red fixed points of theories all of whose terms are nonrenormalizable are free theories.

3.2. *Short-Distance Behaviour: The Beta-Function and Asymptotic Freedom*

Instead of considering the flow as energy μ decreases (length L increases), we can of course consider flowing in the other direction: as μ increases (L decreases). Again, the jargon is borrowed from light: we can consider the flow towards *the ultra-violet*. Looking again at eq. 3.1 (which is eq. 2.4), we see that it is terms/interactions for which $D < 0$ for which the physical coupling constant goes to zero as $L \to 0$; since for these terms, the physical coupling constant scales like L to a positive power.

Of course, the coupling constant being zero means the interaction is not "seen" (cf. Section 2.5.1). The behaviour we see is that of the free, i.e. non-interacting, theory. This is called *asymptotic freedom*. And as in (a) in Section 3.1.4, this free theory may be fixed, i.e. not moved, by the flow. If so, it is an *ultra-violet fixed point*.

On the other hand, if $D = 0$, then according to eq. 3.1, the physical coupling constant scales like L to the zeroth power; that is, it is constant. More precisely: we need to study the range under which eq. 3.1, or some more precise equation, is valid, and what happens to the physical coupling constant(s) beyond that range.

So these cases, $D < 0$ and $D = 0$, are very different; accordingly, there is jargon to distinguish them. Recall that in Section 2.6.1, we called a term for which $D \leq 0$ 'renormalizable'. But we now distinguish the two cases. If $D < 0$ (the "happy case"), we say the theory is *super-renormalizable*. If $D = 0$, we say the theory is ("merely") renormalizable. But if in this latter case, a more subtle analysis shows that the coupling constant goes to zero as $L \to 0$, we will still say the theory is *asymptotically free*. That is: this buzz-word is not reserved for super-renormalizable theories.

We can summarize, using the jargon we have just introduced, like we did at the end of Section 3.1.4. Namely: asymptotically free theories, in particular super-renormalizable theories, have a free theory as an ultra-violet fixed point.

Note that the idea of a ultra-violet fixed point is more general than asymptotic freedom, in that the renormalization group flow could have a non-free theory as an ultra-violet fixed point. The various possibilities for what happens to $g(\mu)$ as μ tends to infinity are often described in terms of the *beta-function*, which is defined by

(3.2) $\beta(g) := dg / d\,(\ln \mu) \equiv \mu(dg / d\mu).$

Here $\ln \mu$ is the logarithm of μ. So the β-function is the rate of increase of g with respect to a logarithmically damped measure $\ln \mu$ of the energy: since logarithm rises very slowly, it is in effect an amplified rate of increase of g with respect energy – amplified by multiplying by the energy itself.

So as μ tends to infinity, there are three possibilities for $g(\mu)$ having a finite limit, i.e. an ultra-violet fixed point. Since there is a fixed point, all will involve $\lim_{\mu \to \infty} \beta = 0$. But g might have zero as its limit (as discussed: asymptotic freedom). Or g might have some non-zero value g_* as its limit. This is called *asymptotic safety*. Or g might be a constant, g_* say, independent of μ; so that g does not, colloquially speaking, tend to a limit – it is already there. This situation occurs in theories which are *conformally invariant*.[11]

To summarize, the three possibilities for $g(\mu)$ having a finite limit at short distances are:

(a) *asymptotic freedom*: $\lim_{\mu \to \infty} \beta = 0$; $\lim_{\mu \to \infty} g = 0$;
(b) *asymptotic safety*: $\lim_{\mu \to \infty} \beta = 0$; $\lim_{\mu \to \infty} g = g_* \neq 0$.
(c) *conformal invariance*: $\beta \equiv 0$ i.e. g is constant, independent of μ.

Compare Figure 1; where the ultra-violet fixed point is the dot. In Fig. 1(a), as μ grows the negative β drives g down to 0. In Fig. 1(b), as μ grows, a positive β drives g up towards the fixed point, while a negative β drives it down. Finally in Fig. 1(c), g is constant independent of μ.[12]

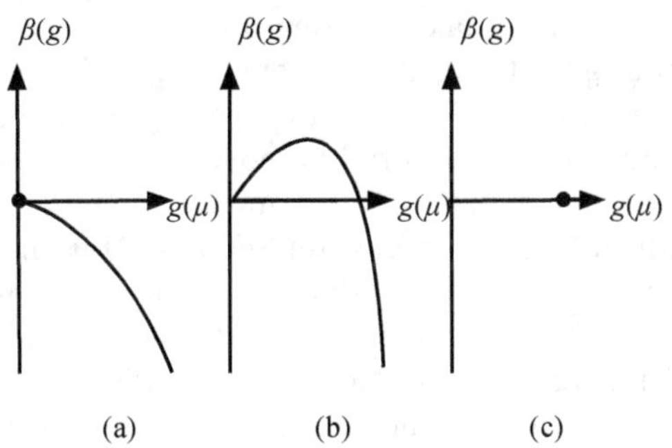

Figure 1: UV fixed points

So much for the general ideas. Let us ask, as we did in Section 2.6.2: How do the quantum field theories which are our best descriptions of the electromagnetic, weak and strong forces, get classified? There, we reported that all three (i.e. QED, QFD and QCD) are renormalizable – in that Section's *inclusive* sense, that $D \leq 0$: i.e. $D < 0$ *or* $D = 0$. More exactly, they are perturbatively renormalizable, since as emphasized there, the theories have not yet been rigorously defined.

Now that we distinguish the two cases, $D < 0$ vs. $D = 0$, there is: bad news and good news – indeed, there are two pieces of each. First, the bad

[11] This means, roughly speaking, that the theory is symmetric under any change of scale (a dilation). This extra symmetry makes conformally invariant theories easier to study in various ways (especially if spacetime has only one spatial dimension); and thus important to us, even though they do not describe the known forces.

[12] Of course, one can give a more fine-grained classification than Figure 1's (a)-(c): cf. e.g. the options in Weinberg (1995a, Section 18.3, pp. 130f.).

news: (that is, in addition to the prevailing lack of rigour). First: None of the theories is super-renormalizable. They are "merely" renormalizable; so we need a more subtle analysis of their short-distance behaviour. Because the three theories are only defined perturbatively, it is hard to make such an analysis. But there is every reason to believe that for QED, there is more bad news; (this is the second piece of bad news). Namely: QED's is badly behaved at short distances. That is: in QED, as L decreases, the coupling constant, i.e. the charge of the electron, at first looks constant – but it then grows and grows. There is good reason to think it tends to infinity, as $L \to 0$.

On the other hand: for QCD, the corresponding analysis yields good news – good short-distance behaviour. That is: There is every reason to believe that QCD is asymptotically free. So at very short distances, quarks do not feel each other's weak or strong force.[13] Besides, there may be some good news about gravity. In this paper, I have of course ignored gravity, apart from saying in Section 3.1.3 that people expect quantum gravity to force a breakdown of quantum field theory. One main reason for that is the fact that the quantum field theory obtained by quantizing general relativity is not renormalizable: and thereby, on Section 2's traditional approach, not acceptable. But there is evidence that this theory is asymptotically safe, i.e. that the physical coupling constant has a finite limit at high energies, case (b) above; (Weinberg 1979, Section 3, pp. 798f).

This good news prompts a broader point, which was foreshadowed at the end of Section 3.1.3's discussion of effective theories. Namely, asymptotic freedom suggests these theories *can* be rigorously defined. This is not to suggest that success is over the next hill: if attainable, it is some way off – but asymptotic freedom gives us grounds for optimism.[14] If so, this would count against the effective field theory vision, that (regardless of gravity) there is a succession of theories, each accurately describing physics in its energy range, but inaccurate beyond it.

[13] Wilczek's Nobel lecture (2005) is a beautiful and masterly introduction to asymptotic freedom, especially in QCD. QFD is, unfortunately, not asymptotically free. Its high energy behaviour is complicated: for details cf. e.g. Horejsi (1996), Moffat (2010); thanks to Nic Teh for these references.

[14] Besides, we can show that if a theory rigorously exists, then its asymptotic freedom can be ascertained perturbatively: so there is no threat of future success undermining our present grounds for optimism. For this pleasant surprise, cf. Gross (1999), p. 571.

3.3. *The Perspective from the Theory of Condensed Matter*

No account of the modern approach to renormalization, however brief, would be complete without some discussion of the role therein of ideas from the theory of condensed matter. ('Condensed matter' is short for 'liquid or solid'.) Ideas from this field have been invaluable. To convey something of these insights, I shall make just three main points: that continuous phase transitions correspond to infra-red fixed points of the renormalization group flow (Section 3.3.1); that renormalization group methods enable us to calculate correctly the critical exponents of such transitions (Section 3.3.2); and finally, that in a condensed matter system, there is a natural lower limit to the cut-off d and length L (Section 3.3.3).[15]

At the outset, I should clarify my usage. I will contrast 'theory of condensed matter', with 'quantum field theory', understood as I have done hitherto in this paper: viz. as describing high energy physics, especially the fundamental forces – electromagnetism, the weak force and the strong force. But I stress that the mathematics of quantum field theory is used endemically to describe condensed matter. For example, one often describes a solid or liquid with a quantum field (say: energy or momentum, or electric field): this amounts to assigning a quantum operator to each point of space or spacetime – thus abstracting away from the atomic constitution of matter. I will briefly return to this in Section 3.3.3.

3.3.1. *Continuous Phase Transitions: Scale-Invariance*

In both classical and quantum physics, understanding condensed matter is usually harder than understanding gases, since the mutual proximity of the atoms makes for stronger interactions, and so for problems that are harder to solve (cf. Section 2.2.1's discussion of intractable Hamiltonians). So it is little wonder that most of the early successes of statistical mechanics – which is the core theory for understanding condensed matter – concerned gases. But the last half-century has seen great strides in understanding liquids and solids, in both classical and quantum physics.

Most relevant to us is the topic of *phase transitions*. These are transitions between macroscopically distinguishable states (also called: *phases*), such as the liquid, solid or gaseous states themselves. So melting and freezing, boiling and condensing, are all phase transitions; but so is the change in the direction of magnetization of some material, under the

[15] Condensed matter is, fortunately, more familiar than quantum fields. Among many approachable references for this material, let me pick out just Kadanoff's masterly surveys (2009, 2013), Batterman (2010), and Menon and Callender (2013).

influence of a changing magnetic field. Here, I will consider only a special kind of phase transition, called *continuous* or *second-order*, or a *critical point*. The idea is that in such a phase transition, the material "looks the same" at whichever length scale you examine it (*scale-invariance*): this phenomenon of scale-invariance does not occur in the more usual *first-order* phase transitions.

Thus consider water boiling, i.e. liquid water being heated and becoming steam. Usually – for example in a kettle – the phase transition is first-order: there is a body of water that is almost entirely liquid but for a few bubbles of steam rising in it, and a body of steam (and air), but for a few droplets of liquid water. If we think of the density as the quantity whose varying values encode the difference between the phases, liquid and solid, there is no scale-invariance. For the two regions where liquid and gas predominate are separated by the bubbling surface; and besides, on a smaller scale, the bubbles of steam in the liquid (and the droplets of water in the gas) have some characteristic mean size (and variance).

But there is a special temperature and pressure, called the *critical point*, at which the distinction between liquid water and steam breaks down, and there *is* scale-invariance. That is: scrutinizing a volume of liquid water, we see that it contains bubbles of steam of widely varying sizes in roughly equal proportions; and scrutinizing such a bubble, we see that *it* contains yet smaller droplets of water, which are themselves of widely varying sizes in roughly equal proportions; and if we scrutinize one of those droplets, we find it contains yet smaller bubbles... and so on, through many orders of magnitude of size, until we reach molecular dimensions, where, of course, the alternation between the phases breaks down.[16]

Thus the critical point involves a "tower of self-similarity," where zooming in to see finer detail presents the same picture as we saw at larger length-scales. More precisely: it presents the same sort of picture, in a statistical sense. That is: the exact position, size and other properties of the bubbles of steam, at any level, is of course a matter of myriadly complicated happenstance. But the statistical properties of the bubbles at different levels match, in the sense that: if we specify the degree of magnification (zooming in: the change of length-scale) between two levels, there is a corresponding re-scaling of the bubbles' other quantities, such as expected size, density etc., which maps the means and variances of

[16] This critical point for water happens at a temperature of 374 degrees Celsius (647 = 374 + 273 degrees Kelvin). The water-steam mixture takes on a cloudy appearance so that images are blurred; and thus the phenomenon is called 'critical opalescence'. As we will see in Section 3.3.2, it also happens for other liquid-gas mixtures.

bubbles' quantities at the first level to those at the second. In short: there is statistical self-similarity, under changes of length-scale, through many orders of magnitude, until we reach molecular dimensions.

Many other phase transitions, in many other materials, can occur in this curious, statistically self-similar, way in which the idea of a boundary between the phases breaks down; (unsurprisingly, this requires special external conditions, like temperatures, pressures etc.). For example, it can happen in a magnet. The analogue to the alternation between bubbles of steam and droplets of liquid water is the alternation between the magnetization in two different spatial directions, for example "up" and "down." At the critical point (requiring a special temperature, the Curie temperature), a region of the magnet with magnetization predominantly up turns out to contain "islands" whose magnetization is predominantly down, but each such island contains islands whose magnetization is predominantly up ... and so on.

We can already see how the idea of a critical point connects with several of the notions in Section 3.1, especially the renormalization group flow, infra-red fixed points and universality (Section 3.1.4). Zooming out our description of a system to longer distances corresponds to flowing to lower energies (decreasing μ) in a quantum field theory. Scale-invariance means that the description does not change as we zoom out further. So such a description corresponds to an infra-red fixed point of the renormalization group flow.

Furthermore, we can strengthen my phrase 'corresponds to' to *is*, if we make the notion of 'description' of the condensed matter system more precise as the set of physical coupling constants that occur in the Hamiltonian that describes the system at the distance-scale concerned. (Similarly, with Lagrangian instead of Hamiltonian; but for short, I will just say 'Hamiltonian'.)

That is: we can set up the idea of a space of theories of condensed matter systems, and a flow on this space, just as we did at the start of Section 3. Given a postulated microscopic Hamiltonian describing the atomic or molecular interactions, the zooming out process is then a matter of successively coarse-graining the description, i.e. defining collective variables, to give an effective Hamiltonian. The standard example, viz. the Ising model, is very intuitive. The system is a regular array of sites (in one or more dimensions: a line or a lattice); with each of which is associated a variable taking one of two values, +1 or −1. The microscopic Hamiltonian encodes the idea that equal values for neighbouring spins are "preferred" (i.e. have lower energy). One coarse-grains by taking a majority vote of the values in a block of, say, ten spins; thus defining a new array, with ten times fewer sites, described by a new effective Hamiltonian. One then defines a flow on the space of Hamiltonians by iterating this coarse-graining; for details, cf. the *maestro*'s exposition (Kadanoff 2013, Section 6.4, pp. 170–172). Thus a Hamiltonian that is unchanged by zooming out, i.e. is scale-invariant, is precisely an infra-red fixed point of this flow.

Finally, note that the notion of universality carries over directly to the context of condensed matter physics. It means that two disparate physical systems, with very different microscopic Hamiltonians, can be described at long length-scales in the same way. (This of course sounds exactly like philosophers' notion of multiple realizability: cf. Section 4.3.) Indeed: if the two microscopic Hamiltonians flow to the same infra-red fixed point, then the two systems are described at their respective critical points, by the very same Hamiltonian. (And similarly, for descriptions in the neighbourhood of the critical point: the descriptions of the two systems are close but not identical.) Having the same Hamiltonian can make the two systems have exactly equal values for corresponding quantities. This quantitative equality, despite very disparate microscopic constitutions, can be very striking – as we will see in the next Subsection.

3.3.2. *Critical Exponents: The Correlation Length*

One aspect of the quantitative behaviour of materials at and near a continuous phase transition, is the fact that the values of various quantities are given by power-laws, i.e. by the value of some other quantity, raised to some power. More specifically: critical points occur only at a specific temperature, the *critical temperature* T_c; and near the critical point, the value of some quantity, $v(Q)$ say, is given by a power of the difference between the actual temperature T and T_c, or by a power of some similar quantity such as the ratio of this difference to T_c:

(3.3) $v(Q) \sim |T - T_c|^p$ or $v(Q) \sim |(T - T_c) / (T_c)|^p$

where p is the power.

It is worth giving some examples of how the same power law (same p) can govern the critical points of very disparate systems. This will show how striking universality can be: which will be relevant to the philosophical discussion in Section 4.3. Besides, since the renormalization group framework can correctly predict these power laws for countless such systems, it will show the amazing success of the framework. My examples will be drawn from condensed matter, not quantum field theory, since this will be technically less demanding: water and steam are familiar, while quarks and gluons are not. But I stress that this Section's themes – phase transitions, critical points, universality and the renormalization group successfully predicting the power laws – nowadays also form a large subject within quantum field theories like QCD; (e.g. Kogut and Stephanov 2004). But note: the details in the rest of this Subsection are *not* needed in my closing philosophical discussion (Section 4), and so can be skipped; nor will Section 4 need these themes in the form they take within quantum field theories.

In eq. 3.3, Q might be a quantity whose value is zero at the critical point (so that p is positive). For example, Q might be:

(i) the difference $\rho_{\text{liquid}} - \rho_{\text{gas}}$ between the densities of liquid water and of steam; or

(ii) the average magnetization m of a piece of iron.

Or Q might be a quantity that diverges (i.e. goes to infinity) at the critical point (so that p is negative). For example, Q might be

(i′) the isothermal compressibility κ_T of water: a measure of how compressible it is, i.e. how the volume changes as a function of pressure, at fixed temperature T (formally: $\kappa_T = -\partial V / \partial p \mid_T$); or

(ii′) the magnetic susceptibility of iron: a measure of how magnetizable it is, i.e. how its magnetization changes as a function of an external applied magnetic field B, at fixed temperature T; (formally: $\chi_T = -\partial m / \partial B \mid_T$).

In these four examples, it is obvious enough that (i), the difference $\rho_{\text{liquid}} - \rho_{\text{gas}}$, vanishes at the critical point. For as I sketched in Section 3.3.1, each body of liquid water contains much gas, which contains much liquid, and so on ... and *vice versa*. A similar argument can be made that (ii) vanishes, and (i′) and (ii′) diverge, at their respective critical points. But we do not need the details of why that is so, in order to make universality vivid.

All I need is to report the striking, indeed amazing, fact that the two members of each pair ((i) and (ii), (i′) and (ii′)) *obey the same law*. That is: (i) and (ii) obey the same power-law, namely with $p \approx 0.35$. In fact this p is written as β. So we have

(3.4) $\rho_{\text{liquid}} - \rho_{\text{gas}} \sim |T - T_c|^{\beta}$ and $m \sim |T - T_c|^{\beta}$; with $\beta \approx 0.35$.

Furthermore, this power-law, with almost the same value of β, describes corresponding phase transitions in many other systems. Here is an example like (i): as one reduces the temperature of a sample of liquid helium below 2 K (two degrees above absolute zero, i.e. –271 degrees Celsius) another phase of helium (a superfluid phase: so-called because it flows without viscosity) condenses out, rather like liquid water condensing out of steam. This transition has a power law like eq. 3.4 with $\beta \approx 0.34$. Another example like (ii) is the magnetization of nickel: here, $\beta \approx 0.36$.

Similarly, (i′) and (ii′) obey the same power-law, with nearly equal values of their exponent p: namely p ~ –1.2. By convention, this p is written as $-\gamma$, so that γ is positive. So we have

(3.5) $\kappa_T \sim |T - T_c|^{-\gamma}$ and $\chi_T \sim |T - T_c|^{-\gamma}$; with $\gamma \approx 1.2$.

Again, this power-law, with almost the same value of γ, describes corresponding phase transitions in many other systems. Here is an example like (i′): a mixture of two organic fluids (viz. trimethylpentane and nitroethane) has a critical point rather like that of water-and-steam: the compressibility κ_T obeys a power law like eq. 3.5, with $\gamma \approx 1.24$. And for helium, i.e. our previous example like (i): the compressibility has $\gamma \approx 1.33$. Another example like (ii′) is the magnetic susceptibility of nickel: here, $\gamma \approx 1.33$.

β and γ are called *critical exponents*. By convention, critical exponents are defined to be positive; so that if the quantity concerned diverges at the critical point, the exponent/index in the power-law is minus the critical exponent. There are several others. All of them encode some power-law behaviour of quantities at or near a critical point: and each of them takes very nearly the same value for critical points occurring in materials and processes that are microscopically very disparate. So again, we see a striking universality of systems' quantitative (albeit arcane!) behaviour.

I need not go into details, except to describe how two critical exponents, written η and ν, encode the power-law behaviour near the critical point – not of straightforwardly macroscopic quantities like the density, compressibility etc. so far mentioned – but of *measures of the microscopic correlations*. Sketching this will bring us back to Section 3.3.1's central idea of scale-invariance.

One of course expects correlations between the states at different locations in a material to decrease, as one considers more widely separated locations. Indeed, this is so: and at or near critical points, the decrease is often given by a power of the separation. Thus suppose we define as our measure of correlation, the expectation (probabilistic average) of the product of local variables (often called 'order parameters') at the two locations, separated by distance r. This is often written $g(r)$.[17] At the critical temperature T_c, $g(r)$ obeys a power law, for r large compared with inter-molecular distances. Namely:

[17] It is usually defined in terms of local variables φ by: (i) assuming spatial homogeneity so that any two locations a distance r apart yield the same expectation, and one may as well choose the origin $\mathbf{0}$ and some location $\mathbf{r}$ which is r units away, so that one would write $g(r) = \langle \varphi(\mathbf{0}) \cdot \varphi(\mathbf{r}) \rangle$, where $\langle\ \rangle$ indicates expectation; and (ii) subtracting away any background constant correlation in φ, the better to study fluctuations, so that one writes $g(r) := \langle \varphi(\mathbf{0}) \cdot \varphi(\mathbf{r}) \rangle - |\langle \varphi \rangle|^2$.

(3.6) $G(r) \sim 1 / r^{d-2+\eta}$

where (i) d is the spatial dimension of the material (usually 3; but e.g. 2 for a monoatomic or monomolecular layer, or a theoretical model of such), and (ii) η is typically between 0 and 0.1. (In my examples, it is: (a) for the mixture of two organic liquids and for helium, 0.02; (b) for iron, 0.07; (c) for nickel, 0.04.)

Near T_c, $g(r)$ obeys, not a simple power law in r, but an exponential decrease:

(3.7) $G(r) \sim \exp(-r / \xi)$;

which defines the length $r = \xi$ over which $g(r)$ decreases by about 66% ($e \approx 2.7$ so that $\exp(-\xi / \xi) \equiv e^{-1} \approx 1/3$). This means that φ fluctuates in correlated blocks of various sizes up to a length ξ, but blocks with a side-length much bigger than ξ are very rare.

It is here that we see again the *scale-invariance* of the critical point – and so its being an infra-red fixed point of the renormalization group flow. For as T_c is approached from above or from below, ξ grows without limit. Indeed, its behaviour is described by (yet another!) power law: with a negative exponent so as to capture the divergence, as with the $-\gamma$ for (i′) and (ii′) above. That is: near T_c (i.e. $|T - T_c| / T_c << 1$), we have

(3.8) $\xi \sim |T - T_c|^{-\nu}$

where ν is typically about 2/3. Again, the approximate value 2/3 covers a whole class of phenomena: for our four examples of the critical points for the two organic liquids, helium, iron and nickel, the range of ν is from 0.62 (the liquids) to 0.69 (iron).[18]

Finally, a note of caution. My praise of the modern approach to renormalization, with its successful calculation of critical exponents (and, needless to say: so much else!), might give the impression that only with the renormalization group did physics cotton on to the broad idea of coarse-graining a microscopic description to give a macroscopic one, or of iterating a scheme for coarse-graining, so as to understand scale-invariant phenomena, and in particular to calculate critical exponents.

That is *not so*. Before the renormalization group analyses were developed, various approaches to understanding these phenomena had considerable, albeit partial, success. That is of course hardly surprising, since the ideas of coarse-graining, and iterating a scheme for coarse-graining, are so

[18] This discussion, especially the data cited, is based on Binney et al. (1992, pp. 7–8, 18–20, 22).

natural. But it is worth emphasizing: not just so as to honour the previous work, but also to avoid philosophers getting the false impression that only with the renormalization group did physics pick up on the idea of multiple realizability. In short: the renormalization group is of course a scientific triumph, and a stimulus to philosophical relection – but that does *not* imply that it harbours entirely novel morals for philosophy. Section 4 will expand on this irenic theme.[19]

3.3.3. *Short Distances: A Natural Lower Limit*

This Section has stressed the analogy between the renormalization group framework for quantum field theory and for condensed matter. But I should end by noting two main differences; the second will be more important. (Of course, there are also many differences of detail.)

First: I have spoken impartially of the renormalization group flow going (i) towards long distances, the infra-red (Sections 3.1 and 3.3.1), and (ii) towards short distances, the ultra-violet (Section 3.2). But broadly speaking, a flow towards the infra-red is often defined by (iterated) coarse-graining. And since coarse-graining involves a loss of microscopic information, such a flow is in general many-one, i.e. irreversible: this underpinned Section 3.3.1's idea of two disparate microscopic Hamiltonians flowing to the same infra-red fixed point. So there is no well-defined inverse flow towards the ultra-violet. (And in the jargon of group theory, one should thus say 'renormalization semi-group', not 'renormalization group'; but the latter name has stuck.) This point is not meant to suggest there is a problem, or even a puzzle, about such irreversible flows. The exact definition of the flow is up to the theorist, and of course varies from theory to theory, and according to the theorist's purposes. Thus the flow towards the infra-red is not always defined by coarse-graining (loss of information); so in some cases it is indeed reversible, and there is a well-defined flow to the ultra-violet.[20]

---

[19] Two broad previous approaches are mean field theory and Landau theory. For a glimpse of what these are and their predictive limitations, in the context of condensed matter, cf. e.g. Kadanoff (2009, Sections 2,3; 2013, Sections 1.2.4–4, pp. 147–164) and Binney et al. (1992, pp. 22, 176–177, 183–184). In particular, mean field theory implies values for the critical exponents I have discussed, as follows: $\beta = 0.5$, $\gamma = 1$, $\eta = 0$, $\nu = 0.5$. As to the false impression that within physics only the renormalization group describes multiple realizability: I think this impression is fostered by some philosophers' praise of the renormalization group; e.g. Batterman (2002, pp. 37–44), Morrison (2012, p. 156, p. 160).

[20] Cf. e.g. Kadanoff (2013, Section 8.5, p. 181). I say there is 'not even a puzzle'; but it can be confusing – witness the exchanges between Cao and Fisher which pepper Cao (1999a).

The second difference is about the ultra-violet regime. Namely: quantum field theory lacks, but condensed matter has, a natural lower limit to the lengths d and L. In more detail: quantum field theories, being *field* theories, assume a spacetime continuum. Although I pointed out that we can make sense of such theories without spacetime being a continuum, namely as describing physics at low enough energies, large enough distances (Sections 3.1.2 and 3.1.3), these theories are undoubtedly conceived and presented as involving a continuum. After all, that was the rationale for the core task, on the traditional approach to renormalization in Sections 2.3 and 2.4: to assign bare coupling constants at each value of the cut-off d, so as to recover the physical coupling constants (cf. eq. 2.3). Similarly for L and μ. In a theory assuming a spacetime continuum, there is no *a priori* obstacle to arbitrarily short-distance, high-energy phenomena – L can be arbitrarily small, μ arbitrarily high.

But in condensed matter, the inter-atomic or inter-molecular distance gives a natural lower limit to the length-scales at which we trust our analyses. Thus recall that in Section 3.3.1, I said the critical point involves scale-invariance, 'until we reach molecular dimensions, where, of course, the alternation between the phases breaks down'. More generally: we of course expect our definition of the renormalization group flow, by which one "vector" of physical coupling constants (or: one Hamiltonian or Lagrangian) replaces another as the energy scale increases (distance scale decreases) to break down at molecular or atomic scales. (Or of course at larger length-scales, if some new physics "kicks in" there, which our definition of the flow has disregarded: an example would be spatially periodic inhomogeneities with, say, a mesoscopic period.) Similarly of course, for theoretical models rather than real samples. The classic example is modelling a crystal as a spatially periodic lattice of sites representing atoms or nuclei. In such a model, the lattice spacing provides the natural lower limit. To sum up: at this limit, the renormalization group transformation – the step along the flow towards the ultra-violet that "zooms in" by an increment – cannot be iterated.

Mention of lattice models in condensed matter prompts a final remark, to complete my description of quantum field theory. Namely, lattice models are also crucial there: but with a lattice of sites throughout a region of spacetime, rather than space. We can easily state two broad reasons why.

(1) Since quantum field theories are complicated, one often cannot solve one's problem analytically (even with perturbation theory, power series etc.) and so needs to solve it numerically, i.e. on a computer: which involves discretizing the field variables, i.e. assigning them to sites of a lattice rather than the continuum of spacetime points.

(2) I enthused about QCD's asymptotic freedom: that the theory becomes very simple, indeed free, at arbitrarily high energies. But consider the other side of this coin: as the energy is lowered, the interaction becomes stronger. Of course it was realized already in the 1930s that the interaction holds the nucleus together, overcoming the electrostatic repulsion among the protons. Hence its name, 'strong', given long before QCD and asymptotic freedom were found. This strength means a large physical coupling constant, and so difficulties for a perturbative expansion (cf. Section 2.2.1's discussion of ε and α): difficulties that led in the 1950s and 1960s to scepticism that any perturbative quantum field theory could describe the interaction. Again there is jargon: the increasing strength at larger distances is called *infra-red slavery*, and the ever-tighter binding of quarks as they try to separate (which prevents us seeing a lone quark) is called *confinement*; (e.g. Wilczek 2005, Sections 1.1, 3.1). But jargon aside, the predicament that the theory is complicated and very hard to calculate in prompts lattice models and the hunt for numerical solutions.

4. Nagelian Reflections

So much by way of reviewing renormalization, in both the traditional and modern approaches. In this final Section, I turn briefly to philosophy. I will confine myself to urging a moral which honours Ernest Nagel. Namely: the material in Sections 2 and 3 fits well with Nagel's account of reduction; and indeed, Hempel's deductive-nomological account of explanation. In particular, the idea of universality (Sections 3.1.4 and 3.3.2) is a case of the philosophical idea of multiple realizability, and does not cause problems for Nagelian reduction.

That is good news for me, since in previous work I joined others in endorsing Nagel's account, and in arguing that multiple realizability is no threat to Nagelian reductions (2011, Sections 3 and 4; and for renormalization, 2014). I will not repeat any of the details of those arguments here.[21]

[21] Not that Nagel's doctrines about explanation, reduction and inter-theoretic relations need *my* endorsement: especially since, happily, his stock-price is rising. For example, one recent defence of Nagel's account, and similar accounts such as Schaffner's (1967, 1976), is Dizadji-Bahmani, Frigg and Hartmann (2010); and Schaffner has recently given a masterly review of this literature, and defence of his own account (2013). As to multiple realizability not threatening Nagelian reduction, I endorse Sober (1999).

I will just sketch Nagel's account (Section 4.1) and describe how the material I have reviewed illustrates it (Section 4.2). Finally in Section 4.3 I will urge that universality is just multiple realizability.

But before setting out, I note as an appetizer for another occasion, two philosophical topics prompted by the material in Sections 2 and 3.

(1) *Real infinities*? Should we acquiesce in the traditional approach's acceptance of infinite bare coupling constants (Section 2.4)? I noted there that most physicists find such physically real, albeit unmeasurable, infinities intellectually uncomfortable; and for all the achievements of the modern approach (especially Sections 3.1, 3.3.2), this approach does *not* expunge these infinities. So what should a philosopher say about such infinities: are they acceptable?[22]

(2) *Instrumentalism*? The dwindling of non-renormalizable terms in the infra-red prompted the effective field theory programme, with its vision of a succession of theories, each accurately describing one energy range, but inaccurate beyond it – suggesting an instrumentalist attitude (Section 3.1.3). But of course, this dwindling (and allied technical considerations) do not make such an attitude compulsory, especially if we think there are good prospects for getting an empirically adequate and rigorously defined interacting quantum field theory: prospects that are certainly enhanced by QCD's asymptotic freedom (cf. the end of Section 3.2). So what attitude should we adopt? The answer is bound to be controversial, since it is an example of the broad, and maybe unending, philosophical debate between varieties of instrumentalism and of realism.[23]

4.1. *Nagel Endorsed*

The traditional philosophical jargon is that one theory T_1 *reduces* to, or *is reducible to*, another T_2 if, roughly speaking, T_1 is a *part* of T_2. To explicate

---

[22] In some of his work, Batterman has advocated physically real infinities (e.g. 2005, pp. 235–236). But beware: this is a different topic. For it rests on different considerations than infinite bare coupling constants in renormalization theory: viz. (i) singular limits, and (ii) the need to take a limit, so as to describe or explain a distinctive phenomenon, e.g. the thermodynamic limit so as to get a phase transition. For my own assessment of his arguments about (i) and (ii), cf. my (2011a, Section 3; 2014, Section 5.1).

[23] For more discussion, with an emphasis on renormalization and the idea of "reductionism," cf.: (i) Cao (1993), Schweber (1993), Cao and Schweber (1993), Cao (1997, Section 11.4 pp. 339–352); (ii) the brief contrasting discussions by Weinberg (1999, pp. 246–250) and Redhead (1999, pp. 38–40); (iii) responses to Cao and Schweber by Hartmann (2001) and Castellani (2002); and more recently, (iv) Bain (2013, 2013a) and my (2014, Section 5.3).

this,[24] I first introduce mnemonic subscripts, to avoid confusion between T_1 and T_2.

For the reducing theory, I will use *b*: 'b' is for bottom/basic/best; 'bottom' and 'basic' connoting microscopic and fundamental, and 'best' connoting a successor theory. And for the reduced theory, I will use *t*: 't' is for top/tangible/tainted; 'top' and 'tangible' connoting macroscopic and observable, and 'tainted' connoting a predecessor theory. Thus I will say that a theory $T_b \equiv T_{\text{bottom/basic/best}}$ reduces $T_t \equiv T_{\text{top/tangible/tainted}}$; and that T_t reduces to, or is reducible to, T_b.

Nagel's main idea is to take theories as linguistic entities, viz. sets of sentences closed under deducibility, and hold that T_t is reduced to T_b by:

(1) *Deducibility*: being shown to be logically deducible from T_b, usually together with some judiciously chosen definitions; and
(2) *Constraints*: the deduction satisfying some informal constraints, e.g. about the derivation counting an explanation of its conclusion.

Before I give some details about (1) and (2), in Sections 4.1.1 and 4.1.2 respectively, my first and most important comment is that taking theories as sets of sentences does *not* imply that the language of the theories, or the notion of deducibility, be formalized (let alone first-order). Nagel and his supporters (including Schaffner et al. listed in footnote 21) of course know that scientific theories are not formalized, and are hardly likely to be, even in mathematized subjects like physics. But that does not prevent one undertaking to assess whether there are relations of reduction between theories (in Nagel's sense; or indeed, in another one). The informality merely makes one's specification of the theories, definitions and deducibility relation, somewhat vague; and so one's ensuing assessments are correspondingly tentative.[25]

[24] Of course, we can and should admit that 'reduction', used as a relation between theories, is vague (my 2011, Section 1.1.2). But this is undoubtedly the core idea.

[25] I agree that the tenor of Nagel's writings – and of his times – also suggests a narrower, and I admit implausible, account requiring that theories, definitions and deductions, be formalized. (Thanks to Bob Batterman and Jim Weatherall for emphasizing this.) I say a bit more about this, and the rival "semantic view" of theories in my (2011, Sections 3, 4; 2014, Section 1.2). I should also note that, on the other hand, some writers have for various reasons criticized the notion of a scientific theory, even if theories are not taken as linguistic entities; and have even invoked quantum field theories in their support. I reply to this line of thought, in the companion paper (2014, Section 4.1).

4.1.1. *Deducibility*

The syntactic conception of theories immediately gives a notion of T_t being a *part* of T_b, viz. when T_t's sentences, i.e. theorems, are a subset of T_b's.

In logic, with its formalized languages, this is called T_t being a *sub-theory* of T_b. But the notion of course applies to unformalized languages. One just has to recognize that which sentences are in a given theory is almost always vague, even when the theory's name is very specific (e.g. 'equilibrium classical statistical mechanics of N point-particles in a cube with side-length l', rather than 'classical statistical mechanics'). And to avoid merely verbal disputes, one must adopt a swings-and-roundabout attitude about whether a given T_t is part of a given T_b. For it all depends on how one makes the two theories, i.e. their sets of sentences, precise; (cf. my 2011, Section 3.1.2; and Schaffner (2013, Section II p. 7, Section VI, p. 29)).

So although scientific theories are of course unformalized, I will continue to use logical jargon, saying e.g. that a theory has a set of sentences, a language has a set of predicates etc. I will also set aside the questions of what a theory's underlying logic is (first-order vs. second-order etc.) and what its mathematical apparatus is (containing set theory, calculus etc.), since for unformalized theories these questions also are vague. So I just refer non-commitally to 'deduction'.

However, one still needs to avoid confusion arising from the same predicate, or other non-logical symbol, occurring in both theories, but with different intended interpretations. This can be addressed by taking the theories to have disjoint non-logical vocabularies, and then defining T_t to be a *definitional extension* of T_b, iff one can add to T_b a set D of definitions, one for each of T_t's non-logical symbols, in such a way that T_t becomes a sub-theory of the augmented theory $T_b \cup D$. That is: In the augmented theory, we can deduce every theorem of T_t.

Note that here again, I mean the word 'definition' as it is used in logic: roughly speaking, a definition states that a given "new" non-logical symbol e.g. a predicate, the *definiens*, is co-extensive with some (maybe very long) open sentence, the *definiendum*. There is no formal requirement that a definition be faithful to some previous meaning (or some other properties) of the definiens: I shall return to this in Section 4.1.2.A.

This is the core formal part of Nagelian reduction: (1) *Deducibility*, in the above notation. The definitions are usually called 'bridge laws' (or 'bridge principles' or 'correspondence rules').[26]

[26] The standard references are Nagel (1961, pp. 351–363; and 1979); cf. also Hempel (1966, Chapter 8). Discussions of bridge laws are at Nagel (1961, pp. 354–358; 1979, pp. 366–368) and Hempel (1966, pp. 72–75, pp. 102–105). My notation *Deducibility* thus covers the

4.1.2. *Informal Constraints on Reduction*

4.1.2.1. *Deducibility Is Too Weak*

Various philosophers have said that even if T_t is a definitional extension of T_b, there can be non-formal aspects of T_t that are not encompassed by (are not part of) the corresponding aspects of T_b; and that, at least in some cases, these aspects seem essential to T_t's functioning as a scientific theory. Nagel agrees with the general point that deducibility is too weak, and (1961, pp. 358–363) adds some informal conditions, mainly motivated by the idea that the reducing theory T_b should explain the reduced theory T_t; and following Hempel, he conceives explanation in deductive-nomological terms. Thus he says, in effect, that T_b reduces T_t iff:

(i) T_t is a definitional extension of T_b; and
(ii) in each of the definitions of T_t's terms, the *definiens* in the language of T_b must play a role in T_b; so it cannot be, for example, a heterogeneous disjunction.

Nagel's proposal (ii) has got some good press; e.g. Sklar (1967, pp. 119–123), Nickles (1973, pp. 190–194). But some philosophers neglect this proposal and-or object that it does not secure what is needed. The most common objection appeals to *multiple realizability*. The leading idea is that the *definiens* of a multiply realizable property shows it to be too "disjunctive" to be suitable for scientific explanation, or to enter into laws. And *pace* Nagel's proposal (ii) above, some philosophers think that multiple realizability prompts a non-Nagelian account of reduction.

I reject this sort of objection. I believe multiple realizability gives no argument against definitional extension; nor even against stronger notions of reduction like Nagel's, that add further constraints additional to deducibility, e.g. about explanation. That is: I believe that such constraints are entirely compatible with multiple realizability. I maintain (2011, Sections 3.1.1 and 4.1.1) that this was shown persuasively by Sober (1999; he mostly targets Putnam (1975) and Fodor (1974)). I will not repeat that rebuttal. Let me just sum up: there is no tension between Nagelian reduction and the fact of multiple realizability.

---

conditions Nagel calls 'derivability' (about logical consequence) and 'connectability' (about definitions); cf. the discussion in Schaffner (2013, Section I). In Section 4.1.2 I will mention revisions by authors such as Schaffner.

4.1.2.2. *Deducibility Is too Strong*

The other traditional criticism of Nagelian reduction goes in the "opposite direction." Instead of saying that Deducibility is too weak, it says that Deducibility is too strong. The idea is that in many cases where T_b reduces T_t, T_b *corrects*, rather than implies, T_t.[27]

But as an objection to Nagel, this is an *ignoratio elenchi*. For Nagel himself made this point. He said that a case in which T_t's laws are a close approximation to what strictly follows from T_b *should* count as reduction. He called this *approximative reduction* (1979, pp. 361–363, 371–373). (Cf. also Hempel 1965, pp. 344–346; 1966, pp. 75–77.) In a similar vein, Schaffner (1967, p. 144; 1976, p. 618) requires a strong analogy between T_t and its corrected version, i.e. what strictly follows from T_b.

Against this, critics have complained that it is too programmatic, since it gives no general account of when an approximation or analogy is close enough. But in Nagel's and Schaffner's defence, I would say that (a) we should not expect, and (b) we do not need, any such general account.[28] What *matters*, both scientifically and conceptually, is that in a given case we can deduce that T_t's proposition (whether a description of particular fact, or a general theorem/law) is approximately true; and that we can quantify how good the approximation is.

[27] This objection is made by e.g. Kemeny and Oppenheim (1956), and Feyerabend (1962). One standard example is Newtonian gravitation theory (T_b) and Galileo's law of free fall (T_t). This T_t says that bodies near the earth fall with constant acceleration. This T_b says that as they fall, their acceleration increases, albeit by a tiny amount. But surely T_b reduces T_t. And similarly in many famous and familiar examples of reduction in physics: wave optics corrects geometric optics, relativistic mechanics corrects Newtonian mechanics etc.

These examples put limiting relations between theories centre-stage. And quite apart from assessing Nagel, such relations have long been a major focus for philosophers of the physical sciences: as also in recent years, cf. e.g. Batterman (2002, 2013), Norton (2012). My own views about these relations are in (2011a): which takes as one of its examples, phase transitions, especially Section 3.3.1's topic of critical points; (cf. also Bouatta and Butterfield (2011, 2015).

[28] I am not alone in this defence; cf. Nickles (1973, p. 189, 195), Schaffner (2013, Section III) and Dizadji-Bahmani, Frigg and Hartmann (2010: Section 3.1, p. 409f.). In a similar vein, the latter argue (their Section 5) that Nagelian reduction does not need to settle once for all whether bridge laws state "mere" correlations, law-like correlations or property-identities (which are Nagel 1961's three options, at pp. 354–357). I entirely agree, *pace* authors like Kim (1999, p. 134; 2006, p. 552) and Causey (1972, pp. 412–421).

4.2. *Renormalizability Deduced at Low Energies as a Family of Nagelian Reductions*

I now conclude by discussing how renormalization illustrates Nagelian reduction. I will confine myself to the topic announced in Section 1: how the modern approach, with its explanation that our good fortune in having renormalizable theories is generic (Section 3.1), accords with Nagelian reduction.

As I said in Section 1, I think philosophers should take note, not just of this specific achievement, but of the general idea of a *space of theories*. This fosters a novel and more ambitious kind of explanatory project than the familiar ones of explaining an individual event, or a single law, or (as in Section 4.1) a theory as a part of, or a good approximation to, another. Namely: to explain a feature of a whole class of theories in a unified way in terms of the structure of the space of theories.

But I will not develop this general theme, but instead concentrate on my main claim, as follows:

> The deduction from a given theory T_b that describes (Perhaps using non-renormalizable terms) high-energy physics, of a renormalizable theory T_t describing low energy physics, is a Nagelian reduction. Besides: for different pairs of theories T_b and T_t, varying across the space of quantum field theories, the reductive relation is similar, thanks to a shared definition of the renormalization group flow, i.e. of the renormalization scheme.

I will fill this out with five short remarks, rehearsing previous material. As regards the philosophical assessment of Nagel's account of reduction, the most important of these remarks are (3), about approximative reduction, and (4), about unity among a family of reductions.

(1) I have specified a theory by a Hamiltonian or Lagrangian. Recall the vector of physical coupling constants $g_1(\mu), \ldots, g_n(\mu)$ which I first introduced in Section 2.5.1, and took as a point in a space of theories at the start of Section 3.

(2) A renormalization scheme that defines a flow towards lower energies (a scheme for coarse-graining so as to eliminate higher-energy degrees of freedom) amounts to the set of definitions D, in logicians' broad sense (Section 4.1.1), needed to make the deduction of T_t from T_b go through.

(3) Since at low energies any non-renormalizable terms in T_b still make non-zero, albeit dwindling, contributions to the theory's predictions (probabilities for various processes), we have here an approximative reduction (Section 4.1.2.2); though the approximation gets better and better as the energy decreases.

(4) A given renormalization scheme (definition of the flow) works to show that many theories T_b lead to a renormalizable low-energy theory T_t. This unity is striking. Hence this Section's title's use of the word 'family': since 'family' connotes resemblance, which 'set' does not.

(5) Agreed: no single renormalization scheme works to prove that all possible theories have dwindling non-renormalizable contributions in the infra-red. And as I have often admitted: the proofs concerned are often not mathematically rigorous. But the various renormalization schemes that have been devised do not contradict one another, and in fact mesh in various ways. So it is fair to say that a generic quantum field theory has dwindling non-renormalizable contributions in the infra-red. As I said in Section 3.1: a satisfying explanation of our good fortune.[29]

So to sum up: the modern approach to renormalization gives a stunning case of explaining something that is otherwise – i.e. on the traditional approach – a coincidence. The coincidence is, in short, that the theory in question, e.g. quantum electrodynamics, has a feature, viz. renormalizability, that seems crucial to it "making sense." This feature also holds of other theories; indeed, it holds of the whole standard model of particle physics, which combines quantum electrodynamics with quantum theories of the weak and strong forces. Thus the coincidence is so grand and important that it seems like manna from heaven; or more prosaically, it seems that renormalizability is in some sense an *a priori* selection-principle for quantum theories. But adopting the modern approach, we can deduce that what seemed manna from heaven is in a sense to be expected.

4.3. *Universality Is Multiple Realizability*

My final point is a brief ancillary one. In Section 3.1.4, I introduced 'universality' as jargon for the idea that dissimilar theories might have similar infra-red behaviour: in particular, the same infra-red fixed point. In Section 3.3.2, we saw vivid examples of this. The dissimilar theories were of utterly different quantities in utterly different systems: e.g. a density difference in a mixture of water and steam, or in a mixture of two phases of liquid helium; or the magnetization of a piece of iron or nickel. And the similar behaviour was quantitatively exact, albeit arcane: the same values

[29] That a satisfying explanation should show the *explanandum* to be generic has of course been a theme in cosmological speculations about explaining initial conditions: not just nowadays, but over the centuries; cf. McMullin (1993).

of critical exponents in power laws describing near-critical behaviour. In the examples just mentioned, it was the critical exponent β in eq. 3.4. Corresponding remarks apply to the exponents γ, η and ν, in eq. 3.5, 3.6 and 3.8 respectively.

These examples obviously illustrate multiple realizability in the sense introduced in Section 4.1.2.1. To take just my first example: having a value of β equal to about 0.35 as in eq. 3.4 is a multiply realizable property. A water-steam mixture has it; so does a mixture of two phases of liquid helium. Similarly for Section 4.1.2.1's idea of a functional definition of a property, viz. specifying a pattern of relations to other properties: clearly, having $\beta \approx 0.35$ is a functional property.

But recall that I join Sober (1999) in seeing no tension between Nagelian reduction and multiple realizability, with the consequent need for functional definitions. So I also see no trouble for Nagelian reduction in universality: e.g. in critical exponents, whether calculated using the renormalization group or using older approaches like mean field theory (cf. footnote 19). In particular, I see no trouble for Section 4.2's main claim that to coarse-grain a class of theories towards their common infra-red fixed point – zooming out our descriptions in each theory, towards longer distances – is to give a unified family of Nagelian reductions.

University of Cambridge
Trinity College
e-mail: jb56@hermes.cam.ac.uk

University of Cambridge
Department of Applied Mathematics and Theoretical Physics
e-mail: n.bouatta@damtp.cam.ac.uk

ACKNOWLEDGEMENTS

JB thanks: Sebastien Rivat, Nic Teh, and seminar audiences in Cambridge for discussion and comments on early versions (2012); Jonathan Bain, Bob Batterman, Tian Yu Cao, Elena Castellani, and Jim Weatherall for generous comments on a later draft (2013); and audiences at Columbia University, New York and Munich (2013). Thanks also to the editors and a very perceptive referee. While I have incorporated most of these comments, I regret that I could not act on them all. I thank the *Journal of Philosophy* for allowing some overlap (in Sections 3.1.2–3.1.4 and 4.2) with the companion paper (2014). This work was supported in part by a grant from the Templeton World Charity Foundation, whose support is gratefully acknowledged.

REFERENCES

Aitchison, I. (1985). Nothing's plenty: the vacuum in quantum field theory. *Contemporary Physics* **26**, 333–391.

Applequist, T., Carazzone, J. (1975). Infra-red singularities and massive fields. *Physical Review D* **11**, 2856–2861.

Baez, J. (2006). *Renormalizability.* http://math.ucr.edu/home/baez/renormalizability.html

Baez, J. (2009). *Renormalization made easy.* http://math.ucr.edu/home/baez/renormalization.html.

Bain, J. (2013). Effective Field Theories. In: R. Batterman (ed.), *The Oxford Handbook of Philosophy of Physics*, pp. 224–254. Oxford: Oxford University Press.

Bain, J. (2013a). Emergence in Effective Field Theories. *European Journal for Philosophy of Science* **3**, 257–273.

Batterman, R. (2002). *The Devil in the Details.* Oxford: Oxford University Press.

Batterman, R. (2005). Critical phenomena and breaking drops: infinite idealizations in physics. *Studies in History and Philosophy of Modern Physics B* **36**, 225–244.

Batterman, R. (2010). Reduction and Renormalization. In: A. Huttemann and G. Ernst (eds.), *Time, Chance, and Reduction: Philosophical Aspects of Statistical Mechanics*, pp. 159–179. Cambridge: Cambridge University Press.

Batterman, R. (2013). The tyranny of scales. In: *The Oxford Handbook of the Philosophy of Physics*, pp. 255–286. Oxford: Oxford University Press.

Binney, J., Dowrick, N., Fisher, A., Newman, M. (1992). *The Theory of Critical Phenomena: an introduction to the renormalization group.* Oxford: Oxford University Press.

Bouatta, N., Butterfield, J. (2011). Emergence and Reduction Combined in Phase Transitions. In: J. Kouneiher, C. Barbachoux and D. Vey (eds.), *Proceedings of Frontiers of Fundamental Physics 11*, pp. 383–403. Available at: http://philsci-archive.pitt.edu/8554/ and at: http://arxiv.org/abs/1104.1371.

Bouatta, N., Butterfield, J. (2015). On emergence in gauge theories at the 't Hooft limit. *European Journal for Philosophy of Science* **5**, 55–87.

Brown, L. (1993). *Renormalization: from Lorentz to Landau (and Beyond).* Berlin: Springer.

Butterfield, J. (2004). Some Aspects of Modality in Analytical Mechanics. In: P. Weingartner and M. Stoeltzner (eds.), *Formale Teleologie und Kausalitat in der Physik*, pp. 160–198. Paderborn: Mentis. http://philsci-archive.pitt.edu/archive/00001192/.

Butterfield, J. (2011). Emergence, Reduction and Supervenience: A Varied Landscape. *Foundations of Physics* **41**, 920–960.

Butterfield, J. (2011a). Less is Different: Emergence and Reduction Reconciled. *Foundations of Physics* **41**, 1065–1135.

Butterfield, J. (2014). Reduction, emergence and renormalization. *The Journal of Philosophy* **111**, 5–49.

Cao, T.Y. (1993). New philosophy of renormalization: From the renormalization group to effective field theories. In: Brown ed. (1993), pp. 87–133.

Cao, T.Y. (1997). *Conceptual Developments of Twentieth Century Field Theories.* Cambridge: Cambridge University Press.

Cao, T.Y. (1999). *Conceptual Foundations of Quantum Field Theory.* Cambridge: Cambridge University Press.

Cao, T.Y. (1999a). Renormalization group: An interesting yet puzzling idea. In: Cao (ed.) (1999), pp. 268–286.

Cao, T.Y., Schweber, S. (1993). The conceptual foundations and the philosophical aspects of renormalization theory. *Synthese* **97**, 33–108.

Castellani, E. (2002). Reductionism, emergence and effective field theories. *Studies in the History and Philosophy of Modern Physics* **33**, 251–267.

Causey, R. (1972). *Attribute-identities and micro-reductions, Journal of Philosophy* **67**, 407–422.

Dizadji-Bahmani, F., Frigg R., Hartmann S. (2010). Who's afraid of Nagelian reduction? *Erkenntnis* **73**, 393–412.

Feyerabend, P. (1962). Explanation, reduction and empiricism. In: H. Feigl and G. Maxwell (eds.), *Minnesota Studies in Philosophy of Science* **3**, pp. 28–97. Minnesota: University of Minnesota Press.

Feynman, R. (1985). *QED.* Princeton: Princeton University Press.

Fodor, J. (1974). Special Sciences (Or: the disunity of science as a working hypothesis). *Synthese* **28**, 97–115.

Giere, R. (1995). *Science Without Laws.* Chicago: University of Chicago Press.

Gross, D. (1999). Renormalization groups. In: P. Deligne (ed.), *Quantum Fields and Strings. A Course for Mathematicians*, pp. 551–596. American Mathematical Society.

Hartmann, S. (2001). Effective field theories, reductionism and scientific explanation. *Studies in History and Philosophy of Modern Physics* **32**, 267–304.

Hempel, C. (1965). *Aspects of Scientific Explanation.* New York: The Free Press.

Hempel, C. (1966). *Philosophy of Natural Science.* Upper Saddle River: Prentice-Hall.

Hesse, M. (1965). *Forces and Fields.* London: Macmillan.

Horejsi, J. (1996). Electroweak interactions and high-energy limit. *Czechoslovak Journal of Physics* **47**, 951–977.

Jaffe, A. (1999). Where does quantum field theory fit into the big picture? In: Cao (ed.) (1999). pp. 136–146.

Jaffe, A. (2008). Quantum theory and relativity. In: R. Doran, C. Moore and R. Zimmer, (eds.), *Contemporary Mathematics* (*Group Representations, Ergodic Theory, and Mathematical Physics: A Tribute to George W. Mackey*), pp. 209–246. Available at http://arthurjaffe.org.

Kadanoff, L. (2009). More is the same: Mean field theory and phase transitions. *Journal of Statistical Physics* **137**, 777–797.

Kadanoff, L. (2013). Theories of matter: infinities and renormalization. In: R. Batterman (ed.), *The Oxford Handbook of the Philosophy of Physics*, pp. 141–188. Oxford: Oxford University Press.

Kemeny, J., Oppenheim P. (1956). On reduction. *Philosophical Studies* **7**, 6–19.

Kim, J. (1999). Making sense of emergence. *Philosophical Studies* **95**, 3–36. Reprinted in: Bedau and Humphreys (eds.). Page reference to the reprint.

Kim, J. (2006). Emergence: Core Ideas and Issues. *Synthese* **151**, 547–559.

Kogut, J., Stephanov, M. (2004). *The Phases of Quantum Chromodynamics.* Cambridge: Cambridge University Press.

Lange, M. (2002). *An Introduction to the Philosophy of Physics.* Oxford: Blackwell.

Lautrup, B., Zinkernagel, H. (1999). g-2 and the trust in experimental results. *Studies in the History and Philosophy of Modern Physics* **30**, 85–110.

McMullin, E. (1993). Indifference principle and anthropic principle in cosmology. *Studies in the History and Philosophy of Science* **24**, 359–389.

McMullin, E. (2002). Origins of the field concept in physics. *Physics in Perspective* **4**, 13–39.

Menon, T., Callender, C. (2013). Turn and face the ch-ch-changes: Philosophical questions raised by phase transitions. In: R. Batterman (ed.), *The Oxford Handbook of the Philosophy of Physics*, pp. 189–223. Oxford: Oxford University Press.

Moffat, J. (2010). Ultraviolet Complete Electroweak Model Without a Higgs Particle. *European Physics Journal Plus* **126**, 53.

Morrison, M. (2012). Emergent physics and micro-ontology. *Philosophy of Science* **79**, 141–166.

Nagel, E. (1961). *The Structure of Science: Problems in the Logic of Scientific Explanation.* New York: Harcourt.

Nagel, E. (1979). Issues in the logic of reductive explanations. In: *Teleology Revisited and Other Essays in the Philosophy and History of Science*. Columbia: Columbia University Press. Reprinted in: Bedau and Humphreys, eds. (2008). Page reference to the reprint.

Nickles, T. (1973). Two concepts of inter-theoretic reduction. *Journal of Philosophy* **70**, 181–201.

Norton, J. (2012). Approximation and idealization: Why the difference matters. *Philosophy of Science* **74**, 207–232.

Putnam, H. (1975). Philosophy and our mental life. In: *Mind, Language and Reality*, pp. 291–303. Cambridge: Cambridge University Press.

Redhead, M. (1999). Quantum field theory and the philosopher. In: Cao (ed.) (1999). pp. 34–40.

Schaffner, K. (1967). Approaches to reduction. *Philosophy of Science* **34**, 137–147.

Schaffner, K. (1976). Reductionism in biology: Problems and prospects. In: R. Cohen et al. (eds). *PSA 1974*, pp. 613–632.

Schaffner, K. (2013). Ernest Nagel and reduction. *Journal of Philosophy* **109**, 534–565.

Schweber, S. (1993). Changing conceptualization of renormalization theory. In: Brown (1993), pp. 135–166.

Schweber, S. (1994). *QED and the Men who Made It.* Princeton: Princeton University Press.

Sklar, L. (1967). Types of intertheoretic reduction. *British Journal for the Philosophy of Science* **18**, 109–124.

Sober, E. (1999). The multiple realizability argument against reductionism. *Philosophy of Science* **66**, 542–564.

Symanzik, K. (1973). Infra-red singularities and small-distance behaviour analysis. *Communications in Mathematical Physics* **34**, 7–36.

Teller, P. (1989). *Infinite renormalization. Philosophy of Science* **56**, *238–257. Reprinted with minor corrections as Chapter 7 of his An Interpretive Introduction to Quantum Field Theory (1995).* Princeton: Princeton University Press.

Weinberg, S. (1995). *The Quantum Theory of Fields*, volume 1. Cambridge: Cambridge University Press.

Weinberg, S. (1995a). *The Quantum Theory of Fields*, volume 2. Cambridge: Cambridge University Press.

Weinberg, S. (1999). What is quantum field theory and what did we think it was. In: Cao ed. (1999), pp. 241–251. Also at: arxiv: hep-th/9702027.

Wightman, A. (1999). The usefulness of a general theory of quantized fields. In: Cao (ed.) (1999). pp. 41–46.

Wilczek, F. (2005). Asymptotic freedom: From paradox to paradigm (Nobel Prize Lecture 2004). *Proceedings of the National Academy of Science* **102**, 8403–8413.

Wilson, K. (1979). Problems in Physics with Many Scales of Length. *Scientific American* **241**, 158–179.

Zuchowski (2013). For electrodynamic consistency. *Studies in the History and Philosophy of Modern Physics* **44**, 135–142.

www.ingramcontent.com/pod-product-compliance
Lightning Source LLC
LaVergne TN
LVHW010623300926
849162LV00001B/39